Linear Algebra

an introduction

Linear Algebra

an introduction

PAUL J. KNOPP

University of Houston

HAMILTON PUBLISHING COMPANY

Santa Barbara, California

Copyright © 1974, by John Wiley & Sons, Inc.
Published by **Hamilton Publishing Company,**
a Division of John Wiley & Sons, Inc.

Library of Congress Cataloging in Publication Data:

Knopp, Paul J.   1934–
    Linear algebra: an introduction

    1. Algebra, Linear. I. Title.
QA184.K66 1974   512′.5   73–10499
ISBN 0-471-49550-6

Printed in the United States of America

10 9 8 7 6 5 4 3 2 1

to my wife
Margaret

Preface

This book presents the essential ideas of elementary linear algebra to students at the freshman-sophomore level. There are many students who need linear algebra, but who will not need calculus or differential equations; and even for those who do need both calculus and linear algebra, a case can be made that linear algebra should precede calculus or at least be taken simultaneously with the first semester of calculus. For these and other reasons, this book does not follow the practices of many other linear algebra texts.

First, no calculus experience is required for this book. Consequently, no function spaces are offered as examples of vector spaces, and no applications to differential equations are presented. The primary objects of study are the vector spaces $\mathbb{R}^n$, linear transformations between these spaces, real matrices, and, in the last chapter, quadratic functions. Discussions have been confined to these topics for several reasons. One is that freshmen and sophomores generally find other vector spaces bewildering. Another reason is that I am convinced that a teacher has done an outstanding teaching job if he gives students at this level a good understanding of linear functions and ample practice with the appropriate computational skills. A third reason is my belief that an understanding of linear functions between the spaces $\mathbb{R}^n$ can make many topics in calculus more accessible.

A second difference between this book and some others is its attitude toward mathematical proofs. Many feel that linear algebra should initiate students into the deductive style of higher level mathematics courses. This view has then influenced their style of presentation. My conviction is that the ideas of linear algebra are valuable for many students who will never progress to higher level mathematics courses or who will never find a need for the formal style of communication usually associated with courses in abstract mathematics. For this reason I have attempted to illustrate ideas and facts with numerical examples and then to discuss them in a more general setting, thus employing an inductive presentation rather than a deductive one. This, I believe, is appropriate for most students at this level. I have not labeled the reasons for a proposition as a proof even though in most cases they actually constitute a proof. On the other hand, I have not been content only with an appeal to numerical examples. The numerical examples are followed by a general formulation to help students make the transition from the particular to the general.

A third departure from the current practice is that set theoretic notation and language has been avoided wherever possible. This was done to make the text accessible to the average student reader. An instructor can easily introduce as much set theoretic notation as he wishes.

Even though the presentation is more inductive than is generally the case in linear algebra texts, still the book can be used for a rather standard course. The instructor will have to supply a few arguments which depend upon mathematical induction, and he will have to insist upon the level of formality he desires in the students' work. Nevertheless, it is my view that students will more quickly learn to state their ideas correctly if they are taught to check them with numerical examples; they must have experience with many examples before they can understand abstractions.

The text is written with the freshman in mind. It is not intended only for mathematics majors, nor is it intended only for nonmathematics majors; it is designed for any freshman student who needs the ideas and techniques of elementary linear algebra. Except for the last chapter, the text does not assume even analytic geometry beyond having some familiarity with obtaining equations for lines and planes. It does not even assume many of the familiar topics from high school algebra. It does assume, however, that the student can do arithmetic with real numbers. Consequently, as far as prerequisites are concerned, the text may fairly be designated a freshman text. The presentations are direct, and the Brief Exercises allow the student to test his understanding of the material immediately preceding them.

This is a mathematics text in that mathematical ideas are the backbone of the book; it is not concerned solely with techniques. Because of this, it is suitable for a first linear algebra course in colleges or in a transfer-oriented curriculum in two-year colleges.

For a one-semester course at the freshman-sophomore level, the first five chapters and the first three sections of Chapter 6 are probably sufficient. If the instructor wishes to include some material on the dot product from Chapter 8, then he may wish to omit the last two sections of Chapter 5 and the final section of Chapter 6. Chapters 7, 8, 9, and 10 are included so that a sophomore-junior level course can also be offered using the text.

Many people have helped me during the preparation of the manuscript. Dennison Brown used an early version in one of his classes, and Richard Sinkhorn gave me continued encouragement for the project. A number of reviewers were helpful. I want to thank the following gentlemen for valuable criticisms and suggestions: Meyer W. Belovicz, Joseph T. Buckley, Jack E. Forbes, Walter J. Gleason, David G. Mead, Richard C. Metzler, and Dieter Schmidt. Also thanks to Joseph D. Thibodeaux for helping me check the arithmetic in the manuscript. But especially I am indebted to my close friend Victor M. Manjarrez for using an early version in class and for many thoughtful conversations. He has greatly influenced my attitude toward mathematical education. The generosity of Mrs. Kathy Magann with her time in doing a fine emergency typing job was crucial in carrying the project to its conclusion. Also, a sincere word of gratitude to my friend Albert F. Thomasson for his encouragement throughout the years.

To my family I owe the greatest thanks. My wife, Margaret, cheerfully encouraged me during the writing, and my children, Cynthia, Elizabeth, and John, were especially understanding in "playing somewhere else." Without their help and support this book could never have been completed.

Finally, the gentlemen of Hamilton Publishing Company gave me invaluable aid and friendly support during the final stages of the writing and during the production of the book.

Houston, Texas   *Paul J. Knopp*

About the Author

Paul J. Knopp was born in San Antonio, Texas, on Janurary 3, 1934. He received his B.S. at Spring Hill College in 1957, his A.M. at Harvard University in 1958, and his Ph.D. at the University of Texas in 1962. He has taught at Spring Hill College (1959), the University of Missouri (1962–1964), and the University of Houston (1964–   ). In 1970 he was a recipient of the Teaching Excellence Award at the University of Houston. In 1971–1973 he served as Associate Director and later as Executive Director of the Committee on the Undergraduate Program in Mathematics (CUPM). He is a member of the Mathematical Association of America and of the American Mathematical Society. He has published research articles dealing with problems in matrix theory and analysis.

Contents

1 Introduction to Systems of Linear Equations   1

SECTION 1-1
Systems of Equations and the Use of Matrices to Solve Them   2

SECTION 1-2
Row Echelon Form   18

SECTION 1-3
Notation for Systems of Linear Equations and for Matrices   35

SECTION 1-4
Some Applications of Systems of Linear Equations   40

2 Vector Spaces   45

SECTION 2-1
Introduction to Vector Spaces   46

SECTION 2-2
Subspaces of Vector Spaces   57

SECTION 2-3
Some Geometry in $\mathbb{R}^2$ and $\mathbb{R}^3$ 63

3 Linear Independence, Bases, Dimension   85

SECTION 3-1
Linear Combinations   86

SECTION 3-2
Span of a Set   92

SECTION 3-3
Vector Equations of Planes in $\mathbb{R}^3$, Subspaces of $\mathbb{R}^2$ and $\mathbb{R}^3$ 100

SECTION 3-4
Linear Independence   114

SECTION 3-5
Existence of Bases   123

SECTION 3-6
Dimension   128

4   Linear Transformations and Matrices   141

SECTION 4-1
Preliminaries About Functions   142

SECTION 4-2
Linear Transformations   152

SECTION 4-3
Matrix Representation of Linear Transformations from $\mathbb{R}^n$ to $\mathbb{R}^m$ 166

SECTION 4-4
Arithmetic of Linear Transformations and Matrices   176

SECTION 4-5
Some Applications   196

5   Nonsingular Linear Transformations and Nonsingular Matrices   203

SECTION 5-1
Introduction to Nonsingular Linear Transformations and Matrices;
   A Method for Finding the Inverse of a Nonsingular Matrix   204

SECTION 5-2
Elementary Matrices   213

SECTION 5-3
Nonsingular Matrices   221

6 The Determinant Function 229

SECTION 6-1
Expansion of the Determinant Along a Row of the Matrix 230

SECTION 6-2
Techniques for Computing the Determinant 241

SECTION 6-3
Matrix of Cofactors, Another Method for Finding the Inverse
   of a Nonsingular Matrix 255

SECTION 6-4
Some Facts About the Determinant 266

7 Kernel, Image, Nullity, and Rank of Linear Transformations and Matrices 271

SECTION 7-1
Row Rank and Column Rank of a Matrix 272

SECTION 7-2
Kernel and Image of Linear Transformations, Dimension Theorem 280

SECTION 7-3
Systems of Equations 292

8 Distance in $\mathbb{R}^n$ 303

SECTION 8-1
Introductory Ideas on Distance, Dot Product 304

SECTION 8-2
Geometric Interpretation of Dot Products 311

SECTION 8-3
Orthogonal and Orthonormal Bases 322

SECTION 8-4
The Cross Product in $\mathbb{R}^3$ 339

9   Change of Basis, Equivalence of Matrices   345

SECTION 9-1
Ordered Bases, Coordinates of Vectors Relative to Ordered
   Bases, Matrix Representation of Linear Transformations
   Relative to Ordered Bases   346

SECTION 9-2
Change of Basis   355

SECTION 9-3
Different Matrices for the Same Linear Transformation,
   Equivalence of Matrices   365

10   Eigenvalues, Eigenvectors, Quadratic Functions   373

SECTION 10-1
Eigenvalues and Eigenvectors   374

SECTION 10-2
Quadratic Functions   389

SECTION 10-3
Conic Sections   401

References   415

Answers to Selected Problems   417

Index   431

Linear Algebra

an introduction

1

Introduction to Systems of Linear Equations

Systems of Equations and the Use of Matrices to Solve Them

PREVIEW   *In this section we discuss methods for finding solutions to systems of linear equations. The procedure involves three different operations: (a) interchange of two equations, (b) multiplication of an equation by a nonzero real number, and (c) addition of a multiple of one equation to another. Finally, we show how our computations can be handled in a more convenient way by using matrices.*

First we study systems of linear equations. An example is

$$x - y = 2$$
$$3x + 2y = 5.$$

The letters x and y represent unknown numbers (unknowns). The numbers 2 and 5 on the right-hand side of the equations are called constants. The numbers multiplying x and y are called the coefficients of x and y. To determine the coefficients, write the equations with a plus sign between the terms. Thus the coefficients of x are 1 in the first equation and 3 in the second. Since $x - y = x + (-1)y$, -1 is the coefficient of y in the first equation, and 2 is the coefficient of y in the second.

We have written the unknowns on the left-hand side of the equations and the constants on the right-hand side, but this is only a convention. We could just as well have written $x = 2 + y$ or $-y = 2 - x$ or $x - y - 2 = 0$. These are different forms of the same equation. However, we will usually write the unknowns on the left of the equality and the constants on the right.

Let us consider another example.

$$2x_1 + 3x_2 - x_3 = 1$$
$$x_1 \qquad + x_3 = 5$$

This is a system of two equations in the three unknowns x_1, x_2, and x_3. The coefficients of x_1 are 2 and 1, the coefficients of x_2 are 3 and 0, and the coefficients of x_3 are -1 and $+1$ in the first and second equations, respectively. The constants are 1 and 5. In this example we distinguish between different unknowns with subscripts on the same letter x, whereas in the first example we used different letters for different unknowns. Both procedures are common; ordinarily we use different letters if there are only a few unknowns and subscripts if there are many.

Let us consider a third example.

$$5x + 2y - z = 0$$

$$x + 3y + 4z = 0$$

$$x + y - 2z = 0$$

This is a system of three equations in three unknowns. This system is called *homogeneous* because the constants are all zero; if at least one constant is different from zero, then the system is called *nonhomogeneous*.

The goal of this section is to give procedures for finding all the solutions for a system of equations. A solution for

$$x - y = 2$$

$$3x + 2y = 5$$

(1)

consists of two numbers, one for x and one for y, which satisfy all the equations. Here, for example, $x = \frac{9}{5}$ and $y = -\frac{1}{5}$ constitute a solution. (In fact, this is the only solution.) A solution for

$$2x_1 + 3x_2 - x_3 = 1$$

$$x_1 + x_3 = 5$$

(2)

consists of three numbers, one for each unknown. For example, $x_1 = 1$, $x_2 = 1$, $x_3 = 4$ is a solution. Another is $x_1 = 0$, $x_2 = 2$, $x_3 = 5$.

Rather than writing $x = \frac{9}{5}$ and $y = -\frac{1}{5}$, we abbreviate by saying that $(\frac{9}{5}, -\frac{1}{5})$ is a solution for the system (1). The first number is used for the first unknown and the second number for the second unknown. Using this kind of abbreviation, we say that (1,1,4) and (0,2,5) are solutions for system (2).

We call $(\frac{9}{5}, -\frac{1}{5})$ an *ordered* pair because the order is important: the order indicates which number corresponds to which unknown. Similarly (1,1,4) is an ordered triple. For a system having four unknowns, a solution is an ordered quadruple; for five unknowns we need an ordered quintuple; for more unknowns we call solutions 6-tuples, 7-tuples, etc.

Now we explain how to find all the solutions for any system of linear equations. Some systems have only one solution, others have many, whereas still others have none. We begin with an example.

Example Consider the system

$$x - y = 2$$

$$2x - 2y = 4$$

$$3x + 2y = 7.$$

We describe a procedure for obtaining all the solutions of this system.

Eliminate x from the second and third equations by adding -2 times the first equation to the second and by adding -3 times the first equation to the third.

$$x - y = 2$$
$$0x + 0y = 0$$
$$5y = 1$$

Interchange the position of the second and third equations.

$$x - y = 2$$
$$5y = 1$$
$$0x + 0y = 0$$

Multiply the third equation by $\frac{1}{5}$.

$$x - y = 2$$
$$y = \tfrac{1}{5}$$
$$0x + 0y = 0$$

Eliminate y from the first equation by adding the second equation to the first.

$$x = \tfrac{11}{5}$$
$$y = \tfrac{1}{5}$$
$$0x + 0y = 0$$

In the example above we go from one system to another by using three operations:

(1) interchange of two equations,
(2) multiplication of an equation by a nonzero real number, and
(3) addition of a multiple of one equation to another.

Using these operations we can find the solutions for (solve) any system of linear equations. We describe a **systematic procedure** which always works. (In this description we label the unknowns x_1, x_2, and so on.)

Step 1: Isolate $1 \cdot x_1$ in the first equation as follows:
(a) locate an equation in which x_1 appears,
(b) place this equation first,
(c) obtain $1 \cdot x_1$ in the first equation by multiplying the first equation by the appropriate number, and
(d) eliminate x_1 from all the other equations by adding multiples of the first equation to the other equations.

Step 2: Isolate $1 \cdot x_2$ in the second equation as follows:
    (a) locate a different equation in which x_2 appears,
    (b) place this equation second,
    (c) obtain $1 \cdot x_2$ in the second equation by multiplying the second equation by the appropriate number, and
    (d) eliminate x_2 from all other equations by adding multiples of the second equation to the others.

At the end of Step 2, x_1 is isolated in the first equation, and x_2 is isolated in the second.

Continue this process until you run out of unknowns or equations. The next example illustrates the procedure.

Example   In this example we use the systematic procedure to find the solutions for the system

$$3x + 2y + 3z = 5$$
$$x - y + 2z = 7.$$

Multiply the first equation by $\frac{1}{3}$.

$$x + \tfrac{2}{3}y + z = \tfrac{5}{3}$$
$$x - y + 2z = 7$$

Add -1 times the first equation to the second.

$$x + \tfrac{2}{3}y + z = \tfrac{5}{3}$$
$$-\tfrac{5}{3}y + z = \tfrac{16}{3}$$

Multiply the second equation by $-\frac{3}{5}$.

$$x + \tfrac{2}{3}y + z = \tfrac{5}{3}$$
$$y - \tfrac{3}{5}z = -\tfrac{16}{5}$$

Add $-\frac{2}{3}$ times the second equation to the first.

$$x + \tfrac{7}{5}z = \tfrac{19}{5}$$
$$y - \tfrac{3}{5}z = -\tfrac{16}{5}$$

We can get many solutions by using different values for z. For $z = 0$, we get $x = \frac{19}{5}$ and $y = -\frac{16}{5}$. For $z = 1$, we get $x = \frac{12}{5}$ and $y = \frac{19}{5}$. In fact, we can describe all solutions by solving for x and y in terms of z.

$$x = \tfrac{19}{5} - \tfrac{7}{5}z$$
$$y = -\tfrac{16}{5} + \tfrac{3}{5}z$$

Then the triple (x,y,z) is a solution if and only if

$$(x,y,z) = (\tfrac{19}{5} - \tfrac{7}{5}z, -\tfrac{16}{5} + \tfrac{3}{5}z, z).$$

All possible solutions are obtained by using all possible real numbers for z and substituting them for z in this formula.

The three operations used in solving a system of equations change the equations but not the solutions. (Some exercises at the end of the section explain why.) Because of this, it is not absolutely essential to use the systematic procedure described above. We can use these operations in any order. Usually we choose the order which involves the simplest arithmetic. The importance of the systematic procedure is that it shows that we can solve any system of equations. Also, because it is systematic, it is a procedure which can be used on a computer.

Brief Exercises

1. Follow the systematic procedure illustrated above to find the solutions for the following systems of linear equations.

(a) $2x + y = 5$
$\quad\;\; x - y = 4$

(b) $3x_1 + \;\; x_2 = 2$
$\quad\;\; x_1 + 2x_2 = 1$

(c) $2x + 3y - \;\; z = 1$
$\quad 4x + 2y + 2z = 4$
$\quad\; x \quad\;\; + z = \tfrac{5}{4}$

2. Can you solve the systems of equations in Exercise 1 with simpler arithmetic by using a procedure different from the systematic one? If so, do so.

Selected Answers: **1(a)** $(3, -1)$ **1(b)** $(\tfrac{3}{5}, \tfrac{1}{5})$ **1(c)** $(\tfrac{5}{4} - z, -\tfrac{1}{2} + z, z)$

The unknowns do not enter into the process of solving a system of linear equations. The computations are done with the coefficient and the constants; the unknowns only keep track of the coefficients. We capitalize on this by introducing a more efficient notation. Rather than writing the entire system of equations, we display only the coefficients and constants in a rectangular pattern. For example,

$$
\begin{array}{c}
x - \;\; y = 2 \\
2x - 2y = 4 \\
3x + 2y = 7
\end{array}
\quad \text{is replaced by} \quad
\begin{pmatrix}
1 & -1 & 2 \\
2 & -2 & 4 \\
3 & 2 & 7
\end{pmatrix}.
$$

The first equation is represented by $(1 \quad -1 \quad 2)$, the second by $(2 \quad -2 \quad 4)$, and the third by $(3 \quad 2 \quad 7)$.

Such a rectangular display is called a *matrix*. The rows of the matrix are the numbers displayed on a horizontal line, and the columns are those numbers on a vertical line. In this case there are three rows; the first row is (1 $\quad$ −1 $\quad$ 2), the second row is (2 $\quad$ −2 $\quad$ 4), and the third row is (3 $\quad$ 2 $\quad$ 7). Furthermore, there are three columns; the first is

$$\begin{pmatrix} 1 \\ 2 \\ 3 \end{pmatrix}.$$

The second column is

$$\begin{pmatrix} -1 \\ -2 \\ 2 \end{pmatrix}$$

and the third column is

$$\begin{pmatrix} 2 \\ 4 \\ 7 \end{pmatrix}.$$

We indicate the size of a matrix by the number of rows and the number of columns, in that order. In this case the matrix is a 3 × 3 (read "three by three") matrix.

The matrix

$$\begin{pmatrix} 1 & -1 \\ 2 & -2 \\ 3 & 2 \end{pmatrix}$$

is the *coefficient matrix* of the system of equations, and the matrix

$$\begin{pmatrix} 1 & -1 & 2 \\ 2 & -2 & 4 \\ 3 & 2 & 7 \end{pmatrix}$$

is the *augmented matrix* of the system, because the coefficient matrix is "augmented" by the column of constants.

In the next two examples we solve some systems of equations using the augmented matrix.

Example In this example we display the calculations used to solve the system of equations

$$x - y = 2$$
$$2x - 2y = 4$$
$$3x + 2y = 7$$

along with the same calculations done with the augmented matrix. Notice how operations with equations correspond to operations with rows of the matrices.

$$
\begin{aligned}
x - y &= 2 \\
2x - 2y &= 4 \\
3x + 2y &= 7
\end{aligned}
\qquad
\begin{pmatrix}
1 & -1 & 2 \\
2 & -2 & 4 \\
3 & 2 & 7
\end{pmatrix}
$$

Add -2 times the first equation to the second.

Add -2 times the first row to the second.

$$
\begin{aligned}
x - y &= 2 \\
0x + 0y &= 0 \\
3x + 2y &= 7
\end{aligned}
\qquad
\begin{pmatrix}
1 & -1 & 2 \\
0 & 0 & 0 \\
3 & 2 & 7
\end{pmatrix}
$$

Add -3 times the first equation to the third.

Add -3 times the first row to the third.

$$
\begin{aligned}
x - y &= 2 \\
0x + 0y &= 0 \\
5y &= 1
\end{aligned}
\qquad
\begin{pmatrix}
1 & -1 & 2 \\
0 & 0 & 0 \\
0 & 5 & 1
\end{pmatrix}
$$

Interchange the second and third equations.

Interchange the second and third rows.

$$
\begin{aligned}
x - y &= 2 \\
5y &= 1 \\
0x + 0y &= 0
\end{aligned}
\qquad
\begin{pmatrix}
1 & -1 & 2 \\
0 & 5 & 1 \\
0 & 0 & 0
\end{pmatrix}
$$

Multiply the second equation by $\frac{1}{5}$.

Multiply the second row by $\frac{1}{5}$.

$$
\begin{aligned}
x - y &= 2 \\
y &= \tfrac{1}{5} \\
0x + 0y &= 0
\end{aligned}
\qquad
\begin{pmatrix}
1 & -1 & 2 \\
0 & 1 & \tfrac{1}{5} \\
0 & 0 & 0
\end{pmatrix}
$$

Add 1 times the second equation to the first.

Add 1 times the second row to the first.

$$
\begin{aligned}
x &= \tfrac{11}{5} \\
y &= \tfrac{1}{5} \\
0x + 0y &= 0
\end{aligned}
\qquad
\begin{pmatrix}
1 & 0 & \tfrac{11}{5} \\
0 & 1 & \tfrac{1}{5} \\
0 & 0 & 0
\end{pmatrix}
$$

When using the matrix notation we retain rows of zeros so that the size of the matrix is not changed.

Example In this example we solve the system of equations

$$x + 2y - z = 1$$
$$2x - y + z = 3.$$

Again we show the calculations on the equations along side those for the augmented matrix. As before, each operation with the equations has its counterpart as an operation with rows of the matrix.

$$x + 2y - z = 1$$
$$2x - y + z = 3$$

$$\begin{pmatrix} 1 & 2 & -1 & 1 \\ 2 & -1 & 1 & 3 \end{pmatrix}$$

Add -2 times the first equation to the second.

Add -2 times the first row to the second.

$$x + 2y - z = 1$$
$$- 5y + 3z = 1$$

$$\begin{pmatrix} 1 & 2 & -1 & 1 \\ 0 & -5 & 3 & 1 \end{pmatrix}$$

Multiply the second equation by $-\frac{1}{5}$.

Multiply the second row by $-\frac{1}{5}$.

$$x + 2y - z = 1$$
$$y - \tfrac{3}{5}z = -\tfrac{1}{5}$$

$$\begin{pmatrix} 1 & 2 & -1 & 1 \\ 0 & 1 & -\frac{3}{5} & -\frac{1}{5} \end{pmatrix}$$

Add -2 times the second equation to the first.

Add -2 times the second row to the first.

$$x \qquad + \tfrac{1}{5}z = \tfrac{7}{5}$$
$$y - \tfrac{3}{5}z = -\tfrac{1}{5}$$

$$\begin{pmatrix} 1 & 0 & \frac{1}{5} & \frac{7}{5} \\ 0 & 1 & -\frac{3}{5} & \frac{1}{5} \end{pmatrix}$$

It is a simple matter to reconstruct the final system of equations from the last matrix and to find that (x,y,z) is a solution if and only if $(x,y,z) = (\tfrac{7}{5} - \tfrac{1}{5}z, -\tfrac{1}{5} + \tfrac{3}{5}z, z)$.

Brief Exercises

Solve the following systems of equations using matrices.

1. $2x + y = 1$ **2.** $3x + 2y = 5$ **3.** $4x + y + z = 1$

   $x - y = 2$ $x - y = 0$ $x + y - z = 4$

Answers: **1** $(1,-1)$ **2** $(1,1)$ **3** $(-1 - \tfrac{2}{3}z, 5 + \tfrac{5}{3}z, z)$

As you see, we can solve a system of equations using only the augmented matrix. One advantage is that there is less to write. Therefore, from now on, we solve systems of equations using the augmented matrix. In doing so it is awkward to use phrases such as "multiply the third row by $\frac{1}{5}$" or "interchange the second and third rows." So we introduce a notation for the operations with rows of a matrix.

To indicate the interchange of the first and third rows, for example, we use $R_{1,3}$; interchanging the second and third rows is indicated by $R_{2,3}$. The subscripts on R tell which rows are being interchanged. In general, this operation is denoted by $R_{i,j}$; the letters i and j show which rows are to be interchanged. (There is no compelling reason for using i and j for this, although it seems to be a tradition in linear algebra.)

To indicate multiplication of the third row by the number 2, we use $R_3(2)$; multiplying the first row by 5 is indicated by $R_1(5)$. The subscript on R tells which row is being multiplied by the number in the parenthesis. Generally this type of row operation is denoted by $R_i(x)$; the letter i shows which row is being multiplied by x.

Finally, adding 3 times the second row to the first is indicated by $R_{1,2}(3)$; adding -6 times the first row to the third is indicated by $R_{3,1}(-6)$. In general, this type of row operation is indicated by $R_{i,j}(x)$; this means we add x times the jth row to the ith row. The order of the subscripts is essential; $R_{1,2}(1)$ and $R_{2,1}(1)$, for example, are different operations.

Definition 1-1_____

The row operations (row transformations) on a matrix are:

(**1**) interchange of the ith and jth rows, indicated by $R_{i,j}$,

(**2**) multiplication of the ith row by the nonzero number x, denoted by $R_i(x)$, and

(**3**) addition of x times the jth row to the ith row, denoted by $R_{i,j}(x)$. (Here x may be 0, but i and j must be different.)

_____

We illustrate these row operations in the next few examples.

Example   We solve the system of equations

$$2x + y - \ z = 2$$

$$x + y + \ z = 6$$

$$x - y + 2z = 7$$

using the augmented matrix and the notation for the row operations.

$$\begin{pmatrix} 2 & 1 & -1 & 2 \\ 1 & 1 & 1 & 6 \\ 1 & -1 & 2 & 7 \end{pmatrix} \xrightarrow{R_{1,2}} \begin{pmatrix} 1 & 1 & 1 & 6 \\ 2 & 1 & -1 & 2 \\ 1 & -1 & 2 & 7 \end{pmatrix}$$

$$\xrightarrow[R_{3,1}(-1)]{R_{2,1}(-2)} \begin{pmatrix} 1 & 1 & 1 & 6 \\ 0 & -1 & -3 & -10 \\ 0 & -2 & 1 & 1 \end{pmatrix} \xrightarrow{R_2(-1)} \begin{pmatrix} 1 & 1 & 1 & 6 \\ 0 & 1 & 3 & 10 \\ 0 & -2 & 1 & 1 \end{pmatrix}$$

$$\xrightarrow[R_{3,2}(2)]{R_{1,2}(-1)} \begin{pmatrix} 1 & 0 & -2 & -4 \\ 0 & 1 & 3 & 10 \\ 0 & 0 & 7 & 21 \end{pmatrix} \xrightarrow{R_3(\frac{1}{7})} \begin{pmatrix} 1 & 0 & -2 & -4 \\ 0 & 1 & 3 & 10 \\ 0 & 0 & 1 & 3 \end{pmatrix}$$

$$\xrightarrow[R_{2,3}(-3)]{R_{1,3}(2)} \begin{pmatrix} 1 & 0 & 0 & 2 \\ 0 & 1 & 0 & 1 \\ 0 & 0 & 1 & 3 \end{pmatrix}$$

Thus the triple $(2,1,3)$ is the only solution.

In the example above we have indicated

$$\xrightarrow[R_{3,1}(-1)]{R_{2,1}(-2)}$$

between the second and third matrices. The operations should be thought of as being done in the order shown, i.e., first $R_{2,1}(-2)$ and then $R_{3,1}(-1)$.

Example Consider the following system of four linear equations in three unknowns.

$$3x + y + 2z = 4$$
$$x - y + z = 2$$
$$2x + y - z = 1$$
$$x + y - z = -2$$

This system has no solutions, but this is not clear at the beginning. Only after some calculations can you be sure there are no solutions.

$$\begin{pmatrix} 3 & 1 & 2 & 4 \\ 1 & -1 & 1 & 2 \\ 2 & 1 & -1 & 1 \\ 1 & 1 & -1 & -2 \end{pmatrix} \xrightarrow{R_{1,2}} \begin{pmatrix} 1 & -1 & 1 & 2 \\ 3 & 1 & 2 & 4 \\ 2 & 1 & -1 & 1 \\ 1 & 1 & -1 & -2 \end{pmatrix}$$

$$\xrightarrow[\substack{R_{3,1}(-2) \\ R_{4,1}(-1)}]{R_{2,1}(-3)} \begin{pmatrix} 1 & -1 & 1 & 2 \\ 0 & 4 & -1 & -2 \\ 0 & 3 & -3 & -3 \\ 0 & 2 & -2 & -4 \end{pmatrix} \xrightarrow[R_4(\frac{1}{2})]{R_3(\frac{1}{3})} \begin{pmatrix} 1 & -1 & 1 & 2 \\ 0 & 4 & -1 & -2 \\ 0 & 1 & -1 & -1 \\ 0 & 1 & -1 & -2 \end{pmatrix}$$

$$\xrightarrow{R_{2,3}} \begin{pmatrix} 1 & -1 & 1 & 2 \\ 0 & 1 & -1 & -1 \\ 0 & 4 & -1 & -2 \\ 0 & 1 & -1 & -2 \end{pmatrix} \xrightarrow[\substack{R_{3,2}(-4) \\ R_{4,2}(-1)}]{R_{1,2}(1)} \begin{pmatrix} 1 & 0 & 0 & 1 \\ 0 & 1 & -1 & -1 \\ 0 & 0 & 3 & 2 \\ 0 & 0 & 0 & -1 \end{pmatrix}$$

$$\xrightarrow{R_3(\frac{1}{3})} \begin{pmatrix} 1 & 0 & 0 & 1 \\ 0 & 1 & -1 & -1 \\ 0 & 0 & 1 & \frac{2}{3} \\ 0 & 0 & 0 & -1 \end{pmatrix} \xrightarrow[R_4(-1)]{R_{2,3}(1)} \begin{pmatrix} 1 & 0 & 0 & 1 \\ 0 & 1 & 0 & -\frac{1}{3} \\ 0 & 0 & 1 & \frac{2}{3} \\ 0 & 0 & 0 & 1 \end{pmatrix}$$

This matrix is the augmented matrix for the system

$$x + 0y + 0z = 1$$
$$0x + y + 0z = -\tfrac{1}{3}$$
$$0x + 0y + z = \tfrac{2}{3}$$
$$0x + 0y + 0z = 1.$$

It is clear that no triple (x,y,z) is a solution to this system, since no numbers x, y, and z can satisfy the last equation. Consequently the original system has no solutions.

If you review the computations you see that we could have stopped when we obtained $(0 \ \ 0 \ \ 0 \ \ -1)$ as the last row in our matrix, because that represents the equation $0x + 0y + 0z = -1$. If you did not notice that the row $(0 \ \ 0 \ \ 0 \ \ -1)$

represents an equation which has no solution, then you can continue the calculations as we did; the impossible equation becomes apparent when you write out the system of equations after your final computation.

Example   Consider the system of equations

$$2x - y + 3z = 1$$
$$x + y - z = 2$$
$$3x - 3y + 7z = 0$$
$$-4x + 5y - 11z = 1.$$

We find the solutions to this system.

$$\begin{pmatrix} 2 & -1 & 3 & 1 \\ 1 & 1 & -1 & 2 \\ 3 & -3 & 7 & 0 \\ -4 & 5 & -11 & 1 \end{pmatrix} \xrightarrow{R_{1,2}} \begin{pmatrix} 1 & 1 & -1 & 2 \\ 2 & -1 & 3 & 1 \\ 3 & -3 & 7 & 0 \\ -4 & 5 & -11 & 1 \end{pmatrix}$$

$$\xrightarrow[\substack{R_{2,1}(-2) \\ R_{3,1}(-3) \\ R_{4,1}(4)}]{} \begin{pmatrix} 1 & 1 & -1 & 2 \\ 0 & -3 & 5 & -3 \\ 0 & -6 & 10 & -6 \\ 0 & 9 & -15 & 9 \end{pmatrix} \xrightarrow{R_2(-\frac{1}{3})} \begin{pmatrix} 1 & 1 & -1 & 2 \\ 0 & 1 & -\frac{5}{3} & 1 \\ 0 & -6 & 10 & -6 \\ 0 & 9 & -15 & 9 \end{pmatrix}$$

$$\xrightarrow[\substack{R_{1,2}(-1) \\ R_{3,2}(6) \\ R_{4,2}(-9)}]{} \begin{pmatrix} 1 & 0 & \frac{2}{3} & 1 \\ 0 & 1 & -\frac{5}{3} & 1 \\ 0 & 0 & 0 & 0 \\ 0 & 0 & 0 & 0 \end{pmatrix}$$

This is the augmented matrix for the system

$$x + \tfrac{2}{3}z = 1$$
$$y - \tfrac{5}{3}z = 1.$$

(We omit the equations $0x + 0y + 0z = 0$ since every vector (x,y,z) satisfies this equation; however, we leave zero rows in our matrix to maintain the same size matrix throughout the calculations.) Consequently (x,y,z) is a solution if and only if $(x,y,z) = (1 - \tfrac{2}{3}z, 1 + \tfrac{5}{3}z, z)$.

Brief Exercises

Solve the following system of equations by using row operations on the augmented matrix for the system.

1. $2x - y = 1$
$x + y = 3$

2. $3x + y + z = 1$
$x \quad\quad - z = 2$

3. $2x + 2y = 4$
$x - y = 6$

Answers: **1** $(\frac{4}{3}, \frac{5}{3})$ **2** $(2 + z, -5 - 4z, z)$ **3** $(4, -2)$

One more point about the use of matrix notation: for a homogeneous system of linear equations, the last column of the augmented matrix has all zero entries. Since a row operation does not change a column of zeros, there is no reason to carry along the column of zeros through the computations. Consequently, when we deal with a homogeneous system, we can omit the column of zeros and apply the row operations to the coefficient matrix. We illustrate this in the next example.

Example   Consider the homogeneous system of equations

$$5x + y - z + w = 0$$
$$x + y + 2z - w = 0$$
$$3x - y + z + 3w = 0.$$

We apply the row operations to the coefficient matrix.

$$
\begin{pmatrix} 5 & 1 & -1 & 1 \\ 1 & 1 & 2 & -1 \\ 3 & -1 & 1 & 3 \end{pmatrix}
\xrightarrow{R_{1,2}}
\begin{pmatrix} 1 & 1 & 2 & -1 \\ 5 & 1 & -1 & 1 \\ 3 & -1 & 1 & 3 \end{pmatrix}
$$

$$
\xrightarrow[R_{3,1}(-3)]{R_{2,1}(-5)}
\begin{pmatrix} 1 & 1 & 2 & -1 \\ 0 & -4 & -11 & 6 \\ 0 & -4 & -5 & 6 \end{pmatrix}
\xrightarrow{R_2(-\frac{1}{4})}
\begin{pmatrix} 1 & 1 & 2 & -1 \\ 0 & 1 & \frac{11}{4} & -\frac{3}{2} \\ 0 & -4 & -5 & 6 \end{pmatrix}
$$

$$
\xrightarrow[R_{3,2}(4)]{R_{1,2}(-1)}
\begin{pmatrix} 1 & 0 & -\frac{3}{4} & \frac{1}{2} \\ 0 & 1 & \frac{11}{4} & -\frac{3}{2} \\ 0 & 0 & 6 & 0 \end{pmatrix}
\xrightarrow{R_3(\frac{1}{6})}
\begin{pmatrix} 1 & 0 & -\frac{3}{4} & \frac{1}{2} \\ 0 & 1 & \frac{11}{4} & -\frac{3}{2} \\ 0 & 0 & 1 & 0 \end{pmatrix}
$$

$$
\xrightarrow[R_{2,3}(-\frac{11}{4})]{R_{1,3}(\frac{3}{4})}
\begin{pmatrix} 1 & 0 & 0 & \frac{1}{2} \\ 0 & 1 & 0 & -\frac{3}{2} \\ 0 & 0 & 1 & 0 \end{pmatrix}
$$

Since we are dealing with a homogeneous system, this is the coefficient matrix for

$$x \qquad + \tfrac{1}{2}w = 0$$
$$y \qquad - \tfrac{3}{2}w = 0$$
$$z \qquad = 0.$$

Thus the solutions are those quadruples (x,y,z,w) which can be expressed in the form $(x,y,z,w) = (-\tfrac{1}{2}w, \tfrac{3}{2}w, 0, w)$.

SUMMARY *In this section we begin a discussion of finding solutions for systems of linear equations. We introduce the coefficient and augmented matrices as devices for working with systems of equations. We also introduce the row operations on a matrix:*

(1) *interchange of the ith and jth rows, denoted by $R_{i,j}$,*
(2) *multiplication of the ith row by the nonzero scalar x, denoted by $R_i(x)$,*
(3) *addition of x times the jth row to the ith row, denoted by $R_{i,j}(x)$, $(i \neq j)$.*

Exercises for Section 1-1

1. Find the solutions for each of the systems listed below by using row operations on the augmented matrix for each system.

(a) $3x - y = 1$
$\quad x + y = 1$

(b) $4x + y - z + w = 2$
$\quad x - y + z - 2w - 1$

(c) $3x + y + z - w + u - v = 1$
$\quad 4x + 3y + 2z + 5w - 6u + 3v = 6$
$\quad x \quad - z + w \quad - v = 7$

(d) $3x + 2y + z = 2$
$\quad x - y + 2z = -5$
$\quad 5x + 5y \quad = 8$

(e) $2x - y + z = 1$
$\quad x + y - z = 0$
$\quad 3x + 2y + 4z = 8$

(f) $3x + 3y + 4z - 5w + 3v = 5$
$\quad x + y \quad + 2w - v = 2$
$\quad y + 2z - w + 2v =$

(g) $2x + 3y + 4z + 2w = 5$
$\quad x - y + 3z + 6w = 2$
$\quad x + y + z + w = 1$

More Challenging Exercises

After the first example we stated that it is essential to know that the solutions for a system of linear equations remain the same if the three row operations are used. Exercises 2 and 3 are designed to help you understand why this is so.

It is easy to see that the solutions are not changed if two equations are interchanged, because a solution for the equations when they are placed in one order must still satisfy them when they are placed in another order. So in Exercises 2 and 3 we concentrate on the other two operations.

2. (a) Give an argument which shows that a solution of $3x - y = 2$ must be a solution of $6x - 2y = 4$, and vice versa.
   (b) Give an argument which shows that a solution of $3x - y = 2$ must be a solution of $3rx - ry = 2r$, and vice versa, provided $r \neq 0$.
   (c) Why is it that the solution set for a system of linear equations remains unchanged if one equation is multiplied by a nonzero real number?

3. (a) Suppose we are given the system

$$5x + 3y + 2z = 6$$
$$x - y + z = 2$$

and suppose we change it by adding two times the second to the first to obtain

$$7x + y + 4z = 10$$
$$x - y + z = 2.$$

Explain how to obtain the first system from the second by the same kind of operation.

   (b) Give an argument which shows that a solution of

$$5x + 3y + 2z = 6$$
$$x - y + z = 2$$

must be a solution of

$$7x + y + 4z = 10$$
$$x - y + z = 2$$

and vice versa.

   (c) Give an argument which shows that a solution to the system

$$5x + 3y + 2z = 6$$
$$x - y + z = 2$$

must be a solution of

$$(5 + r)x + (3 - r)y + (2 + r)z = 6 + 2r$$
$$x - y + z = 2$$

and vice versa. (To get the second system we add r times the second equation to the first equation in the first system.)

(d) Give an argument which shows that the solution set for a system of m linear equations in n unknowns remains the same if we add r times the jth equation to the ith equation.

In Exercises 4 and 5 we ask you to show that the row operations are reversible.

4. Let A be the matrix

$$A = \begin{pmatrix} 1 & 2 & 5 & 4 \\ -3 & 1 & 3 & 2 \\ 4 & 6 & -7 & 8 \end{pmatrix}.$$

(a) Suppose we use the row operation $R_{2,3}$ on A to obtain the matrix

$$B = \begin{pmatrix} 1 & 2 & 5 & 4 \\ 4 & 6 & -7 & 8 \\ -3 & 1 & 3 & 2 \end{pmatrix}.$$

How can we obtain A by applying a row operation to the matrix B?

(b) Suppose we use the row operation $R_3(5)$ on A to obtain the matrix

$$C = \begin{pmatrix} 1 & 2 & 5 & 4 \\ -3 & 1 & 3 & 2 \\ 20 & 30 & -35 & 40 \end{pmatrix}.$$

How can we obtain A by applying a row operation to C?

(c) Suppose we use the row operation $R_{2,1}(3)$ on A to obtain the matrix

$$D = \begin{pmatrix} 1 & 2 & 5 & 4 \\ 0 & 7 & 18 & 14 \\ 4 & 6 & -7 & 8 \end{pmatrix}.$$

How can we obtain the matrix A by applying a row operation to D?

5. Suppose A is a matrix.

(a) If the matrix B is obtained from A by applying the row operation $R_{i,j}$ to A, how can we obtain A from B using a row operation?

(b) If the matrix C is obtained from A by applying the row operation $R_i(x)$, $x \neq 0$, how can we obtain A from C using a row operation?

(c) If the matrix D is obtained from A by applying the row operation $R_{i,j}(x)$, $i \neq j$, how can we obtain A from D using a row operation?

6. Suppose A and B are matrices having the same size (i.e., same number of rows and same number of columns), and suppose B has been obtained from A by a succession of row operations. Is it always possible to obtain A from B by a succession of row operations?

7. Suppose A, B, and C are matrices having the same size. If B can be obtained from A, and if C can be obtained from B by row operations, is it possible to obtain C from A by row operations?

Definition 1-2_____

If A and B are matrices and one can be obtained from the other by row operations, then A and B are said to be *row equivalent*.

_____

SECTION 1-2
Row Echelon Form

PREVIEW   *In solving a system of linear equations, we attempt to isolate each unknown in a different equation. The matrix counterpart of this is the process of transforming a matrix to row echelon form. This section explains what the row echelon form is and how to obtain it.*

The procedure for solving a system of linear equations given in Section 1-1 results in the first unknown appearing only in the first equation, the second unknown only in the second, and so on. This can also be described in matrix terms. Our first goal is to describe the form of the agumented matrix after the process of solving the equations has been completed.

In the following examples we want to produce the numbers 0 and 1 in various positions of a matrix. We count the rows of a matrix from the top and the columns from the left, and we refer to a position in the matrix by giving first the row and then the column in which it appears. For example, the (2,4) position is in the second row and the fourth column. The number in the (2,4) position is the (2,4) entry.

Example   Consider the system

$$
\begin{aligned}
x - y + 3z &= 4 \\
3x + y + z &= 2 \\
4x \quad\;\; + 4z &= 6.
\end{aligned}
$$

The augmented matrix for this system is

$$\begin{pmatrix} 1 & -1 & 3 & 4 \\ 3 & 1 & 1 & 2 \\ 4 & 0 & 4 & 6 \end{pmatrix}.$$

If we want the unknown x to appear only in the first equation and with 1 as co-efficient, then, in matrix language, we must produce the number 1 in the (1,1) position and the number 0 in the (2,1) and (3,1) positions. We can do this by using $R_{2,1}(-3)$ and $R_{3,1}(-4)$ to obtain

$$\xrightarrow[R_{3,1}(-4)]{R_{2,1}(-3)} \begin{pmatrix} 1 & -1 & 3 & 4 \\ 0 & 4 & -8 & -10 \\ 0 & 4 & -8 & -10 \end{pmatrix}.$$

The next step is to isolate y in the second equation. So in the matrix we produce the number 1 in the (2,2) position and the number 0 in every other position in the second column. Here we use $R_2(\tfrac{1}{4})$, $R_{1,2}(1)$, and $R_{3,2}(-4)$ to obtain

$$\xrightarrow[R_{3,2}(-4)]{\substack{R_2(\frac{1}{4}) \\ R_{1,2}(1)}} \begin{pmatrix} 1 & 0 & 1 & \frac{3}{2} \\ 0 & 1 & -2 & -\frac{5}{2} \\ 0 & 0 & 0 & 0 \end{pmatrix}.$$

The third step is to isolate z in the third equation. But this cannot be done because the last row has the number 0 in the (3,3) position. The matrix above is in row echelon form.

Example Consider the matrix

$$\begin{pmatrix} 2 & 2 & 0 & 3 & 1 \\ 3 & 4 & 2 & 1 & 2 \\ 1 & 1 & -1 & 6 & 1 \end{pmatrix}.$$

In this example we operate on this matrix with row operations as though it is the augmented matrix for a system of linear equations. We follow the matrix counterpart of the systematic procedure for solving systems of equations.

We begin with $R_1(\tfrac{1}{2})$ to produce the number 1 in the (1,1) position.

$$\xrightarrow{R_1(\frac{1}{2})} \begin{pmatrix} 1 & 1 & 0 & \frac{3}{2} & \frac{1}{2} \\ 3 & 4 & 2 & 1 & 2 \\ 1 & 1 & -1 & 6 & 1 \end{pmatrix}$$

Then $R_{2,1}(-3)$ and $R_{3,1}(-1)$ produce the number 0 in the (2,1) and (3,1) positions.

$$\xrightarrow[R_{3,1}(-1)]{R_{2,1}(-3)} \begin{pmatrix} 1 & 1 & 0 & \frac{3}{2} & \frac{1}{2} \\ 0 & 1 & 2 & -\frac{7}{2} & \frac{1}{2} \\ 0 & 0 & -1 & \frac{9}{2} & \frac{1}{2} \end{pmatrix}$$

Since we already have the number 1 in the (2,2) position and the number 0 in the (3,2) position, we only need $R_{1,2}(-1)$ to produce the number 0 in the (1,2) position.

$$\xrightarrow{R_{1,2}(-1)} \begin{pmatrix} 1 & 0 & -2 & 5 & 0 \\ 0 & 1 & 2 & -\frac{7}{2} & \frac{1}{2} \\ 0 & 0 & -1 & \frac{9}{2} & \frac{1}{2} \end{pmatrix}$$

Use $R_3(-1)$ to produce the number 1 in the (3,3) position and then $R_{1,3}(2)$ and $R_{2,3}(-2)$ to obtain the number 0 in the (1,3) and (2,3) positions.

$$\xrightarrow[R_{2,3}(-2)]{\substack{R_3(-1) \\ R_{1,3}(2)}} \begin{pmatrix} 1 & 0 & 0 & -4 & -1 \\ 0 & 1 & 0 & \frac{11}{2} & \frac{3}{2} \\ 0 & 0 & 1 & -\frac{9}{2} & -\frac{1}{2} \end{pmatrix}$$

The process ends here; this matrix is in row echelon form.

The examples above illustrate the features of a matrix in row echelon form. To describe these features we need some terminology.

Definition 1-3

If every entry in a row of a matrix is the number zero, then that row is a *zero row*. A row is a *nonzero row* if it contains at least one nonzero entry. The same applies to columns: a *zero column* has all entries equal to 0; a *nonzero column* has at least one nonzero entry.

First, for a matrix to be in row echelon form, *all zero rows must be below the non-zero rows.* In the first example, the zero row was already last, so it was not necessary to place it there. However, if a zero row arises during our calculations, we place it below the nonzero rows with a row operation of the type $R_{i,j}$, an interchange of two rows.

Second, *the first nonzero entry in each nonzero row must be the number 1.* This can be arranged by using a row operation of the type $R_i(x)$, multiplication of a row by a nonzero number. We refer to this number 1 as the *initial* or *leading* 1 in that row.

Third, each initial 1 has been used to "clear" the column in which it appears; i.e., *all other entries in that column are zero.* This can be done by row operations of the type $R_{i,j}(x)$, adding a multiple of one row to another.

The last feature about row echelon form concerns the relative positions of the leading 1's in the nonzero rows. The preceding examples may lead you to expect the leading 1's to be in the (1,1), (2,2), (3,3), etc., positions. That does not always happen. We can, however, arrange the matrix so that as you inspect the rows beginning at the top and proceeding down, *each initial 1 is to the right of the initial 1 of the preceding row.* In other words, the first row has the leading 1 farthest to the left, the second row has the leading 1 second farthest to the left, and so on.

If a matrix has all these features, it is in row echelon form. The reason for the terminology is that the rows give the appearance of being in echelon. An echelon is a troop formation in which the various units are arranged in parallel rows but perhaps unaligned and in a steplike fashion. The position of the leading 1's is the reason for the use of the word "echelon."

Here are some examples of matrices which are not in row echelon form.

Example   The following matrix is not in row echelon form.

$$\begin{pmatrix} 2 & 0 & 1 & 3 \\ 0 & 1 & 0 & 1 \\ 0 & 0 & 0 & 0 \\ 0 & 0 & 1 & 0 \end{pmatrix}$$

There are many reasons for this. First, there is a nonzero row below a zero row. Second, the first nonzero entry in the first row is the number 2 and not the number 1. Third, the leading 1 in the third column has not been used to produce the number 0 in the (1,3) position. Thus the matrix above fails to be in row echelon form for three different reasons. It can be brought into row echelon form by the following sequence of operations: $R_{1,4}(-1)$, $R_1(\frac{1}{2})$, and $R_{3,4}$.

Example   Here is another example of a matrix which is not in row echelon form.

$$\begin{pmatrix} 0 & 1 & 0 & 1 \\ 1 & 0 & 0 & 0 \\ 0 & 0 & 1 & 0 \\ 0 & 0 & 0 & 0 \end{pmatrix}$$

The reason is that the leading 1 of the first row is to the right of the leading 1 of the second row. This matrix can be put into row echelon form by the operation $R_{1,2}$.

Definition 1-4_____

A matrix is in *row echelon* form provided:

(1) the zero rows are below the nonzero rows,
(2) each nonzero row has a leading 1,
(3) each leading 1 is the only nonzero entry in its column, and
(4) each leading 1 is positioned to the right of the leading 1 in the preceding row.

Brief Exercises

Determine which of the following matrices are in row echelon form. If they are not, give reasons why they are not.

1. $\begin{pmatrix} 1 & 0 & 0 \\ 0 & 1 & 0 \\ 0 & 1 & 1 \end{pmatrix}$
  2. $\begin{pmatrix} 1 & 0 & 1 \\ 0 & 1 & 0 \\ 0 & 0 & 0 \end{pmatrix}$

3. $\begin{pmatrix} 1 & 0 & 0 & 1 \\ 0 & 1 & 0 & 2 \\ 0 & 0 & 0 & 0 \\ 0 & 0 & 1 & 0 \end{pmatrix}$
  4. $\begin{pmatrix} 1 & 0 & 0 & 0 \\ 0 & 0 & 0 & 1 \\ 0 & 0 & 0 & 0 \\ 0 & 0 & 0 & 0 \end{pmatrix}$

Answers: **1** No **2** Yes **3** No **4** Yes

We have just described the row echelon form for a matrix. Now we explain why every matrix can be transformed to this form by row operations. We give a systematic procedure which is the matrix form of the procedure to solve a system of equations given in Section 1-1.

Example We transform the matrix

$$\begin{pmatrix} 2 & 1 & 0 \\ 3 & 1 & 2 \end{pmatrix}$$

to row echelon form.

Since the first column is nonzero, we must have a leading 1 in the first column. We have two choices: use $R_1(\frac{1}{2})$ to obtain

$$\begin{pmatrix} 1 & \frac{1}{2} & 0 \\ 3 & 1 & 2 \end{pmatrix}$$

or use $R_2(\frac{1}{3})$ and $R_{1,2}$ to obtain

$$\begin{pmatrix} 1 & \frac{1}{3} & \frac{2}{3} \\ 2 & 1 & 0 \end{pmatrix}.$$

Our choice is to use $R_1(\frac{1}{2})$ to obtain

$$\begin{pmatrix} 1 & \frac{1}{2} & 0 \\ 3 & 1 & 2 \end{pmatrix}.$$

Next we "clear" the first column of other nonzero entries by using $R_{2,1}(-3)$ to obtain

$$\begin{pmatrix} 1 & \frac{1}{2} & 0 \\ 0 & -\frac{1}{2} & 2 \end{pmatrix}.$$

Now we produce the next leading 1. Clearly we can get a leading 1 in the second row by using $R_2(-2)$ to obtain

$$\begin{pmatrix} 1 & \frac{1}{2} & 0 \\ 0 & 1 & -4 \end{pmatrix}.$$

Then we use the leading 1 in the (2,2) position to clear the second column of other nonzero entries with $R_{1,2}(-\frac{1}{2})$.

$$\begin{pmatrix} 1 & 0 & 2 \\ 0 & 1 & -4 \end{pmatrix}$$

The matrix is now in row echelon form.

Example In this example we illustrate a systematic procedure for transforming the following matrix to row echelon form.

$$\begin{pmatrix} 1 & 1 & 2 & 1 \\ 2 & 2 & 1 & 3 \\ 0 & 0 & 4 & 1 \end{pmatrix}$$

The first row already has a leading 1; we use it to clear the first column of other nonzero entries.

$$\xrightarrow{R_{2,1}(-2)} \begin{pmatrix} 1 & 1 & 2 & 1 \\ 0 & 0 & -3 & 1 \\ 0 & 0 & 4 & 1 \end{pmatrix}$$

The next leading 1 must be produced from the second and third rows; by inspecting the matrix above, you can see that the next leading 1 will occur in the (2,3) position.

We use $R_2(-\frac{1}{3})$ to obtain

$$\xrightarrow{R_2(-\frac{1}{3})} \begin{pmatrix} 1 & 1 & 2 & 1 \\ 0 & 0 & 1 & -\frac{1}{3} \\ 0 & 0 & 4 & 1 \end{pmatrix}.$$

Now we use the leading 1 in the (2,3) position to clear the third column of other nonzero entries.

$$\xrightarrow[R_{3,2}(-4)]{R_{1,2}(-2)} \begin{pmatrix} 1 & 1 & 0 & \frac{5}{3} \\ 0 & 0 & 1 & -\frac{1}{3} \\ 0 & 0 & 0 & \frac{7}{3} \end{pmatrix}$$

The next leading 1 must be produced from the last row. Use $R_3(\frac{3}{7})$ to obtain

$$\xrightarrow{R_3(\frac{3}{7})} \begin{pmatrix} 1 & 1 & 0 & \frac{5}{3} \\ 0 & 0 & 1 & -\frac{1}{3} \\ 0 & 0 & 0 & 1 \end{pmatrix}.$$

Use the leading 1 in the (3,4) position to clear the fourth column of other nonzero entries.

$$\xrightarrow[R_{2,3}(\frac{1}{3})]{R_{1,3}(-\frac{5}{3})} \begin{pmatrix} 1 & 1 & 0 & 0 \\ 0 & 0 & 1 & 0 \\ 0 & 0 & 0 & 1 \end{pmatrix}$$

This matrix is in row echelon form.

The systematic procedure can be described as follows:

Step 1: Obtain a leading 1 which is farthest to the left, place it in the first row, and use it to clear the other nonzero entries in its column.

Step 2: Obtain a leading 1 which is second farthest to the left, place it in the second row, and use it to clear the other nonzero entries in its column.

Continue this process as long as necessary.

If you have a choice of obtaining a leading 1 from different entries in the same column, choose the entry which is highest in that column. (In the last example, when we had the matrix

$$\begin{pmatrix} 1 & 1 & 2 & 1 \\ 0 & 0 & -3 & 1 \\ 0 & 0 & 4 & 1 \end{pmatrix}$$

we could have used $R_2(-\frac{1}{3})$ or we could have used $R_3(\frac{1}{4})$ followed by $R_{2,3}$. We chose $R_2(-\frac{1}{3})$ because we could produce the leading 1 from -3, and -3 is higher in that column than 4.) The reason we choose the higher entry is to make the procedure unambiguous.

Consequently, any matrix can be transformed to row echelon form; the systematic procedure gives a recipe for doing it.

Brief Exercises

Follow the systematic procedure in transforming the following matrices to row echelon form.

1.
$$\begin{pmatrix} 1 & 2 \\ 1 & 1 \end{pmatrix}$$

2.
$$\begin{pmatrix} 3 & 1 & 2 \\ 1 & 2 & 1 \end{pmatrix}$$

3.
$$\begin{pmatrix} 1 & 1 & 0 & 1 \\ 2 & 1 & 3 & 0 \\ 0 & 1 & 2 & 1 \\ -4 & 1 & 1 & 1 \end{pmatrix}$$

4.
$$\begin{pmatrix} 2 & 0 & 1 & 3 \\ 4 & 1 & 2 & 1 \\ 1 & 1 & 3 & 2 \end{pmatrix}$$

Answers:

1
$$\begin{pmatrix} 1 & 0 \\ 0 & 1 \end{pmatrix}$$

2
$$\begin{pmatrix} 1 & 0 & \frac{3}{5} \\ 0 & 1 & \frac{1}{5} \end{pmatrix}$$

3
$$\begin{pmatrix} 1 & 0 & 0 & 0 \\ 0 & 1 & 0 & 0 \\ 0 & 0 & 1 & 0 \\ 0 & 0 & 0 & 1 \end{pmatrix}$$

4
$$\begin{pmatrix} 1 & 0 & 0 & \frac{2}{5} \\ 0 & 1 & 0 & -5 \\ 0 & 0 & 1 & \frac{11}{5} \end{pmatrix}$$

The preceding examples and exercises show that you can transform any matrix to row echelon form by following the systematic procedure. However, if you use different row operations and arrive at a matrix in row echelon form, then it is the same matrix which the systematic procedure gives. The practical significance is that you may alter the systematic procedure to avoid messy arithmetic. We illustrate this in the next examples.

Example   Consider the matrix

$$\begin{pmatrix} 1 & 1 & 2 & 1 \\ 2 & 2 & 1 & 3 \\ 0 & 0 & 4 & 1 \end{pmatrix}.$$

In the last example we transformed this to the row echelon form

$$\begin{pmatrix} 1 & 1 & 0 & 0 \\ 0 & 0 & 1 & 0 \\ 0 & 0 & 0 & 1 \end{pmatrix}.$$

Here we use a different sequence of row operations to make the arithmetic easier and still arrive at the same row echelon form.

Use the leading 1 in the first row to clear the first column of other nonzero entries (just as in the previous example).

$$\xrightarrow{R_{2,1}(-2)} \begin{pmatrix} 1 & 1 & 2 & 1 \\ 0 & 0 & -3 & 1 \\ 0 & 0 & 4 & 1 \end{pmatrix}$$

Rather than using $R_2(-\frac{1}{3})$, we use $R_{2,3}(1)$ to avoid fractions.

$$\xrightarrow{R_{2,3}(1)} \begin{pmatrix} 1 & 1 & 2 & 1 \\ 0 & 0 & 1 & 2 \\ 0 & 0 & 4 & 1 \end{pmatrix}$$

Now we use the leading 1 in the (2,3) position to clear the third column of other nonzero entries.

$$\xrightarrow[R_{3,2}(-4)]{R_{1,2}(-2)} \begin{pmatrix} 1 & 1 & 0 & -3 \\ 0 & 0 & 1 & 2 \\ 0 & 0 & 0 & -7 \end{pmatrix}$$

Then use $R_3(-\frac{1}{7})$ to get a leading 1 in the third row.

$$\xrightarrow{R_3(-\frac{1}{7})} \begin{pmatrix} 1 & 1 & 0 & -3 \\ 0 & 0 & 1 & 2 \\ 0 & 0 & 0 & 1 \end{pmatrix}$$

Finally we obtain the row echelon form

$$\xrightarrow[R_{2,3}(-2)]{R_{1,3}(3)}\begin{pmatrix} 1 & 1 & 0 & 0 \\ 0 & 0 & 1 & 0 \\ 0 & 0 & 0 & 1 \end{pmatrix}.$$

Example In this example we transform the following matrix to row echelon form. We use a procedure which makes the calculations easier than they would be if we followed the systematic procedure. We start with the matrix

$$\begin{pmatrix} -3 & -3 & 2 & -3 & 1 & 3 \\ 2 & 2 & -2 & 0 & 1 & 3 \\ 1 & 1 & 4 & 5 & 2 & 1 \\ 3 & 3 & 2 & 5 & 1 & 1 \\ 1 & 1 & 2 & 3 & 3 & 4 \\ 2 & 2 & 1 & 3 & 2 & 1 \end{pmatrix}$$

$$\xrightarrow{R_{1,3}}\begin{pmatrix} 1 & 1 & 4 & 5 & 2 & 1 \\ 2 & 2 & -2 & 0 & 1 & 3 \\ -3 & -3 & 2 & -3 & 1 & 3 \\ 3 & 3 & 2 & 5 & 1 & 1 \\ 1 & 1 & 2 & 3 & 3 & 4 \\ 2 & 2 & 1 & 3 & 2 & 1 \end{pmatrix}$$

$$\xrightarrow[\substack{R_{5,1}(-1) \\ R_{6,1}(-2)}]{\substack{R_{2,1}(-2) \\ R_{3,1}(3) \\ R_{4,1}(-3)}}\begin{pmatrix} 1 & 1 & 4 & 5 & 2 & 1 \\ 0 & 0 & -10 & -10 & -3 & 1 \\ 0 & 0 & 14 & 12 & 7 & 6 \\ 0 & 0 & -10 & -10 & -5 & -2 \\ 0 & 0 & -2 & -2 & 1 & 3 \\ 0 & 0 & -7 & -7 & -2 & -1 \end{pmatrix}.$$

We can avoid fractions by using $R_{6,5}(-4)$; this gets the number 1 into the third column.

$$\xrightarrow{R_{6,5}(-4)}
\begin{pmatrix}
1 & 1 & 4 & 5 & 2 & 1 \\
0 & 0 & -10 & -10 & -3 & 1 \\
0 & 0 & 14 & 12 & 7 & 6 \\
0 & 0 & -10 & -10 & -5 & -2 \\
0 & 0 & -2 & -2 & 1 & 3 \\
0 & 0 & 1 & 1 & -6 & -13
\end{pmatrix}$$

$$\xrightarrow{R_{2,6}}
\begin{pmatrix}
1 & 1 & 4 & 5 & 2 & 1 \\
0 & 0 & 1 & 1 & -6 & -13 \\
0 & 0 & 14 & 12 & 7 & 6 \\
0 & 0 & -10 & -10 & -5 & -2 \\
0 & 0 & -2 & -2 & 1 & 3 \\
0 & 0 & -10 & -10 & -3 & 1
\end{pmatrix}$$

$$\xrightarrow[\substack{R_{5,2}(2) \\ R_{6,2}(10)}]{\substack{R_{1,2}(-4) \\ R_{3,2}(-14) \\ R_{4,2}(10)}}
\begin{pmatrix}
1 & 1 & 0 & 1 & 26 & 53 \\
0 & 0 & 1 & 1 & -6 & -13 \\
0 & 0 & 0 & -2 & 91 & 188 \\
0 & 0 & 0 & 0 & -65 & -132 \\
0 & 0 & 0 & 0 & -11 & -23 \\
0 & 0 & 0 & 0 & -63 & -129
\end{pmatrix}$$

$$\xrightarrow{R_{4,5}(-6)}
\begin{pmatrix}
1 & 1 & 0 & 1 & 26 & 53 \\
0 & 0 & 1 & 1 & -6 & -13 \\
0 & 0 & 0 & -2 & 91 & 188 \\
0 & 0 & 0 & 0 & 1 & 6 \\
0 & 0 & 0 & 0 & -11 & -23 \\
0 & 0 & 0 & 0 & -63 & -129
\end{pmatrix}$$

$$\xrightarrow[\substack{R_{5,4}(11)\\R_{6,4}(63)}]{\substack{R_{1,4}(-26)\\R_{2,4}(6)\\R_{3,4}(-91)}}
\begin{pmatrix}
1 & 1 & 0 & 1 & 0 & -103\\
0 & 0 & 1 & 1 & 0 & 23\\
0 & 0 & 0 & -2 & 0 & -358\\
0 & 0 & 0 & 0 & 1 & 6\\
0 & 0 & 0 & 0 & 0 & 43\\
0 & 0 & 0 & 0 & 0 & 249
\end{pmatrix}$$

$$\xrightarrow{R_5(\frac{1}{43})}
\begin{pmatrix}
1 & 1 & 0 & 1 & 0 & -103\\
0 & 0 & 1 & 1 & 0 & 23\\
0 & 0 & 0 & -2 & 0 & -358\\
0 & 0 & 0 & 0 & 1 & 6\\
0 & 0 & 0 & 0 & 0 & 1\\
0 & 0 & 0 & 0 & 0 & 249
\end{pmatrix}$$

$$\xrightarrow[\substack{R_{4,5}(-6)\\R_{6,5}(-249)}]{\substack{R_{1,5}(103)\\R_{2,5}(-23)\\R_{3,5}(358)}}
\begin{pmatrix}
1 & 1 & 0 & 1 & 0 & 0\\
0 & 0 & 1 & 1 & 0 & 0\\
0 & 0 & 0 & -2 & 0 & 0\\
0 & 0 & 0 & 0 & 1 & 0\\
0 & 0 & 0 & 0 & 0 & 1\\
0 & 0 & 0 & 0 & 0 & 0
\end{pmatrix}$$

$$\xrightarrow[\substack{R_{2,3}(-1)}]{\substack{R_3(-\frac{1}{2})\\R_{1,3}(-1)}}
\begin{pmatrix}
1 & 1 & 0 & 0 & 0 & 0\\
0 & 0 & 1 & 0 & 0 & 0\\
0 & 0 & 0 & 1 & 0 & 0\\
0 & 0 & 0 & 0 & 1 & 0\\
0 & 0 & 0 & 0 & 0 & 1\\
0 & 0 & 0 & 0 & 0 & 0
\end{pmatrix}$$

We have succeeded in arriving at row echelon form without encountering fractions. Contrast this with the systematic procedure.

Since a matrix can be transformed to one and only one matrix in row echelon form with row operations, you may use any choice of row operations to simplify your calculations. We record this fact for future reference. We do not give any reasons for this, because the reasons involve technicalities which lead us away from systems of linear equations.

Proposition 1-1_____

Any matrix can be transformed by row operations to one and only one matrix in row echelon form.

_____

Brief Exercises

Transform the following matrices to row echelon form using row operations. Compare these calculations with those of the systematic procedure.

1. $\begin{pmatrix} 1 & 2 \\ 1 & 1 \end{pmatrix}$

2. $\begin{pmatrix} 3 & 1 & 2 \\ 1 & 2 & 1 \end{pmatrix}$

3. $\begin{pmatrix} 1 & 1 & 0 & 1 \\ 2 & 1 & 3 & 0 \\ 0 & 1 & 2 & 1 \\ -4 & 1 & 1 & 1 \end{pmatrix}$

4. $\begin{pmatrix} 2 & 0 & 1 & 3 \\ 4 & 1 & 2 & 1 \\ 1 & 1 & 3 & 2 \end{pmatrix}$

Answers:

1 $\begin{pmatrix} 1 & 0 \\ 0 & 1 \end{pmatrix}$

2 $\begin{pmatrix} 1 & 0 & \frac{3}{5} \\ 0 & 1 & \frac{1}{5} \end{pmatrix}$

3 $\begin{pmatrix} 1 & 0 & 0 & 0 \\ 0 & 1 & 0 & 0 \\ 0 & 0 & 1 & 0 \\ 0 & 0 & 0 & 1 \end{pmatrix}$

4 $\begin{pmatrix} 1 & 0 & 0 & \frac{2}{5} \\ 0 & 1 & 0 & -5 \\ 0 & 0 & 1 & \frac{11}{5} \end{pmatrix}$

SUMMARY   *A matrix is in row echelon form if:*

(1) *the zero rows are below the nonzero rows,*

(2) *each nonzero row has a leading 1,*

(3) *each leading 1 is the only nonzero entry in its column, and*
(4) *each leading 1 is positioned to the right of the leading 1 of the preceding row.*

Every matrix can be transformed to one and only one matrix in row echelon form by row operations. Because of this, one may follow either the systematic procedure, or he may choose row operations in an order which makes the calculations simpler.

Exercises for Section 1-2

1. Determine whether the following matrices are in row echelon form.

(a) $\begin{pmatrix} 1 & 0 & 0 & 0 \\ 0 & 0 & 1 & 1 \\ 0 & 0 & 0 & 1 \end{pmatrix}$

(c) $\begin{pmatrix} 1 & 2 & 5 & 0 & 3 \\ 0 & 0 & 0 & 1 & 1 \\ 0 & 0 & 0 & 0 & 0 \end{pmatrix}$

(b) $\begin{pmatrix} 1 & 2 & 0 & 0 \\ 0 & 0 & 1 & 1 \\ 0 & 0 & 0 & 0 \end{pmatrix}$

(d) $\begin{pmatrix} 1 & 1 & 0 \\ 0 & 0 & 1 \\ 1 & 0 & 0 \end{pmatrix}$

2. Transform the following matrices to row echelon form.

(a) $\begin{pmatrix} 4 & 1 & 3 & 0 \\ 0 & 1 & -1 & 2 \\ 1 & 5 & 1 & 1 \end{pmatrix}$

(c) $\begin{pmatrix} 1 & -1 & 2 & 0 \\ 3 & 0 & -1 & 2 \\ 1 & -4 & 0 & 3 \end{pmatrix}$

(b) $\begin{pmatrix} 2 & 1 & -1 & 3 & 2 \\ 5 & -1 & 4 & -1 & 1 \end{pmatrix}$

(d) $\begin{pmatrix} 2 & 0 & -1 & 3 \\ 5 & 2 & 8 & -1 \\ 0 & 1 & 1 & 0 \\ 1 & 2 & 0 & 1 \end{pmatrix}$

3. Transform the following matrices to row echelon form in the systematic way described above and then in a way which involves simpler arithmetic.

(a) $\begin{pmatrix} 3 & 0 & 2 & 1 \\ 0 & 5 & 1 & 0 \\ 2 & 1 & 1 & 0 \\ 1 & 0 & 3 & 4 \end{pmatrix}$

(b) $\begin{pmatrix} 2 & 1 & 0 & 1 & 3 \\ 1 & -2 & 1 & 5 & 0 \\ 4 & 1 & 8 & 3 & 1 \\ 1 & 0 & 0 & 1 & 0 \end{pmatrix}$

More Challenging Exercises

4. In this exercise consider the system of equations

$$5x + 2y + 3z + w = 1$$
$$x - y + z - w = 2$$
$$3x + 4y + z + 3w = -3.$$

(a) Transform the augmented matrix for this system to row echelon form. The equations obtained from the row echelon form allow us to solve for x and y in terms of z and w. Do it.

(b) Begin with the augmented matrix for the original system again, and try to solve for x and z in terms of y and w. Do this with the matrix notation; this means you should try to get initial ones appearing in the first and third columns.

(c) Again begin with the augmented matrix for the original system and attempt to solve for y and w in terms of x and z. Is it possible? How can you phrase this part of the problem in matrix terms?

5. Consider the system of equations

$$x - 3y + 2z = 1$$
$$2x - 6y + z = 2.$$

Can you solve for x and y in terms of z? For x and z in terms of y? For y and z in terms of x?

6. (a) Construct a system of three linear equations in four unknowns with which it is possible to solve for any two unknowns in terms of the remaining two.

(b) Construct another system of three linear equations in the four unknowns with which it is possible to solve for x and y in terms of z and w, but it is not possible to solve for x and w in terms of y and z.

7. Suppose you are given a system of six linear equations in seven unknowns. In matrix terms, how many nonzero rows can there be in the row echelon form of the augmented matrix? Can the solution set consist of exactly one 7-tuple, or must there be infinitely many solutions? What is the situation if there are six equations and eight unknowns? What if there are n equations and more than n unknowns; can there be just one solution or must there be infinitely many?

8. Suppose we are told that the solution set for a system of three linear equations in the four unknowns x, y, z, and w is formed by all 4-tuples of the form $(x, y, 3 + 2x - y, 4 + x - 2y)$. In this exercise we show how to find systems of equations with this solution set.

Our approach is to construct an augmented matrix in a form akin to row echelon form which has this solution set. Note that the solution set can be expressed by

$$z = 3 + 2x - y$$
$$w = 4 + x - 2y.$$

These equations can be written in the following way.

$$-2x + y + z \quad\quad = 3$$
$$-x + 2y + \quad\quad w = 4$$

The augmented matrix for this system of equations is

$$\begin{pmatrix} -2 & 1 & 1 & 0 & 3 \\ -1 & 2 & 0 & 1 & 4 \\ 0 & 0 & 0 & 0 & 0 \end{pmatrix}.$$

We can then obtain many systems of equations having the same solution set by applying elementary row operations to this matrix.

9. In each of the following, construct a system of four equations in three unknowns having the solution set indicated. It will be acceptable to give the augmented matrix for the system rather than the system itself.

(a) $(x, 2x, 3 - x)$ (c) $(y + z, y, z)$

(b) $(2, 3 - z, z)$ (d) $(3,1,4)$

10. In Exercise 9, was it possible to decide by inspection that the solution set belonged to a homogeneous system?

It is also possible to operate with the columns of matrices as we have with the rows.

Definition 1-5
The *column operations* on a matrix are:
(1) interchange of the ith and jth columns, denoted by $C_{i,j}$,
(2) multiplication of the ith column by a nonzero real number x, denoted by $C_i(x)$, and
(3) addition of x times the jth column to the ith column, denoted by $C_{i,j}(x)$ $(i \neq j)$.

11. Apply the column operations $C_{1,2}$, $C_{4,1}$, $C_{3,2}$, $C_2(-2)$, $C_4(-3)$, $C_{1,2}(-2)$, $C_{3,2}(-3)$, and $C_{4,2}(-4)$ to the following matrix.

$$\begin{pmatrix} 2 & 1 & 3 & 4 \\ -1 & 2 & 0 & 3 \\ 4 & 1 & 3 & 2 \end{pmatrix}$$

Corresponding to the row echelon form for matrices there is also a column echelon form.

Definition 1-6_____

A matrix is in *column echelon form* provided:
(1) the zero columns are placed to the right of the nonzero columns,
(2) each nonzero column has a leading 1,
(3) each leading 1 is the only nonzero entry in its row, and
(4) each leading 1 is positioned lower than the leading 1 in the preceding column.

12. Determine whether the following matrices are in column echelon form. Give reasons for your answers.

(a) $\begin{pmatrix} 1 & 0 & 0 \\ 0 & 1 & 0 \\ 0 & 0 & 0 \end{pmatrix}$

(c) $\begin{pmatrix} 1 & 0 & 0 & 1 \\ 1 & 0 & 0 & 0 \\ 0 & 0 & 1 & 0 \end{pmatrix}$

(b) $\begin{pmatrix} 1 & 0 & 0 \\ 0 & 0 & 0 \\ 0 & 1 & 0 \end{pmatrix}$

(d) $\begin{pmatrix} 1 & 0 & 0 \\ 0 & 1 & 0 \\ 2 & 1 & 0 \\ 1 & 1 & 0 \end{pmatrix}$

There is a proposition concerning column echelon form which parallels that for row echelon form.

Proposition 1-2_____

Any matrix can be transformed by column operations to one and only one matrix in column echelon form.

13. Transform the following matrices to column echelon form.

(a) $\begin{pmatrix} 1 & 1 \\ 2 & 3 \end{pmatrix}$ **(b)** $\begin{pmatrix} -1 & 1 \\ 0 & 2 \end{pmatrix}$ **(c)** $\begin{pmatrix} 1 & 2 & 1 \\ 0 & 1 & 2 \end{pmatrix}$ **(d)** $\begin{pmatrix} 1 & 1 & 2 & 1 \\ 3 & 1 & 0 & 2 \\ 1 & 0 & -1 & 3 \end{pmatrix}$

14. Transform the matrices listed in Exercise 13 above into row echelon form. Conclude that sometimes the row echelon form and the column echelon form of a matrix are the same and sometimes they are different.

Definition 1-7_____
If we are given a matrix A, we can form a new matrix A^T using the first column of A as the first row of A^T, the second column of A as the second row of A^T, and so on. The matrix A^T is called *the transpose of A*.
_____

15. Take the transpose of the following matrices.

(a) $\begin{pmatrix} 1 & 1 & 2 \\ 1 & 2 & 3 \end{pmatrix}$ **(b)** $\begin{pmatrix} 3 & 1 \\ 0 & 1 \end{pmatrix}$ **(c)** $\begin{pmatrix} 5 & 1 & 2 & 3 \\ 1 & -4 & 3 & 2 \\ 1 & 1 & 3 & 1 \end{pmatrix}$

16. Transform the matrices in Exercise 1 into column echelon form in two different ways: (1) use column operations and (2) use row operations on the transpose to obtain the row echelon form, and then take the transpose of the matrix in row echelon form.

SECTION 1-3
Notation for Systems of Linear Equations and for Matrices

In the previous pages we discussed systems of linear equations having relatively few unknowns. If we have a large number of unknowns, say 17 for example, we list them by writing $x_1, x_2, \ldots, x_{17}$ or by writing $x_1, \ldots, x_{17}$ rather than by writing every unknown explicitly in the list. Inserting dots is a way to shorten long lists.

In a similar way we use three dots in writing linear equations. A linear equation in the two unknowns x_1 and x_2 must have the form

$$a_1 x_1 + a_2 x_2 = y.$$

The letters a_1, a_2, and y represent numbers which are known, and the letters x_1 and x_2 represent numbers which are not known. A linear equation in the three unknowns x_1, x_2, and x_3 has the form

$$a_1 x_1 + a_2 x_2 + a_3 x_3 = y.$$

To represent a linear equation in, say, the 17 unknowns $x_1, \ldots, x_{17}$, we write

$$a_1 x_1 + a_2 x_2 + \cdots + a_{17} x_{17} = y$$

or

$$a_1 x_1 + \cdots + a_{17} x_{17} = y.$$

Here, as before, $a_1, a_2, \ldots, a_{17}$, and y represent known quantities, and x_1, $x_2, \ldots, x_{17}$ represent unknown quantities.

There are times when we want to talk about a linear equation in general, i.e., we want to make statements about every linear equation, no matter how many or how few unknowns it has. We therefore want a symbol which can represent every possible linear equation. Since every linear equation has a certain number of unknowns, we use a letter to represent the number of unknowns; here we use the letter n. The unknowns are therefore represented by

$$x_1, x_2, \ldots, x_n.$$

The device of using the letter n allows us to represent any number of unknowns in this way.

A linear equation in the n unknowns $x_1, x_2, \ldots, x_n$ is represented by

$$a_1 x_1 + a_2 x_2 + \cdots + a_n x_n = y$$

where, as before, $a_1, a_2, \ldots, a_n$, and y are known quantities. For $n = 1$, this equation is to be interpreted $a_1 x_1 = y$. For $n = 2$, it should be read as $a_1 x_1 + a_2 x_2 = y$. Thus, by using a letter, such as n, for the number of unknowns, we can represent any linear equation by the one general equation

$$a_1 x_1 + \cdots + a_n x_n = y.$$

(Whether we write the second quantity in a list or the second term in a sum depends upon the context; if it seems that writing a second term is necessary for clarity, then we do it. Otherwise we omit it.)

Suppose that we want to discuss a system of three linear equations in four unknowns. It is quite common to use the same letter for all the coefficients in all the equations. To distinguish between the coefficients we use two subscripts, the first to indicate the equation in which it appears, and the second to indicate the unknown it multiplies. It is also necessary to use subscripts on the constants to indicate the equation in which they appear. The first equation is represented by

$$a_{11} x_1 + a_{12} x_2 + a_{13} x_3 + a_{14} x_4 = y_1.$$

The second equation is written as

$$a_{21} x_1 + a_{22} x_2 + a_{23} x_3 + a_{24} x_4 = y_2,$$

and the third is given by

$$a_{31} x_1 + a_{32} x_2 + a_{33} x_3 + a_{34} x_4 = y_3.$$

When we wish to speak of a system of linear equations in general, we use one letter for the number of unknowns and another for the number of equations. Here we use n for the number of unknowns and m for the number of equations. We symbolize a system of m linear equations in the n unknowns $x_1, \ldots, x_n$ by

$$a_{11}x_1 + a_{12}x_2 + \cdots + a_{1n}x_n = y_1$$

$$a_{21}x_1 + a_{22}x_2 + \cdots + a_{2n}x_n = y_2$$

$$\vdots \qquad \vdots \qquad \qquad \vdots$$

$$a_{m1}x_1 + a_{m2}x_2 + \cdots + a_{mn}x_n = y_m.$$

The a's and the y's are understood as being known, and the x's are unknown.

If we want to speak of a typical equation in this system, then we speak of the ith equation

$$a_{i1}x_1 + a_{i2}x_2 + \cdots + a_{in}x_n = y_i$$

or the jth equation

$$a_{j1}x_1 + a_{j2}x_2 + \cdots + a_{jn}x_n = y_j$$

or something similar. Notice that the first subscript indicates the number of the equation.

Just as we need a way to describe a general linear equation and a general system of linear equations, we also need a way to describe any matrix. Every matrix has a certain number of rows and a certain number of columns, so we use one letter to represent the number of rows and one to represent the number of columns. For example, we often use m for the number of rows and n for the number of columns; in that case we say we have an $m \times n$ matrix. An example of the way we represent a general $m \times n$ matrix is

$$A = \begin{pmatrix} a_{11} & a_{12} & \cdots & a_{1n} \\ a_{21} & a_{22} & \cdots & a_{2n} \\ \vdots & \vdots & & \vdots \\ a_{m1} & a_{m2} & \cdots & a_{mn} \end{pmatrix}.$$

The a's are called the entries of the matrix. We use two subscripts to indicate the position of each entry in the matrix; the first subscript gives the number of the row and the second gives the number of the column. So, for example, a_{43} stands for the number in the fourth row and third column. We sometimes refer to this position as the (4,3) position or spot in the matrix. If we have more than nine equations or unknowns, then we either use a comma or a space to separate the subscripts; so $a_{14,5}$ or $a_{14,5}$ would be used for the number in the fourteenth row and fifth column.

As an example of the use of this general notation, consider the 3×4 matrix

$$\begin{pmatrix} 2 & 1 & 0 & 4 \\ 1 & -1 & 5 & 1 \\ 2 & 8 & 6 & 7 \end{pmatrix}.$$

The symbols in the general matrix are to be interpreted in the following way to represent the matrix above. Here $m = 3$, $n = 4$, and

$$\begin{array}{llll} a_{11} = 2 & a_{12} = 1 & a_{13} = 0 & a_{14} = 4 \\ a_{21} = 1 & a_{22} = -1 & a_{23} = 5 & a_{24} = 1 \\ a_{31} = 2 & a_{32} = 8 & a_{33} = 6 & a_{34} = 7. \end{array}$$

In a similar fashion any matrix is represented by the general matrix A displayed above.

Using the general notation, the ith row of A is

$$(a_{i1} \quad a_{i2} \quad \cdots \quad a_{in})$$

and the jth column of A is

$$\begin{pmatrix} a_{1j} \\ a_{2j} \\ \cdot \\ \cdot \\ \cdot \\ a_{mj} \end{pmatrix}.$$

With these preliminary remarks, we summarize what we have discussed about systems of linear equations and matrices using the general notation.

Definition 1-8_____

An $m \times n$ matrix is a rectangular array of numbers having the pattern

$$\begin{pmatrix} a_{11} & a_{12} & \cdots & a_{1n} \\ a_{21} & a_{22} & \cdots & a_{2n} \\ \cdot & \cdot & & \cdot \\ \cdot & \cdot & & \cdot \\ \cdot & \cdot & & \cdot \\ a_{m1} & a_{m2} & \cdots & a_{mn} \end{pmatrix}.$$

The ith row of this matrix is

$$(a_{i1} \quad a_{i2} \quad \cdots \quad a_{in}).$$

The jth column is

$$\begin{pmatrix} a_{1j} \\ a_{2j} \\ \cdot \\ \cdot \\ \cdot \\ a_{mj} \end{pmatrix} .$$

An $m \times n$ matrix has m rows and n columns. The a's are called the entries of the matrix. Positions of entries in a matrix are specified by the row and the column in which the entry is placed. The (i,j) position in the matrix is in the ith row and the jth column.

Definition 1-9
A linear equation in the n unknowns $x_1, \ldots, x_n$ is an equation which can be written in the form

$$a_1 x_1 + a_2 x_2 + \cdots + a_n x_n = y.$$

The a's are called the coefficients of the x's, and y is called the constant. The numbers $a_1, \ldots, a_n$, and y are assumed known.

Definition 1-10
A system of m linear equations in the n unknowns $x_1, \ldots, x_n$ is one which can be written in the form

$$a_{11} x_1 + a_{12} x_2 + \cdots + a_{1n} x_n = y_1$$
$$a_{21} x_1 + a_{22} x_2 + \cdots + a_{2n} x_n = y_2$$
$$\vdots$$
$$a_{m1} x_1 + a_{m2} x_2 + \cdots + a_{mn} x_n = y_m.$$

The a's are the coefficients of the unknowns, and the y's are the constants. The coefficients and the constants are assumed to be known, and the x's are assumed not known.

The coefficient matrix for this system is

$$\begin{pmatrix} a_{11} & a_{12} & \cdots & a_{1n} \\ a_{21} & a_{22} & \cdots & a_{2n} \\ \cdot & \cdot & & \cdot \\ \cdot & \cdot & & \cdot \\ \cdot & \cdot & & \cdot \\ a_{m1} & a_{m2} & \cdots & a_{mn} \end{pmatrix} .$$

The augmented matrix for this system is

$$\begin{pmatrix} a_{11} & a_{12} & \cdots & a_{1n} & y_1 \\ a_{21} & a_{22} & \cdots & a_{2n} & y_2 \\ \cdot & \cdot & & \cdot & \cdot \\ \cdot & \cdot & & \cdot & \cdot \\ \cdot & \cdot & & \cdot & \cdot \\ a_{m1} & a_{m2} & \cdots & a_{mn} & y_m \end{pmatrix} .$$

SECTION 1-4
Some Applications of Systems of Linear Equations

Systems of linear equations arise in many different areas. It would be impossible to give an instance of every place where they occur; we give a few so that you can see that they are used in situations outside mathematics.

Example   Some simple electrical circuits can be analyzed with linear equations. Suppose we have the circuit indicated in Figure 1-1. The dimensions used in this

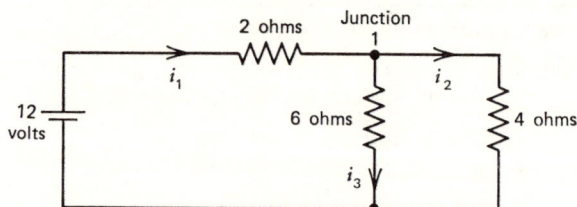

Figure 1-1

example are ohms for resistors, amperes for current, and volts for potential. There are three facts about circuits which we must use:

(1) Ohm's law: the potential difference across a resistor is the product of the current and the value of the resistor;
(2) Kirchhoff's law of current: the sum of the currents flowing from a junction equals the sum of the currents flowing into the junction;
(3) Kirchhoff's law of voltages: the algebraic sum of the voltages in any closed loop is zero.

At junction 1 we use Kirchhoff's law of current to obtain the equation

$$i_1 = i_2 + i_3 \qquad \text{or} \qquad i_1 - i_2 - i_3 = 0.$$

Using Ohm's law we have that the voltage drop across the 2 ohm resistor is $2i_1$, across the 6 ohm resistor is $6i_3$, and across the 4 ohm resistor is $4i_2$. Using Kirchhoff's law of voltage in the left-hand loop we have

$$2i_1 + 6i_3 = 12.$$

In the right-hand loop we get

$$4i_2 - 6i_3 = 0.$$

In the large loop we get

$$2i_1 + 4i_2 = 12.$$

Therefore, we have the system of linear equations

$$
\begin{aligned}
i_1 - i_2 - i_3 &= 0 \\
2i_1 \qquad\;\; + 6i_3 &= 12 \\
4i_2 - 6i_3 &= 0 \\
2i_1 + 4i_2 \qquad\;\; &= 12.
\end{aligned}
$$

The solution to this system, as you can verify, is $i_1 = \frac{30}{11}$, $i_2 = \frac{18}{11}$, and $i_3 = \frac{12}{11}$. Thus we have found the currents by solving a system of linear equations.

Example Suppose four foods are used to make a meal. The foods have the following units per ounce of vitamin B and vitamin C, respectively.

	Vitamin B	Vitamin C
Food 1	8	2
Food 2	4	6
Food 3	7	10
Food 4	3	6

Is it possible to combine these foods into a meal of 16 oz with 96 units of vitamin B and 8 units of vitamin C?

If x_1, x_2, x_3, and x_4 represent the number of ounces of Foods 1, 2, 3, and 4, respectively, we obtain the equations

$$
\begin{aligned}
x_1 + x_2 + x_3 + x_4 &= 16 \\
8x_1 + 4x_2 + 7x_3 + 3x_4 &= 96 \\
2x_1 + 6x_2 + 10x_3 + 6x_4 &= 80.
\end{aligned}
$$

The solution to this system of equations is

$$
\begin{aligned}
x_1 &= 4 + x_3 \\
x_2 &= 28 - 9x_3 \\
x_4 &= -16 + 7x_3.
\end{aligned}
$$

The requirements of the problem show, of course, that none of the quantities are negative. Thus we must have

$$x_1 = 4 + x_3 \geq 0 \quad \text{or} \quad x_3 \geq -4$$
$$x_2 = 28 - 9x_3 \geq 0 \quad \text{or} \quad \tfrac{28}{9} \geq x_3$$
$$x_4 = -16 + 7x_3 \geq 0 \quad \text{or} \quad x_3 \geq \tfrac{16}{7}.$$

Since $\tfrac{28}{9}$ is approximately 3.1, and $\tfrac{16}{7}$ is approximately 2.2, there is a way to satisfy the requirements of this diet by choosing the quantity of Food 3 between 2.2 and 3.1 oz and by choosing the quantity of the other foods as given by the solution equations. Thus we have been able to find a diet meeting the original specifications by solving a system of linear equations and some inequalities imposed by the nature of the problem.

Example In a large company there are departments which produce items for sale (call these production departments), and there are departments which supply supportive services to the company but do not produce items for sale (call these service departments). A service department may be the janitorial department, maintenance, the cafeteria, etc. The company is interested in obtaining a cost analysis.

For our example, suppose the company has five Departments, 1, 2, 3, 4, and 5. Suppose that 1 and 2 are service departments and that 3, 4, and 5 are production departments. Each department has certain direct costs: salaries, supplies, raw materials, etc. Furthermore, for the purpose of determining to what extent each department contributes to each product, each department has the costs which the service departments charge for their services. The service departments are allowed to charge all of their direct costs to the departments they service, but they may not charge more than their total costs; i.e., the service departments are not allowed to "make money." The portion of the costs they charge the various departments is determined by the company's administration. (For example, the janitorial department might charge on the basis of the square footage in each office, or by the number of people in each office, or in some other way.) Finally, a service department does not charge itself for its own services.

We illustrate one such situation numerically. Let $x_1, \ldots, x_5$ denote the total costs of Departments $1, \ldots, 5$, respectively.

	Direct costs	1	2	3	4	5
1	$3,000	0	$\tfrac{1}{4}x_2$	0	0	0
2	$4,000	$\tfrac{1}{6}x_1$	0	0	0	0
3	$6,000	$\tfrac{1}{3}x_1$	$\tfrac{1}{4}x_2$	0	0	0
4	$8,000	$\tfrac{1}{3}x_1$	$\tfrac{1}{4}x_2$	0	0	0
5	$5,000	$\tfrac{1}{6}x_1$	$\tfrac{1}{4}x_2$	0	0	0

Each row shows the costs incurred by that department. For example, Department 1 has $3,000 in direct costs and $\frac{1}{4}$ of the costs of Department 2; Department 3 has $6,000 in direct costs, $\frac{1}{3}$ of the costs of Department 1, and $\frac{1}{4}$ of the costs of Department 2. This information can also be stored in a system of equations

$$x_1 = 3,000 \qquad\quad + \tfrac{1}{4}x_2$$
$$x_2 = 4,000 + \tfrac{1}{6}x_1$$
$$x_3 = 6,000 + \tfrac{1}{3}x_1 + \tfrac{1}{4}x_2$$
$$x_4 = 8,000 + \tfrac{1}{3}x_1 + \tfrac{1}{4}x_2$$
$$x_5 = 5,000 + \tfrac{1}{6}x_1 + \tfrac{1}{4}x_2.$$

The solution to this system of equations is, approximately, $x_1 = \$4,200$, $x_2 = \$4,700$, $x_3 = \$8,600$, $x_4 = \$10,600$, and $x_5 = \$6,900$.

Example   Suppose a simple economy consists of three kinds of industries, I_1, I_2, and I_3, and suppose that they need only the three products of these three industries for continued operation. Each industry uses the products of itself and the other two in a certain ratio as given by the following matrix; we indicate the products of the respective industries by P_1, P_2, and P_3.

$$
\begin{array}{c c}
 & \begin{matrix} P_1 & P_2 & P_3 \end{matrix} \\
\begin{matrix} I_1 \\ I_2 \\ I_3 \end{matrix} &
\begin{pmatrix} \frac{1}{6} & \frac{1}{3} & \frac{1}{3} \\[4pt] \frac{1}{4} & \frac{1}{3} & \frac{1}{2} \\[4pt] \frac{7}{12} & \frac{1}{3} & \frac{1}{6} \end{pmatrix}
\end{array}
$$

Suppose that the income to I_1 is c_1, to I_2 is c_2, and to I_3 is c_3. The manufacturers would prefer not to have to borrow any money, and definitely not from anyone outside the group. Consequently they are interested in maintaining an equilibrium condition in which the amount of money each receives for the goods he produces will equal the amount he must pay for the goods he purchases. This condition can be formulated as follows.

$$\tfrac{1}{6}c_1 + \tfrac{1}{3}c_2 + \tfrac{1}{3}c_3 = c_1$$
$$\tfrac{1}{4}c_1 + \tfrac{1}{3}c_2 + \tfrac{1}{2}c_3 = c_2$$
$$\tfrac{7}{12}c_1 + \tfrac{1}{3}c_2 + \tfrac{1}{6}c_3 = c_3$$

This gives rise to a homogeneous system of equations with the coefficient matrix

$$
\begin{pmatrix} \frac{5}{6} & -\frac{1}{3} & -\frac{1}{3} \\[4pt] -\frac{1}{4} & \frac{2}{3} & -\frac{1}{2} \\[4pt] -\frac{7}{12} & -\frac{1}{3} & \frac{5}{6} \end{pmatrix}.
$$

The row echelon form for this matrix is

$$\begin{pmatrix} 1 & 0 & -\frac{14}{17} \\ 0 & 1 & -\frac{18}{17} \\ 0 & 0 & 0 \end{pmatrix}.$$

This means that $c_1 = \frac{14}{17}c_3$ and $c_2 = \frac{18}{17}c_3$. If the manufacturers want to achieve this equilibrium condition, then each must adjust his prices so that their incomes are in the ratio $(c_1:c_2:c_3) = (14:18:17)$.

While this example may seem artificial at first glance, if we think of including all the manufacturers in a nation, assuming that they deal only with others within the nation, and if we assume a fixed money supply, then the model takes on a greater semblance of reality. The example above is an extremely simple illustration of what is called the Leontief closed model. There is also a model for an economy in which goods produced can be sold outside the given set of industries.

2

Vector Spaces

Introduction to Vector Spaces

PREVIEW *This section gives an introduction to the idea of a vector space. We discuss in some detail the vector spaces $\mathbb{R}^2$, $\mathbb{R}^3$, $\mathbb{R}^4$, and $\mathbb{R}^n$. We list some computational properties in these spaces, concentrating on those which are used to define a vector space in general. The section ends with the definition of a vector space over $\mathbb{R}$.*

In linear algebra we need two kinds of objects, real numbers and vectors.

Vectors have long been used in physics and engineering to study things having a size and a direction. For example, in describing a push, you need to tell how hard you were pushed and in which direction you were pushed; when you are standing at the edge of a cliff it is essential to know whether you are pushed with a force of $\frac{1}{2}$ lb or 250 lb, and also whether you are pushed toward the edge or away from the edge of the cliff. To describe a force accurately it is necessary to use something more than just a real number.

Other things, however, can be described very nicely with a number. Temperature, for example, is described by a real number relative to some standard scale; to say your temperature is 98.6°F describes your temperature accurately. As another example, we know that water boils at 212°F or at 100°C. The boiling point of water is therefore given either by the number 212 or the number 100, depending upon whether we use the Fahrenheit or the Celsius scale. Height can also be described by a real number once a scale is chosen, e.g., feet, inches, centimeters. Quantities of this sort are called scalar quantities.

Because of these considerations, it is quite common in the physical sciences for a "vector" to be defined as a "directed quantity" and for "scalar" to be used as a synonym for "real number." This is because these are the situations in which the physical scientists use the terms vector and scalar.

Other areas of study, also, have some situations which can be described by a real number and some which cannot. Quite often they borrow from the physical sciences: scalar is used for real number, scalar quantity is used for things which can be described using a real number, and vector is used for quantities which are not scalars. In economics, for example, it is not sufficient to give one number to describe the total financial picture of a family at a particular moment. One needs many numbers; e.g., cash on hand, savings, installment loan debts, car debt, house debt, investments, income, etc. The composite picture of the family's finances could be described by a list of numbers, each relating to a different aspect of the financial picture; this list of numbers constitutes the family's "financial vector."

Another example arises in the health checkup. The doctor takes various readings, and this list of numbers gives some indication of your state of health; it is your "health vector."

In constructing a mathematical description of a vector, we want our term to apply to as many different areas as possible. So we do not use the physical definition "directed quantity." It is not clear, for example, how to assign a direction to a family's financial vector or to a health vector. Rather than use any of the particular situations as the basis for our definition, we use computational properties to define mathematical vectors.

We follow the usual procedure of using scalar as a synonym for real number, particularly when a distinction between a real number and a vector is important.

Definition 2-1_____
The word *scalar* is used to mean *real number*. When we refer to the totality of all real numbers, or to the set of all real numbers, it is convenient to use a special symbol rather than continually to repeat the phrase "the set of all real numbers." We use the symbol $\mathbb{R}$.

_____

Brief Exercises

1. Give some examples of situations which are described adequately by one number and may reasonably be called scalars.

2. Give some examples of situations which cannot be described adequately by using just one number. (Hint: automobiles, motorcycles, air contaminants, water pollutants, household finances, measurements for clothes, etc.)

Now we give examples of the mathematical concept of a vector.

Example Let $\mathbb{R}^2$ denote the collection of all ordered pairs of real numbers. Examples of ordered pairs are $(0,0)$, $(2,\frac{1}{3})$, $(-\frac{1}{4},\pi)$, etc. These are the possible solutions of systems of equations having two unknowns. We call these ordered pairs "vectors." We also use phrases such as "vectors in two-dimensional space" or "two-dimensional vectors." We sometimes refer to $\mathbb{R}^2$ as "two-dimensional space" or as "2-space." The numbers in a pair are called components or coordinates. Two pairs must have their first components equal and their second components equal to be called equal pairs. So, for example, $(2,3) \neq (3,3)$ since their first components are different.

We do arithmetic with vectors in $\mathbb{R}^2$. We add two vectors by adding their first components and their second components.

$$(1,5) + (2,4) = (1+2, 5+4) = (3,9)$$

The sum of the vectors $(1,5)$ and $(2,4)$ is thus $(3,9)$. Other vectors are added in the same way: $(-1,3) + (4,-2) = (3,1)$; $(1,2) + (0,0) = (1,2)$; and $(-1,3) + (1,-3) = (0,0)$.

We also multiply a vector in $\mathbb{R}^2$ by a scalar by multiplying each component of the vector by the scalar: $4(2,3) = (4 \cdot 2, 4 \cdot 3) = (8,12)$, $2(3,-1) = (6,-2)$, $(-1)(3,2) = (-3,-2)$, and $0(1,5) = (0,0)$. (This is also referred to as "scalar multiplication." Be careful not to misinterpret this phrase as multiplication of scalars.)

These agreements about adding vectors and multiplying them by scalars allow us to do some arithmetic with vectors. This arithmetic is different from arithmetic of real numbers because, for example, there is no multiplication of vectors, nor is there division of vectors. But there are many similarities.

Brief Exercises

1. Perform the following arithmetic in $\mathbb{R}^2$.

$(2,1) + (3,4)$	$[(2,1) + (3,-2)] + (5,6)$	$(3,1) + (?,?) = (0,0)$
$(3,4) + (2,1)$	$(2,1) + [(3,-2) + (5,6)]$	$(?,?) + (-4,2) = (0,0)$
$(1,-2) + (-4,3)$	$(3,4) + (0,0)$	$(-1)(4,3)$
$(-4,3) + (1,-2)$	$(-6,2) + (0,0)$	$(-1)(-6,7)$

2. Are the following equal?

$$3(4,3) + 2(4,3) = 5(4,3)$$
$$-6(2,3) + 3(2,3) = 3(2,3)$$
$$6(2,3) + 6(-3,1) = 6[(2,3) + (-3,1)]$$
$$(-4)(1,2) + (-4)(5,8) = (-4)[(-1,2) + (5,8)]$$

3. Solve the following vector equations.

(a) $(2,1) + (?,?) = (6,4)$ (c) $(6,8) + 3(?,?) = (1,5)$

(b) $(5,-3) + (?,?) = (2,1)$

Selected Answers: **3(a)** $(4,3)$ **3(b)** $(-3,4)$ **3(c)** $(-\frac{5}{3},-1)$

We also describe vector arithmetic using letters for components of vectors. For example, we can use (a,b) and (c,d) to represent any vectors in $\mathbb{R}^2$. Then we describe addition of vectors by the equation

$$(a,b) + (c,d) = (a + c, b + d). \tag{1}$$

The particular case $(1,5) + (2,4) = (3,9)$ is covered by equation (1) by interpreting $a = 1$, $b = 5$, $c = 2$, and $d = 4$. In fact, equation (1) covers any case of addition of vectors in $\mathbb{R}^2$ by having the letters a, b, c, and d stand for different numbers.

We can also use different letters for the same result. If we use (x_1,x_2) and (y_1,y_2) for vectors in $\mathbb{R}^2$, then vector addition is described by

$$(x_1,x_2) + (y_1,y_2) = (x_1 + y_1, x_2 + y_2).$$

So, for example, $(1,5) + (2,4) = (3,9)$ is obtained with $x_1 = 1$, $x_2 = 5$, $y_1 = 2$, and $y_2 = 4$.

To express multiplication of the vector (a,b) in $\mathbb{R}^2$ by the scalar r, we use the equation

$$r(a,b) = (ra,rb).$$

So the equation $2(3,-1) = (6,-2)$ is obtained by interpreting $r = 2$, $a = 3$, and $b = -1$.

Definition 2-2

The collection of all ordered pairs of real numbers is denoted by $\mathbb{R}^2$. We call these pairs "vectors." Addition of vectors in $\mathbb{R}^2$ and multiplication by scalars are defined as follows.

$$(a,b) + (c,d) = (a + c, b + d)$$
$$r(a,b) = (ra,rb)$$

We call these vectors "vectors in two-dimensional space," or "vectors in 2-space." We call $\mathbb{R}^2$ "two-dimensional space" or "2-space." The number a is the first component or coordinate of (a,b), and b is the second component or coordinate.

Another example of vectors in mathematics is $\mathbb{R}^3$ which we discuss in the next example.

Example Let $\mathbb{R}^3$ denote the collection of all ordered triples of real numbers. Examples of such ordered triples are $(1,0,-5)$, $(3,4,2)$, $(-5,\frac{1}{17},\sqrt{3})$, etc. We call these ordered triples "vectors," "vectors in three-dimensional space," "vectors in 3-space," or "three-dimensional vectors."

Two ordered triples are equal provided their corresponding components are equal.

If (x_1,x_2,x_3) and (y_1,y_2,y_3) are vectors in $\mathbb{R}^3$, we define vector addition by

$$(x_1,x_2,x_3) + (y_1,y_2,y_3) = (x_1 + y_1, x_2 + y_2, x_3 + y_3).$$

This means the sum is obtained by adding the first components, by adding the second components, and by adding the third components. Thus, for example, $(2,1,3) + (-3,4,1) = (-1,5,4)$ and $(5,0,-3) + (-1,2,1) = (4,2,-2)$.

If (x_1,x_2,x_3) is a vector in $\mathbb{R}^3$, we define multiplication of (x_1,x_2,x_3) by the scalar r with the equation

$$r(x_1,x_2,x_3) = (rx_1,rx_2,rx_3).$$

This means that a scalar multiple of (x_1,x_2,x_3) is obtained by multiplying each component by the scalar. Thus, for example, $3(2,1,1) = (6,3,3)$ and $-5(3,-2,1) = (-15,10,-5)$.

Definition 2-3

The collection of all ordered triples of real numbers is denoted by $\mathbb{R}^3$. We call these triples "vectors," "vectors in three-dimensional space," or "vectors in 3-space." We call $\mathbb{R}^3$ "three-dimensional space" or "3-space." Addition of vectors in $\mathbb{R}^3$ and multiplication by scalars are defined as follows.

$$(x_1,x_2,x_3) + (y_1,y_2,y_3) = (x_1 + y_1, x_2 + y_2, x_3 + y_3)$$
$$r(x_1,x_2,x_3) = (rx_1,rx_2,rx_3)$$

The numbers x_1, x_2, and x_3 are the first, second, and third components or co-ordinates of (x_1,x_2,x_3).

Brief Exercises

1. Perform the following calculations.

$(2,1,0) + (3,1,4)$ $3(5,6,-3) + (1,5,1)$

$(5,2,-1) + (3,-4,2)$ $(1,5,1) + 3(5,6,-3)$

$[3(6,-3,2) + 4(8,1,2)] + 5(1,1,1)$

2. Are the following true?

$$6(2,1,0) + (-5)(2,1,0) = (2,1,0)$$
$$3(-1,3,4) + 3(1,2,1) = 3[(-1,3,-4) + (1,2,1)]$$

3. Solve the equations

(a) $(2,1,3) + (?,?,?) = (8,1,-1)$ (b) $2(1,1,1) + 5(?,?,?) = (1,-1,4)$.

Selected Answers: 3(a) $(6,0,-4)$ 3(b) $(-\frac{1}{5},-\frac{3}{5},\frac{2}{5})$

Notice that the ways in which we add vectors and multiply by scalars in $\mathbb{R}^2$ and in $\mathbb{R}^3$ are very similar. The similarity does not stop at the definitions. There are many facts about arithmetic in $\mathbb{R}^2$ and $\mathbb{R}^3$ which are the same. Indeed, the reason that we use the term vector for both is precisely because they share some computational properties. We list these properties in the next example.

Example   Here are some common rules of computations in $\mathbb{R}^2$ and in $\mathbb{R}^3$. In listing these rules or properties we use r and s for scalars, (a,b), (c,d), and (e,f) for vectors in $\mathbb{R}^2$, and (x_1,x_2,x_3), (y_1,y_2,y_3), and (z_1,z_2,z_3) for vectors in $\mathbb{R}^3$.

(1) (Closure or stability under addition)

The sum $(a,b) + (c,d)$ is a vector in $\mathbb{R}^2$.

The sum $(x_1,x_2,x_3) + (y_1,y_2,y_3)$ is a vector in $\mathbb{R}^3$.

(2) (Associativity of addition)

$[(a,b) + (c,d)] + (e,f) = (a,b) + [(c,d) + (e,f)]$.

$[(x_1,x_2,x_3) + (y_1,y_2,y_3)] + (z_1,z_2,z_3) = (x_1,x_2,x_3) + [(y_1,y_2,y_3) + (z_1,z_2,z_3)]$.

(3) (Commutativity of addition)

$(a,b) + (c,d) = (c,d) + (a,b)$.

$(x_1,x_2,x_3) + (y_1,y_2,y_3) = (y_1,y_2,y_3) + (x_1,x_2,x_3)$.

(4) (Existence of a zero vector)

There is precisely one zero vector $(0,0)$ in $\mathbb{R}^2$; its distinguishing feature is that $(a,b) + (0,0) = (a,b)$.

There is precisely one zero vector $(0,0,0)$ in $\mathbb{R}^3$; its distinguishing feature is that $(x_1,x_2,x_3) + (0,0,0) = (x_1,x_2,x_3)$.

(5) (Existence of a negative for each vector)

Each vector (a,b) has precisely one negative $(-a,-b)$; its distinguishing feature is that $(a,b) + (-a,-b) = (0,0)$.

Each vector (x_1,x_2,x_3) has precisely one negative $(-x_1,-x_2,-x_3)$; its distinguishing feature is that $(x_1,x_2,x_3) + (-x_1,-x_2,-x_3) = (0,0,0)$.

(6) (Closure or stability under scalar multiplication)

The scalar multiple $r(a,b)$ is also a vector in $\mathbb{R}^2$.

The scalar multiple $r(x_1,x_2,x_3)$ is also a vector in $\mathbb{R}^3$.

(7) $r[s(a,b)] = (rs)(a,b) = s[r(a,b)]$

$r[s(x_1,x_2,x_3)] = (rs)(x_1,x_2,x_3) = s[r(x_1,x_2,x_3)]$.

(8) $(r + s)(a,b) = r(a,b) + s(a,b)$

$(r + s)(x_1,x_2,x_3) = r(x_1,x_2,x_3) + s(x_1,x_2,x_3)$.

(9) $r[(a,b) + (c,d)] = r(a,b) + r(c,d)$

$r[(x_1,x_2,x_3) + (y_1,y_2,y_3)] = r(x_1,x_2,x_3) + r(y_1,y_2,y_3)$.

(10) $1(a,b) = (a,b)$

$1(x_1,x_2,x_3) = (x_1,x_2,x_3)$.

These are properties common to elements of $\mathbb{R}^2$ and elements of $\mathbb{R}^3$ which are the basis of our convention to call the elements vectors.

Brief Exercises

All of the properties listed in the examples above, except for properties 1 and 6, make a statement about numerical computations. Write out what each statement (except 1 and 6) means if the symbols above are given the following numerical values.

1. In $\mathbb{R}^2$, $(a,b) = (3,4)$, $(c,d) = (-1,2)$, $(e,f) = (-1,1)$, $r = 2$, and $s = -3$.

2. In $\mathbb{R}^2$, $(a,b) = (15,-2)$, $(c,d) = (6,-3)$, $(e,f) = (-2,3)$, $r = -1$, and $s = 2$.

3. In $\mathbb{R}^2$, $(a,b) = (4,-4)$, $(c,d) = (2,-3)$, $(e,f) = (7,1)$, $r = 8$, and $s = -5$.

4. In $\mathbb{R}^3$, $(x_1,x_2,x_3) = (3,4,-2)$, $(y_1,y_2,y_3) = (1,1,0)$, $(z_1,z_2,z_3) = (-5,1,3)$, $r = 7$, and $s = -2$.

5. In $\mathbb{R}^3$, $(x_1,x_2,x_3) = (5,8,-6)$, $(y_1,y_2,y_3) = (-1,2,3)$, $(z_1,z_2,z_3) = (4,4,-4)$, $r = 8$, and $s = -1$.

We give another example of vectors.

Example Let $\mathbb{R}^4$ denote the collection of all ordered quadruples. Typical elements are $(2,1,3,0)$, $(\pi^2,\sqrt{18},5,-3)$, etc. We call these vectors, too. We agree that $(x_1,x_2,x_3,x_4) = (y_1,y_2,y_3,y_4)$ if and only if $x_1 = y_1$, $x_2 = y_2$, $x_3 = y_3$, and $x_4 = y_4$.

We define addition of two ordered quadruples (x_1,x_2,x_3,x_4) and (y_1,y_2,y_3,y_4) in $\mathbb{R}^4$ in a way similar to addition of vectors in $\mathbb{R}^2$ and $\mathbb{R}^3$.

$$(x_1,x_2,x_3,x_4) + (y_1,y_2,y_3,y_4) = (x_1 + y_1,\, x_2 + y_2,\, x_3 + y_3,\, x_4 + y_4)$$

We also define multiplication of (x_1,x_2,x_3,x_4) by the scalar r through the equation

$$r(x_1,x_2,x_3,x_4) = (rx_1,rx_2,rx_3,rx_4).$$

As you may suspect, the ordered quadruples in $\mathbb{R}^4$ with these operations of addition and scalar multiplication satisfy all the computational properties listed for $\mathbb{R}^2$ and $\mathbb{R}^3$ in the previous example. Because these rules are valid for ordered quadruples in $\mathbb{R}^4$ with the particular addition and scalar multiplication just defined, we call them vectors.

We can continue giving the same kinds of examples of vectors by adding one more component in each example. But doing this would be more tedious than any of us could stand. Fortunately, we can use the notation of *n*-tuples to handle all these examples once and for all.

Definition 2-4

Suppose n is a positive integer. An n-tuple of real numbers is represented by $(x_1, x_2, \ldots, x_n)$. Let $\mathbb{R}^n$ denote the collection of all n-tuples of real numbers.

Equality of two n-tuples $(x_1, x_2, \ldots, x_n)$ and $(y_1, y_2, \ldots, y_n)$ is defined by requiring that $x_1 = y_1$, $x_2 = y_2$, $\ldots$, $x_n = y_n$, i.e., that corresponding coordinates or components are equal.

Addition of vectors is defined "componentwise," "component by component," "coordinatewise," or "coordinate by coordinate."

$$(x_1, x_2, \ldots, x_n) + (y_1, y_2, \ldots, y_n) = (x_1 + y_1, x_2 + y_2, \ldots, x_n + y_n)$$

Multiplication of $(x_1, x_2, \ldots, x_n)$ by the scalar r is also defined component by component.

$$r(x_1, x_2, \ldots, x_n) = (rx_1, rx_2, \ldots, rx_n)$$

Brief Exercises

1. The following questions are posed for elements of $\mathbb{R}^6$.

   (a) $(5,1,3,2,-1,6) + (8,2,1,5,4,3) = ?$
   (b) $(2,1,3,8,6,8) + (?,?,?,?,?,?) = (2,1,3,8,6,8)$
   (c) $(5,8,-7,6,4,-3) + (?,?,?,?,?,?) = (0,0,0,0,0,0)$
   (d) $3(4,1,6,2,-1,-3) = ?$

2. Solve the following.

   (a) $3(8,1,5,2,3) + (?,?,?,?,?) = (1,2,3,4,5)$
   (b) $4(1,1,0,1,3) + 2(?,?,?,?,?) = (3,2,0,4,2)$

3. Formulate some exercises in $\mathbb{R}^7$ and $\mathbb{R}^8$ similar to the ones above and solve them.

Answers: **1(a)** $(13,3,4,7,3,9)$ **1(b)** $(0,0,0,0,0,0)$ **1(c)** $(-5,-8,7,-6,-4,3)$ **1(d)** $(12,3,18,6,-3,-9)$ **2(a)** $(-23,-1,-12,-2,-4)$ **2(b)** $(-\frac{1}{2},-1,0,0,-5)$

The n-tuples of $\mathbb{R}^n$ satisfy all the computational rules listed earlier for $\mathbb{R}^2$ and $\mathbb{R}^3$. We could give another list of these same properties using n-tuples rather than pairs, triples, or quadruples. This is not necessary, for it should be clear that we are repeating the same computational rules, one time using two components, another time using three components, another using four, etc. Fortunately, we can record these computational properties in a way which expresses all the cases we have discussed at the same time. If we do this, then we need a notation for vectors which can represent pairs, triples, quadruples, and n-tuples equally well. So we choose a notation which overlooks the fact that pairs have two coordinates, triples have three, and so forth. We use $\mathbf{X}$ to represent (x_1,x_2), (x_1,x_2,x_3), (x_1,x_2,x_3,x_4), or $(x_1,x_2,\ldots,x_n)$ depending whether we are referring to $\mathbb{R}^2$, $\mathbb{R}^3$, $\mathbb{R}^4$, or $\mathbb{R}^n$. Similarly we use $\mathbf{Y}$ for (y_1,y_2), (y_1,y_2,y_3), (y_1,y_2,y_3,y_4), or $(y_1,y_2,\ldots,y_n)$, and we use $\mathbf{Z}$ for (z_1,z_2), (z_1,z_2,z_3), (z_1,z_2,z_3,z_4), or $(z_1,z_2,\ldots,z_n)$. Using this notation we can combine facts about $\mathbb{R}^2$, $\mathbb{R}^3$, $\mathbb{R}^4$, and $\mathbb{R}^n$ into one list. Because the notation does not make the coordinates explicit, it is called a "coordinate-free" notation, i.e., a notation which is free of coordinates.

Proposition 2-1───────────────────────────────────

The following computational properties are valid in $\mathbb{R}^n$. We use $\mathbf{X}$, $\mathbf{Y}$, and $\mathbf{Z}$ to stand for vectors in $\mathbb{R}^n$, and we use r and s to stand for scalars (properties concerned with addition are labeled A.1 through A.5; those concerned with scalar multiplication are labeled SM.1 through SM.5).

A.1 (Closure or stability of $\mathbb{R}^n$ under addition)
   If $\mathbf{X}$ and $\mathbf{Y}$ are in $\mathbb{R}^n$, then $\mathbf{X} + \mathbf{Y}$ is in $\mathbb{R}^n$.

A.2 (Associativity of addition)
   $\mathbf{X} + (\mathbf{Y} + \mathbf{Z}) = (\mathbf{X} + \mathbf{Y}) + \mathbf{Z}$.

A.3 (Commutativity of addition)
   $\mathbf{X} + \mathbf{Y} = \mathbf{Y} + \mathbf{X}$.

A.4 (Existence of a zero vector)
   There is precisely one vector $\mathbf{0}$ in $\mathbb{R}^n$ having the property that $\mathbf{X} + \mathbf{0} = \mathbf{X}$ for every $\mathbf{X}$ in $\mathbb{R}^n$.

A.5 (Existence of a negative for each vector)
   For each $\mathbf{X}$ in $\mathbb{R}^n$, there is precisely one vector $-\mathbf{X}$ in $\mathbb{R}^n$ having the property that $\mathbf{X} + (-\mathbf{X}) = \mathbf{0}$.

SM.1 (Closure or stability of $\mathbb{R}^n$ under scalar multiplication)
   If $\mathbf{X}$ is in $\mathbb{R}^n$ and r is a scalar, then $r\mathbf{X}$ is in $\mathbb{R}^n$.

SM.2 $r[s\mathbf{X}] = (rs)\mathbf{X} = s[r\mathbf{X}]$.

SM.3 $(r + s)\mathbf{X} = r\mathbf{X} + s\mathbf{X}$.

SM.4 $r(\mathbf{X} + \mathbf{Y}) = r\mathbf{X} + r\mathbf{Y}$.

SM.5 $1 \cdot \mathbf{X} = \mathbf{X}$.

──

The reason we have concentrated on these computational properties is that they constitute the mathematical description which has been found most useful. Notice that Proposition 2-1 does not describe each single vector. Rather it describes an entire system consisting of real numbers (scalars) and another set of mathematical objects which can be added and multiplied according to the rules listed in Proposition 2-1. Only when we have this whole system do we call the individual mathematical objects "vector." We use the term "vector space over $\mathbb{R}$" to refer to this whole system consisting of the real scalars (that is the reason for the phrase "over $\mathbb{R}$") and a set of mathematical objects with addition and multiplication by scalars satisfying the basic properties. Consequently, the word vector does not have to refer only to *n*-tuples. It can refer to other objects provided they can be added and multiplied by scalars in accordance with the basic rules.

One technical point: a set may have no members, in which case it is called the empty set, the null set, or the void set. A set which contains at least one member is called nonempty, nonnull, or nonvoid. We agree to deal only with nonempty sets for vector spaces.

Definition 2-5

A *vector space* over $\mathbb{R}$ is a nonempty collection $\mathscr{V}$ of objects, which we denote by **X**, **Y**, **Z**, etc., which can be added and multiplied by real scalars in accordance with the following properties (properties concerning addition are listed A.1 through A.5; those concerning scalar multiplication are listed SM.1 through SM.5):

A.1 If **X** and **Y** are in $\mathscr{V}$, then $\mathbf{X} + \mathbf{Y}$ is in $\mathscr{V}$.

A.2 $\mathbf{X} + (\mathbf{Y} + \mathbf{Z}) = (\mathbf{X} + \mathbf{Y}) + \mathbf{Z}$.

A.3 $\mathbf{X} + \mathbf{Y} = \mathbf{Y} + \mathbf{X}$.

A.4 There is precisely one zero vector **0** in $\mathscr{V}$; its distinguishing property is that $\mathbf{X} + \mathbf{0} = \mathbf{X}$ for every **X** in $\mathscr{V}$.

A.5 For each vector **X** in $\mathscr{V}$ there is precisely one vector $-\mathbf{X}$ in $\mathscr{V}$ having the property that $\mathbf{X} + (-\mathbf{X}) = \mathbf{0}$.

SM.1 If **X** is in $\mathscr{V}$ and r is a scalar, then $r\mathbf{X}$ is in $\mathscr{V}$.

SM.2 $r[s\mathbf{X}] = (rs)\mathbf{X} = s[r\mathbf{X}]$.

SM.3 $(r + s)\mathbf{X} = r\mathbf{X} + s\mathbf{X}$.

SM.4 $r(\mathbf{X} + \mathbf{Y}) = r\mathbf{X} + r\mathbf{Y}$.

SM.5 $1\mathbf{X} = \mathbf{X}$.

The objects in $\mathscr{V}$ are called vectors.

With the operations of addition and multiplication by scalars presented in Definition 2-4, $\mathbb{R}^n$ is a vector space over $\mathbb{R}$.

One final note: the notation of Definition 2-4 allows the possibility of $\mathbb{R}^1$. Strictly speaking, we should denote the 1-tuples by (x_1), (y_1), etc. This would allow us to make a distinction between the vector (or 1-tuple) denoted, for example, by (5) and the scalar 5. However, this distinction in notation is not usual, and we follow standard practice whenever possible.

Brief Exercises

1. Particularize the statements labeled A.2, A.3, A.5, SM.2, SM.3, SM.4, and SM.5 in Definition 2-5 to the situation in $\mathbb{R}^5$ having $\mathbf{X} = (1,2,1,0,3)$, $\mathbf{Y} = (5,4,1,2,-1)$, $\mathbf{Z} = (3,1,-2,-1,1)$, $r = 2$, and $s = -3$.

2. Do the same exercise in $\mathbb{R}^6$ with $\mathbf{X} = (1,0,1,0,1,0)$, $\mathbf{Y} = (4,2,-1,3,1,2)$ $\mathbf{Z} = (-1,-3,0,-2,1,1)$, $r = 4$, and $s = 3$.

SUMMARY   *Various ways of using the term vector are discussed. The term scalar is used in this book as a synonym for real number. Then the standard coordinate operations of vector addition and scalar multiplication in* $\mathbb{R}^2$, $\mathbb{R}^3$, $\mathbb{R}^4$, *and finally* $\mathbb{R}^n$ *are introduced. Ten computational facts common to all these examples are discussed, and then these ten facts are used as the basis for a mathematical definition of a vector space over the real numbers* $\mathbb{R}$.

Exercises for Section 2-1

1. Do the following computations in $\mathbb{R}^2$.

   (a) $(5,1) + 3(-2,4) + 2(5,1)$ **(c)** $2(1,1) + 3(1,1) - 5(1,1)$
   (b) $(8,1) - 6(8,2)$ **(d)** $4(2,3) - 6(2,3) + 2(2,3)$

2. Solve for **X** in $\mathbb{R}^2$.

   (a) $2\mathbf{X} + (1,1) = (7,3)$ **(c)** $-4\mathbf{X} + (1,2) = (3,4)$
   (b) $3\mathbf{X} - (5,-2) = (-6,2)$

3. Do the following computations in $\mathbb{R}^3$.

   (a) $(2,1,3) + 4(1,1,-2) - 5(1,2,1)$ **(c)** $5(2,1,3) - 8(1,6,1)$
   (b) $(4,1,3) + 6(-1,-1,0)$

4. Solve for **X** in $\mathbb{R}^3$.

   (a) $4\mathbf{X} - (3,1,2) = (1,1,4)$ **(c)** $3\mathbf{X} + (4,4,2) = (1,1,2)$
   (b) $-2\mathbf{X} + (1,3,1) = (2,1,8)$

5. Solve for **X** in $\mathbb{R}^4$.

   (a) $3\mathbf{X} + \ (4,1,8,2) = (6,-1,7,2)$ **(c)** $-6\mathbf{X} + 4(2,1,8,1) = (-3,4,-7,6)$
   (b) $5\mathbf{X} + 3(1,1,2,1) = (5,8,6,2)$

The following are systems of equations in vector unknowns.

6. Solve for **X** and **Y** in $\mathbb{R}^2$.   **8.** Solve for **X**, **Y**, and **Z** in $\mathbb{R}^3$.

   $2\mathbf{X} + 3\mathbf{Y} = (2,1)$ $2\mathbf{X} - \mathbf{Y} + 2\mathbf{Z} = (1,0,1)$
   $4\mathbf{X} - \ \ \mathbf{Y} = (1,1)$ $\mathbf{X} + \mathbf{Y} + 3\mathbf{Z} = (2,1,-1)$
                                   $3\mathbf{X} - \mathbf{Y} + \ \ \mathbf{Z} = (1,-1,2)$

7. Same exercise as 6.   **9.** Same exercise as 8.

   $4\mathbf{X} - 3\mathbf{Y} = (-1,2)$ $\mathbf{X} + \ \ \mathbf{Y} + 3\mathbf{Z} = (1,0,1)$
    $\mathbf{X} + \ \ \mathbf{Y} = (5,1)$ $\mathbf{X} + 2\mathbf{Y} + 4\mathbf{Z} = (-1,2,0)$
                              $2\mathbf{X} + \ \ \mathbf{Y} + 5\mathbf{Z} = (1,2,3)$

More Challenging Exercises

10. Use the properties listed in Definition 2-5 to show that if **X** is a vector in a vector space $\mathscr{V}$ over $\mathbb{R}$, then $0\mathbf{X} = \mathbf{0}$ and $(-1)\mathbf{X} = -\mathbf{X}$. (Hint: consider $1\mathbf{X} + 0\mathbf{X}$ for the first part. For the second, consider $1\mathbf{X} + (-1)\mathbf{X}$.)

SECTION 2-2

Subspaces of Vector Spaces

PREVIEW *If one vector space contains another, the second is called a subspace of the first. Here the concept of a subspace is introduced. We show that it is not necessary to check all ten properties listed in the definition of a vector space to determine whether a subset of a vector space is a subspace. It suffices to check just two: A.1, closure or stability relative to addition, and SM.1, closure or stability relative to scalar multiplication.*

We often encounter the situation in which one vector space contains another one; both satisfy all the properties listed in Definition 2-5. The "smaller" one is called a *subspace* of the "larger" one. Let us begin with two examples.

Example We have already seen in Section 2-1 that $\mathbb{R}^2$ is a vector space over $\mathbb{R}$. In this example we are interested in showing that $\mathbb{R}^2$ contains other vector spaces.

Let us consider the set of solutions for the system of equations

$$3x - y = 0$$
$$6x - 2y = 0.$$

Throughout this example we refer to this set, so it is convenient to use a symbol to stand for it. We use $\mathscr{W}$ in this example to take the place of the longer phrase "the set of solutions for the system of equations

$$3x - y = 0$$
$$6x - 2y = 0.\text{''}$$

Since this is a system of linear equations in two unknowns, solutions of the system are ordered pairs. If we use the notation $\mathbb{R}^2$ introduced in Definition 2-2 for all ordered pairs, then the solution to this—and any system of equations in two unknowns—can be thought of as vectors in $\mathbb{R}^2$. We show in this example that $\mathscr{W}$ is a vector space contained in $\mathbb{R}^2$.

First we solve the system. By solving for y in the first equation we find that the vector (x,y) is a solution if and only if

$$(x,y) = (x,3x) = x(1,3).$$

This equation means that $\mathcal{W}$ consists precisely of the multiples of $(1,3)$. We claim that $\mathcal{W}$ satisfies all ten properties listed in Definition 2-5.

For reasons which will be clear in a moment, we consider the properties labeled A.1 and SM.1 first. To show that $\mathcal{W}$ satisfies property A.1, we must show that if we add two multiples of $(1,3)$, we get another multiple of $(1,3)$. To do this, suppose $a(1,3)$ is one multiple and $b(1,3)$ is another. Then

$$a(1,3) + b(1,3) = (a + b)(1,3).$$

Since $(a + b)(1,3)$ is a multiple of $(1,3)$, we have verified that $\mathcal{W}$ satisfies property A.1.

To show that $\mathcal{W}$ satisfies property SM.1 we must show that if a multiple of $(1,3)$ is multiplied by some scalar, then the result is another multiple of $(1,3)$. To do this, let $a(1,3)$ be the multiple of $(1,3)$ and r the scalar. Then

$$r[a(1,3)] = (ra)(1,3).$$

Since $(ra)(1,3)$ is a multiple of $(1,3)$, $\mathcal{W}$ satisfies property SM.1.

The reason we considered A.1 and SM.1 first is that they always need to be verified to show that a set is a subspace. But the pleasant fact is that the other properties never need to be verified. Properties A.2, A.3, SM.2, SM.3, SM.4, and SM.5 make statements about all vectors in $\mathbb{R}^2$, so they are true even if **X**, **Y**, and **Z** are interpreted as multiples of $(1,3)$.

The two properties remaining to be considered are A.4 and A.5. The statements we need to verify are:

> There is precisely one vector **0** in $\mathcal{W}$ having the property that $\mathbf{X} + \mathbf{0} = \mathbf{X}$ for every **X** in $\mathcal{W}$,

and

> For each **X** in $\mathcal{W}$ there is precisely one vector $-\mathbf{X}$ in $\mathcal{W}$ having the property that $\mathbf{X} + (-\mathbf{X}) = \mathbf{0}$.

In other words, we must show that the zero vector **0** is a multiple of $(1,3)$—which it is: $(0,0) = 0 \cdot (1,3)$—and we must show that the negative of a multiple of $(1,3)$ is also a multiple of $(1,3)$—which is so because $-[a(1,3)] = (-a)(1,3)$. However, we are also interested in showing why properties A.4 and A.5 follow automatically once A.1 and SM.1 are verified. The reason is that $0 \cdot \mathbf{X} = \mathbf{0}$ and $(-1)\mathbf{X} = -\mathbf{X}$. You can see that these equations are valid either by thinking of **X** as an ordered pair or from Exercise 10 in Section 2-1. So if the multiples of $(1,3)$ satisfy SM.1, then the zero vector must be a multiple of $(1,3)$, and the negative of a multiple of $(1,3)$ is also a multiple of $(1,3)$.

We have considered this example somewhat at length. The important points are:

(1) some vector spaces contain other vector spaces; and
(2) in verifying that a collection of vectors in some vector space satisfies all the properties for a vector space, it is enough to verify A.1 and SM.1.

Example We give another example of a subspace of a vector space.

Let $\mathscr{W}$ be the collection of all vectors (x,y,z,w) in $\mathbb{R}^4$ which satisfy the equality $x - y = 3z + 4w$. The vector $(1,2,3,4)$ is not in $\mathscr{W}$, for example, because $1 - 2 \neq 3(3) + 4(4)$; the vector $(2,1,3,-2)$ is in $\mathscr{W}$ since $2 - 1 = 3(3) + 4(-2)$.

The purpose of this example is to show that $\mathscr{W}$ is a subspace of $\mathbb{R}^4$.

To show that $\mathscr{W}$ is stable under addition, i.e., satisfies property A.1, we need to show that if (a,b,c,d) and (p,q,r,s) are elements of $\mathscr{W}$, then $(a + p, b + q, c + r, d + s)$ is too; i.e., we need to show that if $a - b = 3c + 4d$ and $p - q = 3r + 4s$, then $(a + p) - (b + q) = 3(c + r) + 4(d + s)$. Since the last equation is obtained by adding the first two, $\mathscr{W}$ is stable under addition.

To show that $\mathscr{W}$ is stable under scalar multiplication, i.e., satisfies property SM.1, suppose (a,b,c,d) is in $\mathscr{W}$ and r is a scalar; so $a - b = 3c + 4d$. We need to show that (ra,rb,rc,rd) is in $\mathscr{W}$; i.e., $ra - rb = 3rc + 4rd$. This equality follows from the first by multiplying the first by the scalar r.

As in the previous example, the remaining properties are automatically satisfied. Properties A.2, A.3, SM.2, SM.3, SM.4, and SM.5 are satisfied just because we are dealing with vectors in the vector space $\mathbb{R}^4$. Properties A.4 and A.5 are satisfied because we know SM.1 is satisfied. The zero vector is in $\mathscr{W}$ since $0 \cdot (2,1,3,-2) = (0,0,0,0)$ is a scalar multiple of a vector in $\mathscr{W}$. The negative of any vector (x,y,z,w) in $\mathscr{W}$ is a multiple of (x,y,z,w) since $(-x,-y,-z,-w) = (-1)(x,y,z,w)$. Consequently, $\mathscr{W}$ is a subspace of $\mathbb{R}^4$.

From these examples you should be able to see the general situation. We have a vector space $\mathscr{V}$ to begin with (e.g., $\mathbb{R}^2$, $\mathbb{R}^4$, or $\mathbb{R}^n$) and some set $\mathscr{W}$ of vectors in $\mathscr{V}$ identified by some means (e.g., as solution set of some system of equations). For $\mathscr{W}$ to be called a subspace of $\mathscr{V}$, $\mathscr{W}$ must satisfy all the properties listed in Definition 2-5 for a vector space; i.e., it must be a vector space in its own right.

Definition 2-6

Suppose that $\mathscr{V}$ is a vector space over $\mathbb{R}$ and that $\mathscr{W}$ is a nonvoid subset of $\mathscr{V}$. Then $\mathscr{W}$ is a (vector) subspace of $\mathscr{V}$ if and only if $\mathscr{W}$ satisfies all the properties of a vector space listed in Definition 2-5 with addition and scalar multiplication in $\mathscr{W}$ being those used in $\mathscr{V}$.

As the examples indicate, to check whether a nonvoid subset $\mathscr{W}$ of a vector space $\mathscr{V}$ is a subspace, it is sufficient to check A.1 and SM.1. The reasons are the same as the ones in the examples: properties A.2, A.3, SM.2, SM.3, SM.4, and SM.5 are automatic because we are in a vector space $\mathscr{V}$. Properties A.4 and A.5 follow from A.1 and SM.1 since $0\mathbf{X} = \mathbf{0}$ and $(-1)\mathbf{X} = -\mathbf{X}$. These observations are important, and we record them for future reference.

Proposition 2-2

Suppose that $\mathscr{V}$ is a vector space over $\mathbb{R}$ and that $\mathscr{W}$ is a nonvoid subset of $\mathscr{V}$. Then $\mathscr{W}$ is a subspace of $\mathscr{V}$ if and only if $\mathscr{W}$ has properties A.1 and SM.1.
A.1   If $\mathbf{X}$ and $\mathbf{Y}$ are in $\mathscr{W}$, then $\mathbf{X} + \mathbf{Y}$ is in $\mathscr{W}$.
SM.1 If $\mathbf{X}$ is in $\mathscr{W}$ and r is a scalar, then $r\mathbf{X}$ is in $\mathscr{W}$.

It is important to understand how to apply Proposition 2-2. To show that A.1 is satisfied by a set $\mathscr{W}$, you must show that, whatever vectors $\mathbf{X}$ and $\mathbf{Y}$ are chosen from $\mathscr{W}$, their sum $\mathbf{X} + \mathbf{Y}$ must also be in $\mathscr{W}$. Consequently, it is not enough to pick one numerical vector for $\mathbf{X}$ and one numerical vector for $\mathbf{Y}$ and show that, for those particular choices of $\mathbf{X}$ and $\mathbf{Y}$, $\mathbf{X} + \mathbf{Y}$ is in $\mathscr{W}$. A similar comment holds for SM.1; it is not sufficient to pick a particular number for r and a particular numerical vector $\mathbf{X}$ from $\mathscr{W}$ and show that $r\mathbf{X}$ is in $\mathscr{W}$. For this reason, when we test A.1 and SM.1 in the last example and in the next, we use (a,b,c,d) for $\mathbf{X}$ and (p,q,r,s) for $\mathbf{Y}$ rather than specific numerical values.

   In the next example we give some further illustrations of subspaces of a vector space.

Example   In this example we take various subsets of $\mathbb{R}^4$ and determine which are subspaces of $\mathbb{R}^4$. Let $\mathscr{W}_1$ be the set of vectors (x,y,z,w) satisfying $3x - 2y = 5$; let $\mathscr{W}_2$ be the set of vectors (x,y,z,w) satisfying $5x - 2y + 3z + w = 0$; and let $\mathscr{W}_3$ be the set of vectors (x,y,z,w) satisfying both $5x + 2y = z$ and $4x - z + w = 0$.

   The set $\mathscr{W}_1$ is not a subspace of $\mathbb{R}^4$, because $\mathscr{W}_1$ is not stable with respect to addition or scalar multiplication. In trying to show A.1 is satisfied in $\mathscr{W}_1$, take (a,b,c,d) and (p,q,r,s) in $\mathscr{W}_1$; then $3a - 2b = 5$ and $3p - 2q = 5$; but then $(a + p, b + q, c + r, d + s)$ is not in $\mathscr{W}_1$ since $3(a + p) - 2(b + q) = 10 \neq 5$. Therefore, A.1 is not satisfied, and $\mathscr{W}_1$ is not a subspace. As a matter of fact, $\mathscr{W}_1$ does not satisfy A.4, A.5, or SM.1, either. But failing to satisfy even one property is enough to keep $\mathscr{W}_1$ from being a subspace.

   The set $\mathscr{W}_2$ is a subspace of $\mathbb{R}^4$. First show that $\mathscr{W}_2$ is stable under addition. If (a,b,c,d) and (p,q,r,s) are in $\mathscr{W}_2$, i.e., $5a - 2b + 3c + d = 0$ and $5p - 2q + 3r + s = 0$, then

$$5(a + p) - 2(b + q) + 3(c + r) + (d + s) = 0,$$

and so $(a + p, b + q, c + r, d + s)$ is in $\mathscr{W}_2$. Now we show that $\mathscr{W}_2$ is stable under scalar multiplication. If r is a scalar and (a,b,c,d) is in $\mathscr{W}_2$, i.e., $5a - 2b + 3c + d = 0$, then $5ra - 2rb + 3rc + rd = 0$ and (ra,rb,rc,rd) is in $\mathscr{W}_2$. Thus $\mathscr{W}_2$ is a subspace of $\mathbb{R}^4$.

The set $\mathscr{W}_3$ is a subspace of $\mathbb{R}^4$. First we show that $\mathscr{W}_3$ is stable under addition. Suppose (a,b,c,d) and (p,q,r,s) are in $\mathscr{W}_3$. This means that $5a + 2b = c$, $4a - c + d = 0$, $5p + 2q = r$, and $4p - r + s = 0$. Then $5(a + p) + 2(b + q) = c + r$ and $4(a + p) - (c + r) + (d + s) = 0$, and so $(a + p, b + q, c + r, d + s)$ is in $\mathscr{W}_3$. Next we show that $\mathscr{W}_3$ is stable under scalar multiplication. If r is in $\mathbb{R}$ and (a,b,c,d) is in $\mathscr{W}_3$, i.e., $5a + 2b = c$ and $4a - c + d = 0$, then $5ra + 2rb = rc$ and $4ra - rc + rd = 0$; this shows that (ra,rb,rc,rd) is in $\mathscr{W}_3$. Thus $\mathscr{W}_3$ is a subspace of $\mathbb{R}^4$.

Brief Exercises

1. In each of the following, determine whether $\mathscr{W}$ is a subspace of $\mathbb{R}^2$.

   (a) $\mathscr{W}$ is the solution set for the system (of one equation)
   $$2x + y = 1.$$

   (b) $\mathscr{W}$ is the solution set for the system
   $$5x + y = 0$$
   $$x - y = 0.$$

   How many vectors are in $\mathscr{W}$?

   (c) $\mathscr{W}$ is the set of vectors (x,y) satisfying $3x + y = 0$.
   (d) $\mathscr{W}$ is the set of vectors (x,y) satisfying $4x - 3y = 2$.

2. In each of the following, determine whether $\mathscr{W}$ is a subspace of $\mathbb{R}^3$.

   (a) $\mathscr{W}$ is the set of vectors (x,y,z) satisfying $3x - y + z = 2$.
   (b) $\mathscr{W}$ is the set of vectors (x,y,z) satisfying $2x + y - z = 0$ and $4x - y + 3z = 0$.
   (c) $\mathscr{W}$ is the set of vectors (x,y,z) satisfying $4x + 2y = 3z$ and $x - y + z = 2$.

3. In Exercises 1 and 2, some of the sets are not subspaces. In these cases can you change the constants in the equations to make the sets subspaces?

Answers: **1(a)** No **1(b)** Yes; exactly one $(0,0)$ **1(c)** Yes **1(d)** No **2(a)** No **2(b)** Yes **2(c)** No **3** Yes; change the constants to 0

SUMMARY *The concept of a subspace of a vector space is introduced; it is a nonempty subset of a vector space which is a vector space itself. This means it satisfies all ten properties for a vector space listed in Definition 2-5. Fortunately,* *it is not necessary to verify all ten properties to determine whether a subset of a vector space is a subspace. The only properties which must be checked are A.1, closure or stability with respect to addition, and SM.1, closure or stability with respect to scalar multiplication.*

Exercises for Section 2-2

1. Determine whether the following are subspaces of $\mathbb{R}^3$.

    (a) the set of vectors having the form $(x,y,0)$
    (b) the set of vectors (x,y,z) satisfying $x + 3y + 2z = 0$
    (c) the set of vectors (x,y,z) satisfying $x - y = 2$
    (d) the set of vectors (x,y,z) satisfying $x + y + z = 0$ and $x - 3y + 2z = 0$
    (e) the set of vectors (x,y,z) satisfying $x + y - z = 2$
    (f) the set of vectors (x,y,z) satisfying the following equations

$$2x + y + z = 0$$

$$x - y - z = 0$$

$$3x + y - z = 0$$

2. Explain why $\mathbb{R}^2$ is not a subspace of $\mathbb{R}^3$.

Exercises 3, 4, and 5 are intended to show that the solution set for a system of equations is a subspace if and only if the system is homogeneous.

3. Let $\mathscr{W}$ be the solution set for the system of equations

$$2x + 3y = k_1$$

$$x - y = k_2,$$

where k_1 and k_2 are some given numbers. Explain why $\mathscr{W}$ is a subspace of $\mathbb{R}^2$ if and only if $k_1 = k_2 = 0$.

4. If $\mathscr{W}$ is the solution set for the system of equations

$$3x + y - z = k_1$$

$$x - y + z = k_2,$$

where k_1 and k_2 are some given numbers, explain why $\mathscr{W}$ is a subspace of $\mathbb{R}^3$ if and only if $k_1 = k_2 = 0$.

5. In general let $\mathscr{W}$ be the solution set for the "general" system of equations

$$a_{11}x_1 + \cdots + a_{1n}x_n = b_1$$
$$a_{21}x_1 + \cdots + a_{2n}x_n = b_2$$
$$\vdots \qquad\qquad \vdots \qquad \vdots$$
$$a_{m1}x_1 + \cdots + a_{mn}x_n = b_m.$$

Explain why $\mathscr{W}$ is a subspace of $\mathbb{R}^n$ if and only if $b_1 = b_2 = \cdots = b_n = 0$, i.e., if and only if the system is homogeneous.

6. Explain why the set consisting only of $(0,0)$ is a subspace of $\mathbb{R}^2$. Explain why the set consisting only of $(0,0,0)$ is a subspace of $\mathbb{R}^3$. Finally, explain why the set consisting only of the zero vector **0** in a vector space $\mathscr{V}$ is a subspace of $\mathscr{V}$. This is called the *zero subspace* of $\mathscr{V}$.

7. Find a subset of $\mathbb{R}^2$ which is stable under scalar multiplication but not stable under addition.

8. Find a subset of $\mathbb{R}^2$ which is stable under addition but not stable under scalar multiplication.

SECTION 2-3
Some Geometry in $\mathbb{R}^2$ and $\mathbb{R}^3$

PREVIEW   *We introduce a way to visualize vectors of $\mathbb{R}^2$ and $\mathbb{R}^3$ geometrically as points in a plane or 3-space and as directed line segments or arrows in a plane or 3-space. We give a geometric interpretation of addition and scalar multiplication of vectors, and we discuss finding a vector equation of a line in a plane and in 3-space.*

In Sections 2-1 and 2-2 we introduced the concept of a vector space over $\mathbb{R}$ and subspaces of a vector space. There we emphasized the arithmetic of vectors. Here we show that vectors in $\mathbb{R}^2$ and $\mathbb{R}^3$ can be pictured geometrically in a plane or three-dimensional space, and that vector addition and scalar multiplication can be visualized geometrically. Unfortunately our method of visualizing $\mathbb{R}^2$ and $\mathbb{R}^3$ does not extend to $\mathbb{R}^n$ if n is greater than 3. Nevertheless the pictures in $\mathbb{R}^2$ and $\mathbb{R}^3$ frequently give clues to ideas which work in general.

First, consider vectors and vector operations in $\mathbb{R}^2$. Suppose that L_1 and L_2 are two lines in a plane with L_1 perpendicular to L_2 (see Figure 2-1), and suppose

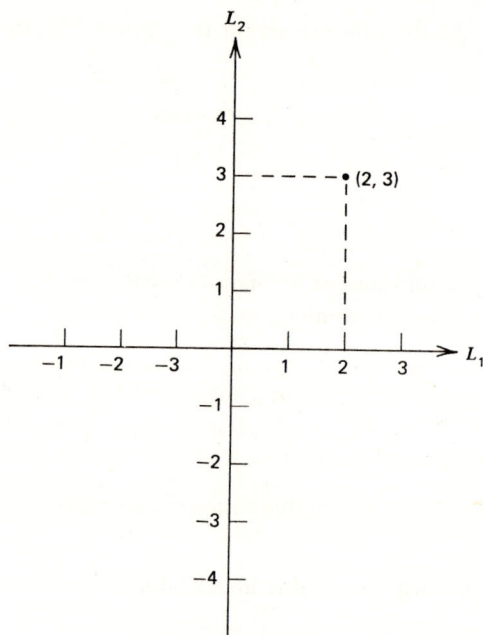

Figure 2-1

the points of these lines are numbered starting with the number 0 at the intersection of the two lines, using the same unit on both lines. The numbers increase to the right on L_1 and upward on L_2.

We can now picture ordered pairs as points. For example, the ordered pair (2,3) is represented by the point which is two units to the right and three units up from O, the origin, as shown in Figure 2-1. The first coordinate tells how many units you move horizontally from the origin, and the second coordinate tells how many units you go vertically to locate the point. Figure 2-2 shows the points for the ordered pairs $(1,-2)$, $(-3,1)$, and $(-1,-2)$. Notice that a negative first coordinate is counted to the left and that a negative second coordinate is counted downward.

Instead of talking about the ordered pair (2,3) and the point determined by (2,3), we usually just speak of the point (2,3). It is generally clear from the context whether we are referring to the geometric point (2,3) or to the numerical ordered pair (2,3).

If we reverse the process, we can take a point and find the ordered pair it is matched with. For example, in Figure 2-3 we illustrate the process of starting with the point $(-2,3)$ and finding the coordinates -2 and 3. Draw a line parallel to L_2 through $(-2,3)$; it crosses L_1 at -2, the first coordinate. To find the second

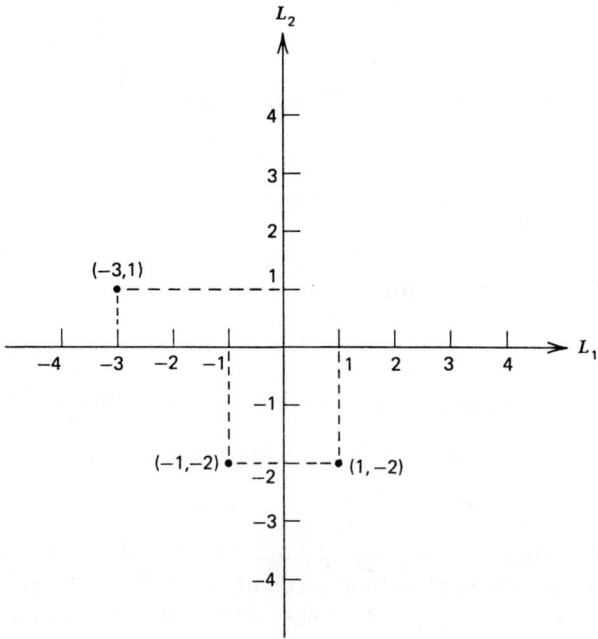

Figure 2-2

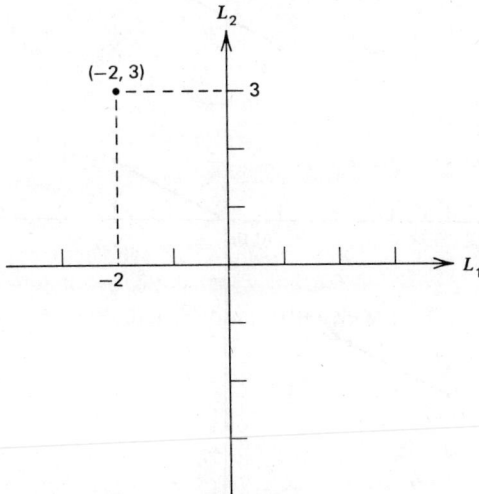

Figure 2-3

coordinate, draw a line parallel to L_1 through $(-2,3)$; it crosses L_2 at 3, the second coordinate.

One more point: it is customary to denote a "general" ordered pair with the symbol (x,y). Because it is such a widespread custom to denote the first coordinate by x and the second by y, the lines L_1 and L_2 of our figures are more usually called the x-axis and y-axis, respectively. From now on, we follow the standard procedure of labeling L_1 with x and L_2 with y in our diagrams.

Brief Exercises

1. Draw a diagram and locate the points with the following coordinates.

$(2,1)$, $(1,2)$, $(-3,4)$, $(4,-3)$, $(-2,-3)$, $(1,-3)$, $(1,-2)$, $(1,-1)$, $(0,2)$, $(4,0)$, $(-3,0)$, $(0,-1)$

There is another way to picture ordered pairs. The pair $(2,1)$ can be pictured as an arrow beginning at the point $(0,0)$ and ending at the point $(2,1)$. It can also be pictured as an arrow beginning at the point $(-1,3)$, for example, and ending at the point $(1,4)$ (see Figure 2-4). In fact, it can be pictured as an arrow beginning at

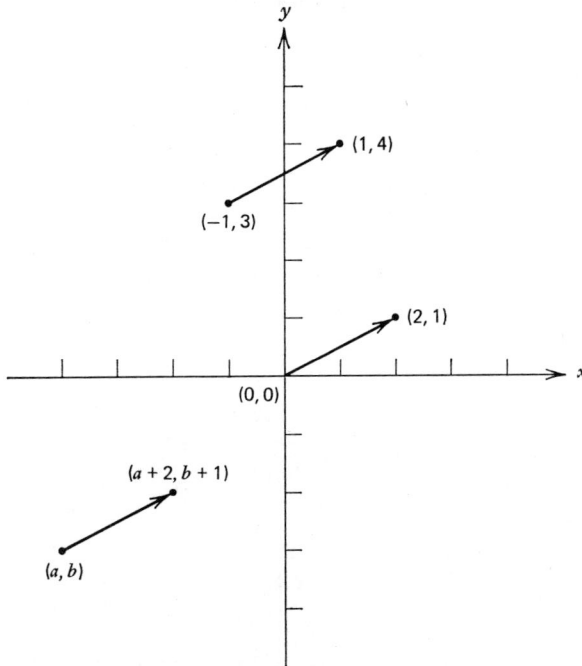

Figure 2-4

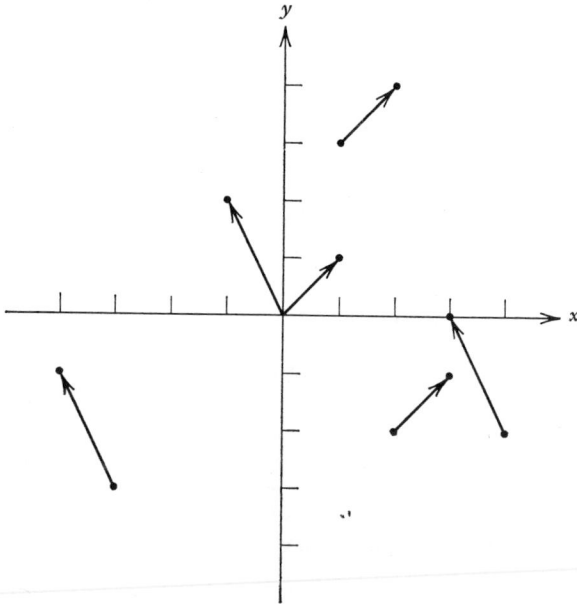

Figure 2-5

any point you please—say (a,b)—and ending at $(a + 2, b + 1)$, i.e., ending at the point obtained by adding $(2,1)$ to (a,b). Some express this by saying that $(2,1)$ can be pictured as any arrow obtained by moving the arrow which begins at $(0,0)$ and ends at $(2,1)$ parallel to itself anywhere in the plane.

Any vector in $\mathbb{R}^2$ or ordered pair can be pictured in a similar way. In Figure 2-5 we illustrate the vectors $(1,1)$ and $(-1,2)$ placed at various positions.

Picturing vectors as arrows helps us to visualize addition of vectors and multiplication of vectors by scalars in a geometric fashion. For example, $(2,3) + (1,-1) = (3,2)$. Picture $(2,3)$ as the arrow from $(0,0)$ to $(2,3)$, and $(1,-1)$ as beginning at $(2,3)$ and ending at $(3,2)$. This is shown in Figure 2-6.

It is not necessary that $(2,3)$ begin at $(0,0)$. In Figure 2-7 the sum $(2,3) + (1,-1)$ is visualized again; place $(2,3)$ so it begins at $(-1,2)$ and ends at $(1,5)$. Place $(1,-1)$ so that it begins at $(1,5)$ and ends at $(2,4)$. The sum is thus the vector beginning at $(-1,2)$ and ending at $(2,4)$, namely the vector $(3,2)$.

In general we can picture $\mathbf{X} + \mathbf{Y}$ by placing $\mathbf{X}$ somewhere and then placing the beginning of $\mathbf{Y}$ at the end of $\mathbf{X}$. Then $\mathbf{X} + \mathbf{Y}$ is pictured as beginning at the initial point of $\mathbf{X}$ and ending at the end of $\mathbf{Y}$, as shown in Figure 2-8.

For a visualization of scalar multiplication, consider the example $2(1,2) = (2,4)$. This is pictured in Figure 2-9 with $(1,2)$ and $(2,4)$ beginning at $(0,0)$. Also

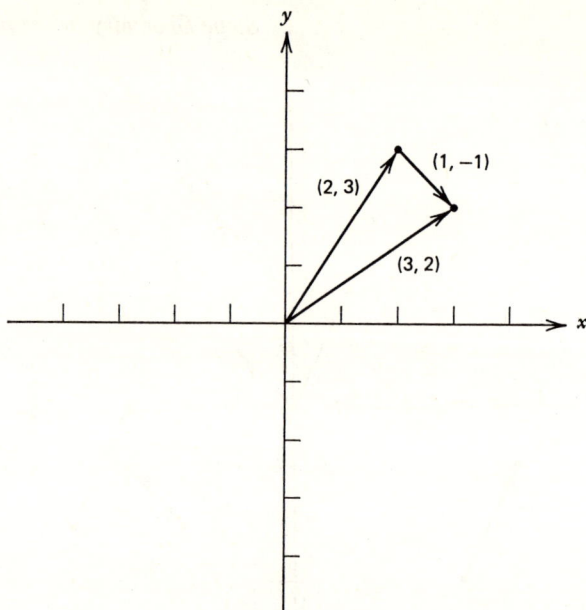

Figure 2-6

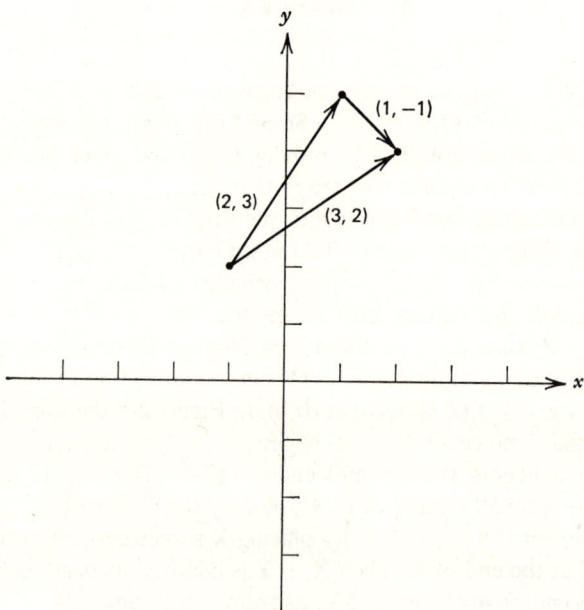

Figure 2-7

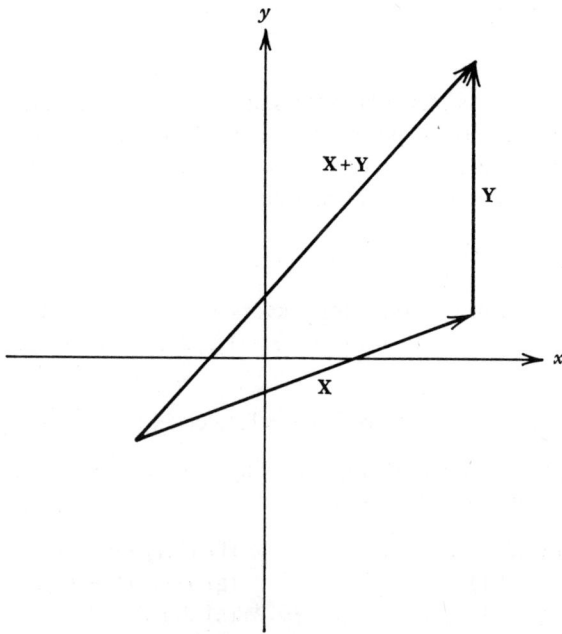

Figure 2-8

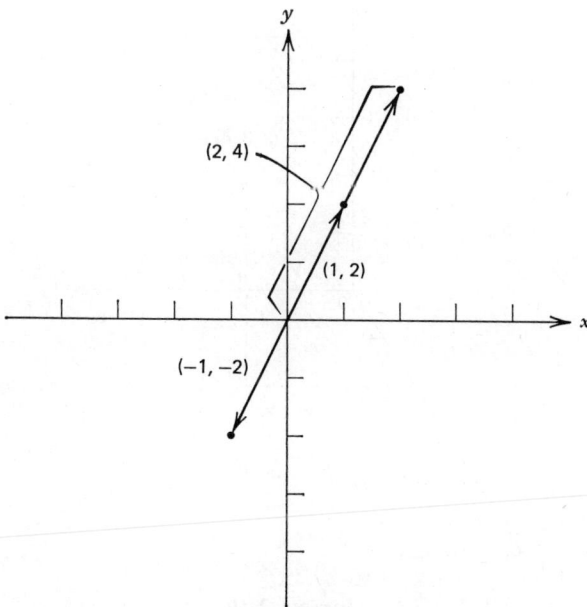

(2, 4)

(1, 2)

(−1, −2)

Figure 2-9

$-1(1,2) = (-1,-2)$ is pictured in the same figure. From this you can see that if a nonzero vector **X** and the scalar multiple t**X** of **X** begin at the same point, then they lie along the same line. If t is positive, then t**X** and **X** point in the same direction; if t is negative, they point in opposite directions.

Brief Exercises

1. Draw a diagram of the following vectors as arrows beginning at $(0,0)$.
   $(1,-1)$, $(3,4)$, $(2,3)$, $(-4,2)$, $(-3,-1)$, $2(1,-1)$, $-1(1,-1)$, $-3(1,-1)$, $0(1,-1)$

2. Draw a diagram of the vectors listed in Exercise 1 as arrows beginning at $(-1,1)$.

3. Draw a diagram to illustrate the following computations; arrange your diagrams so that the result begins at $(0,0)$.

   (a) $(1,1) + (3,2)$
   (b) $(1,1) + (-2,1)$
   (c) $(3,2) + (-2,1)$
   (d) $[(1,1) + (2,-1)] + (1,-2)$
   (e) $(1,1) + [(2,-1) + (1,-2)]$

   (f) $(2,1) + (-3,1)$
   (g) $(-3,1) + (2,1)$
   (h) $2(1,2)$
   (i) $-2(1,2)$
   (j) $3(-1,1)$

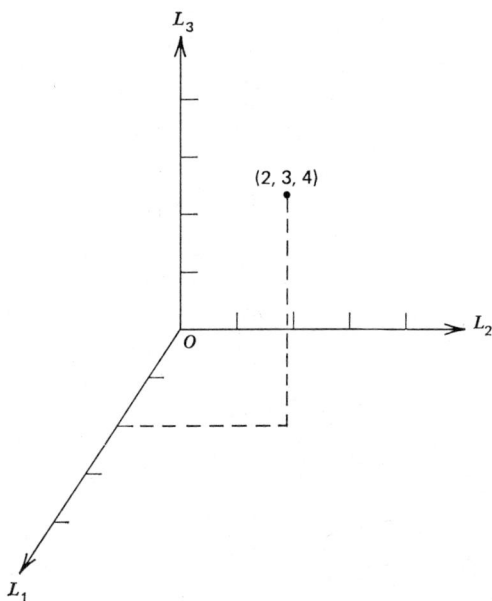

Figure 2-10

Now we turn to a visualization of vectors in $\mathbb{R}^3$. We picture ordered triples either as points or as arrows in ordinary three-dimensional space. We use three mutually perpendicular lines which intersect at the common point labeled O as in Figure 2-10. As we did in $\mathbb{R}^2$, we assume that the points of these lines have been labeled by real numbers starting with the number 0 at the common point of the lines, the origin, and using the same unit on the three lines. The numbers increase in the direction indicated by the arrows in the diagram. We find the point representing the triple $(2,3,4)$ by going 2 units along L_1, then 3 units along L_2, and finally 4 units along L_3. This is illustrated in Figure 2-10.

The same procedure is followed with any other triple; the first coordinate is counted along L_1, the second coordinate along L_2, and the third coordinate along L_3. In Figure 2-11 we illustrate the points corresponding to $(1,-1,2)$, $(-1,1,1)$, and $(2,1,0)$.

As with $\mathbb{R}^2$, it is customary to speak of the point $(1,-1,2)$ rather than the point corresponding to $(1,-1,2)$; we follow this custom from now on. It is also customary to denote the general ordered triple by (x,y,z); for this reason, the lines L_1, L_2, and L_3 are usually called the x-axis, y-axis, and z-axis, respectively. We follow this custom, too.

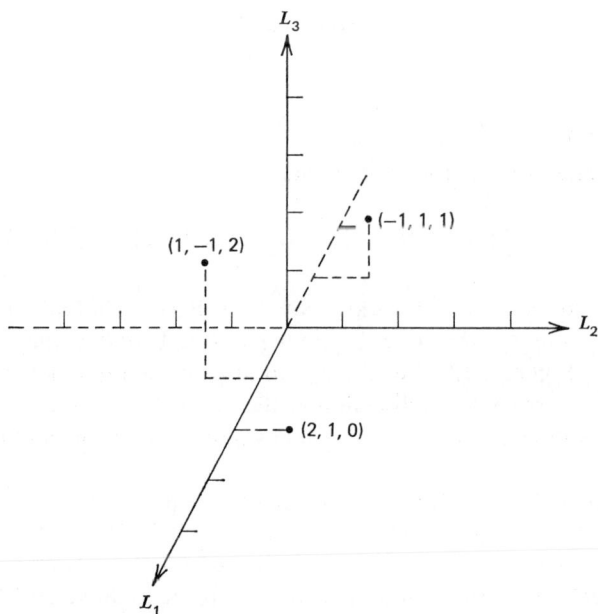

Figure 2-11

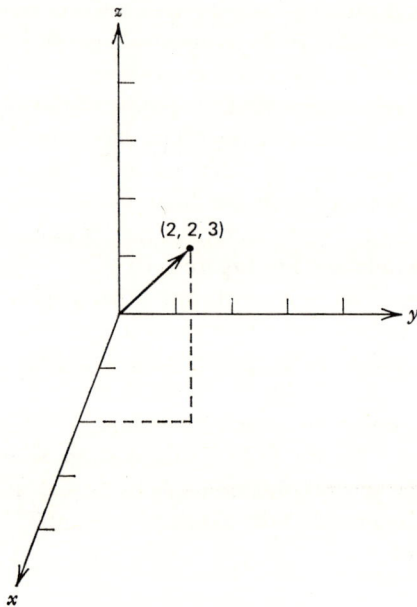

Figure 2-12

Brief Exercises

1. Draw a diagram of the following points.

$$(1,0,1),\ (2,1,-1),\ (1,0,0),\ (0,1,1),\ (2,1,1),\ (0,0,2),\ (1,1,0),\ (3,1,-2)$$

We can also picture ordered triples as arrows, much as we did in $\mathbb{R}^2$. For example, $(2,2,3)$ can be pictured as an arrow beginning at $(0,0,0)$ and ending at $(2,2,3)$. This is illustrated in Figure 2-12. But $(2,2,3)$ can also be pictured as beginning at the point $(1,1,1)$, for example, and ending at the point $(3,3,4)$ (see Figure 2-13). In fact, we can picture $(2,2,3)$ as beginning at any point (a,b,c) and ending at the point $(a + 2, b + 2, c + 3)$.

Addition and scalar multiplication of vectors are pictured in $\mathbb{R}^3$ as they are in $\mathbb{R}^2$. To picture $(1,1,-1) + (2,3,3)$, visualize $(1,1,-1)$ as the arrow from $(0,0,0)$ to $(1,1,-1)$. Place the arrow $(2,3,3)$ so that it begins at $(1,1,-1)$ and ends at $(3,4,2)$. The sum is then the vector from $(0,0,0)$ to $(3,4,2)$ (see Figure 2-14). In general, to picture $\mathbf{X} + \mathbf{Y}$, place $\mathbf{Y}$ so that it begins at the end of $\mathbf{X}$. The sum $\mathbf{X} + \mathbf{Y}$ is pictured by the arrow from the beginning of $\mathbf{X}$ to the end of $\mathbf{Y}$ (see Figure 2-15).

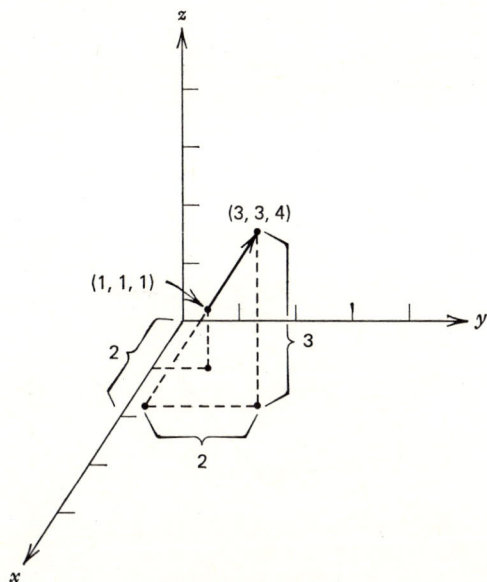

Figure 2-13

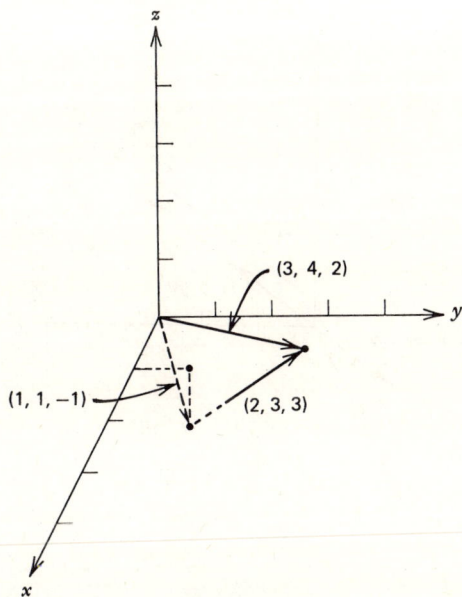

Figure 2-14

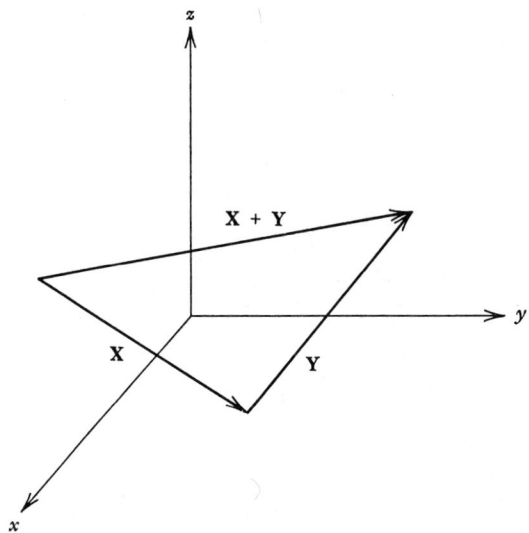

Figure 2-15

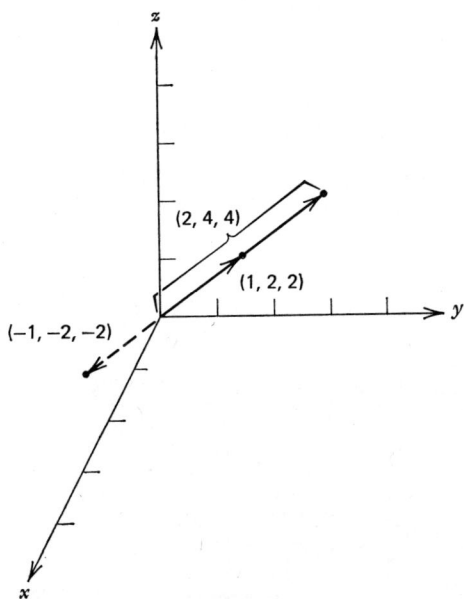

Figure 2-16

In Figure 2-16 we have illustrated the vectors $(1,2,2)$, $2(1,2,2)$, and $(-1)(1,2,2)$. As in the case in $\mathbb{R}^2$, if a vector and a scalar multiple of it begin at the origin, say, then they lie along the same line. If the scalar t is negative, then **X** and t**X** point in opposite directions, whereas if t is positive, then **X** and t**X** point in the same direction.

Brief Exercises

1. Draw a diagram of the following vectors picturing them as arrows beginning at the origin.
$(1,0,1)$, $(2,1,3)$, $(-1,2,1)$, $(0,1,1)$, $(0,0,1)$, $(1,0,1)$, $(0,1,0)$, $3(1,0,1)$, $-2(1,0,1)$, $-1(1,0,1)$

2. Draw a diagram of the following computations.

   (a) $(2,1,0) + (1,0,1)$ **(b)** $2(1,0,-1) + (-1)(1,1,2)$

Now we discuss vector equations of lines in the plane and in 3-space. In this discussion it is convenient to picture vectors as arrows. Our goal is to describe all vectors which begin at the origin and end on a given line (see Figure 2-17).

For example, consider the line passing through $(1,3)$ and $(3,1)$ (see Figure 2-18). It happens that the points $(-1,5)$, $(0,4)$, $(2,2)$, $(4,0)$, and $(5,-1)$ are also on this

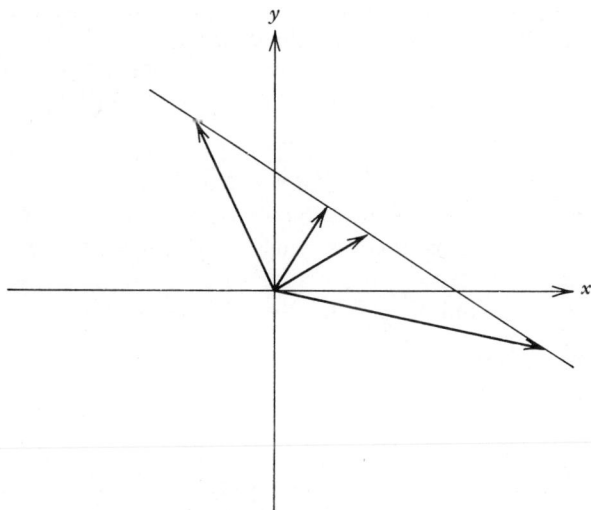

Figure 2-17

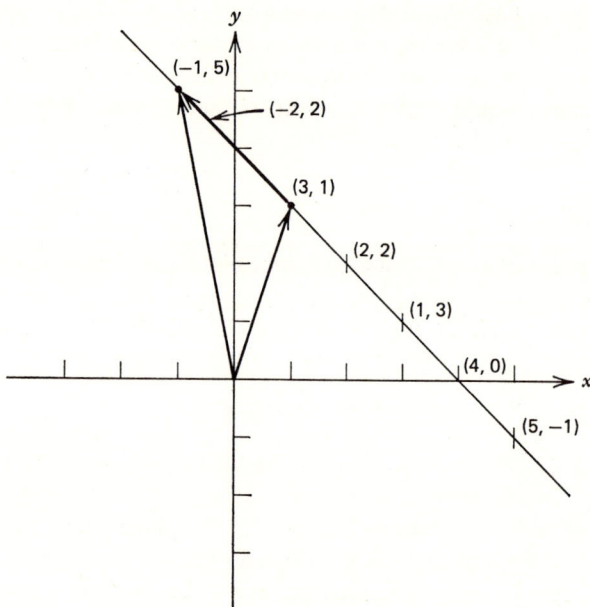

Figure 2-18

line. Our goal is to find a description for all vectors beginning at (0,0) and ending on the line; we use these particular points, $(-1,5)$, $(0,4)$, $(1,3)$, $(2,2)$, $(3,1)$, $(4,0)$, and $(5,-1)$ to illustrate how we obtain this description.

First, the vector from (0,0) to (1,3) begins at the origin and ends on the line. Now consider the vector from (0,0) to $(-1,5)$. We can express $(-1,5)$ in the following way.

$$(-1,5) = (1,3) + (-2,2)$$

The important point to observe here is that we use a known vector $(1,3)$ to get to the line and a vector $(-2,2)$, pictured as lying along the line, to express the vector $(-1,5)$.

We can do the same kind of thing for the other vectors (see Figure 2-19).

$$(0,4) = (1,3) + (-1,1) \qquad (3,1) = (1,3) + (2,-2)$$
$$(1,3) = (1,3) + (0,0) \qquad (4,0) = (1,3) + (3,-3)$$
$$(2,2) = (1,3) + (1,-1) \qquad (5,-1) = (1,3) + (4,-4)$$

From this you can see that we need to find a vector which goes from (0,0) to the line, in this case (1,3), and then add to it vectors which lie along the line or, if placed at the origin, are parallel to the line. How can we find this from the given

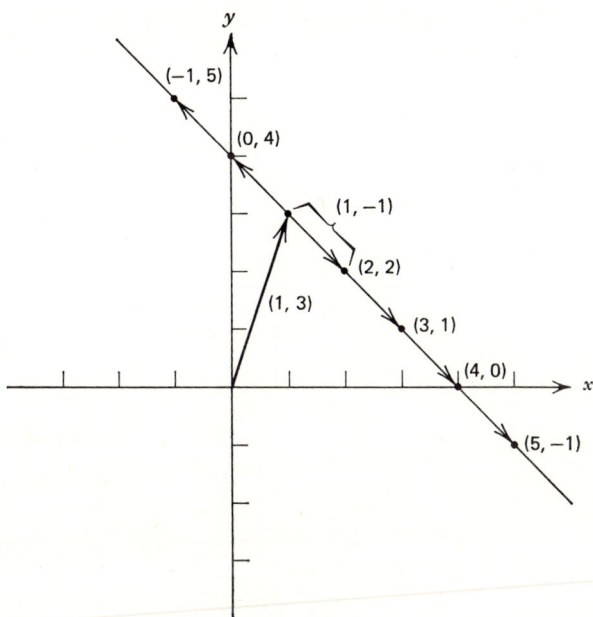

Figure 2-19

pieces of information that the line passes through (1,3) and (3,1)? Certainly the vector (2,−2), pictured as beginning at (1,3) and ending at (3,1), lies along the line. Scalar multiples of this vector can be used to get any vector lying along the same line. If we do this, we can express our previous equalities as follows.

$$(0,4) = (1,3) \quad \tfrac{1}{2}(2,-2) \qquad\qquad (3,1) = (1,3) + 1(2,-2)$$
$$(1,3) = (1,3) + 0(2,-2) \qquad\qquad (4,0) = (1,3) + \tfrac{3}{2}(2,-2)$$
$$(2,2) = (1,3) + \tfrac{1}{2}(2,-2) \qquad\qquad (5,-1) = (1,3) + 2(2,-2)$$

In a similar way we can express any vector from (0,0) to the line in the form (1,3) + t(2,−2) by choosing the proper value for t.

For another example, consider the line passing through (−1,−2) and (3,1). The vector from the point (−1,−2) to the point (3,1) is (4,3), so (4,3) is pictured as lying along the line or as parallel to the line. Any vector from (0,0) to the line can, therefore, be expressed in the form (−1,−2) + t(4,3) by choosing the proper value for t (see Figure 2-20).

We use the expression above in equation form to describe the lines mentioned above. We say $(x,y) = (1,3) + t(2,-2)$ or $\mathbf{X} = (1,3) + t(2,-2)$ is a vector equation of the line passing through (1,3) and (3,1). We say that $(x,y) = (-1,-2) + t(4,3)$ or $\mathbf{X} = (-1,-2) + t(4,3)$ is a vector equation of the line passing through $(-1,-2)$ and (3,1).

These equations are not the only ones which describe these lines. For example, the equation $\mathbf{X} = (2,2) + t(1,-1)$ is a vector equation of the line passing through

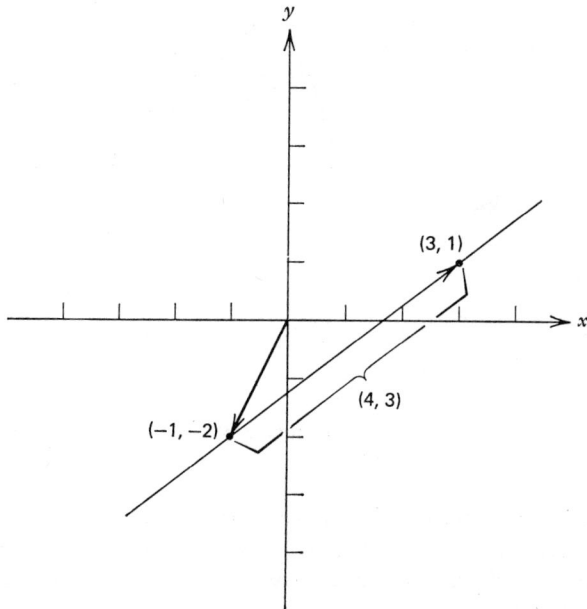

Figure 2-20

(1,3) and (3,1). In this case the equation can be interpreted geometrically as follows: go from the origin to the line with the vector (2,2) and use $(1,-1)$ to go from (2,2) to any other point on the line.

Similarly the line passing through $(-1,-2)$ and (3,1) can be described by the equation $\mathbf{X} = (0,-\tfrac{5}{4}) + t(1,\tfrac{3}{4})$ because $(0,-\tfrac{5}{4})$ is on the line and the vector $(1,\tfrac{3}{4})$ can be pictured as being parallel to the line.

In general, for a vector equation of a line, we need one vector from the origin to some point on the line, call it $\mathbf{X}_0$, and we need a nonzero vector $\mathbf{X}_1$ which is parallel to the line. If we have found two such vectors, then every vector $\mathbf{X}$ from (0,0) to the line can be expressed in the form $\mathbf{X} = \mathbf{X}_0 + t\mathbf{X}_1$. We say this line passes through the point $\mathbf{X}_0$ and is parallel to the vector $\mathbf{X}_1$.

Brief Exercises

1. Write a vector equation of the line passing through the following pairs of points; draw diagrams.

(a) (1,1) and (3,0)   (c) (3,1) and (4,−3)
(b) (−1,1) and (5,1)   (d) (2,−2) and (1,5)

2. Determine the points on the lines above whose first coordinate is 2.

Answers: **1(a)** (1,1) + t(2,−1)   **1(b)** (−1,1) + t(6,0)   **1(c)** (3,1) + t(1,−4)
1(d) (2,−2) + t(−1,7) **2(a)** (2,$\frac{1}{2}$) **2(b)** (2,1) **2(c)** (2,$\frac{7}{4}$) **2(d)** (2,−2)

A vector equation of a line in $\mathbb{R}^3$ is very much like that in $\mathbb{R}^2$. The geometric intuition is the same. If we wish to describe the line passing through the points

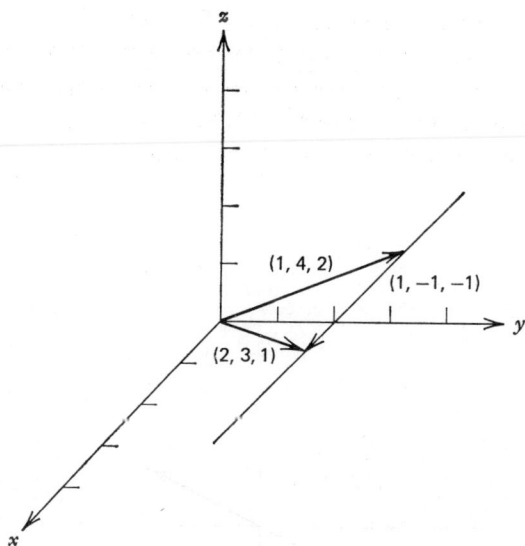

Figure 2-21

(2,3,1) and (1,4,2) (see Figure 2-21), we need a vector from (0,0,0) to the line, say (1,4,2), and we need a vector along the line (or parallel to the line). In this case, the vector (1,−1,−1) can be pictured as lying along the line starting at (1,4,2) and ending at (2,3,1). Then our expression for vectors beginning at the origin and ending on the line has the form $X = (1,4,2) + t(1,-1,-1)$.

This shows that our description of a line in $\mathbb{R}^3$ is like the one we use in $\mathbb{R}^2$; we need a vector X_0 from the origin to the line and a nonzero vector X_1 which is parallel to the line. Then every other vector from (0,0,0) to the line is expressed by $\mathbf{X} = \mathbf{X}_0 + t\mathbf{X}_1$. This line passes through the point $\mathbf{X}_0$ and is parallel to the vector $\mathbf{X}_1$.

Brief Exercises

1. Write a vector equation for the line passing through the following pairs of points.

(a) (1,0,−1) and (2,1,3)   (c) (5,−1,2) and (1,2,−1)

(b) (1,2,−1) and (1,0,3)   (d) (−1,3,1) and (4,0,2)

2. Determine the points on these lines having third coordinate 0.

Answers: **1(a)** (1,0,−1) + t(1,1,4) **1(b)** (1,2,−1) + t(0,−2,4) **1(c)** (5,−1,2) + t(−4,3,−3) **1(d)** (−1,3,1) + t(5,−3,1) **2(a)** $(\frac{5}{4},\frac{1}{4},0)$ **2(b)** $(1,\frac{3}{2},0)$ **2(c)** $(\frac{7}{3},1,0)$ **2(d)** (−6,6,0)

Finally we discuss some geometrical terms for vectors in $\mathbb{R}^2$ or in $\mathbb{R}^3$. In this discussion we always assume the vectors have their beginning at the origin. In $\mathbb{R}^2$ we say that the vector (2,1), for example, determines the line which passes through the origin and is parallel to the vector (2,1): $\mathbf{X} = (0,0) + t(2,1)$. We illustrate this in Figure 2-22. The line can be obtained geometrically by "extending" the arrow

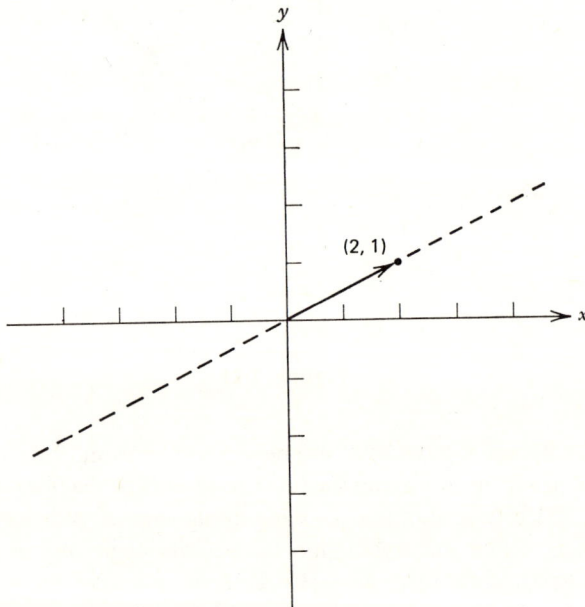

Figure 2-22

in both directions. We say that two vectors are *collinear* if they determine the same line. (Remember, all vectors are placed so they begin at the origin.) If they determine different lines, they are called *noncollinear*.

You can check geometrically that $(2,1)$ and $(4,2)$ are collinear; the vectors $(1,1)$ and $(-2,-2)$ are also collinear. You can also verify that $(2,1)$ and $(3,2)$ are noncollinear; $(5,1)$ and $(2,1)$ are also noncollinear. We are interested in giving an *arithmetic* way to determine whether two vectors are collinear or noncollinear.

As an example, the line passing through $(0,0)$ which is parallel to $(2,1)$ has vector equation $\mathbf{X} = (0,0) + t(2,1)$. In other words, a vector collinear with $(2,1)$ must be a scalar multiple of $(2,1)$. Thus $(4,2)$ and $(2,1)$ are collinear since $(4,2) = 2(2,1)$. Similarly, $(1,1)$ and $(-2,-2)$ are collinear because $(-2,-2) = -2(1,1)$. The vectors $(2,1)$ and $(3,2)$ are noncollinear since $(2,1) \neq t(3,2)$, no matter which value of t is tried. Similarly $(5,1)$ and $(2,1)$ are noncollinear since $(5,1) \neq t(2,1)$, no matter which value of t is used.

The same insight holds in $\mathbb{R}^3$. The vectors $(1,2,1)$ and $(-2,-4,-2)$ are collinear because $(-2,-4,-2) = -2(1,2,1)$, i.e., $(-2,-4,-2)$ is a scalar multiple of $(1,2,1)$.

Consequently, two vectors are collinear if one is a scalar multiple of the other. Otherwise they are noncollinear.

Brief Exercises

1. Determine arithmetically whether the following pairs of vectors are collinear; draw diagrams.

   (a) $(2,0)$ and $(3,0)$ **(c)** $(1,-1)$ and $(2,1)$
   (b) $(1,-2)$ and $(3,-6)$ **(d)** $(4,1)$ and $(-2,-\frac{1}{2})$

2. Do the same exercise for the following pairs of vectors in $\mathbb{R}^3$.

   (a) $(2,1,1)$ and $(-4,-2,-4)$ **(c)** $(3,1,2)$ and $(-1,2,1)$
   (b) $(1,2,1)$ and $(0,1,1)$

SUMMARY   *We show how to visualize vectors of* $\mathbb{R}^2$ *and* $\mathbb{R}^3$ *either as points or as arrows. The arrow representation is used to give a geometric picture of vector addition and of scalar multiplication. Vector equations of a line in* $\mathbb{R}^2$ *and* $\mathbb{R}^3$ *are discussed. A vector* $\mathbf{X}$ *beginning at the origin and ending on the line passing through the endpoint of* $\mathbf{X}_0$ *and parallel to* $\mathbf{X}_1$ *can be described by the equation* $\mathbf{X} = \mathbf{X}_0 + t\mathbf{X}_1$; *various values assigned to t produce different vectors* $\mathbf{X}$. *Finally, two vectors are collinear if and only if one is a scalar multiple of the other.*

Exercises for Section 2-3

1. In each case, write a vector equation of the line passing through the given points in $\mathbb{R}^2$.

   (a) $(2,1)$ and $(3,-1)$ (d) $(7,15)$ and $(0,2)$
   (b) $(-1,5)$ and $(4,2)$ (e) $(1,5)$ and $(1,8)$
   (c) $(1,1)$ and $(4,-3)$

2. Find the points where the following lines intersect the x-axis and the y-axis.

   (a) $(2,1) + t(3,-2)$ (c) $(4,2) + t(3,-1)$
   (b) $(-5,1) + t(1,5)$ (d) $(3,2) + t(-2,2)$

3. Find a vector equation for the line in $\mathbb{R}^3$ passing through the given points.

   (a) $(2,1,5)$ and $(4,1,1)$ (c) $(3,8,-2)$ and $(-1,2,1)$
   (b) $(2,0,3)$ and $(1,-1,0)$

4. In $\mathbb{R}^2$ determine whether the following vectors positioned at the origin are collinear.

   (a) $(1,2)$ and $(-1,2)$ (c) $(-3,0)$ and $(0,2)$
   (b) $(2,4)$ and $(1,2)$

5. In $\mathbb{R}^3$ determine whether the following vectors positioned at the origin are collinear.

   (a) $(2,1,0)$ and $(4,1,-1)$ (c) $(2,4,6)$ and $(-1,-2,-3)$
   (b) $(1,1,2)$ and $(0,1,3)$

6. In $\mathbb{R}^3$ determine whether the given pairs of lines intersect.

   (a) $(2,1,1) + t_1(3,-1,4)$ and $(1,-1,2) + t_2(-1,1,2)$
   (b) $(2,5,4) + t_1(1,1,1)$ and $(-1,4,9) + t_2(2,1,-2)$

7. Draw a diagram of the points in the plane which are expressed by the following.

   (a) $x(2,1) + y(1,3),\ x \geqq 0,\ y \geqq 0$ (c) $x(2,1) + y(1,3),\ x \leqq 0,\ y \geqq 0$
   (b) $x(2,1) + y(1,3),\ x \geqq 0,\ y \leqq 0$ (d) $x(2,1) + y(1,3),\ x \leqq 0,\ y \leqq 0$

8. If $\mathbf{X}$ and $\mathbf{Y}$ are noncollinear vectors in the plane, draw a diagram of the points in the plane expressed by

   (a) $x\mathbf{X} + y\mathbf{Y},\ x \geqq 0,\ y \geqq 0$ (c) $x\mathbf{X} + y\mathbf{Y},\ x \leqq 0,\ y \geqq 0$
   (b) $x\mathbf{X} + y\mathbf{Y},\ x \geqq 0,\ y \leqq 0$ (d) $x\mathbf{X} + y\mathbf{Y},\ x \leqq 0,\ y \leqq 0$

9. Find the point where the diagonals of the parallelogram with vertices at $(5,1)$, $(7,2)$, $(8,6)$, and $(6,5)$ meet. (Hint: write an equation of the line passing

through (5,1) and (8,6) and write an equation of the line passing through (7,2) and (6,5); find where these lines intersect.) (Even better hint: use your high school geometry.)

10. Consider the triangle with vertices at (2,1), (3,4), and (4,0).

   (a) Write a vector equation of the line passing through (3,4) and the midpoint of the line joining (2,1) and (4,0).
   (b) Write a vector equation of the line passing through (2,1) and the midpoint of the line joining (3,4) and (4,0).
   (c) Find where these two medians intersect. Use these results to verify that the medians of a triangle intersect at a point which is $\frac{2}{3}$ of the way from the vertex to the opposite side.

11. Same exercise as Exercise 10 for the triangle in $\mathbb{R}^3$ with vertices at (1,0,1), (2,1,3), and (4,2,3).

12. Describe the solution sets for the following systems of equations.

   (a) $3x + y = 2$
         $x - y = 1$

   (b) $3x + y = 2$
         $6x + 2y = 4$

13. Give an argument that the solution set for a system of linear equations in two unknowns must be empty, one point, a line, or all of $\mathbb{R}^2$. Can you relate the geometry of the solution set with the number of nonzero rows in the row echelon form for the augmented matrix? (Try a few numerical examples to discover a pattern.)

3

Linear Independence, Bases, Dimension

Linear Combinations

PREVIEW   *This section introduces the concept of a linear combination of some vectors. A linear combination is formed by multiplying the given vectors by various scalars and adding. A special kind of linear combination, which we call a proper linear combination, is introduced. Finally, the concept of a linear combination is related to systems of linear equations.*

Addition of vectors and multiplication of vectors by scalars are the two operations in any vector space. The vectors obtained by repeated application of these operations are linear combinations.

For example, in $\mathbb{R}^2$ we can begin with $(1,3)$ and $(2,-1)$ and form various linear combinations.

$$
\begin{aligned}
2(1,3) + 4(2,-1) &= (10,2) \\
5(1,3) - 3(2,-1) &= (-1,18) \\
2(1,3) + 4(2,-1) - (1,3) + 2(2,-1) &= (13,-3) \\
(1,3) + 6(2,-1) &= (13,-3)
\end{aligned}
$$

Thus the vectors $(10,2)$, $(-1,18)$, and $(13,-3)$ have been obtained from $(1,3)$ and $(2,-1)$ by scalar multiplication and vector addition, and so they are linear combinations of $(1,3)$ and $(2,-1)$. These are linear combinations of two vectors. We can, of course, form linear combinations of three, four, or more vectors.

$$2(1,2) + 3(2,3) - 4(3,4) + 2(4,5) - 5(5,6) = (-21,-23).$$

A linear combination is a sum of scalar multiples of vectors. From time to time we need to indicate the general form of a linear combination. A linear combination involving two terms has the form $a_1\mathbf{X}_1 + a_2\mathbf{X}_2$. Here a_1 and a_2 stand for scalars, and $\mathbf{X}_1$ and $\mathbf{X}_2$ stand for any vectors. A linear combination involving three terms has the form $a_1\mathbf{X}_1 + a_2\mathbf{X}_2 + a_3\mathbf{X}_3$. In this expression, as before, a_1, a_2, and a_3 represent scalars, and $\mathbf{X}_1$, $\mathbf{X}_2$, and $\mathbf{X}_3$ represent vectors. Rather than continue this kind of description indefinitely, getting longer and longer sums, we resort to the following device. Any linear combination involves a certain number of terms. Let n denote the number of terms; then the linear combination has the form

$$a_1\mathbf{X}_1 + a_2\mathbf{X}_2 + \cdots + a_n\mathbf{X}_n \quad \text{or} \quad a_1\mathbf{X}_1 + \cdots + a_n\mathbf{X}_n.$$

The dots indicate some terms which we are not writing out; each of the terms not mentioned explicitly still has the form $a\mathbf{X}$.

If $n = 1$, then we interpret the expression above to be $a_1\mathbf{X}_1$. If $n = 2$, then we interpret it to be $a_1\mathbf{X}_1 + a_2\mathbf{X}_2$.

Definition 3-1————————————————————————————————

Let $\mathscr{V}$ be a vector space over $\mathbb{R}$, and let $\mathscr{S}$ be a set of vectors in $\mathscr{V}$. If a vector $\mathbf{X}$ can be expressed as a finite sum of the form $a_1\mathbf{X}_1 + \cdots + a_n\mathbf{X}_n$, where $\mathbf{X}_1, \ldots, \mathbf{X}_n$ are vectors in the set $\mathscr{S}$ and $a_1, \ldots, a_n$ are scalars, then $\mathbf{X}$ is a *linear combination* of vectors in $\mathscr{S}$ or $\mathbf{X}$ is a *linear combination* of the vectors $\mathbf{X}_1, \ldots, \mathbf{X}_n$. If all the scalars $a_1, \ldots, a_n$ equal 0, the linear combination is a *zero linear combination*. If at least one of the scalars is different from 0, the linear combination is a *nonzero linear combination*.

————————————————————————————————

Notice that it is possible to express the zero vector $\mathbf{0}$ as a nonzero linear combination; e.g., in $\mathbb{R}^2$, $(0,0) = (1,2) - (3,6) + (2,4)$.

There is a slightly troublesome point in this definition: it is whether the vectors $\mathbf{X}_1, \ldots, \mathbf{X}_n$ are n different vectors or whether two different symbols, $\mathbf{X}_i$ and $\mathbf{X}_j$ say, with $i \neq j$, may represent the same vector $\mathbf{X}_i = \mathbf{X}_j$. In the definition of linear combination above we do *not* require that the vectors $\mathbf{X}_1, \ldots, \mathbf{X}_n$ be different from each other. But there are situations in which we must require that the vectors be different from each other. We introduce the term proper linear combination for this.

Definition 3-2————————————————————————————————

Let $\mathscr{V}$ be a vector space over $\mathbb{R}$, and let $\mathscr{S}$ be a set of vectors in $\mathscr{V}$. If a vector $\mathbf{X}$ can be expressed as a linear combination $a_1\mathbf{X}_1 + \cdots + a_n\mathbf{X}_n$, where $\mathbf{X}_1, \ldots, \mathbf{X}_n$ are n *different* vectors in $\mathscr{S}$, then $\mathbf{X}$ is a *proper linear combination* of some vectors in $\mathscr{S}$ or $\mathbf{X}$ is a *proper linear combination* of the vectors $\mathbf{X}_1, \ldots, \mathbf{X}_n$.

————————————————————————————————

At the beginning of this section we have

$$2(1,3) + 4(2,-1) - (1,3) + 2(2,-1) = (13,-3)$$

and

$$(1,3) + 6(2,-1) = (13,-3).$$

The first equation shows that $(13,-3)$ is a linear combination of $(1,3)$ and $(2,-1)$, whereas the second shows that $(13,-3)$ can be expressed as a *proper* linear combination of $(1,3)$ and $(2,-1)$ by combining the multiples of $(1,3)$ and of $(2,-1)$ into single terms. In the same way, any linear combination can be expressed as a proper linear combination by combining scalar multiples of each vector. Thus, for example,

$$5(1,0,1) + 6(3,2,-1) + 4(1,0,1) + (5,1,0)$$

can be expressed as a proper linear combination as

$$9(1,0,1) + 6(3,2,-1) + (5,1,0).$$

Now we look at a few more examples.

Example   Here are some linear combinations in $\mathbb{R}^4$. Since

$$5(1,2,-1,0) + 3(1,1,0,1) - 2(3,1,0,0) = (2,11,-5,3)$$

the vector $(2,11,-5,3)$ is a linear combination of the vectors $(1,2,-1,0)$, $(1,1,0,1)$, and $(3,1,0,0)$.

Example   Here are some linear combinations in $\mathbb{R}^3$. Since

$$2(1,-1,1) + 5(-3,2,1) - 4(1,2,0) + 0(1,1,1) = (-17,0,7)$$

the vector $(-17,0,7)$ is a linear combination of the vectors $(1,-1,1)$, $(-3,2,1)$, $(1,2,0)$, and $(1,1,1)$, even though the zero scalar multiplies $(1,1,1)$. But $(-17,0,7)$ is also a linear combination of $(1,-1,1)$, $(-3,2,1)$, and $(1,2,0)$ because

$$2(1,-1,1) + 5(-3,2,1) - 4(1,2,0) = (-17,0,7).$$

As the last example indicates, we use $0\mathbf{X}$ in a linear combination or we omit it if we wish since $0\mathbf{X} = \mathbf{0}$ (see Exercise 10, Section 2-1).

Brief Exercises

1. Compute the following linear combinations.

(a) $3(5,1) - 4(6,8) + 3(2,-3)$
(b) $2(-1,3) + 3(2,-1) + 4(6,2)$
(c) $3(8,1) - 6(2,1) + 5(8,1) + 2(3,1)$
(d) $3(1,0,1) - 6(2,1,3) + 4(1,0,1) + 2(1,-1,3) + 4(2,1,3)$
(e) $2(1,1,-17) + 4(8,1,2) - 3(8,1,2)$
(f) $2(1,1,-1,2) + 3(0,1,2,1) - 6(1,8,2,1) + 4(0,1,2,1) + 3(1,8,2,1)$
$\qquad - 5(1,1,-1,2)$

2. Which of the linear combinations in Exercise 1 are proper? For those which are not proper linear combinations, express them as proper linear combinations of the same vectors by combining multiples of each vector.

Selected Answers: **2(a)** Proper   **2(b)** Proper   **2(c)** $8(8,1) - 6(2,1) + 2(3,1)$
2(d) $7(1,0,1) - 2(2,1,3) + 2(1,-1,3)$ **2(e)** $2(1,1,-17) + 1(8,1,2)$ **2(f)** $-3(1,1,-1,2) + 7(0,1,2,1) - 3(1,8,2,1)$

There is no difficulty in computing linear combinations; it is just scalar multiplication and addition. A slightly more difficult problem, and therefore a more interesting one, is to determine whether a given vector is a linear combination of some others. We illustrate this next.

Example　In this example we determine whether in $\mathbb{R}^2$ the vector $(2,1)$ is a linear combination of $(1,-1)$, $(-3,1)$, and $(0,2)$.

The vector $(2,1)$ is a linear combination of these three vectors if and only if there are scalars x, y, and z so that

$$(2,1) = x(1,-1) + y(-3,1) + z(0,2)$$

or

$$(2,1) = (x - 3y, -x + y + 2z).$$

Since two vectors in $\mathbb{R}^2$ are equal if and only if they have equal components, this vector equality is the same as two scalar equations.

$$
\begin{aligned}
x - 3y \quad\;\;\; &= 2 \\
-x + \; y + 2z &= 1
\end{aligned}
$$

Use row operations on the augmented matrix to find the solution to this system. The row echelon form of the augmented matrix is

$$\begin{pmatrix} 1 & 0 & -3 & -\frac{5}{2} \\ 0 & 1 & -1 & -\frac{3}{2} \end{pmatrix}.$$

This is the augmented matrix for the system

$$
\begin{aligned}
x - 3z &= -\tfrac{5}{2} \\
y - \; z &= -\tfrac{3}{2}
\end{aligned}
$$

or

$$
\begin{aligned}
x &= -\tfrac{5}{2} + 3z \\
y &= -\tfrac{3}{2} + \; z.
\end{aligned}
$$

So, for example, by using $x = -\frac{5}{2}$, $y = -\frac{3}{2}$, and $z = 0$, we can express $(2,1)$ as a linear combination of $(1,-1)$, $(-3,1)$, and $(0,2)$.

$$(2,1) = -\tfrac{5}{2}(1,-1) - \tfrac{3}{2}(-3,1) + 0(0,2)$$

(Of course, there are other choices of scalars corresponding to different choices for z, such as $x = -1$, $y = -1$, $z = \frac{1}{2}$.) Thus $(2,1)$ is a linear combination of $(1,-1)$, $(-3,1)$, and $(0,2)$.

This example shows that deciding whether $(2,1)$ is a linear combination of $(1,-1)$, $(-3,1)$, and $(0,2)$ is the same as finding the solution set for a system of linear equations. This is particularly easy to see if we write vectors in $\mathbb{R}^2$ vertically rather than horizontally; $(2,1)$ is a linear combination of $(1,-1)$, $(-3,1)$, and $(0,2)$ if and only if there are scalars x, y, and z so that

$$ x\begin{pmatrix} 1 \\ -1 \end{pmatrix} + y\begin{pmatrix} -3 \\ 1 \end{pmatrix} + z\begin{pmatrix} 0 \\ 2 \end{pmatrix} = \begin{pmatrix} 2 \\ 1 \end{pmatrix}. $$

This display makes it easy to picture the system of equations which must be solved.

We can use this device of writing vectors vertically to look at systems of equations from another point of view; a system of equations has a solution if the last column of the augmented matrix is a linear combination of the other columns.

Example   Consider the system of equations

$$ \begin{aligned} 3x_1 + 2x_2 - x_3 + x_4 &= 5 \\ 4x_1 - x_2 + 3x_3 &= 1 \\ x_1 + 2x_2 - x_3 + 4x_4 &= 3. \end{aligned} $$

To find the solutions of these equations, we need to find scalars x_1, x_2, x_3, and x_4 so that

$$ x_1\begin{pmatrix} 3 \\ 4 \\ 1 \end{pmatrix} + x_2\begin{pmatrix} 2 \\ -1 \\ 2 \end{pmatrix} + x_3\begin{pmatrix} -1 \\ 3 \\ -1 \end{pmatrix} + x_4\begin{pmatrix} 1 \\ 0 \\ 4 \end{pmatrix} = \begin{pmatrix} 5 \\ 1 \\ 3 \end{pmatrix}. $$

Consequently, the original system of equations has a solution if and only if $(5,1,3)$ is a linear combination of the vectors $(3,4,1)$, $(2,-1,2)$, $(-1,3,-1)$, and $(1,0,4)$. The augmented matrix for this system is

$$ \begin{pmatrix} 3 & 2 & -1 & 1 & 5 \\ 4 & -1 & 3 & 0 & 1 \\ 1 & 2 & -1 & 4 & 3 \end{pmatrix}. $$

In matrix language, the original system of equations has a solution if and only if the last column of the augmented matrix is a linear combination of the first four columns. Notice that we have considered the columns of the matrix as vectors in $\mathbb{R}^3$.

The observations made in the example above are important enough to be recorded in a proposition.

Proposition 3-1 _____

A system of linear equations has a solution if and only if the column of constants in the augmented matrix is a linear combination of the other columns.

Brief Exercises

In Exercises 1, 2, and 3, answer the question. Then express the question and the answer in terms of (a) a system of linear equations and (b) the columns of the augmented matrix for the system.

1. Is $(2,1)$ a linear combination of $(1,1)$, $(1,3)$, and $(-1,4)$?

2. Is $(2,1,-3)$ a linear combination of $(1,1,0)$, $(1,0,1)$, and $(0,1,0)$?

3. Is $(1,1,0)$ a linear combination of $(2,-1,1)$, $(3,-2,1)$, and $(-1,1,0)$?

In Exercises 4, 5, and 6, answer the question. Then phrase the question and its answer in terms of (a) linear combinations and (b) the columns of the augmented matrix.

4. Does the following system of equations have a solution?

$$3x + y - z = 2$$
$$x - y + z = 3$$

5. Does the following system of equations have a solution?

$$2x + y = 1$$
$$x - y = 2$$

6. Does the following system of equations have a solution?

$$4x - 6y + z = 2$$
$$x + 2y + 3z = 1$$
$$4x - y + 2z = 3$$
$$x + y - 3z = 1$$

Answers: **1** $(2,1) = (\frac{5}{2} + \frac{7}{2}z)(1,1) + (-\frac{1}{2} - \frac{5}{2}z)(1,3) + z(-1,4)$ **2** $(2,1,-3) = 5(1,1,0) - 3(1,0,1) - 4(0,1,0)$ **3** No **4** $(\frac{5}{4}, -\frac{7}{4} + z, z)$ **5** $(1,-1)$ **6** No

SUMMARY *The concepts of linear combination and proper linear combination of vectors are introduced and related to systems of equations; a system of equations has a solution if and only if the column of constants (the last column) in the augmented matrix for the system of equations can be expressed as a linear combination of the other columns.*

Exercises for Section 3-1

1. Form the indicated linear combinations of the given vectors.
   (a) In $\mathbb{R}^5$: $-5(1,1,2,3,4) + 16(3,4,5,\sqrt{17},\pi)$
   (b) In $\mathbb{R}^5$: $13(1,15,16,11,0) + 9(2,16,13,9,7)$
   (c) In $\mathbb{R}^2$: $\frac{1}{4}(\frac{2}{3},\frac{1}{5}) + \frac{1}{3}(1,0) + \frac{1}{12}(18,-6)$

2. Write the following linear combinations as proper linear combinations.
   (a) In $\mathbb{R}^4$: $2(1,3,1,4) - 5(6,3,-1,0) + 4(1,3,1,4) + 2(9,0,1,3)$
   (b) In $\mathbb{R}^3$: $(1,0,1) + 2(1,1,0) + 3(1,0,1) - 1(1,1,0)$

3. (a) In $\mathbb{R}^2$ is $(1,1)$ a linear combination of $(1,-1)$ and $(3,2)$?
   (b) In $\mathbb{R}^3$ is $(1,1,0)$ a linear combination of $(1,-1,0)$ and $(3,2,0)$?
   (c) In $\mathbb{R}^4$ is $(1,1,0,3)$ a linear combination of $(1,1,1,0)$, $(2,1,0,4)$, and $(4,1,0,1)$?
   (d) In $\mathbb{R}^3$ is $(1,1,0)$ a linear combination of $(1,0,1)$, $(3,4,5)$, and $(6,8,2)$?

4. Express all the questions in 3 as questions concerning the solution of a system of linear equations.

5. Express all the questions in 3 as questions concerning the columns of the augmented matrix for a system of linear equations.

SECTION 3-2
Span of a Set

PREVIEW   *The linear span of a set of vectors is introduced. If every vector in a vector space $\mathscr{V}$ can be expressed as a linear combination of some vectors in a set $\mathscr{S}$, then $\mathscr{S}$ is called a generating set for $\mathscr{V}$. The collection of all vectors which can be expressed as a linear combination of vectors in $\mathscr{S}$ is called the (linear) span of $\mathscr{S}$.*

The problem discussed in Section 3-1 is to determine whether a given vector is expressible as a linear combination of some other vectors. Now we consider the problem of whether or not *every* vector in a vector space can be expressed as a linear combination of some given vectors.

Example   In this example we want to know whether every vector in $\mathbb{R}^2$ is a linear combination of $(1,2)$, $(-1,1)$, and $(3,4)$. To do this we must determine whether, given a general vector (x,y), it is always possible to find scalars a_1, a_2, and a_3 so that

$$a_1 \begin{pmatrix} 1 \\ 2 \end{pmatrix} + a_2 \begin{pmatrix} -1 \\ 1 \end{pmatrix} + a_3 \begin{pmatrix} 3 \\ 4 \end{pmatrix} = \begin{pmatrix} x \\ y \end{pmatrix}.$$

To solve this system of equations we transform the augmented matrix to its row echelon form.

$$\begin{pmatrix} 1 & 0 & \frac{7}{3} & \frac{1}{3}(x+y) \\ 0 & 1 & -\frac{2}{3} & \frac{1}{3}(y-2x) \end{pmatrix}$$

This is the augmented matrix for

$$a_1 + \tfrac{7}{3}a_3 = \tfrac{1}{3}(x+y)$$
$$a_2 - \tfrac{2}{3}a_3 = \tfrac{1}{3}(y-2x)$$

or

$$a_1 = \tfrac{1}{3}(x+y) - \tfrac{7}{3}a_3$$
$$a_2 = \tfrac{1}{3}(y-2x) + \tfrac{2}{3}a_3.$$

Thus, if (x,y) is given, we can choose a value for a_3 and then determine a_1 and a_2 to express (x,y) as a linear combination of $(1,2)$, $(-1,1)$, and $(3,4)$. For example, for $a_3 = 0$ we get $a_1 = \tfrac{1}{3}(x+y)$ and $a_2 = \tfrac{1}{3}(y-2x)$. For $a_3 = 3$ we get $a_1 = \tfrac{1}{3}(x+y) - 7$ and $a_2 = \tfrac{1}{3}(y-2x) + 2$. So, if $(x,y) = (1,1)$, then $(1,1)$ can be expressed as a linear combination of the given vectors by using $a_1 = \tfrac{2}{3}$, $a_2 = -\tfrac{1}{3}$, and $a_3 = 0$, or by using $a_1 = -\tfrac{19}{3}$, $a_2 = \tfrac{5}{3}$, and $a_3 = 3$. The conclusion we get is that any vector in $\mathbb{R}^2$ is a linear combination of the vectors $(1,2)$, $(-1,1)$, and $(3,4)$.

Example Here we determine whether every vector in $\mathbb{R}^3$ is a linear combination of $(1,0,1)$, $(2,1,3)$, and $(5,2,7)$. To do this we must determine whether, given a general vector (x,y,z), it is always possible to find scalars a_1, a_2, and a_3 so that

$$a_1 \begin{pmatrix} 1 \\ 0 \\ 1 \end{pmatrix} + a_2 \begin{pmatrix} 2 \\ 1 \\ 3 \end{pmatrix} + a_3 \begin{pmatrix} 5 \\ 2 \\ 7 \end{pmatrix} = \begin{pmatrix} x \\ y \\ z \end{pmatrix}.$$

We solve by transforming the augmented matrix to its row echelon form.

$$\begin{pmatrix} 1 & 0 & 1 & x-2y \\ 0 & 1 & 2 & y \\ 0 & 0 & 0 & z-x-y \end{pmatrix}.$$

This is the augmented matrix for the system of equations

$$a_1 \qquad + \ a_3 = x - 2y$$
$$a_2 + 2a_3 = y$$
$$0a_1 + 0a_2 + 0a_3 = z - x - y$$

or

$$a_1 = x - 2y - a_3$$
$$a_2 = y - 2a_3$$
$$0 = z - x - y.$$

The last equation tells us that (x,y,z) is a linear combination of $(1,0,1)$, $(2,1,3)$, and $(5,2,7)$, if its coordinates satisfy $0 = z - x - y$. Many vectors do not satisfy this equation; e.g., $(1,0,0)$, $(1,1,1)$, $(0,0,1)$, etc. Consequently we cannot express every vector (x,y,z) in $\mathbb{R}^3$ as a linear combination of the given vectors. However, if a vector's coordinates do satisfy $z - x - y = 0$, then the first two equations show how to choose the scalars. For example, $(-1,2,1)$ satisfies the third equation; the first two equations give choices for the scalars a_1, a_2, and a_3. One choice is $a_1 = -5$, $a_2 = 2$, and $a_3 = 0$; another choice is $a_1 = 0$, $a_2 = 12$, and $a_3 = -5$.

Brief Exercises

1. Determine whether every vector in $\mathbb{R}^2$ is a linear combination of the given vectors. If the answer is yes, then show how the vectors $(1,1)$, $(-1,3)$, and $(5,-7)$ can be expressed as linear combinations of the given vectors.

    (a) $(1,-2)$ and $(3,2)$ (b) $(1,-1)$, $(3,-1)$, and $(2,4)$

2. Determine whether every vector in $\mathbb{R}^3$ is a linear combination of the given vectors. If so, then express $(2,1,0)$, $(-1,3,5)$, and $(6,-10,8)$ as a linear combination of the given vectors.

    (a) $(1,1,3)$, $(2,1,5)$, $(1,0,0)$ (b) $(1,2,5)$, $(2,1,0)$

Selected Answers: **1(a)** $(x,y) = \frac{1}{8}(2x - 3y)(1,-2) + \frac{1}{8}(2x + y)(3,2)$ **1(b)** $(x,y) = (7a - \frac{1}{2}(x + 3y))(1,-1) + (\frac{1}{2}(x + y) - 3a)(3,-1) + a(2,4)$, for any value of a **2(a)** $(x,y,z) = \frac{1}{2}(5y - z)(1,1,3) + \frac{1}{2}(z - 3y)(2,1,5) + \frac{1}{2}(2x + y - z)(1,0,0)$ **2(b)** No, we cannot express every vector as a linear combination of the given vectors, yet $(2,1,0) = 0(1,2,5) + 1(2,1,0)$

Now we introduce some terminology which makes it easier to describe the results of the examples above.

Definition 3-3

Let $\mathcal{V}$ be a vector space over $\mathbb{R}$, and let $\mathcal{S}$ be a nonvoid subset of $\mathcal{V}$. The *linear span of $\mathcal{S}$* or the *span of $\mathcal{S}$* is the set of all vectors in $\mathcal{V}$ which are linear combinations of some vectors from $\mathcal{S}$. The set $\mathcal{S}$ is a *spanning* or *generating set* for $\mathcal{V}$ (*spans* or *generates $\mathcal{V}$*) if every vector in $\mathcal{V}$ is a linear combination of some vectors in $\mathcal{S}$, or, equivalently, if the span of $\mathcal{S}$ equals $\mathcal{V}$. The vector space $\mathcal{V}$ is *finitely generated* if there is a finite set which generates $\mathcal{V}$.

In the first example in this section, $(1,2)$, $(-1,1)$, and $(3,4)$ form a generating set for $\mathbb{R}^2$. In the second, $(1,0,1)$, $(2,1,3)$, and $(5,2,7)$ do not form a generating set for $\mathbb{R}^3$, but they do generate the subspace consisting of vectors (x,y,z) which satisfy $z - x - y = 0$.

There are many obvious examples of generating sets for the spaces $\mathbb{R}^n$. In $\mathbb{R}^2$ the vectors $(1,0)$ and $(0,1)$ form a generating set since any vector (x,y) in $\mathbb{R}^2$ can be expressed as $(x,y) = x(1,0) + y(0,1)$. Quite often $\mathbf{i}$ and $\mathbf{j}$ are used for $(1,0)$ and $(0,1)$, respectively, and so $(x,y) = x\mathbf{i} + y\mathbf{j}$. In $\mathbb{R}^3$ the vectors $(1,0,0)$, $(0,1,0)$, and $(0,0,1)$ form a generating set since any vector (x,y,z) can be expressed

$$(x,y,z) = x(1,0,0) + y(0,1,0) + z(0,0,1).$$

You often see $\mathbf{i}$, $\mathbf{j}$, and $\mathbf{k}$ used for $(1,0,0)$, $(0,1,0)$, and $(0,0,1)$, respectively, and so

$$(x,y,z) = x\mathbf{i} + y\mathbf{j} + z\mathbf{k}.$$

In $\mathbb{R}^n$ the vectors $(1,0,\ldots,0), (0,1,0,\ldots,0), \ldots, (0,\ldots,0,1)$—the vectors having one coordinate equal to 1 and the other coordinates equal to 0—form a generating set for $\mathbb{R}^n$. This set is called the *natural* or *standard* generating set for $\mathbb{R}^n$, and so it is clear that $\mathbb{R}^n$ is a finitely generated vector space.

Example We determine whether $(1,2,1)$, $(3,4,5)$, and $(-5,-6,-9)$ form a generating set for $\mathbb{R}^3$. To do this we have to determine whether, given any vector (x,y,z) in $\mathbb{R}^3$, it is possible to find scalars a_1, a_2, and a_3 so that

$$a_1 \begin{pmatrix} 1 \\ 2 \\ 1 \end{pmatrix} + a_2 \begin{pmatrix} 3 \\ 4 \\ 5 \end{pmatrix} + a_3 \begin{pmatrix} -5 \\ -6 \\ -9 \end{pmatrix} = \begin{pmatrix} x \\ y \\ z \end{pmatrix}.$$

We solve this system of equations for a_1, a_2, and a_3. The row echelon form of the augmented matrix is

$$\begin{pmatrix} 1 & 0 & 1 & -2x + \frac{3}{2}y \\ 0 & 1 & -2 & x - \frac{1}{2}y \\ 0 & 0 & 0 & z - 3x + y \end{pmatrix}.$$

This is the augmented matrix for

$$a_1 + \quad + a_3 = -2x + \tfrac{3}{2}y$$

$$a_2 - 2a_3 = \quad x - \tfrac{1}{2}y$$

$$0 = z - 3x + y.$$

Thus there is a solution provided $z - 3x + y = 0$. Consequently, the given vectors do not form a generating set for $\mathbb{R}^3$, because many vectors in $\mathbb{R}^3$ do not satisfy this equation, e.g., $(1,0,0)$, $(1,1,1)$, etc. However, the given vectors do generate the subspace consisting of vectors (x,y,z) satisfying $z - 3x + y = 0$.

Example   We find a generating set for two subspaces of $\mathbb{R}^4$.

$\mathscr{W}_1$, the vectors (x,y,z,w) satisfying $x - y + z + w = 0$
$\mathscr{W}_2$, the vectors (x,y,z,w) satisfying $3x + y - z + 2w = 0$ and $x - y + 4z - 2w = 0$

To find a generating set for $\mathscr{W}_1$, solve for one of the unknowns in terms of the others. We solve for y: $y = x + z + w$. Then a vector (x,y,z,w) is in $\mathscr{W}_1$ if and only if it can be expressed in the form (substitute $x + z + w$ for y)

$$(x, x + z + w, z, w) = x(1,1,0,0) + z(0,1,1,0) + w(0,1,0,1).$$

Consequently $(1,1,0,0)$, $(0,1,1,0)$, and $(0,1,0,1)$ form a generating set for $\mathscr{W}_1$.

If we solve for a different unknown, we get a different generating set for $\mathscr{W}_1$. For example, solve for x: $x = y - z - w$. Then a vector (x,y,z,w) is in $\mathscr{W}_1$ if and only if it can be expressed in the form (substitute $y - z - w$ for x)

$$(y - z - w, y, z, w) = y(1,1,0,0) + z(-1,0,1,0) + w(-1,0,0,1).$$

Consequently $(1,1,0,0)$, $(-1,0,1,0)$, and $(-1,0,0,1)$ also form a generating set for $\mathscr{W}_1$. The same kind of thing occurs if we solve for z or w; we obtain other generating sets for $\mathscr{W}_1$.

To find a generating set for $\mathscr{W}_2$, solve for two of the unknowns in terms of the others. In this case we solve for x and y in terms of z and w.

$$x = -\tfrac{3}{4}z$$

$$y = \tfrac{13}{4}z - 2w$$

Then a vector (x,y,z,w) is in $\mathscr{W}_2$ if and only if it can be expressed in the form

$$(-\tfrac{3}{4}z, \tfrac{13}{4}z - 2w, z, w) = z(-\tfrac{3}{4},\tfrac{13}{4},1,0) + w(0,-2,0,1).$$

Thus the vectors $(-\tfrac{3}{4},\tfrac{13}{4},1,0)$ and $(0,-2,0,1)$ form a generating set for $\mathscr{W}_2$.

Brief Exercises

1. Do the sets consisting of the following vectors generate $\mathbb{R}^2$?

    (a) $(1,1)$, $(0,1)$ **(c)** $(3,-1)$, $(-3,1)$

    (b) $(2,2)$, $(1,1)$ **(d)** $(2,1)$, $(1,3)$

2. Do the sets consisting of the following vectors generate $\mathbb{R}^3$? Give reasons.

    (a) $(2,1,0)$ **(c)** $(1,0,1)$, $(0,1,0)$, $(0,1,1)$

    (b) $(1,1,0)$, $(0,1,0)$ **(d)** $(1,0,0)$, $(2,1,0)$, $(0,1,0)$

3. Find a generating set for the following vector spaces.

    (a) the vectors (x,y) in $\mathbb{R}^2$ satisfying $x - y = 0$

    (b) the vectors (x,y) in $\mathbb{R}^2$ satisfying $x + 2y = 0$

    (c) the vectors (x,y,z) in $\mathbb{R}^3$ satisfying $x - y + z = 0$

    (d) the vectors (x,y,z) in $\mathbb{R}^3$ satisfying $x + y - z = 0$ and $x + 3y + z = 0$

Answers: **1(a)** Yes **1(b)** No **1(c)** Yes **1(d)** Yes **2(a)** No **2(b)** No **2(c)** Yes **2(d)** No **3** There is no single answer to these. Possible answers: **3(a)** $(1,1)$ **3(b)** $(-2,1)$ **3(c)** $(1,1,0)$, $(0,1,1)$ **3(d)** $(-2,1,-1)$

We conclude with a fact concerning the linear span of sets. The linear span of a set of vectors in a vector space $\mathscr{V}$ is a subspace of $\mathscr{V}$. We illustrate this in the next example.

Example Suppose $\mathscr{S}$ is the set in $\mathbb{R}^3$ consisting of the vectors $(2,1,0)$ and $(3,4,-2)$. The linear span of $\mathscr{S}$ is a subspace of $\mathbb{R}^3$, and the reason is easy to understand. The span of $\mathscr{S}$ is the set of all linear combinations of $(2,1,0)$ and $(3,4,-2)$. The sum of two such linear combinations is also such a linear combination

$$a(2,1,0) + b(3,4,-2)$$

$$\frac{c(2,1,0) + d(3,4,-2)}{(a + c)(2,1,0) + (b + d)(3,4,-2)}$$

Also, a scalar multiple of a linear combination of these two vectors is again a linear combination of the two vectors.

$$r[a(2,1,0) + b(3,4,-2)] = ra(2,1,0) + rb(3,4,-2)$$

Thus the linear span of $\mathscr{S}$ is a subspace of $\mathbb{R}^3$ because properties A.1 and SM.1 from Definition 2-5 are satisfied (see Proposition 2-2).

The fact illustrated in this example is true in general; if $\mathscr{S}$ is a nonvoid set of vectors in a vector space $\mathscr{V}$, then the linear span of $\mathscr{S}$ is a subspace of $\mathscr{V}$. The reasons are the same as given in the example: properties A.1 and SM.1 are satisfied.

The sum of two linear combinations of vectors in $\mathscr{S}$ is a (perhaps longer) linear combination of vectors in $\mathscr{S}$.

$$
(\underbrace{\hspace{3cm}}_{\substack{\text{linear combination} \\ \text{of vectors in } \mathscr{S}}}) + (\underbrace{\hspace{3cm}}_{\substack{\text{linear combination} \\ \text{of vectors in } \mathscr{S}}})
$$

$$
\underbrace{\hspace{9cm}}_{\substack{\text{linear combination} \\ \text{of vectors in } \mathscr{S}}}
$$

A scalar multiple of a linear combination of vectors in $\mathscr{S}$ is a linear combination of the same vectors.

Proposition 3-2

If $\mathscr{S}$ is a nonvoid set of vectors in a vector space $\mathscr{V}$, then the linear span of $\mathscr{S}$ is a subspace of $\mathscr{V}$.

Brief Exercises

1. In each of the following, if $\mathscr{S}$ is the set consisting of the given vectors, describe the span of $\mathscr{S}$ both arithmetically and geometrically, if possible.

(a) $(1,1)$

(b) $(0,1,1)$, $(2,0,2)$

(c) $(0,1,1)$, $(1,0,1)$

(d) $(1,0,1,0)$, $(0,1,1,0)$, $(1,0,1,1)$

Answers: **1(a)** Multiples of $(1,1)$; line through origin determined by $(1,1)$ **1(b)** Vectors of the form $(2x, y, x + 2y)$; plane passing through $(0,0,0)$, $(0,1,1)$, and $(2,0,2)$ **1(c)** Vectors of the form $(x, y, x + y)$; plane passing through $(0,0,0)$, $(0,1,1)$, and $(1,0,1)$ **1(d)** Vectors of the form $(x + z, y, x + y + z, z)$

> **SUMMARY**   *If every vector in a vector space $\mathscr{V}$ can be expressed as a linear combination of vectors in $\mathscr{S}$, then $\mathscr{S}$ is called a generating set for $\mathscr{V}$. The set of all linear combinations formed from vectors in $\mathscr{S}$ is called the linear span of $\mathscr{S}$. The linear span of $\mathscr{S}$ is a subspace of $\mathscr{V}$ containing $\mathscr{S}$.*

Exercises for Section 3-2

1. Do the given vectors form a generating set for $\mathbb{R}^2$?

(a) $(1,1)$, $(1,2)$, $(3,4)$

(b) $(2,1)$, $(3,2)$

2. Do the given vectors form a generating set for $\mathbb{R}^3$?

(a) $(1,4,2)$, $(2,3,1)$, $(-4,-1,1)$

(b) $(1,0,1)$, $(3,4,5)$, $(6,8,2)$

3. Do the given vectors form a generating set for $\mathbb{R}^4$?

  (a) (1,1,0,1), (1,1,2,1), (1,1,0,0)
  (b) (1,1,2,0), (1,1,2,1), (2,1,1,0), (1,1,0,1), (0,1,0,1)

More Challenging Exercises

4. If $\mathscr{A}$ is the set consisting of (1,0,1) and (0,1,1) and if $\mathscr{B}$ is the set consisting of (1,2,3) and (1,−1,0), does the span of $\mathscr{A}$ equal the span of $\mathscr{B}$?

5. In the 2 × 3 matrix

$$\begin{pmatrix} 5 & 1 & 2 \\ 1 & -1 & 7 \end{pmatrix}$$

we can consider each row as an element of $\mathbb{R}^3$. Label them as $\mathbf{X}_1 = (5,1,2)$ and $\mathbf{X}_2 = (1,-1,7)$. If we apply a row operation, for example $R_{1,2}(-1)$, we obtain the matrix

$$\begin{pmatrix} 4 & 2 & -5 \\ 1 & -1 & 7 \end{pmatrix}.$$

Label the rows of this matrix as $\mathbf{Y}_1 = (4,2,-5)$ and $\mathbf{Y}_2 = (1,-1,7)$. It is easy to see that the **Y**'s are linear combinations of the **X**'s.

$$\mathbf{Y}_1 = \mathbf{X}_1 - \mathbf{X}_2$$

$$\mathbf{Y}_2 = \mathbf{X}_2$$

This is an example of a general fact.

  (a) Give an argument supporting the statement that if the matrix B is obtained from A by row operations, then the rows of B are linear combinations of the rows of A. Consequently the rows of B are in the span of the rows of A.
  (b) Since it is possible to reverse row operations—$R_{i,j}$ reverses $R_{i,j}$; $R_i(x^{-1})$ reverses $R_i(x)$; and $R_{i,j}(-x)$ reverses $R_{i,j}(x)$—does it follow that if B has been obtained from A by row operations, then the rows of A are in the span of the rows of B? Why?
  (c) If A and B are row equivalent, then their rows span the same subspace of $\mathbb{R}^n$. (See Definition 1-2 in Exercises for Section 1-1.)
  (d) Suppose that A and B are matrices and that the rows of each are in the span of the other. Can B be obtained from A by row operations?

6. Give an argument supporting the general statement that if $\mathscr{A}$ and $\mathscr{B}$ are subsets of a vector space $\mathscr{V}$ and if $\mathscr{A}$ is contained in the span of $\mathscr{B}$ and $\mathscr{B}$ is contained in the span of $\mathscr{A}$, then the span of $\mathscr{A}$ equals the span of $\mathscr{B}$.

7. Give an argument to support the statement that if the $m \times n$ matrix B is obtained from the matrix A by column operations, then the columns of A and the columns of B span the same subspace of $\mathbb{R}^m$. (Look at Exercise 5; trying a few numerical examples never hurts.)

SECTION 3-3
Vector Equations of Planes in $\mathbb{R}^3$, Subspaces of $\mathbb{R}^2$ and $\mathbb{R}^3$

PREVIEW *In this section we return to geometrical considerations. First, a geometrical argument is given to show that any two noncollinear vectors form a generating set for $\mathbb{R}^2$. Then vector equations of planes in $\mathbb{R}^3$ are discussed. This leads to deciding whether three vectors at the origin lie in one plane—are coplanar— or whether they do not lie in one plane—are noncoplanar. A geometric argument shows that three noncoplanar vectors form a generating set for $\mathbb{R}^3$. Finally a geometric description is given of all subspaces of $\mathbb{R}^2$ and $\mathbb{R}^3$.*

In this section we do some more geometry, first in $\mathbb{R}^2$. We use the terminology introduced in Section 2-3. At the origin, take any two noncollinear vectors $\mathbf{X}_1$ and $\mathbf{X}_2$. Our aim is to show geometrically that $\mathbf{X}_1$ and $\mathbf{X}_2$ form a generating set for $\mathbb{R}^2$.

For example, let $\mathbf{X}_1 = (1,2)$ and $\mathbf{X}_2 = (-2,-1)$, and let $\mathbf{X} = (-3,5)$. To write $\mathbf{X}$ as a linear combination of $\mathbf{X}_1$ and $\mathbf{X}_2$ we need to find scalars a_1 and a_2 so that

$$a_1 \begin{pmatrix} 1 \\ 2 \end{pmatrix} + a_2 \begin{pmatrix} -2 \\ -1 \end{pmatrix} = \begin{pmatrix} -3 \\ 5 \end{pmatrix}.$$

The solution to this system of equations is $a_1 = \frac{13}{3}$ and $a_2 = \frac{11}{3}$. Thus $(-3,5) = \frac{13}{3}(1,2) + \frac{11}{3}(-2,-1)$. This is pictured in Figure 3-1.

This can be done geometrically, too. The line through the point $(-3,5)$ parallel to the vector $\mathbf{X}_2$ intersects the line determined by $\mathbf{X}_1$ at the point labeled $\frac{13}{3}\mathbf{X}_1$, whereas the line through the point $(-3,5)$ parallel to the vector $\mathbf{X}_1$ intersects the line determined by $\mathbf{X}_2$ at the point labeled $\frac{11}{3}\mathbf{X}_2$. Thus $(-3,5) = \frac{13}{3}(1,2) + \frac{11}{3}(-2,-1)$.

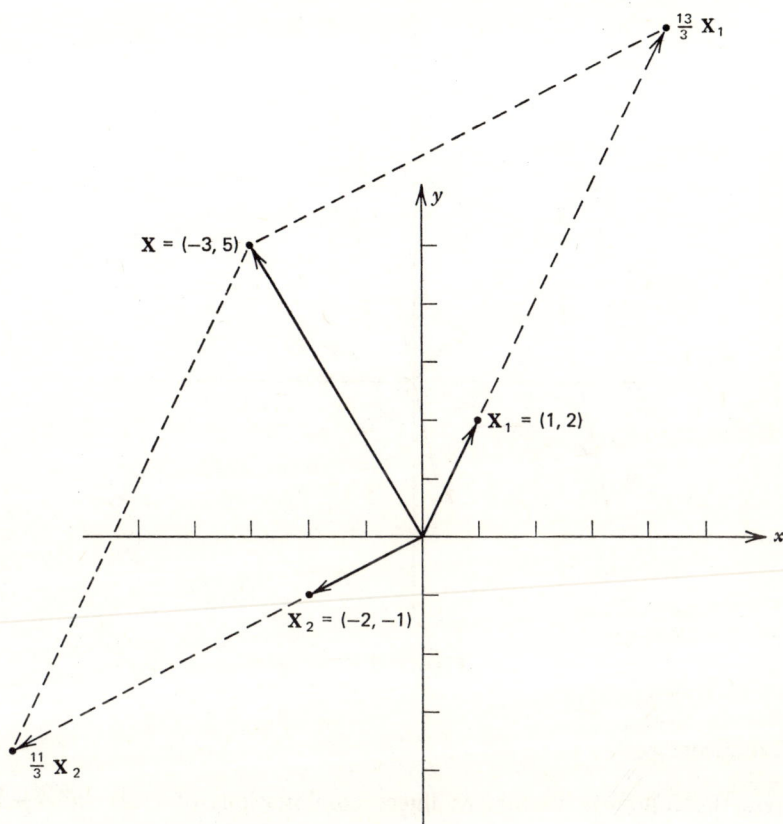

Figure 3-1

This shows how any vector **X** can be expressed geometrically as a linear combination of two noncollinear vectors $\mathbf{X}_1$ and $\mathbf{X}_2$ (see Figure 3-2). Through the endpoint of **X** draw a line parallel to $\mathbf{X}_2$; this intersects the line determined by $\mathbf{X}_1$ at the point P_1; the vector from O to P_1 is some scalar multiple of $\mathbf{X}_1$, say $a_1\mathbf{X}_1$. Similarly, the line parallel to $\mathbf{X}_1$ through the endpoint of **X** intersects the line determined by $\mathbf{X}_2$ at the point P_2; the vector from O to P_2 is some scalar multiple of $\mathbf{X}_2$, say $a_2\mathbf{X}_2$. It is geometrically clear that $\mathbf{X} = a_1\mathbf{X}_1 + a_2\mathbf{X}_2$.

From the numerical example and the geometrical discussion, it is plausible to expect any two noncollinear vectors in $\mathbb{R}^2$ to generate $\mathbb{R}^2$, and that in $\mathbb{R}^2$ there is only one way to represent a vector as a proper linear combination of two noncollinear vectors.

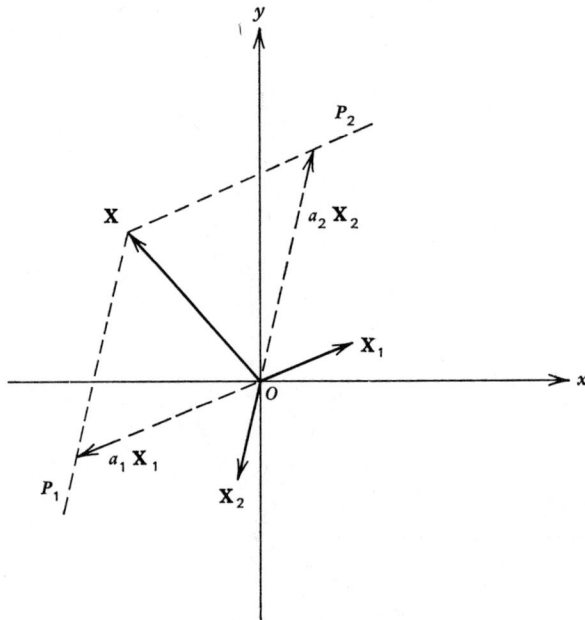

Figure 3-2

Brief Exercises

1. Express the following vectors as linear combinations of $(1,2)$ and $(-2,3)$; draw a diagram for each.

$$(2,1), (3,-2), (2,4)$$

2. In each of the parts of Exercise 1, verify that there is only one way to express those vectors as a proper linear combination of $(1,2)$ and $(-2,3)$.

Selected Answers: **1** $(2,1) = \frac{8}{7}(1,2) - \frac{3}{7}(-2,3)$; $(3,-2) = \frac{5}{7}(1,2) - \frac{8}{7}(-2,3)$; $(2,4) = 2(1,2) + 0(-2,3)$

Now consider vectors in $\mathbb{R}^3$. It is a geometric fact that three points which do not all lie on one line determine a plane. We use this fact to determine a vector equation of a plane in $\mathbb{R}^3$.

For example, three points $(4,2,1)$, $(2,4,2)$, and $(1,3,4)$ do not all lie on one line. We find a vector equation for the plane determined by these three points (see Figure 3-3), i.e., an expression for all vectors beginning at the origin and ending on the plane. Let $\mathbf{X}_0$ be the vector from $(0,0,0)$ to $(4,2,1)$, let $\mathbf{X}_1$ be the vector from

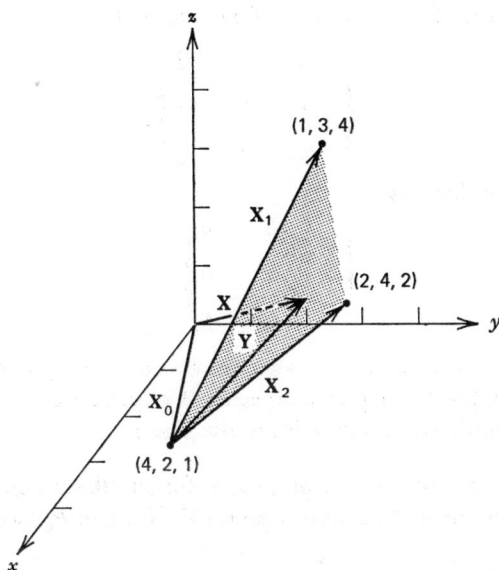

Figure 3-3

(4,2,1) to (1,3,4), and let X_2 be the vector from (4,2,1) to (2,4,2). Thus $X_0 =$ (4,2,1), $X_1 = (-3,1,3)$, and $X_2 = (-2,2,1)$. Then any vector X from (0,0,0) to the plane can be expressed as $X_0 + Y$, where Y is the vector from the point (4,2,1) to the endpoint of X. Because Y lies in the plane, it is a linear combination of X_1 and X_2, say $Y = sX_1 + tX_2$ for properly chosen scalars s and t. The equation

$$X = X_0 + sX_1 + tX_2$$
$$= (4,2,1) + s(-3,1,3) + t(-2,2,1)$$

is the vector equation of this plane. Different numerical choices for s and t produce different vectors with endpoints on the plane. Thus we can decide whether the vector (1,3,1), for example, positioned at the origin has its endpoint in this plane. If it does, then there are scalars s and t so that

$$(1,3,1) = (4,2,1) + s(-3,1,3) + t(-2,2,1).$$

Subtract (4,2,1) from both sides to obtain

$$(-3,1,0) = s(-3,1,3) + t(-2,2,1).$$

The augmented matrix for this system of equations is

$$\begin{pmatrix} -3 & -2 & -3 \\ 1 & 2 & 1 \\ 3 & 1 & 0 \end{pmatrix}$$

and the row echelon form is

$$\begin{pmatrix} 1 & 0 & 0 \\ 0 & 1 & 0 \\ 0 & 0 & 1 \end{pmatrix}.$$

The equations have no solution, as is shown by the last equation. Therefore the point $(1,3,1)$ does not lie in the plane, or, equivalently, the vector $(1,3,1)$ positioned at the origin does not have its endpoint in the plane.

This example shows the general procedure for finding a vector equation of a plane determined by three noncollinear points P_1, P_2, and P_3 (see Figure 3-4). Let

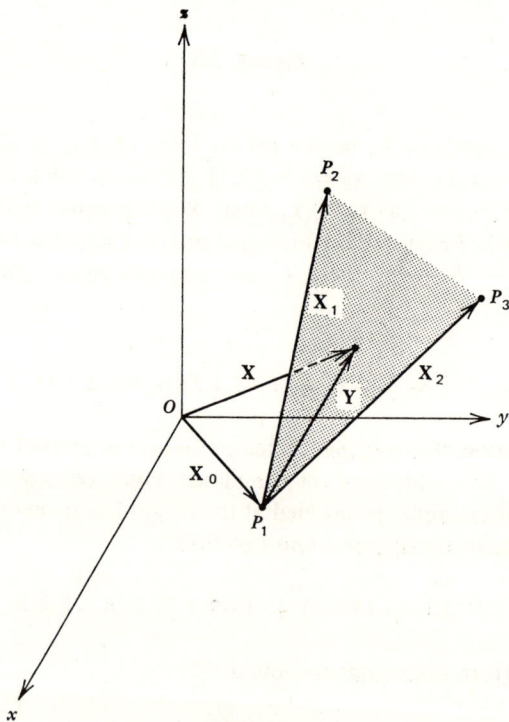

Figure 3-4

X_0 be the vector from the origin to P_1, let X_1 be the vector from P_1 to P_2, and let X_2 be the vector from P_1 to P_3. Then any vector X from the origin to the plane can be expressed as $X = X_0 + Y$, where Y is the vector from the endpoint of X_0 to the endpoint of X. But since Y lies in the plane, it is a linear combination of X_1 and X_2, $Y = sX_1 + tX_2$. Therefore

$$X = X_0 + sX_1 + tX_2$$

is a vector equation of the plane.

Brief Exercises

1. Write a vector equation of the plane passing through the given three points.

   (a) $(1,0,1)$, $(0,1,1)$, $(1,1,0)$ **(b)** $(2,1,4)$, $(3,8,-2)$, $(6,5,7)$

2. Are any of the following points on any of the planes listed in Exercise 1?

$$(2,1,5), \ (1,5,2), \ (11,16,5)$$

3. Write a vector equation of the plane passing through the origin which contains the following two vectors.

   (a) $(1,2,1)$, $(3,0,4)$ **(b)** $(-3,6,0)$, $(1,5,2)$

Answers: **1(a)** $(1,0,1) + s(-1,1,0) + t(0,1,-1)$ **1(b)** $(2,1,4) + s(1,7,-6) + t(4,4,3)$ **2** None of the points are on the lines **3(a)** $s(1,2,1) + t(3,0,4)$ **3(b)** $s(-3,6,0) + t(1,5,2)$

In Section 2-3 we discussed noncollinear vectors in $\mathbb{R}^2$ and $\mathbb{R}^3$. Now we discuss noncoplanar vectors in $\mathbb{R}^3$.

Suppose that three nonzero vectors are positioned at the origin in $\mathbb{R}^3$. These vectors are *coplanar* if they all lie in some plane; they are *noncoplanar* if there is no plane containing all of them. If the vectors are noncoplanar, then none of them is in the plane determined by the other two, or, phrased arithmetically, none is a linear combination of the remaining two.

Now we show that any three noncoplanar vectors in $\mathbb{R}^3$ form a generating set for $\mathbb{R}^3$. For example, consider the three vectors $X_1 = (1,0,1)$, $X_2 = (-1,2,0)$, and $X_3 = (0,-1,3)$. You can verify that they are indeed noncoplanar. We show that every vector in $\mathbb{R}^3$ can be expressed as a linear combination of these three vectors. We do this arithmetically first, and then we give a geometrical argument that the same result is true if X_1, X_2, and X_3 are any noncoplanar vectors in $\mathbb{R}^3$.

To show that the vectors mentioned above form a generating set for $\mathbb{R}^3$, suppose that (x,y,z) is a vector in $\mathbb{R}^3$; we determine whether there are scalars a_1, a_2, and a_3 so that

$$a_1 \begin{pmatrix} 1 \\ 0 \\ 1 \end{pmatrix} + a_2 \begin{pmatrix} -1 \\ 2 \\ 0 \end{pmatrix} + a_3 \begin{pmatrix} 0 \\ -1 \\ 3 \end{pmatrix} = \begin{pmatrix} x \\ y \\ z \end{pmatrix}.$$

The row echelon form for the augmented matrix is

$$\begin{pmatrix} 1 & 0 & 0 & \frac{1}{7}(& 6x + 3y + & z) \\ 0 & 1 & 0 & \frac{1}{7}(& -x + 3y + & z) \\ 0 & 0 & 1 & \frac{1}{7}(-2x - & y + 2z) \end{pmatrix}.$$

Thus any vector (x,y,z) in $\mathbb{R}^3$ is expressible as a linear combination of the three noncoplanar vectors $(1,0,1)$, $(-1,2,0)$, and $(0,-1,3)$; therefore, these three noncoplanar vectors form a generating set for $\mathbb{R}^3$.

Now we give a geometric argument that three noncoplanar vectors form a generating set for $\mathbb{R}^3$. Let $\mathbf{X}_1$, $\mathbf{X}_2$, and $\mathbf{X}_3$ be three noncoplanar vectors situated at the origin (see Figure 3-5), and suppose that $\mathbf{X}$ is a vector in $\mathbb{R}^3$ also positioned at the origin. The plane through the endpoint of $\mathbf{X}$ and parallel to the plane determined by $\mathbf{X}_2$ and $\mathbf{X}_3$ intersects the line determined by $\mathbf{X}_1$ at a point P_1; the vector from O to P_1 is some scalar multiple $a_1\mathbf{X}_1$ of $\mathbf{X}_1$ (see Figure 3-6).

The plane through the endpoint of $\mathbf{X}$ and parallel to the plane determined by $\mathbf{X}_1$ and $\mathbf{X}_2$ intersects the line determined by $\mathbf{X}_2$ at some point P_2; the vector from O to P_2 is some scalar multiple $a_2\mathbf{X}_2$ of $\mathbf{X}_2$ (see Figure 3-7).

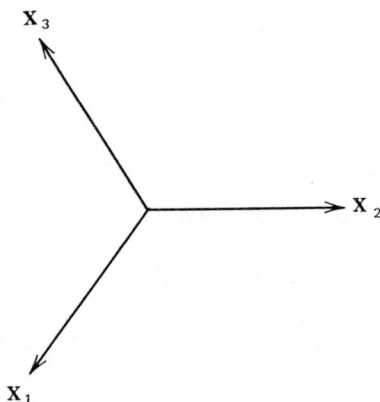

Figure 3-5

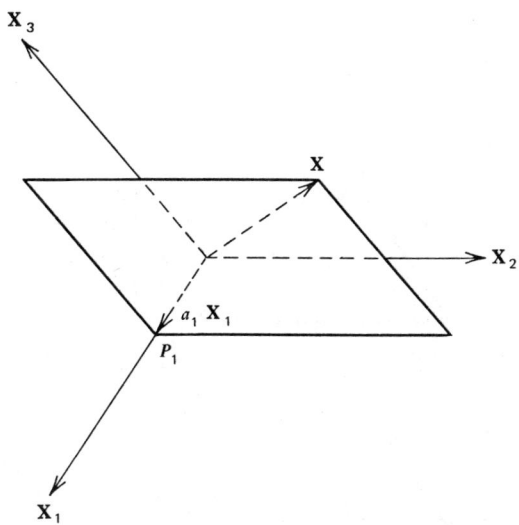

Figure 3-6

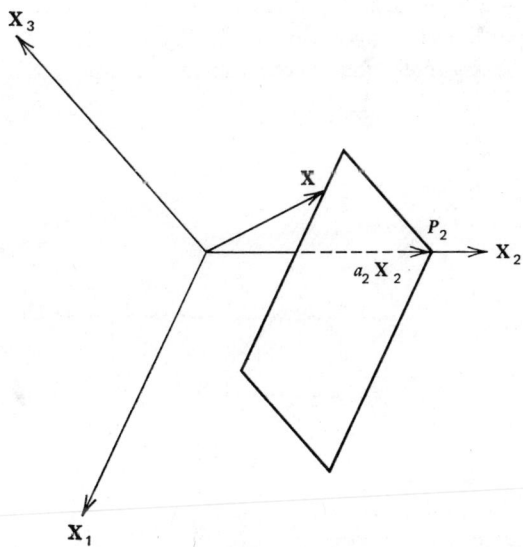

Figure 3-7

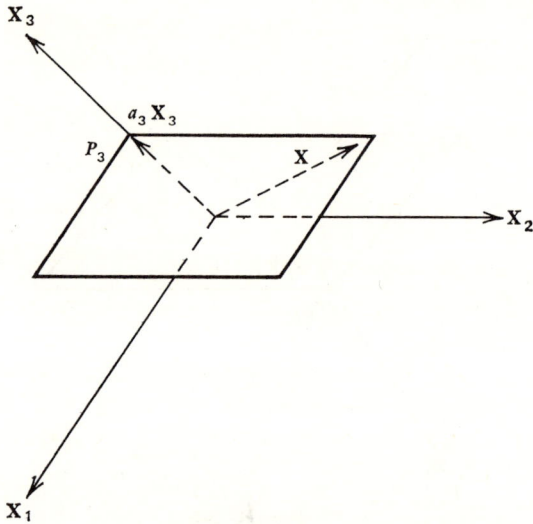

Figure 3-8

Finally, the plane through the endpoint of $\mathbf{X}$ and parallel to the plane determined by $\mathbf{X}_1$ and $\mathbf{X}_2$ intersects the line determined by $\mathbf{X}_3$ at some point P_3; the vector from O to P_3 is some scalar multiple $a_3\mathbf{X}_3$ of $\mathbf{X}_3$ (see Figure 3-8).

Then $\mathbf{X} = a_1\mathbf{X}_1 + a_2\mathbf{X}_2 + a_3\mathbf{X}_3$ (see Figure 3-9). In fact, this is the only way to express $\mathbf{X}$ as a proper linear combination of $\mathbf{X}_1$, $\mathbf{X}_2$, and $\mathbf{X}_3$.

This completes the argument that a set of three noncoplanar vectors in $\mathbb{R}^3$ is a generating set for $\mathbb{R}^3$.

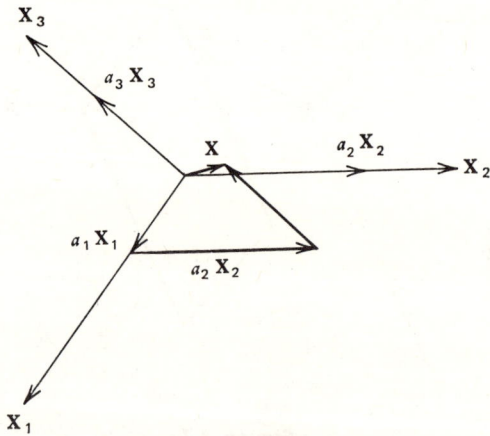

Figure 3-9

Brief Exercises

1. Write the following vectors as a linear combination of the vectors (1,0,1), (2,1,0), and (0,1,2) (check your work).

$$(3,1,4), (1,0,0), (0,1,0), (2,5,2)$$

2. Verify that the four vectors listed in Exercise 1 can be expressed in only one way as a proper linear combination of the vectors (1,0,1), (2,1,0), and (0,1,2).

In Section 2-2 we introduced subspaces of a vector space. Now we give a geometric description of all the subspaces of $\mathbb{R}^2$ and $\mathbb{R}^3$. We use $\mathscr{W}$ for a given subspace.

First consider $\mathbb{R}^2$. If $\mathscr{W}$ contains no nonzero vector, then $\mathscr{W}$ is the zero subspace. If $\mathscr{W}$ contains nonzero vectors, then we distinguish two cases: (1) $\mathscr{W}$ contains two noncollinear vectors and (2) $\mathscr{W}$ does not contain two noncollinear vectors.

　　Case 1: $\mathscr{W}$ contains two noncollinear vectors $\mathbf{X}_1$ and $\mathbf{X}_2$. These form a generating set for $\mathbb{R}^2$, so $\mathscr{W} = \mathbb{R}^2$.

　　Case 2: $\mathscr{W}$ contains nonzero vectors, but does not contain two noncollinear vectors. Take $\mathbf{X}_1$, any nonzero vector in $\mathscr{W}$ (see Figure 3-10). Since

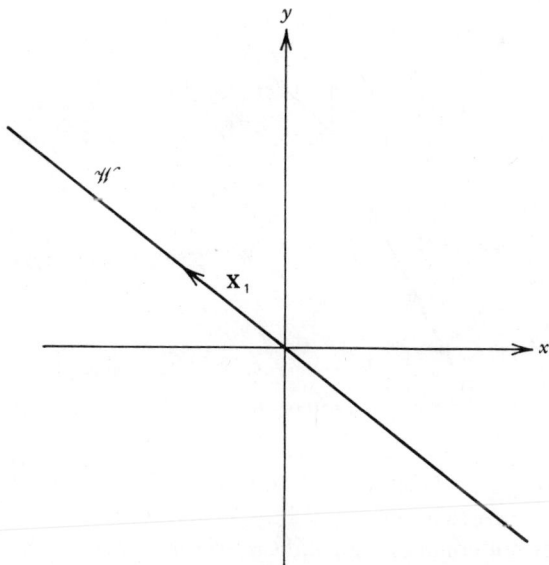

Figure 3-10

every other vector in $\mathscr{W}$ is collinear with $\mathbf{X}_1$, $\mathscr{W}$ is the line through the origin determined by $\mathbf{X}_1$.

Thus the only subspaces of $\mathbb{R}^2$ are the zero subspace, a line through the origin, or $\mathbb{R}^2$ itself. In other words, the subspaces of $\mathbb{R}^2$ are the zero subspace, the span of one nonzero vector, or the span of two noncollinear vectors.

Now we turn to subspaces of $\mathbb{R}^3$. If $\mathscr{W}$ contains no nonzero vector, then $\mathscr{W}$ is the zero subspace. If $\mathscr{W}$ contains some nonzero vectors, then we consider three cases: (1) $\mathscr{W}$ contains three noncoplanar vectors; (2) $\mathscr{W}$ does not contain three noncoplanar vectors, but does contain two noncollinear vectors; and (3) $\mathscr{W}$ does not contain two noncollinear vectors.

Case 1: $\mathscr{W}$ contains three noncoplanar vectors $\mathbf{X}_1$, $\mathbf{X}_2$, and $\mathbf{X}_3$. Then $\mathbf{X}_1$, $\mathbf{X}_2$, and $\mathbf{X}_3$ generate $\mathbb{R}^3$ and so $\mathscr{W} = \mathbb{R}^3$.

Case 2: $\mathscr{W}$ does not contain three noncoplanar vectors, but does contain two noncollinear vectors $\mathbf{X}_1$ and $\mathbf{X}_2$ (see Figure 3-11). Any other vector in

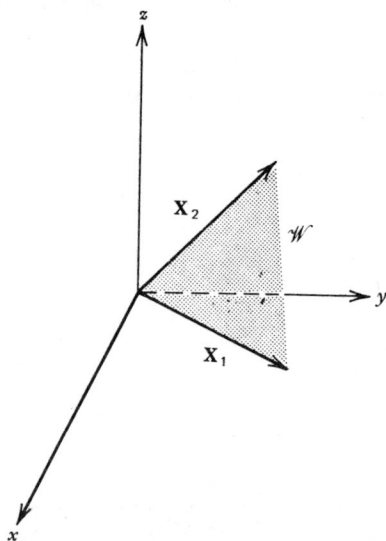

Figure 3-11

$\mathscr{W}$ must be coplanar with $\mathbf{X}_1$ and $\mathbf{X}_2$, and so $\mathscr{W}$ is the plane passing through the origin determined by $\mathbf{X}_1$ and $\mathbf{X}_2$.

Case 3: $\mathscr{W}$ does not contain two collinear vectors, but does contain a nonzero vector $\mathbf{X}_1$ (see Figure 3-12). Any other vector in $\mathscr{W}$ must be collinear with $\mathbf{X}_1$, so $\mathscr{W}$ is the line passing through the origin determined by $\mathbf{X}_1$.

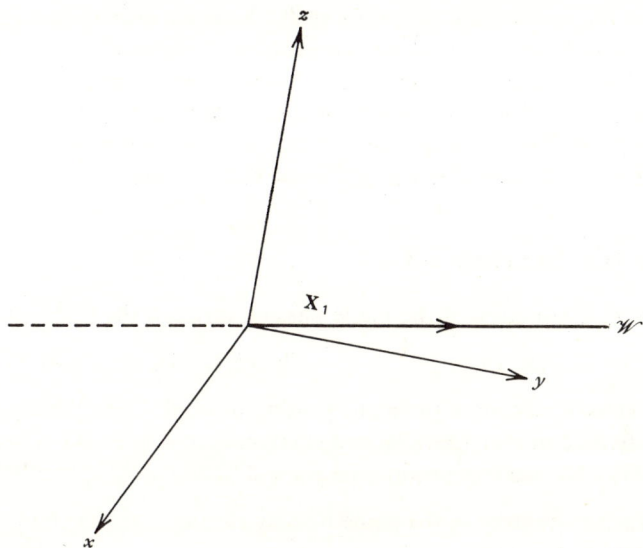

Figure 3-12

Thus we have shown that if $\mathscr{W}$ is a subspace of $\mathbb{R}^3$, then it must be the zero subspace, $\mathbb{R}^3$, a plane passing through the origin, or a line passing through the origin. The subspaces of $\mathbb{R}^3$ are therefore the zero subspace, the span of one non-zero vector, the span of two noncollinear vectors, or the span of three noncoplanar vectors.

Brief Exercises

1. Explain why the vectors in $\mathbb{R}^2$ positioned at the origin with endpoints on a line passing through the origin constitute a subspace of $\mathbb{R}^2$.

2. Explain why the vectors in $\mathbb{R}^3$ positioned at the origin with endpoints on a line passing through the origin form a subspace of $\mathbb{R}^3$.

3. Explain why the vectors in $\mathbb{R}^3$ positioned at the origin with endpoints on a plane passing through the origin form a subspace of $\mathbb{R}^3$.

SUMMARY   *Any two noncollinear vectors in* $\mathbb{R}^2$ *form a generating set for* $\mathbb{R}^2$. *A vector equation of the plane in* $\mathbb{R}^3$ *passing through the noncollinear points* P_1, P_2, *and* P_3 *is given by*

$$\mathbf{X} = \mathbf{X}_0 + s\mathbf{X}_1 + t\mathbf{X}_2$$

where $\mathbf{X}_0$ *is the vector from the origin to* P_1, $\mathbf{X}_1$ *is the vector from* P_1 *to* P_2, $\mathbf{X}_2$ *is the vector from* P_1 *to* P_3, *and s and t are scalars. A geometric argument establishes that any three noncoplanar vectors in* $\mathbb{R}^3$ *form a generating set for* $\mathbb{R}^3$. *Finally the subspaces of* $\mathbb{R}^2$ *are described: the zero subspace, the span of one nonzero vector, and* $\mathbb{R}^2$. *The subspaces of* $\mathbb{R}^3$ *are the zero subspace, the span of one nonzero vector, the span of two noncollinear vectors, and* $\mathbb{R}^3$.

Exercises for Section 3-3

1. Find a vector equation of the plane passing through the three given points.

   (a) $(1,2,1)$, $(2,1,3)$, $(4,1,1)$ (b) $(5,1,-1)$, $(0,-3,4)$, $(1,1,2)$

2. Find a vector equation of the plane passing through $(2,1,4)$ and parallel to the plane described in 1(a). (Parallel planes either do not intersect or they coincide.) (Hint: you only need to position the plane differently using another vector $\mathbf{X}_0$.)

3. Find a vector equation of the plane passing through $(1,2,1)$ and parallel to the plane described in 1(b).

4. Determine whether these planes are parallel.

   (a) $(2,1,1) + s(1,-1,2) + t(1,1,-1)$ and $(1,1,2) + u(-1,2,1) + v(0,1,1)$

   (b) $(4,-1,2) + s(1,0,2) + t(-1,3,4)$ and $(1,0,2) + u(-2,1,1) + v(3,0,-2)$

   (Hint: find an equation for the points common to both planes. There should be no common points, a line of common points, or a plane of common points. To do this arithmetically in (a), determine whether there are scalars s, t, u, and v so that $(2,1,1) + s(1,-1,2) + t(1,1,-1) = (1,1,2) + u(-1,2,1) + v(0,1,1)$.)

5. Do the following pairs of lines intersect? If so, find their intersection.

   (a) $(1,1,0) + t(-1,2,3)$ and $(2,0,1) + s(2,-1,4)$
   (b) $(2,1,1) + t(2,4,-5)$ and $(1,1,2) + s(3,-1,0)$

6. Determine whether the given line and plane intersect. If they do, find their intersection.

   (a) $(2,1,-1) + t(1,2,1)$ and $(-2,3,0) + s(-1,0,2) + u(2,1,0)$
   (b) $(5,1,2) + t(4,-1,2)$ and $(1,1,1) + s(2,1,3) + u(-1,2,1)$

7. In the following find two noncollinear vectors which generate the given subspace of $\mathbb{R}^3$.

   (a) the set of vectors (x,y,z) satisfying $3x + y - z = 0$
   (b) the set of vectors (x,y,z) satisfying $2x + y + 2z = 0$
   (c) the set of vectors (x,y,z) satisfying $5x + 4y - 3z = 0$

8. Describe geometrically the solutions for the following.

(a) $3x + y + z = 2$
    $x - y + 2z = 1$
    $2x + y \quad\;\; = 3$

(c) $3x + y - 2z = 1$
    $6x + 2y - 4z = 2$
    $9x + 3y - 6z = 3$

(b) $3x + y - 2z = 1$
    $x - y \quad\;\; = 2$
    $x + 3y - 2z = -3$

9. After having done Exercise 8, can you give a general argument that the solution set for a system of linear equations in three unknowns is empty, one point, a line, a plane, or all of $\mathbb{R}^3$? Can you relate the type of geometrical configuration of the solution set to the number of nonzero rows in the row echelon form for the augmented matrix?

10. How are the solution sets for the systems below related to each other?

$$x + y - z = 2 \qquad x + y - z = 1$$
$$2x - y + 2z = 5 \qquad 2x - y + 2z = 3$$

Can you generalize your answer to the situation of two systems of m equations in three unknowns whose coefficient matrices are equal but whose augmented matrices are different?

More Challenging Exercises

Definition 3-4_____

Two subspaces $\mathcal{W}_1$ and $\mathcal{W}_2$ of a vector space $\mathcal{V}$ are called *complementary* subspaces if (1) the zero vector $\mathbf{0}$ is the only vector in both subspaces and (2) each vector $\mathbf{X}$ in $\mathcal{V}$ can be expressed as the sum $\mathbf{X} = \mathbf{X}_1 + \mathbf{X}_2$ with $\mathbf{X}_1$ in $\mathcal{W}_1$ and $\mathbf{X}_2$ in $\mathcal{W}_2$.

_____

11. In $\mathbb{R}^2$ let $\mathcal{W}_1$ be the subspace generated by the one vector $(1,2)$. Describe geometrically those subspaces which are complementary to $\mathcal{W}_1$.

12. Same exercise as 11 with $\mathcal{W}_1$ being the subspace generated by the one vector $(-2,3)$.

13. In $\mathbb{R}^3$ let $\mathcal{W}_1$ be the subspace generated by the one vector $(1,0,2)$. Describe geometrically those subspaces which are complementary to $\mathcal{W}_1$.

14. Same exercise as 13 with $\mathcal{W}_1$ the subspace generated by the one vector $(2,-1,3)$.

15. Same exercise as 13 with $\mathscr{W}_1$ the subspace generated by the two vectors $(1,2,1)$ and $(5,-1,0)$.

16. Same exercise as 13 with $\mathscr{W}_1$ the subspace generated by the two vectors $(2,-1,3)$ and $(4,-5,0)$.

17. Suppose $\mathscr{W}_1$ and $\mathscr{W}_2$ are complementary subspaces of a vector space $\mathscr{V}$. Show that each vector $\mathbf{X}$ in $\mathscr{V}$ can be expressed in exactly one way as $\mathbf{X} = \mathbf{X}_1 + \mathbf{X}_2$, where $\mathbf{X}_1$ is in $\mathscr{W}_1$ and $\mathbf{X}_2$ is in $\mathscr{W}_2$.

18. Let $\mathscr{W}_1$ be the solution set in $\mathbb{R}^2$ for the equation $2x + y = 0$. Let $\mathscr{W}_2$ be the subspace spanned by $(2,1)$. Are $\mathscr{W}_1$ and $\mathscr{W}_2$ complementary subspaces?

19. Let $\mathscr{W}_1$ be the solution set in $\mathbb{R}^3$ for the system

$$3x - y + z = 0$$
$$x + y + 2z = 0.$$

Let $\mathscr{W}_2$ be the subspace spanned by the rows of the coefficient matrix. Are $\mathscr{W}_1$ and $\mathscr{W}_2$ complementary subspaces?

SECTION 3-4
Linear Independence

PREVIEW *The concepts of two noncollinear vectors and three noncoplanar vectors need to be extended to sets having more than three vectors. The term used for this more general concept is linear independence. A computational criterion for determining linear independence is given, and a connection between linear independence and solutions of homogeneous systems of equations is established. Finally, we show that if a linearly independent set $\mathscr{S}$ does not span the whole vector space $\mathscr{V}$, then one can enlarge $\mathscr{S}$ with a vector $\mathbf{Y}$ not in $\mathscr{S}$ and yet retain a linearly independent set.*

Two vectors in $\mathbb{R}^2$ and $\mathbb{R}^3$ are noncollinear if and only if neither is a scalar multiple of the other. Three vectors in $\mathbb{R}^3$ are noncoplanar if and only if none of the vectors is a linear combination of the other two. We also speak of two vectors in $\mathbb{R}^n$ being noncollinear. Of course, we cannot picture vectors in $\mathbb{R}^n$, so we use the arithmetic description: neither is a scalar multiple of the other. Similarly, we speak of three vectors in $\mathbb{R}^n$ as being noncoplanar if none is a linear combination of the other two. If n is four or larger, we extend the idea expressed by noncollinear and noncoplanar to four or more vectors. The term used for this is *linear independence* of a set of vectors.

Definition 3-5_____

Let $\mathscr{V}$ be a vector space over $\mathbb{R}$, and let $\mathscr{S}$ be a nonvoid set of vectors in $\mathscr{V}$. Then $\mathscr{S}$ is a *linearly independent set* if and only if (1) $\mathscr{S}$ contains at least one nonzero vector and (2) no vector in $\mathscr{S}$ is a linear combination of some other vectors in $\mathscr{S}$. We say that $\mathscr{S}$ is a *linearly dependent* set if and only if it is not a linearly independent set; i.e., $\mathscr{S}$ either consists of just the zero vector or at least one vector in $\mathscr{S}$ is a linear combination of some other vectors in $\mathscr{S}$.

Example Consider the set consisting of $(1,2)$, $(-1,1)$, and $(2,4)$ in $\mathbb{R}^2$. It is a simple matter to show that $(2,4)$ is a linear combination of the other two: $(2,4) = 2(1,2) + 0(-1,1)$. Therefore the set is a linearly dependent set.

Example Consider the set consisting of the vectors $(1,0,1)$, $(1,1,0)$, and $(0,1,1)$ in $\mathbb{R}^3$. It is easy to show—and we leave it to you—that none of these is a linear combination of the other two. Therefore this set is a linearly independent set.

If we wish to show that a set containing n vectors $\mathbf{X}_1, \ldots, \mathbf{X}_n$ is a linearly independent set, then we need to show that $\mathbf{X}_1$ is not a linear combination of the others, that $\mathbf{X}_2$ is not a linear combination of the others, and so on; we would have to work n different problems. Fortunately, there is a shorter way: if the zero vector $\mathbf{0}$ can be expressed as a nonzero proper linear combination of vectors in $\mathscr{S}$, then $\mathscr{S}$ is a linearly dependent set. Otherwise $\mathscr{S}$ is a linearly independent set. (You must use only *proper* linear combinations; it is always possible to write $\mathbf{0} = \mathbf{X} + (-1)\mathbf{X}$ or $\mathbf{0} = \mathbf{X} + 2\mathbf{X} - 3\mathbf{X}$, etc.; this is not the kind of linear combination that will give information about linearly independent sets.)

Example Consider the set containing the vectors $(1,0,1)$, $(1,1,0)$, and $(0,1,1)$ in $\mathbb{R}^3$. In the previous example you were asked to verify that this is a linearly independent set because none of the vectors is a linear combination of the others. Here we show that the only way $(0,0,0)$ can be expressed as a proper linear combination of these three vectors is with the zero linear combination. To show this, we ask which scalars x_1, x_2, and x_3 must we use to obtain

$$x_1 \begin{pmatrix} 1 \\ 0 \\ 1 \end{pmatrix} + x_2 \begin{pmatrix} 1 \\ 1 \\ 0 \end{pmatrix} + x_3 \begin{pmatrix} 0 \\ 1 \\ 1 \end{pmatrix} = \begin{pmatrix} 0 \\ 0 \\ 0 \end{pmatrix}.$$

The solution to these equations is $x_1 = x_2 = x_3 = 0$. Consequently the zero linear combination must be used to express $(0,0,0)$ as a proper linear combination of the given vectors. This shows that the three vectors form a linearly independent set. If, for example, $(1,0,1)$ were a linear combination of the other two, say

$$(1,0,1) = b(1,1,0) + c(0,1,1),$$

then

$$(0,0,0) = -1(1,0,1) + b(1,1,0) + c(0,1,1),$$

contradicting the result just established that zero scalars must be used to express $(0,0,0)$ as a proper linear combination of the given vectors, since -1 is obviously not 0.

Example In this example we show that the set consisting of the vectors $(1,-1)$, $(1,2)$, and $(-3,0)$ is a linearly dependent set in $\mathbb{R}^2$. To do this we ask which scalars x_1, x_2, and x_3 give

$$x_1 \begin{pmatrix} 1 \\ -1 \end{pmatrix} + x_2 \begin{pmatrix} 1 \\ 2 \end{pmatrix} + x_3 \begin{pmatrix} -3 \\ 0 \end{pmatrix} = \begin{pmatrix} 0 \\ 0 \end{pmatrix}.$$

The row echelon form for the coefficient matrix is

$$\begin{pmatrix} 1 & 0 & -2 \\ 0 & 1 & -1 \end{pmatrix}.$$

This is the coefficient matrix for

$$x_1 \qquad - 2x_3 = 0$$
$$x_2 - \quad x_3 = 0.$$

Of course $x_1 = x_2 = x_3 = 0$ is a solution, but it is not the *only* one; another is $x_1 = 2$, $x_2 = 1$, $x_3 = 1$. Thus we can write

$$(0,0) = 2(1,-1) + 1(1,2) + 1(-3,0).$$

If we add $-2(1,-1)$ to both sides and divide by -2 we get

$$(1,-1) = -\tfrac{1}{2}(1,2) - \tfrac{1}{2}(-3,0),$$

and so $(1,-1)$ is expressed as a linear combination of the other vectors in the set. Thus the set is a linearly dependent set. The reason we can express $(1,-1)$ as a linear combination of $(1,2)$ and $(-3,0)$ is that $(1,-1)$ is multiplied by a nonzero scalar in the equation

$$(0,0) = 2(1,-1) + 1(1,2) + 1(-3,0).$$

For the same reason $(1,2)$ can be expressed as a linear combination of $(1,-1)$ and $(-3,0)$; a similar fact is true for $(-3,0)$.

The next proposition has a statement of the criterion for linear independence.

Proposition 3-3

Suppose $\mathscr{V}$ is a vector space over $\mathbb{R}$ and $\mathscr{S}$ is a nonvoid set of vectors in $\mathscr{V}$. Then $\mathscr{S}$ is a linearly independent set if and only if the only way the zero vector can be expressed as a proper linear combination of some vectors in $\mathscr{S}$ is with a zero linear combination.

Another way to say the same thing is this: $\mathscr{S}$ is a linearly dependent set if and only if the zero vector can be expressed as a nonzero proper linear combination of some vectors in $\mathscr{S}$.

The reasons for this proposition are the same as the ones given in the examples. If $\mathscr{S}$ is a linearly dependent set and a vector $\mathbf{X}_1$ in $\mathscr{S}$ is expressed as a linear combination

$$\mathbf{X}_1 = a_2\mathbf{X}_2 + \cdots + a_n\mathbf{X}_n$$

of other vectors $\mathbf{X}_2, \ldots, \mathbf{X}_n$ in $\mathscr{S}$, then the zero vector can be expressed as a nonzero, proper linear combination

$$\mathbf{0} = (-1)\mathbf{X}_1 + a_2\mathbf{X}_2 + \cdots + a_n\mathbf{X}_n$$

of vectors in $\mathscr{S}$. On the other hand, if the zero vector can be expressed as a nonzero, proper linear combination

$$\mathbf{0} = a_1\mathbf{X}_1 + \cdots + a_n\mathbf{X}_n$$

of vectors in $\mathscr{S}$ (discard all terms of the form $0\mathbf{X}_i$ on the right-hand side), then, solving for $\mathbf{X}_1$,

$$\mathbf{X}_1 = (-a_2/a_1)\mathbf{X}_2 + \cdots + (-a_n/a_1)\mathbf{X}_n,$$

we have a vector $\mathbf{X}_1$ in $\mathscr{S}$ expressed as a linear combination of other vectors $\mathbf{X}_2, \ldots, \mathbf{X}_n$ in $\mathscr{S}$, and so $\mathscr{S}$ is linearly dependent.

In the next examples we illustrate the procedure for determining whether a set is linearly independent.

Example In this example we determine whether $(1,1,2)$ and $(3,4,5)$ form a linearly independent set in $\mathbb{R}^3$. We ask which scalars x_1 and x_2 give

$$x_1\begin{pmatrix} 1 \\ 1 \\ 2 \end{pmatrix} + x_2\begin{pmatrix} 3 \\ 4 \\ 5 \end{pmatrix} = \begin{pmatrix} 0 \\ 0 \\ 0 \end{pmatrix}.$$

The solution is $x_1 = x_2 = 0$. Thus the only way to express $(0,0,0)$ as a proper linear combination of $(1,1,2)$ and $(3,4,5)$ is with a zero linear combination, so the set is linearly independent.

Example   In this example we determine whether $(1,1)$, $(3,2)$, and $(-1,4)$ form a linearly independent set in $\mathbb{R}^2$. To do this, we ask which scalars x_1, x_2, and x_3 give us

$$x_1 \begin{pmatrix} 1 \\ 1 \end{pmatrix} + x_2 \begin{pmatrix} 3 \\ 2 \end{pmatrix} + x_3 \begin{pmatrix} -1 \\ 4 \end{pmatrix} = \begin{pmatrix} 0 \\ 0 \end{pmatrix}.$$

The row echelon form for the coefficient matrix is

$$\begin{pmatrix} 1 & 0 & 14 \\ 0 & 1 & -5 \end{pmatrix}.$$

This is the coefficient matrix for

$$x_1 \qquad + 14x_3 = 0$$

$$x_2 - \quad 5x_3 = 0.$$

There are many solutions other than $x_1 = x_2 = x_3 = 0$; for example, $x_1 = -14$, $x_2 = 5$, and $x_3 = 1$. Since it is possible to express $(0,0)$ as a nonzero proper linear combination of the vectors $(1,1)$, $(3,2)$, and $(-1,4)$, the vectors $(1,1)$, $(3,2)$, and $(-1,4)$ form a linearly dependent set in $\mathbb{R}^2$.

Brief Exercises

1. Determine whether the sets consisting of the following vectors are linearly independent subsets of $\mathbb{R}^2$.
   (a) $(1,1)$, $(2,2)$
   (b) $(1,0)$, $(2,1)$
   (c) $(1,1)$
   (d) $(3,4)$
   (e) $(1,2)$, $(0,3)$, $(5,0)$
   (f) $(1,1)$, $(2,3)$

2. Same exercise in $\mathbb{R}^3$.
   (a) $(1,0,1)$
   (b) $(3,-1,5)$
   (c) $(1,2,1)$, $(1,3,2)$
   (d) $(1,0,1)$, $(0,1,1)$, $(1,0,0)$
   (e) $(2,1,3)$, $(1,0,1)$, $(1,5,1)$, $(2,3,4)$

3. Same exercise in $\mathbb{R}^4$.
   (a) $(1,1,1,1)$
   (b) $(2,0,1,1)$, $(1,-1,2,1)$
   (c) $(2,1,1,3)$, $(4,2,2,6)$
   (d) $(3,1,0,1)$, $(1,0,3,1)$, $(4,1,3,2)$

4. Give an argument to show that a set consisting of exactly one nonzero vector **X** in a vector space $\mathscr{V}$ is a linearly independent set.

Answers: **1(a)** No **1(b)** Yes **1(c)** Yes **1(d)** Yes **1(e)** No **1(f)** Yes **2(a)** Yes **2(b)** Yes **2(c)** Yes **2(d)** Yes **2(e)** No **3(a)** Yes **3(b)** Yes **3(c)** No **3(d)** No

From the preceding examples you can see that determining whether a set of vectors in $\mathbb{R}^m$ is linearly independent is the same as determining whether a homogeneous system of linear equations has only one solution. The reason is that the homogeneous system

$$a_{11}x_1 + a_{12}x_2 + \cdots + a_{1n}x_n = 0$$
$$a_{21}x_1 + a_{22}x_2 + \cdots + a_{2n}x_n = 0$$
$$\vdots \qquad \vdots \qquad \qquad \vdots$$
$$a_{m1}x_1 + a_{m2}x_2 + \cdots + a_{mn}x_n = 0$$

can be rewritten as

$$x_1 \begin{pmatrix} a_{11} \\ a_{21} \\ \vdots \\ a_{m1} \end{pmatrix} + x_2 \begin{pmatrix} a_{12} \\ a_{22} \\ \vdots \\ a_{m2} \end{pmatrix} + \cdots + x_n \begin{pmatrix} a_{1n} \\ a_{2n} \\ \vdots \\ a_{mn} \end{pmatrix} = \begin{pmatrix} 0 \\ 0 \\ \vdots \\ 0 \end{pmatrix}.$$

If any columns of the coefficient matrix are equal, then there is always more than one solution; multiply one of the equal columns by 1, the other by -1, and the remaining columns by 0. If the columns are all different, then there is more than one solution if and only if the columns form a linearly dependent set in $\mathbb{R}^m$.

Brief Exercises

1. Rephrase the exercise of solving the following systems of equations as an exercise about the linear independence of some vectors.

(a) $2x + y = 0$
$\quad\ \ x - y = 0$

(b) $3x_1 + 2x_2 + 5x_3 = 0$
$\quad\ x_1 - \ x_2 + \ x_3 = 0$

(c) $4x - \ y + \ z = 0$
$\quad 3x - 2y + 3z = 0$
$\quad\ \ x + \ y - \ z = 0$
$\quad 3x + \ y + \ z = 0$

Answers: **1(a)** Do $(2,1)$ and $(1,-1)$ form a linearly independent set? **1(b)** Do $(3,1)$, $(2,-1)$, and $(5,1)$ form a linearly independent set? **1(c)** Do $(4,3,1,3)$, $(-1,2,1,1)$, and $(1,3,-1,1)$ form a linearly independent set?

We present one more fact about linearly independent sets; if a linearly independent set does not generate the whole vector space, then it is always possible to include another vector in the set and retain linear independence. For example, the set consisting of the one vector (2,4,3) in $\mathbb{R}^3$ is a linearly independent set. It generates the line through the origin passing through the point (2,4,3), but it does not generate $\mathbb{R}^3$. Pick any vector not on the line; (2,2,0), for example. Then (2,4,3) and (2,2,0) form a linearly independent set, as you can verify easily. Then (2,4,3) and (2,2,0) generate the plane passing through the points (0,0,0), (2,4,3), and (2,2,0), and so they do not generate the whole vector space $\mathbb{R}^3$. Pick any vector not in this plane; for example, from Figure 3-13 it is easy to see that (1,0,2) is such a vector. Then (2,4,3), (2,2,0), and (1,0,2) form a linearly independent set since they are three noncoplanar vectors. These three vectors generate the entire vector space $\mathbb{R}^3$, so it is not possible to include a fourth vector from $\mathbb{R}^3$ in this set and maintain a linearly independent set.

This illustrates the general situation. If $\mathscr{S}$ is a linearly independent set and the vector **Y** is not a linear combination of vectors in $\mathscr{S}$, then we can enlarge $\mathscr{S}$ by including **Y** and retain a linearly independent set.

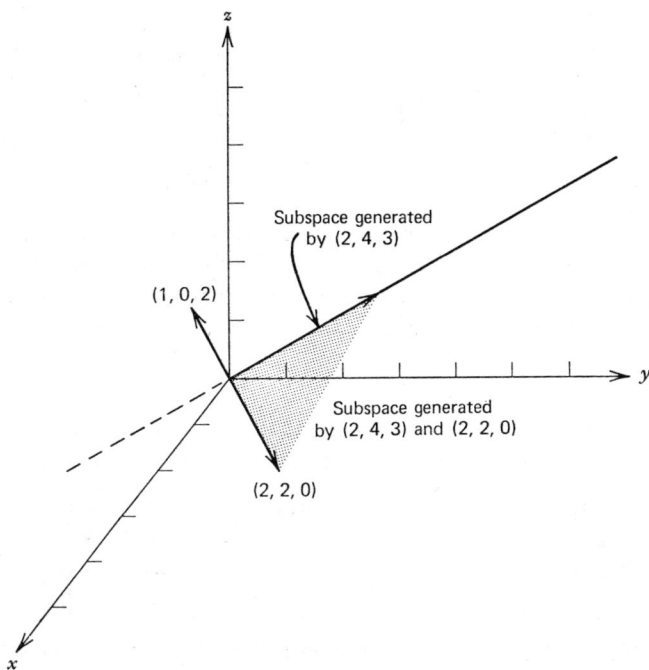

Figure 3-13

Proposition 3-4

Suppose $\mathcal{V}$ is a vector space over $\mathbb{R}$. If $\mathcal{S}$ is a linearly independent subset of $\mathcal{V}$ and $\mathbf{Y}$ is a vector in $\mathcal{V}$ which is not a linear combination of vectors from $\mathcal{S}$, then the set obtained by including $\mathbf{Y}$ with $\mathcal{S}$ is linearly independent.

Because this proposition concerns all vector spaces and not just $\mathbb{R}^3$, we must use the general test for linear independence in Proposition 3-3. So suppose

$$a_1 \mathbf{X}_1 + \cdots + a_n \mathbf{X}_n + b\mathbf{Y} = \mathbf{0}$$

is a proper linear combination with $\mathbf{X}_1, \ldots, \mathbf{X}_n$ from $\mathcal{S}$. We want to know whether the scalars $a_1, \ldots, a_n$, and b must all equal zero. Now $b = 0$, for if $b \neq 0$ we can solve for $\mathbf{Y}$ in terms of $\mathbf{X}_1, \ldots, \mathbf{X}_n$. So, since $b = 0$, the linear combination is

$$a_1 \mathbf{X}_1 + \cdots + a_n \mathbf{X}_n = \mathbf{0}.$$

This is a proper linear combination of vectors in $\mathcal{S}$, a linearly independent set. Thus $a_1 = \cdots = a_n = 0$. Consequently, the only way to express $\mathbf{0}$ as a proper linear combination of $\mathbf{Y}$ and vectors from $\mathcal{S}$ is with a zero linear combination. Thus the set consisting of $\mathbf{Y}$ along with the vectors from $\mathcal{S}$ is a linearly independent set.

Brief Exercises

1. In the following let $\mathcal{S}$ consist of the given vectors. Give a geometric description of those vectors $\mathbf{Y}$ which, along with the vectors from $\mathcal{S}$, form a linearly independent set.

   (a) $(1,0)$ in $\mathbb{R}^2$ (e) $(1,1,2)$ in $\mathbb{R}^3$
   (b) $(3,2)$ in $\mathbb{R}^2$ (f) $(1,0,0), (0,1,0)$ in $\mathbb{R}^3$
   (c) $(1,-1), (2,0)$ in $\mathbb{R}^2$ (g) $(1,0,0), (0,0,1)$ in $\mathbb{R}^3$
   (d) $(2,1,0)$ in $\mathbb{R}^3$ (h) $(1,1,0), (1,0,1)$ in $\mathbb{R}^3$

Answers: **1(a)** Any vector noncollinear with $(1,0)$ **1(b)** Any vector noncollinear with $(3,2)$ **1(c)** None **1(d)** Any vector noncollinear with $(2,1,0)$ **1(e)** Any vector noncollinear with $(1,1,2)$ **1(f)** Any vector noncoplanar with $(1,0,0)$ and $(0,1,0)$ **1(g)** Any vector noncoplanar with $(1,0,0)$ and $(0,0,1)$ **1(h)** Any vector noncoplanar with $(1,1,0)$ and $(1,0,1)$

SUMMARY *A set of vectors $\mathcal{S}$ is linearly independent if $\mathcal{S}$ contains at least one nonzero vector and no vector in $\mathcal{S}$ can be expressed as a linear combination of other vectors in $\mathcal{S}$. This definition is awkward to use in practice, so a more computational test is introduced; $\mathcal{S}$ is linearly independent if and only if the only* *way to express the zero vector as a proper linear combination of vectors in $\mathcal{S}$ is to use a zero linear combination. This test amounts to deciding whether a homogeneous system of linear equations has more than one solution. Finally, we show that if a linearly independent set $\mathcal{S}$ does not span the whole vector space $\mathcal{V}$, then we can find a vector $\mathbf{Y}$ in $\mathcal{V}$ so that the set consisting of $\mathbf{Y}$ along with the vectors in $\mathcal{S}$ is linearly independent.*

Exercises for Section 3-4

In Exercises 1 through 7, determine whether the set consisting of the given vectors is a linearly independent set.

1. $(1,0,-1,0)$, $(-1,1,0,0)$, $(3,0,1,1)$ in $\mathbb{R}^4$

2. $(1,0,0,0)$, $(1,1,0,0)$, $(1,1,1,0)$, $(1,1,1,1)$ in $\mathbb{R}^4$

3. $(1,5,3,6)$, $(2,-1,6,2)$, $(4,1,1,0)$ in $\mathbb{R}^4$

4. $(3,1,1)$, $(1,-1,1)$, $(1,1,1)$ in $\mathbb{R}^3$

5. $(1,1,0)$, $(0,1,1)$, $(2,5,3)$ in $\mathbb{R}^3$

6. $(5,1,3)$, $(2,4,6)$ in $\mathbb{R}^3$

7. $(4,1,3)$, $(2,4,6)$, $(1,1,1)$, $(2,3,1)$ in $\mathbb{R}^3$

More Challenging Exercises

8. Give an argument showing that in a vector space over $\mathcal{V}$ a nonvoid subset of a linearly independent set is also a linearly independent set.

9. Suppose $\mathbf{X}_1$ and $\mathbf{X}_2$ form a linearly independent set in a vector space $\mathcal{V}$, and suppose that (a_1,a_2) and (b_1,b_2) form a linearly independent set in $\mathbb{R}^2$. Show that $a_1\mathbf{X}_1 + a_2\mathbf{X}_2$ and $b_1\mathbf{X}_1 + b_2\mathbf{X}_2$ form a linearly independent set in $\mathcal{V}$. (Take numerical examples for (a_1,a_2) and (b_1,b_2) if you have difficulty getting started.)

10. Suppose $\mathbf{X}_1$, $\mathbf{X}_2$, and $\mathbf{X}_3$ form a linearly independent set in a vector space $\mathcal{V}$, and suppose that (a_1,a_2,a_3), (b_1,b_2,b_3), and (c_1,c_2,c_3) form a linearly independent set in $\mathbb{R}^3$. Show that $a_1\mathbf{X}_1 + a_2\mathbf{X}_2 + a_3\mathbf{X}_3$, $b_1\mathbf{X}_1 + b_2\mathbf{X}_2 + b_3\mathbf{X}_3$, and $c_1\mathbf{X}_1 + c_2\mathbf{X}_2 + c_3\mathbf{X}_3$ form a linearly independent set in $\mathcal{V}$.

11. Suppose $\mathbf{X}_1, \ldots, \mathbf{X}_n$ form a linearly independent set in a vector space $\mathcal{V}$, and suppose that $(a_{11},\ldots,a_{1n}), \ldots, (a_{n1},\ldots,a_{nn})$ form a linearly independent set in $\mathbb{R}^n$. Show that $a_{11}\mathbf{X}_1 + \cdots + a_{1n}\mathbf{X}_n, \ldots, a_{n1}\mathbf{X}_1 + \cdots + a_{nn}\mathbf{X}_n$ form a linearly independent set in $\mathcal{V}$.

12. Give an argument showing that any set of vectors containing the zero vector must be linearly dependent.

13. Give an argument to show that the nonzero rows of a matrix in row echelon form constitute a linearly independent set.

14. Give an argument showing that the rows of a matrix form a linearly dependent set if their sum is the zero vector.

SECTION 3-5
Existence of Bases

PREVIEW *A linearly independent set which generates a vector space $\mathscr{V}$ is called a basis for $\mathscr{V}$. If a nonzero vector space is generated by a finite number of vectors, then it is generated by a finite linearly independent set, i.e., the vector space contains a basis having a finite number of vectors. Furthermore, every vector in the vector space has a unique representation as a proper linear combination of vectors in the basis.*

Section 3-3 shows that both $\mathbb{R}^2$ and $\mathbb{R}^3$ can be generated by a linearly independent set, two noncollinear vectors in $\mathbb{R}^2$, and three noncoplanar vectors in $\mathbb{R}^3$. A linearly independent generating set is called a basis.

Definition 3-6_____
A *basis* for a vector space $\mathscr{V}$ is a linearly independent set which generates $\mathscr{V}$.

The results in Section 3-3 can be stated differently using Definition 3-6. Two noncollinear vectors in $\mathbb{R}^2$ form a basis for $\mathbb{R}^2$. Three noncoplanar vectors in $\mathbb{R}^3$ form a basis for $\mathbb{R}^3$. Every nonzero subspace of $\mathbb{R}^2$ contains a basis with no more than two vectors in it, and every nonzero subspace of $\mathbb{R}^3$ contains a basis with no more than three vectors in it.

The main result of this section is that if a nonzero vector space over $\mathbb{R}$ is generated by a finite set of vectors, then it has a basis. The approach we use is simple: start with the generating set and discard vectors from it one by one, being sure to maintain a generating set each time. When the process ends, you have a linearly independent generating set which is a subset of the original generating set.

Example Consider the set $\mathscr{S}$ consisting of the vectors $(1,1)$, $(2,1)$, $(0,1)$, and $(1,-1)$ in $\mathbb{R}^2$. Clearly $\mathscr{S}$ is a generating set for $\mathbb{R}^2$. We are interested in finding a subset of $\mathscr{S}$ with fewest number of vectors which still generates $\mathbb{R}^2$.

Are there any three vectors in $\mathscr{S}$ which generate $\mathbb{R}^2$? The answer is that any three vectors from $\mathscr{S}$ generate $\mathbb{R}^3$.

Are there any two vectors in $\mathscr{S}$ which generate $\mathbb{R}^2$? The answer is that any two vectors in $\mathscr{S}$ generate $\mathbb{R}^2$.

Finally, do any one-element subsets of $\mathscr{S}$ generate $\mathbb{R}^2$? The answer is that no single vector generates $\mathbb{R}^2$.

Therefore, the smallest subsets of $\mathscr{S}$ which generate $\mathbb{R}^2$ are the two-element subsets. The various two-element subsets are all linearly independent subsets of $\mathscr{S}$, and so these subsets are bases for $\mathbb{R}^2$.

Example   Consider the set $\mathscr{S}$ consisting of the vectors $(2,1,3)$, $(0,1,0)$, $(1,1,0)$, $(2,4,3)$, and $(5,4,3)$ in $\mathbb{R}^3$. It is easy to check that $\mathscr{S}$ is a generating set for $\mathbb{R}^3$. We are interested in finding a generating subset of $\mathscr{S}$ with fewest number of vectors.

First we ask if there are four vectors in $\mathscr{S}$ which generate $\mathbb{R}^3$. The answer, which you can check for yourself, is that any four vectors from $\mathscr{S}$ form a generating set for $\mathbb{R}^3$.

Then we ask if there are three-element subsets of $\mathscr{S}$ which generate $\mathbb{R}^3$. The answer, which again you can check for yourself, is that there are eight three-element subsets of $\mathscr{S}$ which generate $\mathbb{R}^3$. They are the sets consisting of the following vectors.

$$(2,1,3), (0,1,0), (1,1,0)$$

$$(2,1,3), (0,1,0), (5,4,3)$$

$$(2,1,3), (1,1,0), (2,4,3)$$

$$(2,1,3), (2,4,3), (5,4,3)$$

$$(0,1,0), (1,1,0), (2,4,3)$$

$$(0,1,0), (1,1,0), (5,4,3)$$

$$(0,1,0), (2,4,3), (5,4,3)$$

$$(1,1,0), (2,4,3), (5,4,3)$$

Finally we ask if there are any two-element subsets of $\mathscr{S}$ which generate $\mathbb{R}^3$. The answer, of course, is that no two-element subset of $\mathbb{R}^3$ generates $\mathbb{R}^3$. Consequently, a generating subset of $\mathscr{S}$ with fewest elements contains three elements—we listed the eight above. You can check by inspection that the eight sets are linearly independent. Hence we have found eight subsets of $\mathscr{S}$ which are bases for $\mathbb{R}^3$.

These examples indicate the general situation which we record in a proposition.

Proposition 3-5_____

If $\mathscr{V}$ is a nonzero vector space over $\mathbb{R}$ and $\mathscr{S}$ is a finite generating set for $\mathscr{V}$, then $\mathscr{S}$ contains a basis for $\mathscr{V}$.

_____

As we showed in the examples, it is possible to discard vectors one by one from $\mathscr{S}$, being careful to maintain a generating set at each step, until the process stops. We end up with a subset $\mathscr{T}$ of $\mathscr{S}$ which is a generating set, and we cannot remove another vector from $\mathscr{T}$ and still have a generating set. The set $\mathscr{T}$ must be linearly independent; if it were not, then some vector **X** in $\mathscr{T}$ would be a linear combination of the other vectors in $\mathscr{T}$. This would imply that a linear combination of the vectors in $\mathscr{T}$ using **X** could be turned into a linear combination using only the vectors in $\mathscr{T}$ which are different from **X**. Thus, the set obtained by removing **X** from $\mathscr{T}$ would be a generating set for $\mathscr{V}$. This would contradict the way $\mathscr{T}$ was constructed since it has fewer elements than $\mathscr{T}$. Thus $\mathscr{T}$ must be a linearly independent generating set for $\mathscr{V}$, and so $\mathscr{T}$ is a basis for $\mathscr{V}$.

Brief Exercises

1. Decide which of the following are generating sets for $\mathbb{R}^2$ and which are bases for $\mathbb{R}^2$. For those generating sets which are not bases, find all subsets which are bases.

   (a) $(1,2)$, $(2,4)$, $(4,8)$, $(0,1)$ **(c)** $(3,1)$, $(2,-3)$, $(5,-2)$
   (b) $(1,0)$, $(1,1)$

2. Same exercise in $\mathbb{R}^3$ for

   (a) $(2,1,0)$, $(1,3,1)$, $(3,4,1)$, $(0,0,1)$ **(c)** $(1,5,1)$, $(2,0,1)$, $(3,4,5)$, $(-1,2,3)$
   (b) $(1,2,1)$, $(0,1,1)$, $(1,1,0)$

Answers: **1(a)** Generating set; bases are $(1,2)$, $(0,1)$; $(2,4)$, $(0,1)$; $(4,8)$, $(0,1)$ **1(b)** Generating set and basis **1(c)** Generating set; bases are $(3,1)$, $(2,-3)$; $(3,1)$, $(5,-2)$; $(2,-3)$, $(5,-2)$ **2(a)** Generating set; bases are $(2,1,0)$, $(1,3,1)$, $(0,0,1)$; $(2,1,0)$, $(3,4,1)$, $(0,0,1)$; $(1,3,1)$, $(3,4,1)$, $(0,0,1)$ **2(b)** Not a generating set **2(c)** Generating set; all three-element subsets are bases

In Section 3-3 we showed that any vector in $\mathbb{R}^2$ can be expressed as a linear combination of two noncollinear vectors, and that any vector in $\mathbb{R}^3$ can be expressed as a linear combination of three noncoplanar vectors. We mentioned that there is only one way that a vector in $\mathbb{R}^2$ or $\mathbb{R}^3$ can be expressed as a proper linear combination of the noncollinear or the noncoplanar vectors; the discussion there was geometric. Now we give an arithmetic argument for a more general statement.

Proposition 3-6

If $\mathcal{V}$ is a vector space over $\mathbb{R}$, and if $\mathbf{X}_1, \ldots, \mathbf{X}_n$ form a basis for $\mathcal{V}$, then each vector in $\mathcal{V}$ has a unique representation as a proper linear combination of the basis vectors.

This proposition is easy to understand. Suppose that some vector $\mathbf{X}$ in $\mathcal{V}$ can be expressed in two ways as a proper linear combination of $\mathbf{X}_1, \ldots, \mathbf{X}_n$:

$$\mathbf{X} = a_1\mathbf{X}_1 + \cdots + a_n\mathbf{X}_n$$

and

$$\mathbf{X} = b_1\mathbf{X}_1 + \cdots + b_n\mathbf{X}_n.$$

Subtracting the second equation from the first we get

$$\mathbf{0} = (a_1 - b_1)\mathbf{X}_1 + \cdots + (a_n - b_n)\mathbf{X}_n.$$

Since $\mathbf{X}_1, \ldots, \mathbf{X}_n$ form a linearly independent set, the coefficients in the last equation must be zero, so $a_1 = b_1, \ldots, a_n = b_n$, and the two expressions are actually the same.

From the argument above, you can see that the unique representation refers to the scalars which multiply the various basis elements; it does not refer to the order in which the terms $a_i\mathbf{X}_i$ are written in the linear combination. For example, we say that $(3,5)$ has a unique representation as a linear combination of $(1,0)$ and $(0,1)$ even though

$$(3,5) = 3(1,0) + 5(0,1) = 5(0,1) + 3(1,0).$$

The unique representation lies in the fact that the scalar 3 multiplies $(1,0)$ and the scalar 5 multiplies $(0,1)$.

Example   In the second example of this section we observed that the vectors $(2,1,3)$, $(0,1,0)$, and $(1,1,0)$ form a basis for $\mathbb{R}^3$. In this example we illustrate Proposition 3-6 by writing vectors from $\mathbb{R}^3$ as linear combinations of these vectors, and we illustrate a method for determining whether these vectors form a basis for $\mathbb{R}^3$.

To show that the set of vectors given above is a basis we need to show that it is a generating set and that it is a linearly independent set. This might appear to involve two problems. We show that by determining whether the set generates $\mathbb{R}^3$ we can also decide whether the set is linearly independent.

Suppose (x,y,z) is a vector in $\mathbb{R}^3$. To show the given set is a generating set we need to show that there are scalars a_1, a_2, and a_3 so that

$$a_1\begin{pmatrix}2\\1\\3\end{pmatrix} + a_2\begin{pmatrix}0\\1\\0\end{pmatrix} + a_3\begin{pmatrix}1\\1\\0\end{pmatrix} = \begin{pmatrix}x\\y\\z\end{pmatrix}.$$

Solving, we find that $a_1 = \frac{1}{3}z$, $a_2 = y - x + \frac{1}{3}z$, and $a_3 = x - \frac{2}{3}z$. Consequently, the three given vectors constitute a generating set for $\mathbb{R}^3$.

Now, to determine whether the given set is linearly independent, we solve the same problem with $x = y = z = 0$. The solution is clearly $a_1 = a_2 = a_3 = 0$, and so the set is linearly independent. Thus $(2,1,3)$, $(0,1,0)$, and $(1,1,0)$ form a basis for $\mathbb{R}^3$.

To express (x,y,z) as a linear combination of the basis vectors we must have

$$(x,y,z) = \tfrac{1}{3}z(2,1,3) + (y - x + \tfrac{1}{3}z)(0,1,0) + (x - \tfrac{2}{3}z)(1,1,0).$$

Notice that once the vector (x,y,z) is given, the scalars are completely determined. For example,

$$(2,1,4) = \tfrac{4}{3}(2,1,3) + \tfrac{1}{3}(0,1,0) - \tfrac{2}{3}(1,1,0).$$

SUMMARY *A linearly independent generating set for a vector space is a basis for the vector space. If a vector space is generated by a finite number of vectors, then it is possible to find a subset of this generating set which is a basis. Thus, if a vector space is finitely generated, it contains a basis with a finite number of vectors. Further, each vector in the vector space can be represented in only one way as a linear combination of vectors in a basis.*

Exercises for Section 3-5

In Exercises 1 through 9, determine whether the given vectors form a basis.

1. $(2,1)$, $(3,2)$ in $\mathbb{R}^2$

2. $(1,1)$, $(2,1)$, $(-5,3)$ in $\mathbb{R}^2$

3. $(3,4)$, $(1,1)$ in $\mathbb{R}^2$

4. $(2,1,0)$, $(5,2,0)$, $(1,1,0)$ in $\mathbb{R}^3$

5. $(3,4,-2)$, $(1,1,2)$, $(3,-4,1)$ in $\mathbb{R}^3$

6. $(2,1,3)$, $(5,1,6)$, $(1,1,2)$, $(1,-1,3)$ in $\mathbb{R}^3$

7. $(1,0,0,0)$, $(1,1,1,0)$, $(0,0,1,0)$, $(2,0,1,3)$ in $\mathbb{R}^4$

8. $(1,0,-1,0)$, $(-1,1,0,0)$, $(3,0,1,1)$ in $\mathbb{R}^4$

9. $(2,1,4,3,0)$, $(1,5,8,2,1)$, $(2,0,1,1,0)$, $(1,0,1,0,2)$, $(2,1,1,0,0)$ in $\mathbb{R}^5$

10. In each of the following, find a basis for $\mathbb{R}^2$ which is a subset of the generating set consisting of the vectors

    (a) $(2,1)$, $(3,2)$ (b) $(1,1)$, $(2,1)$, $(-5,3)$

11. Same exercise in $\mathbb{R}^3$.

   (a) $(1,1,0)$, $(2,-1,5)$, $(0,1,0)$, $(3,2,1)$
   (b) $(1,1,0)$, $(0,1,0)$, $(0,0,1)$
   (c) $(2,1,4)$, $(3,2,5)$, $(-3,1,1)$, $(4,2,2)$
   (d) $(1,0,0)$, $(0,1,0)$, $(0,0,1)$, $(1,1,1)$, $(1,1,0)$

More Challenging Exercises

12. In Exercise 5(c) of Section 3-2 you were asked to show that if the $m \times n$ matrices A and B are row equivalent, then their rows generate the same subspace of $\mathbb{R}^n$. In Exercise 13 of Section 3-4 you were asked to show that the nonzero rows of a $m \times n$ matrix in row echelon form from a linearly independent set in $\mathbb{R}^n$. Use these facts to show the following. Suppose the subspace $\mathscr{V}$ of $\mathbb{R}^n$ is generated by the vectors $\mathbf{X}_1, \ldots, \mathbf{X}_m$. Form the $m \times n$ matrix A having $\mathbf{X}_1, \ldots, \mathbf{X}_m$ as its first, $\ldots$, mth rows, respectively. Transform A to row echelon form B. Then the nonzero rows of B form a basis for $\mathscr{V}$.

13. Using the results of Exercise 12, find a basis for the subspace of $\mathbb{R}^3$ generated by the vectors $(2,1,3)$, $(1,1,1)$, $(3,2,4)$, and $(4,3,5)$.

14. Same exercise for the subspace of $\mathbb{R}^4$ generated by $(5,1,2,0)$, $(-1,2,3,0)$, $(7,-3,-4,0)$, and $(0,1,0,1)$.

SECTION 3-6
Dimension

PREVIEW   *The central fact established in this section is that if a vector space has one basis which contains n vectors, then any set containing more than n vectors is linearly dependent. Because of this, every other basis for the same space has precisely n vectors, too. To establish this, we show that a homogeneous system of linear equations with more unknowns than equations always has nonzero solutions.*
   Once the central fact is established, the dimension of a (finitely generated) vector space is introduced, and the number of elements in generating sets and in linearly independent sets is related to the dimension of the space. This is used to show that the dimension of a subspace cannot be greater than the dimension of the space. Finally, we show that any linearly independent set can be enlarged to a basis.

The main fact we establish in this section is that if a vector space has a basis with n elements, then every basis for that space also contains exactly n elements. Before we get to this, we need a fact about solutions for homogeneous systems of linear equations.

When solving a homogeneous system of linear equations, we try to solve for each unknown in a different equation. If there are more unknowns than equations, then we cannot solve for every unknown. We can solve for some unknowns in terms of others; by assigning nonzero values to the others, we can construct nonzero solutions.

Example If we start with

$$3x + 2y + z = 0$$
$$4x + 2y - z = 0$$

we can solve for x and y.

$$x = \ 2z$$
$$y = -\tfrac{7}{2}z.$$

Nonzero solutions can be constructed by choosing nonzero values for z: $(2,-\tfrac{7}{2},1)$, $(4,-7,2)$, etc.

Example Consider the system of equations

$$x - 2y + 3z + w = 0$$
$$2x - 4y + \ z - w = 0$$
$$3x - 6y + 4z \qquad = 0.$$

Here we solve for x and z in terms of y and w. The row echelon form leads to the system of equations

$$x - 2y \qquad - \tfrac{4}{5}w = 0$$
$$z + \tfrac{3}{5}w = 0.$$

We rewrite these equations in the following form.

$$x \qquad = 2y + \tfrac{4}{5}w$$
$$z = \qquad \tfrac{3}{5}w.$$

Nonzero solutions can be constructed by choosing at least one nonzero value for y and w such as $(6,1,3,5)$, $(4,0,3,5)$, etc.

These numerical examples show what happens in general. A homogeneous system such as

$$a_{11}x_1 + a_{12}x_2 + \cdots + a_{1n}x_n = 0$$

$$\vdots \qquad \vdots \qquad \qquad \vdots \qquad \vdots$$

$$a_{m1}x_1 + a_{m2}x_2 + \cdots + a_{mn}x_n = 0$$

which has more unknowns than equations can be transformed to a system which looks like

$$x_{c_1} = \boxed{}$$

$$x_{c_2} =$$

$$\vdots$$

$$x_{c_r} =$$

where the right-hand side of each equation is a linear combination of the unknowns we cannot solve for. By assigning nonzero values to the unknowns on the right-hand side and computing the values for the unknowns on the left-hand side, we can construct nonzero solutions for the original system. (By the way, the reason that the unknowns on the left-hand side are labeled the way they are is that we cannot be sure that we are solving for the first r unknowns; look at the last example where we could not solve for the first two unknowns x and y at the same time.)

This is an important fact, and we record it for reference.

Proposition 3-7
A homogeneous system of linear equations having more unknowns than equations always has nonzero solutions.

Brief Exercises

1. Verify Proposition 3-7 by solving the following homogeneous systems of linear equations; find nonzero solutions.

   (a) $3x - y + z = 0$
   $\quad\;\; 5x + y + 3z = 0$

   (c) $5x_1 + 3x_2 - x_3 + 4x_4 = 0$
   $\quad\;\; 6x_1 - 2x_2 + 3x_3 + 5x_4 = 0$

   (b) $4x + y + 2z = 0$
   $\quad\;\; 8x + 2y + 4z = 0$

2. Relate the number of unknowns one can solve for to the number of nonzero rows in the row echelon form of the coefficient matrix.

Selected Answers: **1(a)** $y(1,1,-2)$ **1(b)** $x(1,-4,0) + z(0,-2,1)$
1(c) $x_2(-23,1,0,28) + x_3(17,0,1,-21)$

With this fact about homogeneous systems of equations we can discuss the main point of this section: if a basis for a vector space has exactly n vectors, then any set containing more than n vectors is linearly dependent.

In $\mathbb{R}^2$, for example, suppose that $\mathbf{X}_1 = (1,2)$ and $\mathbf{X}_2 = (0,1)$ form a basis. We claim that any set having three vectors must be linearly dependent. For example, the three vectors $\mathbf{Y}_1 = (5,-2)$, $\mathbf{Y}_2 = (3,-4)$, and $\mathbf{Y}_3 = (-2,1)$ form a linearly dependent set. Suppose

$$b_1\mathbf{Y}_1 + b_2\mathbf{Y}_2 + b_3\mathbf{X}_3 = \mathbf{0}. \tag{1}$$

Since $\mathbf{X}_1$ and $\mathbf{X}_2$ form a basis, each $\mathbf{Y}$ is a linear combination of the $\mathbf{X}$'s.

$$\mathbf{Y}_1 = (5,-2) = 5(1,2) - 12(0,1)$$

$$\mathbf{Y}_2 = (3,-4) = 3(1,2) - 10(0,1)$$

$$\mathbf{Y}_3 = (-2,1) = -2(1,2) + 5(0,1).$$

Substitute these into equation (1).

$$\mathbf{0} = b_1(5(1,2) - 12(0,1)) + b_2(3(1,2) - 10(0,1)) + b_3(-2(1,2) + 5(0,1)).$$

Collecting terms we get

$$\mathbf{0} = (5b_1 + 3b_2 - 2b_3)(1,2) + (-12b_1 - 10b_2 + 5b_3)(0,1).$$

Since $\mathbf{X}_1$ and $\mathbf{X}_2$ form a basis, the scalars in the last equation must be zero.

$$5b_1 + 3b_2 - 2b_3 = 0$$

$$-12b_1 - 10b_2 + 5b_3 = 0.$$

In any case, we get two equations in the three unknowns b_1, b_2, and b_3. Therefore, there are nonzero solutions; for example, $(5,1,14)$ is a nonzero solution. Therefore,

$$5\mathbf{Y}_1 + \mathbf{Y}_2 + 14\mathbf{Y}_3 = \mathbf{0}.$$

This shows that $\mathbf{Y}_1$, $\mathbf{Y}_2$, and $\mathbf{Y}_3$ form a linearly dependent set.

The same kind of argument works in general. Suppose $\mathscr{V}$ is a vector space over $\mathbb{R}$ with $\mathbf{X}_1, \ldots, \mathbf{X}_n$ forming a basis for $\mathscr{V}$, and suppose $\mathscr{S}$ is a set of vectors containing more than n vectors.

Pick any $n + 1$ vectors $\mathbf{Y}_1, \ldots, \mathbf{Y}_{n+1}$ from $\mathscr{S}$. We show that these $n + 1$ vectors form a linearly dependent set.

Suppose $b_1\mathbf{Y}_1 + \cdots + b_{n+1}\mathbf{Y}_{n+1} = \mathbf{0}$. Our attack is the same as above; we express the $\mathbf{Y}$'s in terms of the $\mathbf{X}$'s, substitute these expressions into the equation above, and get a system of n equations in the $n + 1$ unknowns $b_1, \ldots, b_{n+1}$. This system has nonzero solutions for the *b*'s by Proposition 3-7, and so $\mathbf{Y}_1, \ldots, \mathbf{Y}_{n+1}$ form a linearly dependent set; consequently $\mathscr{S}$ is a linearly dependent set.

This is an important fact, so we record it as a proposition.

Proposition 3-8

Suppose $\mathscr{V}$ is a vector space over $\mathbb{R}$ and suppose $\mathbf{X}_1, \ldots, \mathbf{X}_n$ form a basis for $\mathscr{V}$. If $\mathscr{S}$ is a set of more than *n* vectors in $\mathscr{V}$, then $\mathscr{S}$ is a linearly dependent set.

An immediate consequence of this proposition is the fact that any two bases for $\mathscr{V}$ contain exactly the same number of vectors. For suppose $\mathbf{X}_1, \ldots, \mathbf{X}_n$ form a basis for $\mathscr{V}$ and $\mathbf{Y}_1, \ldots, \mathbf{Y}_m$ also form a basis for $\mathscr{V}$. Because $\mathbf{Y}_1, \ldots, \mathbf{Y}_m$ form a linearly independent set and $\mathbf{X}_1, \ldots, \mathbf{X}_n$ form a basis, we know from the proposition above that $m \leq n$. On the other hand, because $\mathbf{X}_1, \ldots, \mathbf{X}_n$ form a linearly independent set and $\mathbf{Y}_1, \ldots, \mathbf{Y}_m$ form a basis, the proposition above shows that $n \leq m$.

Proposition 3-9

If $\mathscr{V}$ is a vector space over $\mathbb{R}$ having a basis consisting of *n* vectors, then every basis for $\mathscr{V}$ has exactly *n* vectors in it.

This proposition gives the reason for the next definition.

Definition 3-7

If $\mathscr{V}$ is the zero vector space, then the *dimension* of $\mathscr{V}$ is zero, dim $\mathscr{V} = 0$. If $\mathscr{V}$ is a nonzero vector space over $\mathbb{R}$ and has a basis with exactly *n* vectors, then the *dimension* of $\mathscr{V}$ is *n*, dim $\mathscr{V} = n$. We also use the phrase "$\mathscr{V}$ is an *n*-dimensional vector space" to mean that dim $\mathscr{V} = n$.

Brief Exercises

1. In $\mathbb{R}^2$ let $\mathbf{X}_1 = (1,0)$, $\mathbf{X}_2 = (0,1)$, $\mathbf{Y}_1 = (2,-3)$, $\mathbf{Y}_2 = (4,2)$, and $\mathbf{Y}_3 = (-1,2)$. Go through the steps of the argument for Proposition 3-8 to show the **Y**'s form a linearly dependent set.

2. Same exercise in $\mathbb{R}^3$ with $\mathbf{X}_1 = (1,0,0)$, $\mathbf{X}_2 = (0,1,0)$, $\mathbf{X}_3 = (0,0,1)$, $\mathbf{Y}_1 = (2,1,5)$, $\mathbf{Y}_2 = (-3,6,1)$, $\mathbf{Y}_3 = (4,1,5)$, and $\mathbf{Y}_4 = (6,-8,2)$.

3. What is the dimension of $\mathbb{R}^2$? of $\mathbb{R}^3$? of $\mathbb{R}^4$? of $\mathbb{R}^n$? Give reasons.

Now it is possible to get information about the number of vectors in generating sets and in linearly independent sets.

As an example, consider $\mathbb{R}^3$. We know that dim $\mathbb{R}^3 = 3$. First, a generating set must have at least three vectors in it, because, by Proposition 3-5, a generating set must contain a basis. Second, a generating set with three vectors in it must be a basis, once again because a generating set contains a basis; since a basis must have three vectors in it, the generating set must be a basis. Third, a linearly independent set can have no more than three vectors; this comes from Proposition 3-8. Fourth, a linearly independent set having three vectors in it must be a basis. If it were not, it must fail to be a generating set; then by Proposition 3-4 we could find a fourth vector to join with these three vectors to form a linearly independent set with four vectors, which is impossible in $\mathbb{R}^3$.

The same facts are true in any vector space $\mathscr{V}$ with dim $\mathscr{V} = n > 0$. (1) A generating set must have at least n vectors in it, because by Proposition 3-5 a generating set contains a basis. (2) A generating set having exactly n vectors in it must be a basis, for the same reasons given for $\mathbb{R}^3$. (3) A linearly independent set has no more than n vectors in it because of Proposition 3-8. (4) A linearly independent set having n vectors in it must be a basis, for if it were not, it would not be a generating set. Then by Proposition 3-4 we could find an $n + 1$-st vector to join with these n vectors to form a linearly independent set with $n + 1$ vectors in it, an impossibility.

Proposition 3-10_____

Suppose that $\mathscr{V}$ is an n-dimensional vector space over $\mathbb{R}$.

(1) A generating set for $\mathscr{V}$ must contain at least n vectors.
(2) A generating set with n vectors is a basis.
(3) A linearly independent set contains at most n vectors.
(4) A linearly independent set with n vectors is a basis.

Some consequences of Propositions 3-8, 3-9, and 3-10 are illustrated in the next examples.

Example The dimension of $\mathbb{R}^2$ is 2, the dimension of $\mathbb{R}^3$ is 3, and, in general, the dimension of $\mathbb{R}^n$ is n. The reason is that the natural basis for $\mathbb{R}^n$ has n elements. (The *natural basis* is formed by $(1,0,\ldots,0)$, $(0,1,0,\ldots,0)$, $\ldots$, $(0,\ldots,0,1)$, the vectors having the number 1 in one component and 0 in all the others.)

Example Do $(1,0,-1,0)$, $(-1,1,0,0)$, and $(3,0,1,1)$ form a basis for $\mathbb{R}^4$? Before this section we solved this problem by asking whether, given any (x,y,z,w) in $\mathbb{R}^4$, there are scalars a_1, a_2, and a_3 so that

$$a_1(1,0,-1,0) + a_2(-1,1,0,0) + a_3(3,0,1,1) = (x,y,z,w).$$

Now it is easy to see that the given vectors do not generate $\mathbb{R}^4$, since four vectors are needed to generate $\mathbb{R}^4$.

Brief Exercises

1. Do the following vectors form a linearly independent set in $\mathbb{R}^2$?

    (a) $(2,1)$ **(c)** $(2,3)$, $(2,-1)$, $(3,4)$

    (b) $(1,1)$, $(3,2)$, $(5,4)$ **(d)** $(1,2)$, $(1,-1)$

2. Give an example of a set in $\mathbb{R}^2$ having five vectors in it and which is not a generating set for $\mathbb{R}^2$.

3. Same exercise in $\mathbb{R}^6$. What is the difference between solving this exercise and solving Exercise 2?

4. How many vectors does it take to generate the solution set for $x - y + z = 0$ in $\mathbb{R}^3$?

5. Same question in $\mathbb{R}^4$ for

$$2x + y - 3z + w = 0$$

$$x + y + z - w = 0$$

$$3x - y + 2z \qquad = 0.$$

Selected Answers: **1(a)** Yes   **1(b)** No   **1(c)** No   **1(d)** Yes   **4** At least two: e.g., $x(1,1,0) + z(0,1,1)$ **5** At least one: $w(-\frac{2}{19}, \frac{12}{19}, \frac{9}{19}, 1)$

In Section 3 we gave geometric arguments to show that every subspace of $\mathbb{R}^2$ has dimension 0, 1, or 2, and that every subspace of $\mathbb{R}^3$ has dimension 0, 1, 2, or 3. A similar situation occurs in $\mathbb{R}^4$; any subspace $\mathscr{W}$ of $\mathbb{R}^4$ has dimension 0, 1, 2, 3, or 4. If $\mathscr{W}$ is the zero subspace, then dim $\mathscr{W} = 0$. If $\mathscr{W}$ contains nonzero vectors, pick a nonzero vector $\mathbf{X}_1$ in $\mathscr{W}$. If $\mathbf{X}_1$ generates $\mathscr{W}$, then dim $\mathscr{W} = 1$. If not, then there is a vector $\mathbf{X}_2$ in $\mathscr{W}$ which is not a multiple of $\mathbf{X}_1$, and so by Proposition 3-4, $\mathbf{X}_1$ and $\mathbf{X}_2$ form a linearly independent set. If $\mathbf{X}_1$ and $\mathbf{X}_2$ span $\mathscr{W}$, then dim $\mathscr{W} = 2$. If not, then there is a vector $\mathbf{X}_3$ in $\mathscr{W}$ which is not a linear combination of $\mathbf{X}_1$ and $\mathbf{X}_2$, and so by Proposition 3-4 again, $\mathbf{X}_1$, $\mathbf{X}_2$, and $\mathbf{X}_3$ form a linearly independent set. If $\mathbf{X}_1$, $\mathbf{X}_2$, and $\mathbf{X}_3$ generate $\mathscr{W}$, then dim $\mathscr{W} = 3$. If not, then there is a vector $\mathbf{X}_4$ in $\mathscr{W}$ which is not a linear combination of $\mathbf{X}_1$, $\mathbf{X}_2$, and $\mathbf{X}_3$; again by Proposition 3-4, $\mathbf{X}_1$, $\mathbf{X}_2$, $\mathbf{X}_3$, and $\mathbf{X}_4$ form a linearly independent set. The process ends here, because in $\mathbb{R}^4$ it is impossible to have more than four vectors in a linearly independent set. Consequently, a subspace of $\mathbb{R}^4$ has dimension 0, 1, 2, 3, or 4.

In general suppose $\mathcal{W}$ is a subspace of an n-dimensional vector space $\mathcal{V}$. We can determine the dimension of $\mathcal{W}$ by the same process that we just described for subspaces of $\mathbb{R}^4$; the process may have more steps, but it must terminate after at most n steps because no linearly independent set in $\mathcal{V}$ has more than n vectors in it. So the process ends at some stage with linearly independent vectors $\mathbf{X}_1, \ldots, \mathbf{X}_m$ chosen from $\mathcal{W}$ with $m \leq n$. Since the process stopped at this point, we cannot find a vector in $\mathcal{W}$ which is not spanned by $\mathbf{X}_1, \ldots, \mathbf{X}_m$; that is, the vectors $\mathbf{X}_1, \ldots, \mathbf{X}_m$ form a basis for $\mathcal{W}$. Therefore dim $\mathcal{W} = m$.

Proposition 3-11_____

If $\mathcal{W}$ is a subspace of an n-dimensional vector space $\mathcal{V}$ over $\mathbb{R}$, then dim $\mathcal{W} \leq n$.

Example Let $\mathcal{W}$ be the vectors (x,y,z,w) in $\mathbb{R}^4$ for which $2x + 3y + z - w = 0$ and $x - 2y + z = 0$. It is easy to check that $\mathcal{W}$ is a subspace of $\mathbb{R}^4$. We find a basis for $\mathcal{W}$ in this example.

Since $\mathcal{W}$ is a subspace of $\mathbb{R}^4$, the dimension of $\mathcal{W}$ is 0, 1, 2, 3, or 4. Now dim $\mathcal{W} \neq 4$ since, for example, the vector $(1,0,0,0)$ is not in $\mathcal{W}$. We could guess about vectors in $\mathcal{W}$ trying to find a basis; however, we illustrate a systematic way to find a basis for $\mathcal{W}$.

The vector (x,y,z,w) is in $\mathcal{W}$ if and only if it is a solution to

$$2x + 3y + z - w = 0$$

$$x - 2y + z \qquad = 0.$$

The solution is

$$x \qquad + \tfrac{3}{7}z - \tfrac{2}{7}w = 0$$

$$y - \tfrac{1}{7}z - \tfrac{1}{7}w = 0$$

or

$$x = -\tfrac{5}{7}z + \tfrac{2}{7}w$$

$$y = \tfrac{1}{7}z + \tfrac{1}{7}w.$$

Thus (x,y,z,w) is in $\mathcal{W}$ if and only if

$$(x,y,z,w) = (-\tfrac{5}{7}z + \tfrac{2}{7}w, \tfrac{1}{7}z + \tfrac{1}{7}w, z, w)$$

$$= z(-\tfrac{5}{7},\tfrac{1}{7},1,0) + w(\tfrac{2}{7},\tfrac{1}{7},0,1).$$

So $(-\tfrac{5}{7},\tfrac{1}{7},1,0)$ and $(\tfrac{2}{7},\tfrac{1}{7},0,1)$ form a basis for $\mathcal{W}$.

Brief Exercises

In each of the following find the dimension of the solution set.

1. $3x - y + z = 0$
$x + y + z = 0$

5. $x + y - z \phantom{{}+ w} = 0$
$\phantom{x + {}}y \phantom{{}- z} + w = 0$
$x - y - z - w = 0$

2. $5x + y - z = 0$

6. $\phantom{x + {}}y \phantom{{}- z} + w = 0$
$x - y - z - w = 0$

3. $4x + y + z = 0$
$x - y + z = 0$
$x + y - z = 0$

7. $x - y - z - w = 0$

4. $4x - y + z - w = 0$
$x + y - z \phantom{{}- w} = 0$
$\phantom{4x + {}}y \phantom{{}- z} + w = 0$
$x - y - z - w = 0$

Answers: **1** 1 **2** 2 **3** 0 **4** 0 **5** 1 **6** 2 **7** 3

Finally, we show that it is always possible to enlarge a linearly independent set to a basis.

In $\mathbb{R}^2$, for example, if we have two linearly independent vectors $\mathbf{X}_1$ and $\mathbf{X}_2$, then we already have a basis. If we have one nonzero vector $\mathbf{X}_1$, then we can pick any $\mathbf{X}_2$ which is noncollinear with $\mathbf{X}_1$; then $\mathbf{X}_1$ and $\mathbf{X}_2$ form a basis for $\mathbb{R}^2$.

In $\mathbb{R}^3$, if we begin with one nonzero vector $\mathbf{X}_1$, we pick $\mathbf{X}_2$ noncollinear with $\mathbf{X}_1$, and then we pick $\mathbf{X}_3$ noncoplanar with $\mathbf{X}_1$ and $\mathbf{X}_2$; this yields a basis consisting of $\mathbf{X}_1$, $\mathbf{X}_2$, and $\mathbf{X}_3$.

The same kind of thing can be done in general. Suppose $\mathscr{V}$ is an n-dimensional vector space over $\mathbb{R}$, and suppose we start with a linearly independent set of m vectors $\mathbf{X}_1, \ldots, \mathbf{X}_m$ where $m < n$. Since $m < n$, the vectors $\mathbf{X}_1, \ldots, \mathbf{X}_m$ cannot generate $\mathscr{V}$. Therefore, there is a vector $\mathbf{X}_{m+1}$ which is in $\mathscr{V}$ and which is not spanned by $\mathbf{X}_1, \ldots, \mathbf{X}_m$. By Proposition 3-4, the vectors $\mathbf{X}_1, \ldots, \mathbf{X}_m, \mathbf{X}_{m+1}$ form a linearly independent set. If $m + 1 = n$, we have a basis. If $m + 1 < n$, we do the same thing again, getting a linearly independent set with $m + 2$ vectors, and so on. Finally the process ends with our having found vectors $\mathbf{X}_{m+1}, \ldots, \mathbf{X}_n$ to join with our original vectors $\mathbf{X}_1, \ldots, \mathbf{X}_m$ to form the basis $\mathbf{X}_1, \ldots, \mathbf{X}_m, \mathbf{X}_{m+1}, \ldots, \mathbf{X}_n$.

Proposition 3-12

Suppose $\mathbf{X}_1, \ldots, \mathbf{X}_m$ form a linearly independent set in the n-dimensional vector space $\mathscr{V}$ over $\mathbb{R}$, and suppose $m < n$. Then there are vectors $\mathbf{X}_{m+1}, \ldots, \mathbf{X}_n$ in $\mathscr{V}$ so that $\mathbf{X}_1, \ldots, \mathbf{X}_m, \mathbf{X}_{m+1}, \ldots, \mathbf{X}_n$ form a basis for $\mathscr{V}$.

Example Consider the set consisting of (1,1,0) and (2,0,1) in $\mathbb{R}^3$. This is clearly a linearly independent set. It is easy to see that no vector in the natural basis for $\mathbb{R}^3$ is in the linear span of (1,1,0) and (2,0,1). Therefore, we can enlarge the set consisting of these two vectors to a basis for $\mathbb{R}^3$ by including any one of the vectors from the natural basis. Therefore, the following three-element sets are bases for $\mathbb{R}^3$.

$$(1,1,0), (2,0,1), (1,0,0)$$
$$(1,1,0), (2,0,1), (0,1,0)$$
$$(1,1,0), (2,0,1), (0,0,1)$$

These bases contain the original linearly independent set with the "extra" vectors coming from the natural basis. The geometrical reason for this is that none of the vectors in the natural basis lies in the plane determined by the two vectors (1,1,0) and (2,0,1).

Brief Exercises

1. Find bases for $\mathbb{R}^2$ which contain the following, if possible.

(**a**) (1,2)

(**b**) (2,3)

(**c**) (1,1), (2,2)

(**d**) (0,1)

2. Same exercise in $\mathbb{R}^3$.

(**a**) (1,0,1)

(**b**) (2,1,1)

(**c**) (1,0,0), (0,1,0)

(**d**) (1,1,1), (2,1,0)

Answers: **1(a)** Any vector noncollinear with (1,2) **1(b)** Any vector noncollinear with (2,3) **1(c)** No basis containing these **1(d)** Any vector noncollinear with (0,1) **2(a)** Any two noncollinear vectors whose plane does not contain (1,0,1) **2(b)** Any two noncollinear vectors whose plane does not contain (2,1,1) **2(c)** Any vector noncoplanar with (1,0,0) and (0,1,0) **2(d)** Any vector noncoplanar with (1,1,1) and (2,1,0)

SUMMARY *A homogeneous system of linear equations with more unknowns than equations always has nonzero solutions. This fact is used to show that any set of more than n vectors in a vector space having a basis with n vectors must be linearly dependent, and hence any two bases have the same number of vectors. This number is the dimension of the vector space. In an n-dimensional vector space, generating sets must have at least n vectors. Finally, every linearly independent set, if not already a basis, can be enlarged to a basis.*

Exercises for Section 3-6

1. Determine whether the following vectors form a basis for $\mathbb{R}^2$.

   (a) $(2,1)$ and $(3,2)$

   (b) $(2,-1)$ and $(4,-2)$

   (c) $(2,-1)$, $(2,1)$, and $(4,0)$

   (d) $(5,1)$ and $(17,\frac{17}{5})$

   (e) $(1,1)$ and $(0,1)$

2. Determine whether the following vectors form a basis for $\mathbb{R}^3$.

   (a) $(1,1,0)$, $(2,1,0)$, and $(3,1,0)$

   (b) $(5,0,1)$, $(1,1,0)$, and $(2,1,1)$

   (c) $(5,1,2)$ and $(3,1,4)$

   (d) $(1,5,1)$, $(2,1,0)$, $(3,1,2)$, and $(4,1,1)$

   (e) $(6,-1,1)$, $(-3,2,5)$, and $(8,6,1)$

3. Determine whether the following vectors form a basis for $\mathbb{R}^4$.

   (a) $(2,1,5,2)$, $(1,-1,3,0)$, $(1,0,2,0)$, and $(0,1,1,1)$

   (b) $(8,1,0,2)$, $(0,3,5,1)$, $(2,4,6,0)$, and $(1,3,5,7)$

4. Find bases for the following subspaces of $\mathbb{R}^3$.

   (a) the vectors (x,y,z) satisfying $2x + y + z = 0$

   (b) the vectors (x,y,z) satisfying $x + y + z = 0$ and $x - y - z = 0$

   (c) the vectors (x,y,z) satisfying $2x + y + z = 0$, $x + y + z = 0$, and $x + z = 0$

5. Find bases for the following subspaces of $\mathbb{R}^4$.

   (a) the vectors (x,y,z,w) satisfying $x + y + w = 0$

   (b) the vectors (x,y,z,w) satisfying $x + y + z - w = 0$ and $x - y + z + 2w = 0$

   (c) the vectors (x,y,z,w) satisfying the equations $x + y + z = 0$, $y + z + w = 0$, and $x + z + w = 0$

   (d) the vectors (x,y,z,w) satisfying the equations $x + y + z = 0$, $y + z + w = 0$, $x + z + w = 0$, and $2x + 2y + 3z + 2w = 0$

6. In each part of Exercise 4, find vectors which can be added to the basis for the subspace to yield a basis for $\mathbb{R}^3$.

7. In each part of Exercise 5, find vectors which can be added to the basis for the subspace to yield a basis for $\mathbb{R}^4$.

More Challenging Exercises

8. Suppose a subspace $\mathscr{W}$ of $\mathbb{R}^3$ contains the vectors $(3,1,2)$ and $(5,-1,1)$. Find a homogeneous system of equations which has $\mathscr{W}$ as its solution set. (Form the matrix A having the given vectors as rows.

$$A = \begin{pmatrix} 3 & 1 & 2 \\ 5 & -1 & 1 \end{pmatrix}$$

Transform A to its row echelon form B.

$$B = \begin{pmatrix} 1 & 0 & \frac{3}{8} \\ 0 & 1 & \frac{7}{8} \end{pmatrix}$$

By Exercise 5(c) of Section 3-2 we know that the rows of B and the rows of A span the same space. Therefore, $\mathscr{W}$ is spanned by $(1,0,\frac{3}{8})$ and $(0,1,\frac{7}{8})$. So (x,y,z) is in $\mathscr{W}$ if and only if $(x,y,z) = x(1,0,\frac{3}{8}) + y(0,1,\frac{7}{8})$. This vector equality is equivalent to one scalar equality: $z = \frac{3}{8}x + \frac{7}{8}y$, or, equivalently, $3x + 7y - 8z = 0$. This is the homogeneous system of equations—consisting of only one equation—which has $\mathscr{W}$ as its solution set.)

9. Use the method outlined in Exercise 8 to find a homogeneous system of equations whose solution set is the space indicated.

(a) the subspace of $\mathbb{R}^3$ generated by $(2,1,1)$
(b) the subspace of $\mathbb{R}^3$ generated by $(2,1,1)$ and $(1,0,1)$
(c) the subspace of $\mathbb{R}^4$ generated by $(1,1,0,1)$
(d) the subspace of $\mathbb{R}^4$ generated by $(1,1,0,1)$ and $(1,0,1,0)$
(e) the subspace of $\mathbb{R}^4$ generated by $(1,1,0,1)$, $(1,0,1,0)$, and $(1,1,0,0)$
(f) the subspace of $\mathbb{R}^4$ generated by $(1,1,0,1)$, $(1,0,1,0)$, $(1,1,0,0)$, and $(3,2,1,1)$

10. Use the results of Exercise 9 to describe a general procedure for finding a homogeneous system of equations whose solution set is a given subspace of $\mathbb{R}^n$. What is the connection between the dimension of the subspace and the number of equations needed? Explain.

11. In the Exercises for Section 3-3 we introduced the concept of complementary subspaces. Use ideas from this chapter to argue that if $\mathscr{W}_1$ is a subspace of the vector space $\mathscr{V}$, then there is a subspace $\mathscr{W}_2$ of $\mathscr{V}$ which is a complementary subspace for $\mathscr{W}_1$. (look at the argument for Proposition 3-12.)

12. If $\mathscr{W}_1$ and $\mathscr{W}_2$ are complementary subspaces in the vector space $\mathscr{V}$ and dim $\mathscr{V} = n$, give an argument to show that

$$\dim \mathscr{V} = \dim \mathscr{W}_1 + \dim \mathscr{W}_2.$$

13. Suppose $\mathscr{V}$ is a vector space over $\mathbb{R}$ with dim $\mathscr{V} = n$, and suppose dim $\mathscr{V} = \dim \mathscr{W}_1 + \dim \mathscr{W}_2$, where $\mathscr{W}_1$ and $\mathscr{W}_2$ are subspaces of $\mathscr{V}$. Are $\mathscr{W}_1$ and $\mathscr{W}_2$ necessarily complementary subspaces? If not, what additional condition can be imposed which insures that they are complementary?

4

Linear
Transformations
and
Matrices

Preliminaries About Functions

PREVIEW *In this section we introduce functions from one vector space to another. We show how to add functions, multiply functions by scalars, and form composition of functions. Finally, we verify that composition of functions is associative.*

One of the most useful ideas in scientific work is that of a mathematical function. The word "function" is used in everyday language to describe the natural or proper action of something, an assigned duty or activity, an official ceremony, in addition to the mathematical usage. In mathematics, however, it expresses relationships or dependence of one thing upon another, how one quantity varies as another varies, or the change of something from one condition to another. Other words used in mathematics for function are map, mapping, and transformation.

For example, the cost of sending a letter by airmail depends upon the weight of the letter, 11¢ per ounce for letters weighing less than 8 oz. We can think of cost as a function of the weight; once the weight is known, the cost is determined. The cost is considered as variable from letter to letter depending upon the weight of the letter. The cost is called the dependent variable, and the weight is called the independent variable. Since the cost is determined once the weight is known, mathematicians call cost a function of one (independent) variable, the weight.

We write "$c(\frac{1}{2}) = 11$" to convey the information that the cost of sending a $\frac{1}{2}$ oz letter is 11¢. Using this notation we have $c(\frac{1}{4}) = 11$, $c(\frac{3}{4}) = 11$, $c(1.1) = 22$, etc. The letter c represents the function (cost in this instance). The number enclosed in the parentheses represents a particular value of the independent variable (weight), and the number on the right-hand side gives the cost for that particular weight.

This example can be extended. For letters heavier than 8 oz, the cost depends upon two factors, the weight and the zone into which it is sent. Some entries from the 1972 table are given below.

Weight not over	Local Zones 1, 2 and 3	Zone 4	Zone 5	Zone 6	Zone 7	Zone 8
1	1.00	1.00	1.00	1.00	1.00	1.00
$1\frac{1}{2}$	1.20	1.22	1.25	1.30	1.40	1.50
2	1.40	1.43	1.51	1.60	1.68	1.77
$2\frac{1}{2}$	1.60	1.65	1.76	1.90	2.02	2.16

So, for example, it costs \$1.30 to send a letter weighing $1\frac{1}{4}$ lb to Zone 6. One way to indicate this in mathematical notation is "$c(1\frac{1}{4}, 6) = 1.30$." The factors which determine the price are listed in the parentheses, and the price for that weight and zone is on the right-hand side. For example, $c(2\frac{1}{8}, 7) = 2.02$, $c(1\frac{1}{8}, 5) = 1.25$. Because, in this instance, two factors are needed to determine the cost, this cost function is a function of two (independent) variables.

In most colleges the scores on college entrance tests are used to determine section assignments in courses such as English, mathematics, foreign languages, etc. For example, suppose your college entrance test has four parts, so that you have four scores. These four numbers or scores are the factors which determine the sections to which you are assigned in English, mathematics, and foreign languages. Your sectioning is thus a function of your scores. This is a function of four (independent) variables—for the purposes of sectioning, the scores are considered to be independent of each other even though there may be some correlation between reading ability, for example, and other skills.—This sectioning function also determines more than one number; it determines a section number in three courses. We therefore think of the sectioning function as determining three things or as determining an ordered triple as section assignments. We call this a vector-valued function of four variables. It is also a vector-valued function of a vector variable, since the scores can be considered an ordered quadruple.

In all the examples just given, the function determines a number or a vector once another number or vector is known. This is typical of the functions considered in this book; we only deal with functions which determine a vector (from a second vector space) once a vector (from a first vector space) is known.

Example   A function f from $\mathbb{R}$ to $\mathbb{R}$ can be specified by the equation $f(x) = x^2 + 3x + 2$. Once the number x is known, the value $x^2 + 3x + 2$ is determined. Thus $f(3) = 3^2 + 3 \cdot 3 = 2 = 20, f(4) = 4^2 + 3 \cdot 4 + 2 = 30, f(-2) = (-2)^2 + 3(-2) + 2 = 0$, etc. This is called a real-valued function of a real variable because, once the real variable x is known, the value $f(x)$ is also a real number.

Example   A function f from $\mathbb{R}$ to $\mathbb{R}^3$ can be specified by the equation $f(x) = (x, 3x^2 - 5, x^3 + x + 1)$. Once the number x is known, the value $(x, 3x^2 - 5, x^3 + x + 1)$ is determined. For example, $f(1) = (1, 3 \cdot 1^2 - 5, 1^3 + 1 + 1) = (1, -2, 3)$, $f(2) = (2, 3 \cdot 2^2 - 5, 2^3 + 2 + 1) = (2, 7, 11), f(-3) = (-3, 3(-3)^2 - 5, (-3)^3 + (-3) + 1) = (-3, 22, -29)$, etc. This is called a vector-valued function of a real variable because once the real variable x is known, the value $f(x)$ is a vector.

Example   A function f from $\mathbb{R}^2$ to $\mathbb{R}^2$ can be specified by $f(x, y) = (x^2 + y^2, 3x + 4y)$. Once (x, y) is known, the value $f(x, y)$ is determined. For example,

$f(1,0) = (1^2 + 0^2, 3 \cdot 1 + 4 \cdot 0) = (1,3), f(2,3) = (2^2 + 3^2, 3 \cdot 2 + 4 \cdot 3) = (13,18), f(-1,4) = ((-1)^2 + 4^2, 3(-1) + 4 \cdot 4) = (17,13)$, etc. This is a vector-valued function of a vector variable because, once the (vector) variable (x,y) is known, the value $f(x,y)$ is a vector.

Definition 4-1

A *function f* from a vector space $\mathscr{V}$ to a vector space $\mathscr{W}$ is a relationship or a rule of correspondence by which each vector in $\mathscr{V}$ is assigned a vector in $\mathscr{W}$. If $\mathbf{X}$ is a vector in $\mathscr{V}$, the value of f at $\mathbf{X}$ is denoted by $f(\mathbf{X})$.

Rather than saying "f is a function from $\mathscr{V}$ to $\mathscr{W}$," it is customary to use the abbreviation $f: \mathscr{V} \to \mathscr{W}$. The *domain* of f is $\mathscr{V}$, and the *range* or *image* of f consists of all vectors in $\mathscr{W}$ which have been assigned to vectors in $\mathscr{V}$.

The words "map," "mapping," and "transformation" are used as synonyms for "function." If one of these is used, then we sometimes say that f maps $\mathbf{X}$ to $f(\mathbf{X})$ or that f transforms $\mathbf{X}$ to $f(\mathbf{X})$.

The examples above show that a function $f: \mathscr{V} \to \mathscr{W}$ is completely determined if, for every $\mathbf{X}$ in $\mathscr{V}$, we can find $f(\mathbf{X})$. This gives a method of describing functions, and it gives a way of determining whether two functions $f: \mathscr{V} \to \mathscr{W}$ and $g: \mathscr{V} \to \mathscr{W}$ are actually the same function, i.e., are equal.

Definition 4-2

The functions $f: \mathscr{V} \to \mathscr{W}$ and $g: \mathscr{V} \to \mathscr{W}$ are equal if, for each $\mathbf{X}$ in $\mathscr{V}, f(\mathbf{X}) = g(\mathbf{X})$. If this is the case, we write $f = g$.

Brief Exercises

1. In each of the following suppose the function $f: \mathbb{R} \to \mathbb{R}$ is determined by the formula given; compute $f(0), f(1), f(-1), f(2), f(-2), f(5)$, and $f(-6)$.

   (a) $f(x) = 3x + 2$ (c) $f(x) = x^2 + 2$
   (b) $f(x) = 6x - 5$ (d) $f(x) = x^2 + 2x - 5$

2. In each of the following, suppose the function $f: \mathbb{R} \to \mathbb{R}^2$ is determined by the formula given; compute $f(0), f(1), f(-1), f(2), f(-2), f(5)$, and $f(-6)$.

   (a) $f(x) = (2x^2 + 3, 5x - 1)$ (c) $f(x) = (5x^2 - 2, 4x^3 + 1)$
   (b) $f(x) = (x^2 - 4x + 2, 3x^2 + x - 2)$

3. In each of the following, suppose the function $\mathbb{R}^2 \to \mathbb{R}^3$ is determined by the formula given; compute $f(1,0), f(0,1), f(2,3), f(-2,4), f(3,-2)$.

   (a) $f(x,y) = (x^2y, x - y, 4x + 2y)$
   (b) $f(x,y) = (2x^2 + y, x + 3y, 2x - y)$
   (c) $f(x,y) = (4x^2 + 3xy^2 - x - y, 2x^2 + 4xy, 5x - 6y + 1)$

Selected Answers: **1(a)** $f(0) = 2$, $f(-2) = -4$ **1(b)** $f(1) = 1$, $f(-2) = -17$
1(c) $f(2) = 6$, $f(5) = 27$ **1(d)** $f(-1) = -6$, $f(-6) = 19$ **2(a)** $f(1) = (5,4)$,
$f(5) = (53,24)$ **2(b)** $f(-2) = (14,8)$, $f(-6) = (62,100)$ **2(c)** $f(-1) = (3,-3)$,
$f(-2) = (18,-31)$ **3(a)** $f(1,0) = (0,1,4)$, $f(2,3) = (12,-1,14)$ **3(b)** $f(0,1) =$
$(1,3,-1)$, $f(-2,4) = (12,10,-8)$ **3(c)** $f(2,3) = (65,32,-7)$, $f(3,-2) =$
$(71,-6,28)$

We can add functions and multiply by scalars. This is illustrated in the next
examples.

Example   Suppose $f : \mathbb{R} \to \mathbb{R}$ and $g : \mathbb{R} \to \mathbb{R}$ are defined by $f(x) = x^2 + 3x - 1$
and $g(x) = 2x^2 - 2x + 3$. We can form a new function $h : \mathbb{R} \to \mathbb{R}$ using f and g;
the values of h are computed by adding the values of f and the values of g. Thus

$$
\begin{aligned}
h(0) &= f(0) &+ g(0) &= -1 + 3 = 2 \\
h(1) &= f(1) &+ g(1) &= 3 + 3 = 6 \\
h(-1) &= f(-1) &+ g(-1) &= -3 + 7 = 4 \\
h(2) &= f(2) &+ g(2) &= 9 + 7 = 16 \\
h(-2) &= f(-2) &+ g(-2) &= -3 + 15 = 12
\end{aligned}
$$

and so forth. We can give the same information in a more abbreviated form: for
each x in $\mathbb{R}$,

$$ h(x) = f(x) + g(x). $$

This equation explains how to find the number which h associates with a given
number x. It is natural to denote this new function by $f + g$.

Example   Suppose $f : \mathbb{R}^2 \to \mathbb{R}^3$ and $g : \mathbb{R}^2 \to \mathbb{R}^3$ are defined by the equations

$$ f(x,y) = (x^2y + 3y, x - y, x^2y^2) $$
$$ g(x,y) = (5x + 2y, x^2 - y, 3x - y^2). $$

We can form a new function $h : \mathbb{R}^2 \to \mathbb{R}^3$ by adding f and g:

$$
\begin{aligned}
h(0,0) &= f(0,0) + g(0,0) = (0,0,0) + (0,0,0) = (0,0,0) \\
h(1,0) &= f(1,0) + g(1,0) = (0,1,0) + (5,1,3) = (5,2,3) \\
h(0,1) &= f(0,1) + g(0,1) = (3,-1,0) + (2,-1,-1) = (5,-2,-1)
\end{aligned}
$$

and so on. This information can be given by stating that for each (x,y) in $\mathbb{R}^2$,

$$h(x,y) = f(x,y) + g(x,y) = (x^2y + 5x + 5y, x^2 + x - 2y, x^2y^2 + 3x - y^2).$$

This equation explains how to find the element of $\mathbb{R}^3$ which h associates with a given element (x,y) of $\mathbb{R}^2$. We denote this function h by $f + g$.

Definition 4-3_____

Suppose $f: \mathscr{V} \to \mathscr{W}$ and $g: \mathscr{V} \to \mathscr{W}$ are functions. The sum $f + g$ is the function whose value at each $\mathbf{X}$ in $\mathscr{V}$ is the sum of the values of f and g at $\mathbf{X}$.

$$(f + g)(\mathbf{X}) = f(\mathbf{X}) + g(\mathbf{X})$$

Now we illustrate multiplication of functions by scalars.

Example If $f: \mathbb{R} \to \mathbb{R}$ is defined by $f(x) = x^2 + 3x - 1$, then we can form a new function $k: \mathbb{R} \to \mathbb{R}$ using f and the scalar 3, for example,

$$k(0) = 3f(0) = 3(-1) = -3$$
$$k(1) = 3f(1) = 3 \cdot 3 = 9$$
$$k(2) = 3f(2) - 3 \cdot 9 = 27$$

etc. This can be symbolized by the following: for each x in $\mathbb{R}$,

$$k(x) = 3f(x).$$

This equation explains how to find the number which k associates with a given number x. We denote this function by $3f$.

Example If $f: \mathbb{R}^2 \to \mathbb{R}^3$ is defined by

$$f(x,y) = (x^2y + 3y, x - y, x^2y^3)$$

then we can form the function $-2f$, for example.

$$(-2f)(1,1) = -2[f(1,1)] = -2(4,0,1) = (-8,0,-2)$$
$$(-2f)(1,0) = -2[f(1,0)] = -2(0,1,0) = (0,-2,0)$$
$$(-2f)(0,1) = -2[f(0,1)] = -2(3,-1,0) = (-6,2,0)$$

Definition 4-4_____

Suppose $f: \mathscr{V} \to \mathscr{W}$ is a function and suppose that r is a scalar. The scalar multiple rf of f is the function whose value at each $\mathbf{X}$ in $\mathscr{V}$ is r times the values of f at $\mathbf{X}$.

$$(rf)(\mathbf{X}) = r[f(\mathbf{X})]$$

Brief Exercises

1. If $f: \mathbb{R} \to \mathbb{R}$ and $g: \mathbb{R} \to \mathbb{R}$ are the functions indicated, compute $(f + g)(0)$, $(f + g)(-1)$, $(f + g)(2)$, and $(f + g)(x)$; also compute $(5f)(0)$, $(5f)(-1)$, $(5f)(2)$, and $(5f)(x)$.

    (a) $f(x) = x^2$, $g(x) = 3x - 5$
    (b) $f(x) = 4x^3 + 2x - 4$, $g(x) = 3x^2 + x - 4$

2. If $f: \mathbb{R}^2 \to \mathbb{R}$ and $g: \mathbb{R}^2 \to \mathbb{R}$ are the functions indicated, compute $(f + g)(0,1)$, $(f + g)(1,1)$, $(f + g)(2,1)$, $(f + g)(x,y)$, $(3f)(0,1)$, $(3f)(1,1)$, $(3f)(2,1)$, and $(3f)(x,y)$.

    (a) $f(x,y) = 5x + 6y + xy^2 - 2$, $g(x,y) = 8xy + 5y^2$
    (b) $f(x,y) = 6x^2 + 6y^2 - 3xy$, $g(x,y) = 2xy - 3x + 2$

3. If $f: \mathbb{R} \to \mathbb{R}^3$ and $g: \mathbb{R} \to \mathbb{R}^3$ are the functions indicated, compute $(f + g)(0)$, $(f + g)(-1)$, $(f + g)(2)$, $(f + g)(x)$, $(-3f)(0)$, $(-3f)(1)$, and $(-3f)(x)$.

    (a) $f(x) = (x^2 - 5, 8x^3 + 3x, -5x^2 + 2)$, $g(x) = (0, 2x, 5)$
    (b) $f(x) = (6x^2 - 2x + 1, x + 2, x + 3)$, $g(x) = (5x, -3, 4x + 2)$

Selected Answers: **1(a)** $(f + g)(0) = -5$, $(f + g)(2) = 5$ **1(b)** $(f + g)(-1) = -12$, $5f(2) = 160$ **2(a)** $(f + g)(2,1) = 37$, $3f(1,1) = 30$ **2(b)** $(f + g)(1,1) = 10$, $3f(1,1) = 27$ **3(a)** $(f + g)(-1) = (-4, -13, 2)$, $-3f(1) = (12, -33, 9)$ **3(b)** $(f + g)(2) = (31, 1, 15)$, $-3f(0) = (-3, -6, -9)$

There is another way to construct a new function from other functions which is important to us. The new function is the *composite* or the *composition*. In this operation, the order in which the functions are used is extremely important, as we illustrate in the next few examples.

Example Suppose $f: \mathbb{R} \to \mathbb{R}$ and $g: \mathbb{R} \to \mathbb{R}$ are defined by $f(x) = x^2$ and $g(y) = y^3 + 3y^2 - 5$. We form a new function, denoted by $g \circ f$, by applying f to x and applying g to the result. So, for example,

$$(g \circ f)(0) = g(f(0)) = g(0) = -5$$
$$(g \circ f)(1) = g(f(1)) = g(1) = -1$$
$$(g \circ f)(2) = g(f(2)) = g(4) = 107$$
$$(g \circ f)(-3) = g(f(-3)) = g(9) = 967$$

and so on. This can be summarized in a more abbreviated notation as follows.

$$(g \circ f)(x) = g(f(x)) = g(x^2) = (x^2)^3 + 3(x^2)^2 - 5 = x^6 + 3x^4 - 5$$

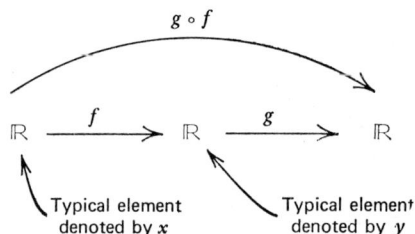

Figure 4-1

This is called "the composition of f by g" or "f followed by g." Notice that there may be another function we can get by composing f and g in the opposite order; the function $f \circ g$ is usually different from $g \circ f$. In this case

$$
\begin{aligned}
(f \circ g)(0) &= f(g(0)) &&= f(-5) = 25 \\
(f \circ g)(1) &= f(g(1)) &&= f(-1) = 1 \\
(f \circ g)(2) &= f(g(2)) &&= f(15) = 225 \\
(f \circ g)(-3) &= f(g(-3)) &&= f(-5) = 25
\end{aligned}
$$

and so on. The function $f \circ g$ can be described more succinctly by the equation

$$
(f \circ g)(y) = f(g(y)) = f(y^3 + 3y^2 - 5) = (y^3 + 3y^2 - 5)^2
$$
$$
= y^6 + 6y^5 + 9y^4 - 10y^3 - 30y^2 + 25.
$$

Since $f \circ g$ and $g \circ f$ are usually different functions, it is important to note the order in which the functions are applied. The formula for $g \circ f$ is $(g \circ f)(x) = g(f(x))$. For this to have meaning, $f(x)$ must be an element to which g can be applied; i.e., $f(x)$ must be an element of the domain of g.

Another point: we have defined f by the equation $f(x) = x^2$ and $g(y) = y^3 + 3y^2 - 5$. The letter x is understood to denote elements in the domain of f in this example, and the letter y is being used for elements in the domain of g. Of course, the domain of f is $\mathbb{R}$, and the domain of g is $\mathbb{R}$, so why use different symbols? The reason is that when we form $g \circ f$ and $f \circ g$, we must distinguish between the

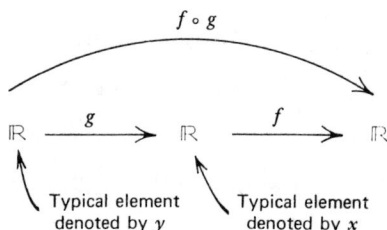

Figure 4-2

domain of f and the domain of g. We do so by using different letters to represent their elements. See Figures 4-1 and 4-2.

Example Let $f: \mathbb{R}^2 \to \mathbb{R}^2$ and $g: \mathbb{R}^2 \to \mathbb{R}$ be defined by $f(x,y) = (x^2 + y, 3x + 4y)$ and $g(x,y) = 3x^2y - xy^2 + y$. Then the function $g \circ f$ can be defined since the range of f is contained in $\mathbb{R}^2$, and $\mathbb{R}^2$ is the domain of g. However, $f \circ g$ cannot be formed since the range of g is contained in $\mathbb{R}$ whereas the domain of f is $\mathbb{R}^2$.

Definition 4-5

Suppose $\mathscr{U}, \mathscr{V}$, and $\mathscr{W}$ are vector spaces, and suppose $f: \mathscr{U} \to \mathscr{V}$ and $g: \mathscr{V} \to \mathscr{W}$ are functions. Then the function $g \circ f: \mathscr{U} \to \mathscr{W}$ is defined by the equation $(g \circ f)(\mathbf{X}) = g(f(\mathbf{X}))$ for each $\mathbf{X}$ in $\mathscr{U}$ (see Figure 4-3).

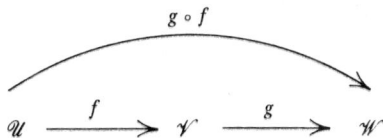

$$g \circ f$$

$$\mathscr{U} \xrightarrow{\ f\ } \mathscr{V} \xrightarrow{\ g\ } \mathscr{W}$$

Figure 4-3

Finally, we show that composition of functions is associative, i.e., $h \circ (g \circ f) = (h \circ g) \circ f$.

Example Suppose the functions f, g, and h are defined by the following formulas:
$f(x) = x^2 + 1$, $g(y) = y - 5$, and $h(z) = z^3 + 2z - 1$. Then

$$(g \circ f)(x) = g(f(x)) = g(x^2 + 1) = (x^2 + 1) - 5 = x^2 - 4$$
$$(h \circ g)(y) = h(g(y)) = h(y - 5) = (y - 5)^3 + 2(y - 5) - 1$$
$$= y^3 - 15y^2 + 75y - 125 + 2y - 10 - 1$$
$$= y^3 - 15y^2 + 77y - 136$$
$$[h \circ (g \circ f)](x) = h((g \circ f)(x)) = h(x^2 - 4)$$
$$= (x^2 - 4)^3 + 2(x^2 - 4) - 1$$
$$= x^6 - 12x^4 + 48x^2 - 64 + 2x^2 - 8 - 1$$
$$= x^6 - 12x^4 + 50x^2 - 73$$
$$[(h \circ g) \circ f](x) = (h \circ g)(x^2 + 1)$$
$$= (x^2 + 1)^3 - 15(x^2 + 1)^2 + 77(x^2 + 1) - 136$$
$$= x^6 + 3x^4 + 3x^2 + 1 - 15x^4 - 30x^2 - 15 + 77x^2 + 77 - 136$$
$$= x^6 - 12x^4 + 50x^2 - 73.$$

So $h \circ (g \circ f)$ and $(h \circ g) \circ f$ are the same function.

The general situation we have in mind is the composition of three functions f, g, and h. For the compositions $h \circ (g \circ f)$ or $(h \circ g) \circ f$ to have any meaning, we need the image of f contained in the domain of g and the image of g contained in the domain of h. So we make this hypothesis (see Figure 4-4).

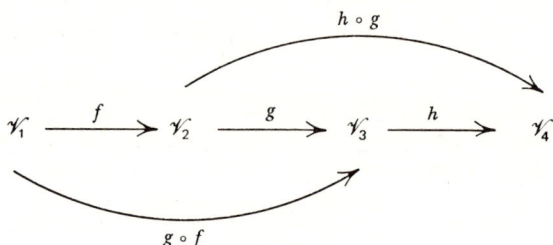

Figure 4-4

Proposition 4-1

If $f: \mathcal{V}_1 \to \mathcal{V}_2$, $g: \mathcal{V}_2 \to \mathcal{V}_3$, and $h: \mathcal{V}_3 \to \mathcal{V}_4$ are functions, then

$$h \circ (g \circ f) = (h \circ g) \circ f.$$

To see this we take an element of $\mathcal{V}_1$, call it $\mathbf{X}$. Then we observe that $[h \circ (g \circ f)](\mathbf{X}) = h(g(f(\mathbf{X})))$ and $[(h \circ g) \circ f](\mathbf{X}) = h(g(f(\mathbf{X})))$ as you easily see by applying Definition 4-5 a number of times.

Brief Exercises

1. In each of the following, compute the composite $g \circ f$.

(a) $f(x) = x^2 + 2x - 1$, $g(y) = 3y - 2$
(b) $f(x) = 3x - 2$, $g(y) = y^2 + 2y - 1$
(c) $f(x) = 4x^2 - 5$, $g(y) = 4y^2 - 5$
(d) $f(x) = (3x, 4x - 2)$, $g(u,v) = u^2 + 3uv^2$
(e) $f(x,y) = (x + 3y, 5x - 2y)$, $g(u,v) = (4u + 3v, 5u - v)$
(f) $f(x,y,z) = xy + 3z$, $g(u) = (u^2 + 1, u^2 - 1)$

2. Verify Proposition 4-1 in the special cases listed below, i.e., verify by calculation that $h \circ (g \circ f) = (h \circ g) \circ f$.

(a) $f(x) = 3x + 5$, $g(y) = 4y + 6$, $h(z) = 5z + 2$
(b) $f(x) = 4x + 6$, $g(y) = 3y + 5$, $h(z) = 5z + 2$
(c) $f(x,y) = 2xy + y^2$, $g(u) = (u + 2, 3u - 1)$, $h(v,w) = (2v - 3w, v + w)$

(d) $f(x_1,x_2,x_3) = (x_1 - x_2, 3x_1 + 2x_2 - x_3)$,
$g(y_1,y_2) = (4y_1 + 3y_2, 5y_1 + y_2, y_1 - y_2)$,
$h(z_1,z_2,z_3) = (2z_1 - z_2 + z_3, -z_3 + z_2 - z_1)$

Selected Answers: **1(a)** $3x^2 + 6x - 5$ **1(d)** $144x^3 - 135x^2 + 36x$ **1(f)** $(x^2y^2 + 6xyz + 9z^2 + 1, x^2y^2 + 6xyz + 9z^2 - 1)$

SUMMARY *A function from a vector space $\mathscr{V}$ to a vector space $\mathscr{W}$ is a rule of correspondence which assigns to each element of $\mathscr{V}$ one specific element of $\mathscr{W}$. The domain of f is $\mathscr{V}$, and the set of elements in $\mathscr{W}$ which are assigned to elements of $\mathscr{V}$ is called the range or image of f. This is symbolized by $f: \mathscr{V} \to \mathscr{W}$. If $f: \mathscr{V} \to \mathscr{W}$ and $g: \mathscr{V} \to \mathscr{W}$ are functions, then we add f and g and we multiply f by a scalar r to obtain new functions $(f + g): \mathscr{V} \to \mathscr{W}$ and $rf: \mathscr{V} \to \mathscr{W}$ by the defining equations*

$$(f + g)(\mathbf{X}) = f(\mathbf{X}) + g(\mathbf{X})$$

$$(rf)(\mathbf{X}) = r[f(\mathbf{X})]$$

for each $\mathbf{X}$ in $\mathscr{V}$. If $\mathscr{U}, \mathscr{V}$, and $\mathscr{W}$ are vector spaces and $f: \mathscr{U} \to \mathscr{V}$ and $g: \mathscr{V} \to \mathscr{W}$ are functions, then the composite of f followed by g is the function $g \circ f: \mathscr{U} \to \mathscr{W}$ defined by requiring that, for each $\mathbf{X}$ in $\mathscr{U}$,

$$(g \circ f)(\mathbf{X}) = g(f(\mathbf{X})).$$

Finally, the composition of functions is associative.

Exercises for Section 4-1

1. Suppose f, g, and h are the functions defined from $\mathbb{R}$ to $\mathbb{R}$ by the equations: $f(x) = x^2 + 2, g(x) = 3x + 5,$ and $h(x) = 4x^2 + x$. Find the following functions.

(a) $f + g$ **(d)** $f \circ h$

(b) $3f + 2g$ **(e)** $h \circ f$

(c) $5g - 2f + 3h$

2. Suppose f and g are functions from $\mathbb{R}$ to $\mathbb{R}^2$ defined by $f(x) = (x^2 + 2, 5x - 1)$ and $g(x) = (3x + 2, 5x^2 - 3)$. Suppose further that $h: \mathbb{R}^2 \to \mathbb{R}^2$ is defined by $h(x,y) = (x + y, x^2 - y)$. Find the following functions.

(a) $2f + 3g$ **(c)** $h \circ (2f + 4g)$

(b) $5f - 3g$ **(d)** $(3h) \circ (f - g)$

3. Give all the details of an argument for Proposition 4-1.

SECTION 4-2
Linear Transformations

PREVIEW   *Linear transformations between the n-tuple spaces are described; a linear transformation $T: \mathbb{R}^n \to \mathbb{R}^m$ has the defining equation*

$$T(x_1, \ldots, x_n) = (a_{11}x_1 + \cdots + a_{1n}x_n, \ldots, a_{m1}x_1 + \cdots + a_{mn}x_n).$$

This describes a linear function in terms of coordinates. If $\mathscr{V}$ and $\mathscr{W}$ are vector spaces, a function $T: \mathscr{V} \to \mathscr{W}$ is additive if $T(\mathbf{X} + \mathbf{Y}) = T(\mathbf{X}) + T(\mathbf{Y})$ for every $\mathbf{X}$ and $\mathbf{Y}$ in $\mathscr{V}$, and T is homogeneous if $T(r\mathbf{X}) = rT(\mathbf{X})$ for every $\mathbf{X}$ in $\mathscr{V}$ and r in $\mathbb{R}$. A function $T: \mathbb{R}^n \to \mathbb{R}^m$ is linear if and only if it is additive and homogeneous. This is a coordinate-free description of linear functions. Then we show that a linear function is determined by its values on a basis. Finally, if $\mathbf{X}_1, \ldots, \mathbf{X}_n$ form a basis for $\mathscr{V}$ and $\mathbf{Y}_1, \ldots, \mathbf{Y}_n$ are in $\mathscr{W}$, then there is exactly one linear transformation $T: \mathscr{V} \to \mathscr{W}$ having the property that $T(\mathbf{X}_1) = \mathbf{Y}_1, T(\mathbf{X}_2) = \mathbf{Y}_2, \ldots, T(\mathbf{X}_n) = \mathbf{Y}_n$.

Some of the simplest functions to deal with are linear functions. They are also extremely important in mathematics and in other areas of study.

In our treatment we use the letters T, S, and R for linear functions rather than f, g, and h; this is only a matter of personal preference. We also use the terms "linear transformation," "linear mapping," and "linear function" interchangeably.

A linear function is always a function between vector spaces. In this book we are interested almost exclusively in linear functions between the n-tuple spaces.

Example   Let the function $T: \mathbb{R} \to \mathbb{R}$ be defined by requiring that, for each x in $\mathbb{R}$, $T(x) = 3x$. This is a linear function.

Another linear function $S: \mathbb{R} \to \mathbb{R}$ is defined by the equation $S(x) = -5x$ for each x in $\mathbb{R}$.

In fact, the linear functions from $\mathbb{R}$ to $\mathbb{R}$ are precisely those functions $T: \mathbb{R} \to \mathbb{R}$ whose defining equation is of the type $T(x) = ax$, where a is some given real number. These are perhaps the simplest examples of linear functions.

Example   In this example we describe linear functions from $\mathbb{R}^2$ to $\mathbb{R}$. These are linear functions $T: \mathbb{R}^2 \to \mathbb{R}$ whose defining equation is of the type $T(x,y) = ax + by$, where a and b are real numbers.

Thus the equation $T(x,y) = 5x - 7y$ defines a linear function from $\mathbb{R}^2$ to $\mathbb{R}$, $T(x,y) = -6x + 2y$ defines another, and $T(x,y) = 5y = 0x + 5y$ defines another.

The equation $f(x,y) = 5x^2 + 3y$ defines a function from $\mathbb{R}^2$ to $\mathbb{R}$, but not a linear function, because $5x^2 + 3y$ does not have the form $ax + by$ no matter how a and b might be chosen.

Notice that the formula $ax + by$ is the same as the left-hand side of a linear equation in two unknowns.

Example In this example we describe linear functions from $\mathbb{R}^3$ to $\mathbb{R}$. These are linear functions $T: \mathbb{R}^3 \to \mathbb{R}$ whose defining equation is of the type $T(x,y,z) = ax + by + cz$, where a, b, and c are real numbers.

Thus the equation $T(x,y,z) = 3x - 4z = 3x + 0y - 4z$ defines a linear function from $\mathbb{R}^3$ to $\mathbb{R}$, the equation $T(x,y,z) = 4x - 2y + z$ defines another, and the equation $T(x,y,z) = 0 = 0x + 0y + 0z$ defines another.

These are the only linear functions from $\mathbb{R}^3$ to $\mathbb{R}$. Notice that the formula $ax + by + cz$ is the same as the left-hand side of a linear equation in three unknowns.

These examples give some insight as to how linear functions from $\mathbb{R}^n$ to $\mathbb{R}$ can be described.

Definition 4-6_____

A linear function $T: \mathbb{R}^n \to \mathbb{R}$ is one defined by an equation of the form

$$T(x_1, \ldots, x_n) = a_1 x_1 + \cdots + a_n x_n$$

where $a_1, \ldots, a_n$ are real numbers.

_____

This shows the close relation between linear functions and linear equations. The formula for a linear function from $\mathbb{R}^n$ to $\mathbb{R}$ is the same as the left-hand side of a linear equation in n unknowns (with the unknowns on the left-hand side). These functions are simple and easy to compute with.

Brief Exercises

1. Are the following functions from $\mathbb{R}$ to $\mathbb{R}$ linear?

(a) $T(x) = 3x$

(b) $T(x) = -4x$

(c) $T(x) = 0$

(d) $T(x) = x - 2$

(e) $T(x) = x^2$

(f) $T(x) = 4x^2 - 2x$

2. Are the following functions from $\mathbb{R}^2$ to $\mathbb{R}$ linear?

(a) $T(x,y) = 2x - y$

(b) $T(x,y) = 0$

(c) $T(x,y) = -2x + y$

(d) $T(x_1, x_2) = x_1 + x_2$

(e) $T(x_1, x_2) = 3x_1 - 2x_2$

(f) $T(x_1, x_2) = 3x_1 x_2$

3. Are the following functions from $\mathbb{R}^3$ to $\mathbb{R}$ linear?
   (a) $T(x,y,z) = 2x - y + z$ (c) $T(x_1,x_2,x_3) = 4x_1 + x_2 - 3x_2$
   (b) $T(x,y,z) = 5x - 7z + 8y$ (d) $T(x_1,x_2,x_3) = 2x_1^2 - x_3$

4. Are the following functions from $\mathbb{R}^4$ to $\mathbb{R}$ linear?
   (a) $T(x,y,z,w) = x - 2y + 3z - w$ (c) $T(x_1,x_2,x_3,x_4) = 0$
   (b) $T(x_1,x_2,x_3,x_4) = 4x_1 + 2x_3$ (d) $T(w_1,w_2,w_3,w_4) = 5w_1 - 2w_2 + w_4$

Answers: **1(a)** Yes **1(b)** Yes **1(c)** Yes **1(d)** No **1(e)** No **1(f)** No **2(a)** Yes **2(b)** Yes **2(c)** Yes **2(d)** Yes **2(e)** Yes **2(f)** No **3(a)** Yes **3(b)** Yes **3(c)** Yes **3(d)** No **4(a)** Yes **4(b)** Yes **4(c)** Yes **4(d)** Yes

So far we have described linear functions from $\mathbb{R}^n$ to $\mathbb{R}$. Now we describe linear functions from $\mathbb{R}^n$ to $\mathbb{R}^2$, to $\mathbb{R}^3$, etc.

Example   In this example we describe linear functions from $\mathbb{R}$ to $\mathbb{R}^2$. A linear function $T: \mathbb{R} \to \mathbb{R}^2$ has a defining equation of the form $T(x) = (ax, bx)$, where a and b are real numbers. Thus, for example, the equation $T(x) = (2x, 3x)$ defines a linear function from $\mathbb{R}$ to $\mathbb{R}^2$, the equation $T(x) = (-6x, 7x)$ defines another, whereas the equation $T(x) = (5x^2, -2x)$ does not define a linear function.

Notice that to describe a function from $\mathbb{R}$ into $\mathbb{R}^2$ we need two "coordinate functions." In this case they are given by the formulas ax and bx. This is the key; if each coordinate function is linear, then the function T is linear.

Example   In this example we describe linear functions from $\mathbb{R}^2$ to $\mathbb{R}^2$. Here we need two coordinate functions of two variables. A linear function $T: \mathbb{R}^2 \to \mathbb{R}^2$ has a defining equation of the form $T(x,y) = (ax + by, cx + dy)$, where a, b, c, and d are real numbers. Notice that the formulas $ax + by$ and $cx + dy$ in each coordinate are formulas which define linear functions from $\mathbb{R}^2$ to $\mathbb{R}$. Another way to view a linear function $T: \mathbb{R}^2 \to \mathbb{R}^2$ is to picture two linear equations in two unknowns.

$$5x + 3y = 6$$

$$x - y = 2$$

Use the left-hand sides as components and you have a linear function: $T(x,y) = (5x + 3y, x - y)$.

Some examples of linear functions from $\mathbb{R}^2$ to $\mathbb{R}^2$ are given by the following defining equations: $T(x,y) = (3x - y, 4x + 2y)$, $T(x,y) = (8x + 7y, 3x)$, and $T(x,y) = (4x + 2y, -x + 3y)$.

Some examples of functions from $\mathbb{R}^2$ to $\mathbb{R}^2$ which are not linear are given by $T(x,y) = (x^2y, x - 2)$ and $T(x,y) = (x + y - 5, x + y)$.

Example   In this example we describe linear functions from $\mathbb{R}^2$ to $\mathbb{R}^3$. In this case we need three coordinate functions, each having two variables. A linear function $T: \mathbb{R}^2 \to \mathbb{R}^3$ has a defining equation of the form $T(x,y) = (ax + by, cx + dy, px + qy)$, where a, b, c, d, p, and q are real numbers. Notice that each coordinate function is a linear function from $\mathbb{R}^2$ to $\mathbb{R}$; in this case we need three of them since the range is contained in $\mathbb{R}^3$.

Another way to view a linear function $T: \mathbb{R}^2 \to \mathbb{R}^3$ is to picture three linear equations in two unknowns.

$$x - y = 6$$

$$4x + 2y = 1$$

$$5x - 4y = 7$$

Use the left-hand sides as components and you have a linear function.

$$T(x,y) = (x - y, 4x + 2y, 5x - 4y)$$

Examples of linear functions $T: \mathbb{R}^2 \to \mathbb{R}^2$ are given by the defining equations $T(x,y) = (2x - y, 3y, 5x - 2y)$ and $T(x,y) = (x, 0, 4x + y)$.

These examples show how to construct linear functions $T: \mathbb{R}^n \to \mathbb{R}^m$. We need m component functions, each one a linear function of n variables. From Definition 4-6 each component function must look like $a_1 x_1 + \cdots + a_n x_n$. As with systems of linear equations, we distinguish between the formulas for different components by using two subscripts. The first component is denoted by $a_{11}x_1 + \cdots + a_{1n}x_n$, the second component by $a_{21}x_1 + \cdots + a_{2n}x_n$, and the mth component by $a_{m1}x_1 + \cdots + a_{mn}x_n$.

Definition 4-7

A linear function $T: \mathbb{R}^n \to \mathbb{R}^m$ has a defining equation of the form

$$T(x_1, \ldots, x_n) = (a_{11}x_1 + \cdots + a_{1n}x_n, \ldots, a_{m1}x_1 + \cdots + a_{mn}x_n).$$

This definition is perhaps more easily remembered if you picture m linear equations in n unknowns.

$$a_{11}x_1 + a_{12}x_2 + \cdots + a_{1n}x_n = b_1$$
$$a_{21}x_1 + a_{22}x_2 + \cdots + a_{2n}x_n = b_2$$
$$\vdots$$
$$a_{m1}x_1 + a_{m2}x_2 + \cdots + a_{mn}x_n = b_m$$

Use the left-hand side as the components of T: the first equation for the first component, the second equation for the second component, and so on. We call this linear function T the one which corresponds to the system of linear equations.

Brief Exercises

1. Are the following functions from $\mathbb{R}^3$ to $\mathbb{R}^2$ linear?

(a) $T(x,y,z) = (3x + 2y - z, 4x - y + 3z)$
(b) $T(x_1,x_2,x_3) = (4x_1 - x_3, 0)$
(c) $T(x_1,x_2,x_3) = (5x_1 + x_2 - x_3, x_1 + x_2^2 + 3x_3)$

2. Are the following functions from $\mathbb{R}^2$ to $\mathbb{R}^4$ linear?

(a) $T(x,y) = (x + y, x - y, 3x - 5y, x + 6y)$
(b) $T(x,y) = (3x - y, 0, 0, 5x + y^2)$
(c) $T(x,y) = (4x, y, 2y, 3x - y)$
(d) $T(x,y) = (4,0,2x,3y)$

3. Given an example of a linear function from $\mathbb{R}$ to $\mathbb{R}^4$, $\mathbb{R}$ to $\mathbb{R}^5$, $\mathbb{R}^2$ to $\mathbb{R}^3$, $\mathbb{R}^2$ to $\mathbb{R}^5$, $\mathbb{R}^4$ to $\mathbb{R}^3$, $\mathbb{R}^6$ to $\mathbb{R}^4$, and $\mathbb{R}^6$ to $\mathbb{R}^3$.

4. Find the linear function corresponding to the following systems of equations.

(a) $2x + 3y - z = 2$
    $x + y - z = 1$

(b) $4x - 3y + 2z = 1$
    $4x + y + z = 2$
    $x \quad\quad + z = 0$

Selected Answers: **1(a)** Yes **1(b)** Yes **1(c)** No **2(a)** Yes **2(b)** No **2(c)** Yes **2(d)** No **4(a)** $T(x,y,z) = (2x + 3y - z, x + y - z)$ **4(b)** $T(x,y,z) = (4x - 3y + 2z, 4x + y + z, x + z)$

The description we have given for linear functions from $\mathbb{R}^n$ to $\mathbb{R}^m$ depends essentially on the fact that the vectors in $\mathbb{R}^n$ and $\mathbb{R}^m$ have components. This description therefore relies on the form of the vectors in the n-tuple space. Sometimes this is useful, whereas at other times it is clumsy, precisely because there are too many components to keep track of. Fortunately, there is a description of linear functions which involves vector addition and multiplication by scalars, the characteristic operations in all vector spaces.

Example   Consider the linear function $T: \mathbb{R}^2 \to \mathbb{R}$ whose defining equation is $T(x_1,x_2) = 3x_1 + 4x_2$. Let $\mathbf{X} = (1,5)$ and $\mathbf{Y} = (-4,3)$. Then $T(\mathbf{X} + \mathbf{Y}) = T(-3,8) = -9 + 32 = 23$. Now $T(\mathbf{X}) = T(1,5) = 23$ and $T(\mathbf{Y}) = T(-4,3) = 0$. Thus $T(\mathbf{X} + \mathbf{Y}) = T(\mathbf{X}) + T(\mathbf{Y})$. Not only is this equation true for $\mathbf{X} = (1,5)$ and

$\mathbf{Y} = (-4,3)$, it is true for every $\mathbf{X}$ and $\mathbf{Y}$. To see this let $\mathbf{X} = (x_1, x_2)$ and $\mathbf{Y} = (y_1, y_2)$. Then

$$T(\mathbf{X} + \mathbf{Y}) = T(x_1 + y_1, x_2 + y_2) = 3(x_1 + y_1) + 4(x_2 + y_2)$$
$$= (3x_1 + 4x_2) + (3y_1 + 4y_2) = T(x_1, x_2) + T(y_1, y_2)$$
$$= T(\mathbf{X}) + T(\mathbf{Y}).$$

When the function T satisfies the equation $T(\mathbf{X} + \mathbf{Y}) = T(\mathbf{X}) + T(\mathbf{Y})$ for every choice of $\mathbf{X}$ and $\mathbf{Y}$, we say that T is *additive*.

Another thing to notice is that (letting $\mathbf{X} = (1,5)$ again) $T(5\mathbf{X}) = T(5,25) = 15 + 100 = 115$. Also $5\,T(\mathbf{X}) = 5\,T(1,5) = 5 \cdot 23 = 115$. Thus $T(5\mathbf{X}) = 5T(\mathbf{X})$. This equation also holds for any $\mathbf{X} = (x_1, x_2)$ and for any scalar r.

$$T(r\mathbf{X}) = T(rx_1, rx_2) = 3rx_1 + 4rx_2$$
$$= r(3x_1 + 4x_2) = rT(x_1, x_2)$$
$$= rT(\mathbf{X})$$

When the function T satisfies the equation $T(r\mathbf{X}) = rT(\mathbf{X})$ for every choice of $\mathbf{X}$ and r, we say that T is *homogeneous*.

Example In this example we show that *any* linear function $T: \mathbb{R}^2 \to \mathbb{R}$ must be additive and homogeneous. Such a function is defined by an equation of the form $T(x,y) = ax + by$, where a and b are some given real numbers. Let $\mathbf{X} = (x_1, x_2)$, $\mathbf{Y} = (y_1, y_2)$, and let r be a scalar. Then

$$T(\mathbf{X} + \mathbf{Y}) = T(x_1 + y_1, x_2 + y_2)$$
$$= a(x_1 + y_1) + b(x_2 + y_2)$$
$$= (ax_1 + bx_2) + (ay_1 + by_2)$$
$$= T(x_1, x_2) + T(y_1, y_2)$$
$$= T(\mathbf{X}) + T(\mathbf{Y}).$$

This shows T must be additive.

$$T(r\mathbf{X}) = T(rx_1, rx_2)$$
$$= arx_1 + brx_2$$
$$= r(ax_1 + bx_2)$$
$$= rT(x_1, x_2) = rT(\mathbf{X})$$

This shows T must be homogeneous. Consequently, as we claimed, any linear function $T: \mathbb{R}^2 \to \mathbb{R}$ must be additive and homogeneous.

Example   In this example we show that any function $T: \mathbb{R}^2 \to \mathbb{R}$ which is additive and homogeneous must be linear. This result, coupled with the result in the last example, shows that the linear functions from $\mathbb{R}^2$ to $\mathbb{R}$ and those which are both additive and homogeneous are the same functions. Here is the argument.

$$
\begin{aligned}
T(x_1, x_2) &= T(x_1(1,0) + x_2(0,1)) \\
&= T(x_1(1,0)) + T(x_2(0,1)) \\
&= x_1 T(1,0) + x_2 T(0,1)
\end{aligned}
$$

(The first equality is true because $(x_1, x_2) = x_1(1,0) + x_2(0,1)$; the second is true because T is additive; the third is true because T is homogeneous.) If we let $a = T(1,0)$ and $b = T(0,1)$, then we have the result that T is defined by the equation $T(x_1, x_2) = ax_1 + bx_2$, and so T is linear.

Definition 4-8_____

Let $\mathscr{V}$ and $\mathscr{W}$ be vector spaces over $\mathbb{R}$, and let $T: \mathscr{V} \to \mathscr{W}$ be a function from $\mathscr{V}$ to $\mathscr{W}$:

(1) T is *additive* if and only if $T(\mathbf{X} + \mathbf{Y}) = T(\mathbf{X}) + T(\mathbf{Y})$ for every choice of $\mathbf{X}$ and $\mathbf{Y}$ in $\mathscr{V}$, and
(2) T is *homogeneous* if and only if $T(r\mathbf{X}) = rT(\mathbf{X})$ for every choice of $\mathbf{X}$ in $\mathscr{V}$ and every scalar r.

_____

In the last three examples we have discussed particular cases of an important general fact; we record it in the next proposition.

Proposition 4-2_____

Let $T: \mathbb{R}^n \to \mathbb{R}^m$ be a function. Then T is linear if and only if it is additive and homogeneous.

_____

The point of Proposition 4-2 is that we can think of linear functions from $\mathbb{R}^n$ to $\mathbb{R}^m$ either in terms of their coordinate description given in Definition 4-7, or we can think of them as the additive and homogeneous functions. However, the description in Definition 4-7 applies only to functions involving n-tuples. The description of linear functions as additive and homogeneous does not involve coordinates—it is a coordinate-free description—and therefore it can be used in situations where the coordinate description cannot be used. In that sense, the coordinate-free description provides a more general description of linear functions.

Definition 4-9

(Alternate definition of linearity) Suppose $\mathscr{V}$ and $\mathscr{W}$ are vector spaces over $\mathbb{R}$, and suppose T is a function from $\mathscr{V}$ into $\mathscr{W}$. T is linear if and only if it is additive and homogeneous.

Brief Exercises

In Exercises 1 through 4 verify that the linear function is additive and homogeneous

1. $T: \mathbb{R} \to \mathbb{R}$

   (a) $T(x) = 2x$ (b) $T(x) = -4x$ (c) $T(x) = ax$

2. $T: \mathbb{R} \to \mathbb{R}^2$

   (a) $T(x) = (3x, 2x)$ (b) $T(x) = (-5x, 7x)$ (c) $T(x) = (ax, bx)$

3. $T: \mathbb{R}^2 \to \mathbb{R}^3$

   (a) $T(x, y) = (2x - y, 3x + 4y, 5x + y)$
   (b) $T(x, y) = (x - 3y, x + 2y, 6x - 7y)$
   (c) $T(x, y) = (ax + by, cx + dy, px + qy)$

4. $T: \mathbb{R}^3 \to \mathbb{R}^2$

   (a) $T(x, y, z) = (x - y + z, 2x + 3y + 4z)$
   (b) $T(x, y, z) = (6x + 2y - z, 4x + 3y - 5z)$
   (c) $T(x, y, z) = (ax + by + cz, px + qy - rz)$

5. Show that a function $T: \mathbb{R} \to \mathbb{R}^2$ which is additive and homogeneous must be linear.

6. Show that a function $T: \mathbb{R} \to \mathbb{R}^3$ which is additive and homogeneous must be linear.

7. Show that a function $T: \mathbb{R}^2 \to \mathbb{R}^2$ which is additive and homogeneous must be linear.

Using the coordinate-free description we show how to apply a linear transformation to a linear combination of vectors. Suppose $\mathscr{V}$ and $\mathscr{W}$ are vector spaces and $T: \mathscr{V} \to \mathscr{W}$ is a linear transformation. Suppose $\mathbf{X}_1$, $\mathbf{X}_2$, $\mathbf{X}_3$, and $\mathbf{X}_4$ are vectors in $\mathscr{V}$. Applying T to the linear combination $\mathbf{X} = 5\mathbf{X}_1 + 7\mathbf{X}_2$, for example, we obtain

$$T(\mathbf{X}) = T(5\mathbf{X}_1 + 7\mathbf{X}_2) = T(5\mathbf{X}_1) + T(7\mathbf{X}_1) \qquad (T \text{ is additive})$$

$$= 5T(\mathbf{X}_1) + 7T(\mathbf{X}_2). \qquad (T \text{ is homogeneous})$$

Applying T to $\mathbf{X} = 4\mathbf{X}_1 - 3\mathbf{X}_2 + \mathbf{X}_3 + 6\mathbf{X}_4$ we obtain

$$T(\mathbf{X}) = T(4\mathbf{X}_1 - 3\mathbf{X}_2 + \mathbf{X}_3 + 6\mathbf{X}_4)$$

$$= T(4\mathbf{X}_1) + T(-3\mathbf{X}_2) + T(\mathbf{X}_3) + T(6\mathbf{X}_4) \qquad (T \text{ is additive})$$

$$= 4T(\mathbf{X}_1) - 3T(\mathbf{X}_2) + T(\mathbf{X}_3) + 6T(\mathbf{X}_4). \qquad (T \text{ is homogeneous})$$

The same thing happens with any linear combination. If $\mathbf{X}_1, \ldots, \mathbf{X}_n$ are in $\mathscr{V}$ and $\mathbf{X} = a_1\mathbf{X}_1 + \cdots + a_n\mathbf{X}_n$, then

$$T(\mathbf{X}) = T(a_1\mathbf{X}_1 + \cdots + a_n\mathbf{X}_n)$$

$$= T(a_1\mathbf{X}_1) + \cdots + T(a_n\mathbf{X}_n) \qquad (T \text{ is additive})$$

$$= a_1T(\mathbf{X}_1) + \cdots + a_nT(\mathbf{X}_n). \qquad (T \text{ is homogeneous})$$

Proposition 4-3

Suppose $\mathscr{V}$ and $\mathscr{W}$ are vector spaces over $\mathbb{R}$ and $T: \mathscr{V} \to \mathscr{W}$ is a linear transformation. If $\mathbf{X}_1, \ldots, \mathbf{X}_n$ are vectors in $\mathscr{V}$, then

$$T(a_1\mathbf{X}_1 + \cdots + a_n\mathbf{X}_n) = a_1T(\mathbf{X}_1) + \cdots + a_nT(\mathbf{X}_n).$$

The importance of Proposition 4-3 is this: if we know $T(\mathbf{X}_1), \ldots, T(\mathbf{X}_n)$, then we can calculate the value of T at any linear combination of $\mathbf{X}_1, \ldots, \mathbf{X}_n$. In particular, if $\mathbf{X}_1, \ldots, \mathbf{X}_n$ form a basis for $\mathscr{V}$, then we can calculate the value of T at any vector $\mathbf{X}$ in $\mathscr{V}$. To do this, express $\mathbf{X}$ as a linear combination of the basis vectors: $\mathbf{X} = a_1\mathbf{X}_1 + \cdots + a_n\mathbf{X}_n$. Then $T(\mathbf{X}) = a_1T(\mathbf{X}_1) + \cdots + a_nT(\mathbf{X}_n)$. This has the consequence that if S and T are linear transformations from $\mathscr{V}$ to $\mathscr{W}$, and if $S(\mathbf{X}_1) = T(\mathbf{X}_1), \ldots, S(\mathbf{X}_n) = T(\mathbf{X}_n)$, then $S = T$, because

$$S(\mathbf{X}) = a_1S(\mathbf{X}_1) + \cdots + a_nS(\mathbf{X}_n)$$

$$= a_1T(\mathbf{X}_1) + \cdots + a_nT(\mathbf{X}_n) = T(\mathbf{X}).$$

Proposition 4-4

Suppose $\mathscr{V}$ and $\mathscr{W}$ are vector spaces over $\mathbb{R}$ and $\mathbf{X}_1, \ldots, \mathbf{X}_n$ form a basis for $\mathscr{V}$. If $S: \mathscr{V} \to \mathscr{W}$ and $T: \mathscr{V} \to \mathscr{W}$ are linear transformations having equal values at each vector in the basis, i.e., $S(\mathbf{X}_1) = T(\mathbf{X}_1), \ldots, S(\mathbf{X}_n) = T(\mathbf{X}_n)$, then $S = T$.

The last point we discuss is this: suppose we want to construct a linear function T from a vector space $\mathscr{V}$ to a vector space $\mathscr{W}$, and suppose that $\mathbf{X}_1, \ldots, \mathbf{X}_n$ form a basis for $\mathscr{V}$. If the vectors $\mathbf{Y}_1, \ldots, \mathbf{Y}_n$ are chosen from $\mathscr{W}$, not necessarily different vectors (i.e., we allow $\mathbf{Y}_1 = \mathbf{Y}_2$, for example), then there is always exactly one linear transformation $T: \mathscr{V} \to \mathscr{W}$ which has the property that $T(\mathbf{X}_1) = \mathbf{Y}_1, \ldots, T(\mathbf{X}_n) = \mathbf{Y}_n$.

Example In this example we consider linear transformations from $\mathbb{R}^3$ to $\mathbb{R}^2$. Suppose that $(1,0,0)$, $(0,1,0)$, and $(0,0,1)$ form the given basis in $\mathbb{R}^3$. Then a linear transformation can be found which sends these basis vectors into any given vectors in $\mathbb{R}^2$.

Suppose we want a linear transformation which sends $(1,0,0)$ into $(2,3)$, sends $(0,1,0)$ into $(1,5)$, and sends $(0,0,1)$ into $(-1,-3)$. This determines the function T, for if (x,y,z) is in $\mathbb{R}^3$, then

$$T(x,y,z) = T(x(1,0,0) + y(0,1,0) + z(0,0,1))$$

$$= xT(1,0,0) + yT(0,1,0) + zT(0,0,1)$$

$$= x(2,3) + y(1,5) + z(-1,-3)$$

$$= (2x + y - z,\ 3x + 5y - 3z).$$

Any values for $T(1,0,0)$, $T(0,1,0)$, and $T(0,0,1)$ could have been chosen; different values give different linear transformations, of course. If, for example, we had wanted $S(1,0,0) = S(0,1,0) = S(0,0,1) = (5,2)$, then the linear transformation S would be defined by

$$S(x,y,z) = x(5,2) + y(5,2) + z(5,2)$$

$$= (5x + 5y + 5z,\ 2x + 2y + 2z).$$

More generally, suppose a linear transformation $R\colon \mathbb{R}^3 \to \mathbb{R}^2$ sends $(1,0,0)$ into (a_1,b_1), sends $(0,1,0)$ into (a_2,b_2), and sends $(0,0,1)$ into (a_3,b_3). The formula for $R(x,y,z)$ is

$$R(x,y,z) = R(x(1,0,0) + y(0,1,0) + z(0,0,1))$$

$$= xR(1,0,0) + yR(0,1,0) + zR(0,0,1)$$

$$= x(a_1,b_1) + y(a_2,b_2) + z(a_3,b_3)$$

$$= (a_1 x + a_2 y + a_3 z,\ b_1 x + b_2 y + b_3 z).$$

This shows that knowing the values of R at $(1,0,0)$, $(0,1,0)$, and $(0,0,1)$ allows us to compute the value of R at any vector (x,y,z).

Example In this example we consider linear transformations from $\mathbb{R}^3$ into $\mathbb{R}^3$. Suppose that $(1,2,0)$, $(1,0,1)$, and $(0,1,1)$ form the given basis in $\mathbb{R}^3$. Suppose we want a linear transformation $T\colon \mathbb{R}^3 \to \mathbb{R}^3$ with the properties that $T(1,2,0) = (4,0,2)$, $T(1,0,1) = (-1,1,2)$, and $T(0,1,1) = (0,-3,1)$. To determine the value of T at a vector $\mathbf{X}$ in $\mathbb{R}^3$, suppose $\mathbf{X} = (x,y,z)$. Write $\mathbf{X}$ as a linear combination of the basis vectors.

$$\mathbf{X} = (x,y,z) = \tfrac{1}{3}(x + y - z)(1,2,0) + \tfrac{1}{3}(2x - y + z)(1,0,1)$$

$$+ \tfrac{1}{3}(-2x + y + 2z)(0,1,1)$$

Then T can be evaluated at $\mathbf{X}$ by

$$
\begin{aligned}
T(\mathbf{X}) &= \tfrac{1}{3}(x + y - z)T(1,2,0) + \tfrac{1}{3}(2x - y + z)T(1,0,1) \\
&\quad + \tfrac{1}{3}(-2x + y + 2z)T(0,1,1) \\
&= \tfrac{1}{3}(x + y - z)(4,0,2) + \tfrac{1}{3}(2x - y + z)(-1,1,2) \\
&\quad + \tfrac{1}{3}(-2x + y + 2z)(0,-3,1) \\
&= (\tfrac{1}{3}(2x + 5y - 5z), \tfrac{1}{3}(8x - 4y - 5z), \tfrac{1}{3}(4x + y + 2z)).
\end{aligned}
$$

If different values had been chosen for $T(1,2,0)$, $T(1,0,1)$, and $T(0,1,1)$, then a different function T would have been determined.

These examples show how linear transformations can be constructed by specifying values on a basis. The same procedure works in general: if $\mathbf{X}_1, \ldots, \mathbf{X}_n$ form a basis for $\mathscr{V}$ and $\mathbf{Y}_1, \ldots, \mathbf{Y}_n$ are in $\mathscr{W}$, then we construct a linear function $T: \mathscr{V} \to \mathscr{W}$ so that $T(\mathbf{X}_1) = \mathbf{Y}_1, \ldots, T(\mathbf{X}_n) = \mathbf{Y}_n$. If $\mathbf{X}$ is in $\mathscr{V}$, express $\mathbf{X}$ as a linear combination of the basis vectors: $\mathbf{X} = a_1\mathbf{X}_1 + \cdots + a_n\mathbf{X}_n$. Define $T(\mathbf{X})$ to be the vector $a_1\mathbf{Y}_1 + \cdots + a_n\mathbf{Y}_n$. There is only one possible value of $T(\mathbf{X})$, since the scalars $a_1, \ldots, a_n$ are uniquely determined (see Proposition 3-6). Therefore we have defined a function. To show that T is linear we show that it is additive and homogeneous. Suppose we take two vectors $\mathbf{X}'$ and $\mathbf{X}''$ in $\mathscr{V}$. Express them as linear combinations of the basis vectors, say

$$
\begin{aligned}
\mathbf{X}' &= a_1'\mathbf{X}_1 + \cdots + a_n'\mathbf{X}_n \\
\mathbf{X}'' &= a_1''\mathbf{X}_1 + \cdots + a_n''\mathbf{X}_n.
\end{aligned}
$$

Then

$$
\mathbf{X}' + \mathbf{X}'' = (a_1' + a_1'')\mathbf{X}_1 + \cdots + (a_n' + a_n'')\mathbf{X}_n.
$$

Our definition of T shows that

$$
\begin{aligned}
T(\mathbf{X}') &= a_1'T(\mathbf{X}_1) + \cdots + a_n'T(\mathbf{X}_n) \\
T(\mathbf{X}'') &= a_1''T(\mathbf{X}_1) + \cdots + a_n''T(\mathbf{X}_n) \\
T(\mathbf{X}' + \mathbf{X}'') &= (a_1' + a_1'')\mathbf{X}_1 + \cdots + (a_n' + a_n'')\mathbf{X}_n.
\end{aligned}
$$

These equations show that $T(\mathbf{X}' + \mathbf{X}'') = T(\mathbf{X}') + T(\mathbf{X}'')$. Furthermore,

$$
r\mathbf{X}' = ra_1'\mathbf{X}_1 + \cdots + ra_n'\mathbf{X}_n
$$

and

$$
\begin{aligned}
T(r\mathbf{X}') &= ra_1'T(\mathbf{X}_1) + \cdots + ra_n'T(\mathbf{X}_n) \\
&= r(a_1'T(\mathbf{X}_1) + \cdots + a_n'T(\mathbf{X}_n)).
\end{aligned}
$$

Consequently $T(r\mathbf{X}') = rT(\mathbf{X}')$. This verifies that T is linear.

Proposition 4-5

Suppose $\mathcal{V}$ and $\mathcal{W}$ are vector spaces over $\mathbb{R}$, $\mathbf{X}_1, \ldots, \mathbf{X}_n$ form a basis for $\mathcal{V}$, $\mathbf{Y}_1, \ldots, \mathbf{Y}_n$ are vectors in $\mathcal{W}$. Then there is a unique linear transformation $T \colon \mathcal{V} \to \mathcal{W}$ which has the property that $T(\mathbf{X}_1) = \mathbf{Y}_1, \ldots, T(\mathbf{X}_n) = \mathbf{Y}_n$.

The remarks preceding the proposition show how to construct T, and Proposition 4-4 shows that there is only one such function.

Brief Exercises

Find the defining equations for the linear transformations specified by the following information.

1. $T \colon \mathbb{R} \to \mathbb{R}$, $T(2) = 3$
2. $T \colon \mathbb{R}^2 \to \mathbb{R}$, $T(1,0) = 1$, $T(0,1) = 4$
3. $T \colon \mathbb{R}^2 \to \mathbb{R}^2$, $T(1,0) = (5,2)$, $T(0,1) = (3,4)$
4. $T \colon \mathbb{R}^2 \to \mathbb{R}^2$, $T(1,1) = (6,-2)$, $T(-1,2) = (3,1)$
5. $T \colon \mathbb{R}^3 \to \mathbb{R}^2$, $T(1,0,1) = (1,5)$, $T(1,1,0) = (1,2)$, $T(0,1,0) = (2,3)$

Answers: **1** $T(x) = \frac{3}{2}x$ **2** $T(x,y) = x + 4y$ **3** $T(x,y) = (5x + 3y, 2x + 4y)$ **4** $T(x,y) = (3x + 3y, \frac{1}{3}(-5x - y))$ **5** $T(x,y,z) = (-x + 2y + 2z, -x + 3y + 6z)$

SUMMARY *The linear transformations from $\mathbb{R}^n$ to $\mathbb{R}^m$ are described in terms of the form of their component functions:*

$$T(x_1, \ldots, x_n) = (a_{11}x_1 + \cdots + a_{1n}x_n, \ldots, a_{m1}x_1 + \cdots + a_{mn}x_n).$$

The form of the components is the same as the left-hand sides of linear equations in n unknowns. Consequently, there is a correspondence between a system of m linear equations in n unknowns and a linear transformation from $\mathbb{R}^n$ to $\mathbb{R}^m$.

    Then a coordinate-free description of linear transformations is given. A function $T \colon \mathcal{V} \to \mathcal{W}$ from one vector space $\mathcal{V}$ to another $\mathcal{W}$ is additive if

$$T(\mathbf{X} + \mathbf{Y}) = T(\mathbf{X}) + T(\mathbf{Y})$$

and it is homogeneous if

$$T(r\mathbf{X}) = rT(\mathbf{X})$$

(these must be true for any choice of $\mathbf{X}$ and $\mathbf{Y}$ in $\mathcal{V}$ and any scalar r). With this terminology a linear function can be described as one which is additive and homogeneous. Since this description depends only on vector addition and scalar multiplication, it is the one we use when coordinates are either not present or not

convenient. With this coordinate-free description of a linear tranformation we establish that a linear transformation is determined by its values on a basis for its domain. Then we show that if $\mathbf{X}_1, \ldots, \mathbf{X}_n$ *form a basis for* $\mathscr{V}$ *and* $\mathbf{Y}_1, \ldots, \mathbf{Y}_n$ *are in* $\mathscr{W}$ *, then there is a unique linear transformation* $T: \mathscr{V} \to \mathscr{W}$ *having* $T(\mathbf{X}_1) = \mathbf{Y}_1, \ldots, T(\mathbf{X}_n) = \mathbf{Y}_n.$

Exercises for Section 4-2

1. Determine whether the following functions are linear functions.

   (a) $f: \mathbb{R}^2 \to \mathbb{R}^2$
      (i) $f(x,y) = (x^2 - y^2, x^2 + 3xy - 4y^2)$
      (ii) $f(x,y) = (x - 2y, \sin xy)$
      (iii) $f(x,y) = (x + 2y, x - 3y)$
   (b) $g: \mathbb{R}^2 \to \mathbb{R}^3$
      (i) $g(x,y) = (2x + 5y, x - y, 3x^2)$
      (ii) $g(x,y) = (2xy, 0, 0)$
      (iii) $g(x,y) = (4x + 17y, x^3 - 3y, 8y + 2x)$
   (c) $h: \mathbb{R}^3 \to \mathbb{R}^2$
      (i) $h(x,y,z) = (x - y, 3\sqrt{2}x + 5\sqrt{2}y)$
      (ii) $h(x,y,z) = (x + 2y, (-17\sqrt{2}/5)x + (34/19)y)$

2. (a) A linear transformation T from $\mathbb{R}^2$ to $\mathbb{R}^3$ has the property that $T(1,1) = (2,1,3)$ and $T(-1,1) = (1,5,4)$. Find a formula for $T(x,y)$.
   (b) A linear transformation T from $\mathbb{R}^2$ to $\mathbb{R}^2$ has the property that $T(2,1) = (1,2)$ and $T(3,4) = (-1,0)$. Find a formula for $T(x,y)$.
   (c) A linear transformation from $\mathbb{R}^2$ to $\mathbb{R}^3$ has the property that $T(-1,2) = (1,2,0)$ and $T(1,-3) = (1,2,0)$. Find a formula for $T(x,y)$.
   (d) A linear transformation T from $\mathbb{R}^3$ to $\mathbb{R}^2$ has the property that $T(1,1,0) = (2,5)$, $T(1,0,1) = (4,3)$, and $T(1,0,0) = (1,3)$. Find a formula for $T(x,y,z)$.
   (e) A linear transformation T from $\mathbb{R}^2$ to $\mathbb{R}^3$ has the property that $T(1,2) = (2,5,7)$ and $T(15,30) = (30,75,105)$. Explain why there is not enough information to determine a formula for $T(x,y)$.

3. (a) Let $T: \mathbb{R}^2 \to \mathbb{R}^3$ be the linear transformation defined by $T(x,y) = (x + y, x - y, 2x + 3y)$. The points on the line $\mathscr{L}$ in $\mathbb{R}^2$ passing through $(1,2)$ and parallel to $(3,4)$ satisfy the equation $(x,y) = (1,2) + t(3,4)$. Find an equation for $T(\mathscr{L})$ obtained by applying T to every $\mathbf{X}$ in $\mathscr{L}$. How do you describe $T(\mathscr{L})$ geometrically?

(b) If $T: \mathbb{R}^3 \to \mathbb{R}^2$ is defined by $T(x,y,z) = (x + y - z, x + z)$, find an equation satisfied by the points of $T(\mathscr{L})$ if $\mathscr{L}$ is the set of points on the line in $\mathbb{R}^3$ passing through $(2,1,3)$ and parallel to $(4,-5,2)$.

(c) If $T: \mathbb{R}^2 \to \mathbb{R}^2$ is defined by $T(x,y) = (0, 2x + 3y)$, find a geometric description of $T(\mathscr{L})$, where $\mathscr{L}$ is the line passing through $(1,1)$ and parallel to $(3,-2)$.

(d) If $T: \mathbb{R}^2 \to \mathbb{R}^2$ is defined by $T(x,y) = (3x + 4y, x - y)$, find a description of the geometric configuration obtained by applying T to every point on

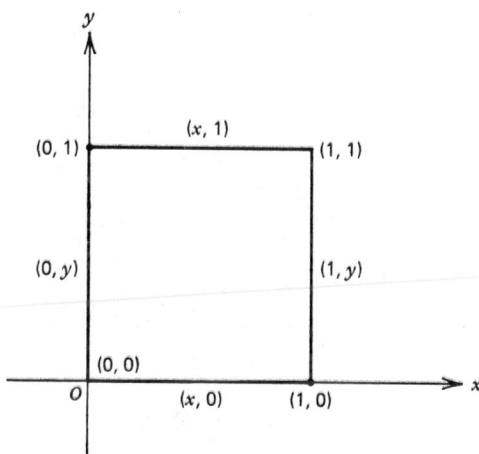

Figure 4-5

the square in Figure 4-5. (Hint: consider the four sides. The points can be described as follows.

$$(x,0), \quad 0 \le x \le 1 \qquad (x,1), \quad 0 \le x \le 1$$
$$(1,y), \quad 0 \le y \le 1 \qquad (0,y), \quad 0 \le y \le 1$$

Apply T to each edge.)

4. Is there a linear transformation of $\mathbb{R}^2$ into $\mathbb{R}^2$ which sends the square $\mathscr{S}_1$ into the quadrilateral $\mathscr{S}_2$ shown in Figure 4-6 (sending P to P', Q to Q', R to R', and S to S')?

5. Does a linear transformation $T: \mathbb{R}^2 \to \mathbb{R}^2$ always send parallel lines into parallel lines?

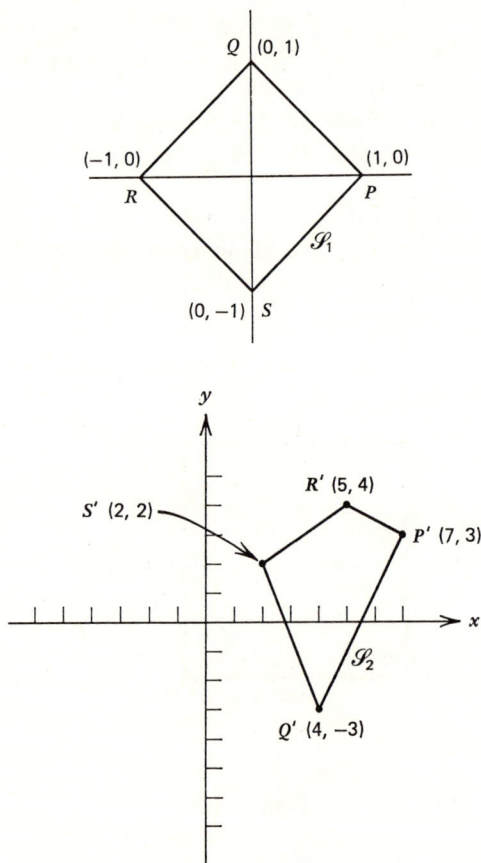

Figure 4-6

SECTION 4-3
Matrix Representation of Linear Transformations from $\mathbb{R}^n$ to $\mathbb{R}^m$

PREVIEW *In this section we show how a linear transformation $T: \mathbb{R}^n \to \mathbb{R}^m$ is represented by an $m \times n$ matrix and how, conversely, every $m \times n$ matrix represents a linear transformation $T: \mathbb{R}^n \to \mathbb{R}^m$.*

In Section 4-2 we introduced linear transformations from $\mathbb{R}^n$ to $\mathbb{R}^m$. We gave a coordinate description and a coordinate-free description. The coordinate description is the one we usually work with when we make numerical computations, whereas the coordinate-free description is usually more helpful when we are discussing linear transformations in general. As we mentioned in the last section, the coordinate description

$$T(x_1, \ldots, x_n) = (a_{11}x_{11} + \cdots + a_{1n}x_n, \ldots, a_{m1}x_1 + \cdots + a_{mn}x_n)$$

can be somewhat clumsy. In this section we show that we can use matrices as a tidier way of writing the coordinate description of a linear transformation.

We begin with some examples.

Example Consider the linear transformation $T: \mathbb{R}^2 \to \mathbb{R}^2$ defined by the equation $T(x,y) = (3x - y, 4x + 2y)$. As we mentioned in Section 4-2, the coordinates for T are the same as the left-hand side of a system of two equations in two unknowns.

$$3x - y = b$$

$$4x + 2y = c$$

The 2×2 matrix we use to represent T is the coefficient matrix of this system of equations.

$$\begin{pmatrix} 3 & -1 \\ 4 & 2 \end{pmatrix}$$

To compute $T(5,-6)$, for example, write $(5,-6)$ vertically to the right of the matrix as shown.

$$\begin{pmatrix} 3 & -1 \\ 4 & 2 \end{pmatrix} \begin{pmatrix} 5 \\ -6 \end{pmatrix}$$

To compute the first component of $T(5,-6)$ use the first row of the matrix and the coordinates of $(5,-6)$. Add the products of corresponding entries

$$\begin{pmatrix} 3 & -1 \\ 4 & 2 \end{pmatrix} \begin{pmatrix} 5 \\ -6 \end{pmatrix} = \begin{pmatrix} 3 \cdot 5 + (-1)(-6) \\ 4 \cdot 5 + 2(-6) \end{pmatrix} = \begin{pmatrix} 21 \\ 8 \end{pmatrix}.$$

To compute the second component use the second row of the matrix in a similar fashion. You notice that when we use this device, the vector $T(5,-6)$ comes out written vertically rather than horizontaly. We call this applying the matrix to $(5,-6)$.

For another example we compute $T(4,-3)$ by applying the matrix to $(4,-3)$.

$$\begin{pmatrix} 3 & -1 \\ 4 & 2 \end{pmatrix} \begin{pmatrix} 4 \\ -3 \end{pmatrix} = \begin{pmatrix} 3 \cdot 4 + (-1)(-3) \\ 4 \cdot 4 + 2(-3) \end{pmatrix} = \begin{pmatrix} 15 \\ 10 \end{pmatrix}$$

In general we compute $T(x,y)$ as follows.

$$\begin{pmatrix} 3 & -1 \\ 4 & 2 \end{pmatrix} \begin{pmatrix} x \\ y \end{pmatrix} = \begin{pmatrix} 3x - y \\ 4x + 2y \end{pmatrix}$$

Another way to obtain the matrix for T is this. Use $T(1,0)$ as the first column of the matrix and $T(0,1)$ as the second column. Since $T(1,0) = (3,4)$ and $T(0,1) = (-1,2)$, we get the same matrix using this method.

Example Consider the linear transformation $T: \mathbb{R}^2 \to \mathbb{R}^3$ defined by $T(x,y) = (5x + 6y, 7x - y, 8x - 3y)$. Another way of finding the matrix for T is to fill in the matrix to produce the equality

$$\begin{pmatrix} & & \\ & ? & \\ & & \end{pmatrix} \begin{pmatrix} x \\ y \end{pmatrix} = \begin{pmatrix} 5x + 6y \\ 7x - y \\ 8x - 3y \end{pmatrix}.$$

It is clear that we must have

$$\begin{pmatrix} 5 & 6 \\ 7 & -1 \\ 8 & -3 \end{pmatrix} \begin{pmatrix} x \\ y \end{pmatrix} = \begin{pmatrix} 5x + 6y \\ 7x - y \\ 8x - 3y \end{pmatrix}.$$

For example, we compute $T(5,8)$ and $T(-3,6)$ by applying the matrix to $(5,8)$ and $(-3,6)$.

$$\begin{pmatrix} 5 & 6 \\ 7 & -1 \\ 8 & -3 \end{pmatrix} \begin{pmatrix} 5 \\ 8 \end{pmatrix} = \begin{pmatrix} 5 \cdot 5 + 6 \cdot 8 \\ 7 \cdot 5 + (-1)8 \\ 8 \cdot 5 + (-3)8 \end{pmatrix} = \begin{pmatrix} 73 \\ 27 \\ 16 \end{pmatrix}$$

$$\begin{pmatrix} 5 & 6 \\ 7 & -1 \\ 8 & -3 \end{pmatrix} \begin{pmatrix} -3 \\ 6 \end{pmatrix} = \begin{pmatrix} 5(-3) + 6 \cdot 6 \\ 7(-3) + (-1)6 \\ 8(-3) + (-3)6 \end{pmatrix} = \begin{pmatrix} 21 \\ -27 \\ -42 \end{pmatrix}$$

Another way to obtain the matrix for T is to use $T(1,0) = (5,7,8)$ as the first column and $T(0,1) = (6,-1,-3)$ as the second column of the matrix.

Example Consider the linear transformation $T: \mathbb{R}^4 \to \mathbb{R}^3$ determined by the equation

$$T(x_1,x_2,x_3,\ x_4) = (2x_1 - 3x_3 + x_4,\ 5x_1 + x_2 - 2x_3 + 3x_4,\ x_1 + x_2 - x_4).$$

The matrix for T is the 3×4 matrix

$$\begin{pmatrix} 2 & 0 & -3 & 1 \\ 5 & 1 & -2 & 3 \\ 1 & 1 & 0 & -1 \end{pmatrix}.$$

We compute $T(-1,3,4,2)$ by applying the matrix to $(-1,3,4,2)$.

$$\begin{pmatrix} 2 & 0 & -3 & 1 \\ 5 & 1 & -2 & 3 \\ 1 & 1 & 0 & -1 \end{pmatrix} \begin{pmatrix} -1 \\ 3 \\ 4 \\ 2 \end{pmatrix} = \begin{pmatrix} 2(-1) + 0\cdot 3 + (-3)4 + 1\cdot 2 \\ 5(-1) + 1\cdot 3 + (-2)4 + 3\cdot 2 \\ 1(-1) + 1\cdot 3 + 0\cdot 4 + (-1)2 \end{pmatrix}$$

$$= \begin{pmatrix} -12 \\ -4 \\ 0 \end{pmatrix}$$

Thus $T(-1,3,4,2) = (-12,-4,0)$.

As in the previous examples, we can obtain the matrix by using $T(1,0,0,0) = (2,5,1)$ as the first column, $T(0,1,0,0) = (0,1,1)$ as the second, $T(0,0,1,0) = (-3,-2,0)$ as the third, and $T(0,0,0,1) = (1,3,-1)$ as the fourth.

These examples show how to determine the matrix for linear transformation $T: \mathbb{R}^n \to \mathbb{R}^m$. If T is given by the equation

$$T(x_1, \ldots, x_n) = (a_{11}x_1 + \cdots + a_{1n}x_1, \ldots, a_{m1}x_1 + \cdots + a_{mn}x_n)$$

then we can compute with T by using the $m \times n$ matrix

$$\begin{pmatrix} a_{11} & a_{12} & \cdots & a_{1n} \\ \cdot & \cdot & & \cdot \\ \cdot & \cdot & & \cdot \\ \cdot & \cdot & & \cdot \\ a_{m1} & a_{m2} & \cdots & a_{mn} \end{pmatrix}.$$

Thus we get

$$\begin{pmatrix} a_{11} & a_{12} & \cdots & a_{1n} \\ a_{21} & a_{22} & \cdots & a_{2n} \\ \cdot & \cdot & & \cdot \\ \cdot & \cdot & & \cdot \\ \cdot & \cdot & & \cdot \\ a_{m1} & a_{m2} & \cdots & a_{mn} \end{pmatrix} \begin{pmatrix} x_1 \\ x_2 \\ \cdot \\ \cdot \\ \cdot \\ x_n \end{pmatrix} = \begin{pmatrix} a_{11}x_1 + a_{12}x_2 + \cdots + a_{1n}x_n \\ a_{21}x_1 + a_{22}x_2 + \cdots + a_{2n}x_n \\ \cdot \\ \cdot \\ \cdot \\ a_{m1}x_1 + a_{m2}x_2 + \cdots + a_{mn}x_n \end{pmatrix}.$$

This matrix is called the matrix for T. The first column is $T(1,0, \ldots ,0)$, the second is $T(0,1,0, \ldots ,0), \ldots$, the nth column is $T(0, \ldots ,0,1)$.

Notice that if $T: \mathbb{R}^n \to \mathbb{R}^m$ is a linear transformation, the matrix for T is an $m \times n$ matrix. The dimension of the domain vector space $\mathbb{R}^n$ determines the number of columns of the matrix, whereas the dimension of the space $\mathbb{R}^m$ determines the number of rows.

Definition 4-10

The matrix constructed above is called the *matrix for the linear transformation T*, or the *matrix representing the linear transformation T*. We often denote this matrix by A_T.

Proposition 4-6

If $T: \mathbb{R}^n \to \mathbb{R}^m$ is a linear transformation, then there is a unique $m \times n$ matrix for T obtained in the manner outlined above.

The matrix determined by T is clearly unique since its columns are the vectors $T(1,0, \ldots ,0)$, $T(0,1,0, \ldots ,0), \ldots$, $T(0, \ldots ,0,1)$.

Brief Exercises

1. Determine the matrices for the linear transformations defined by the following equations.

   (a) $T(x) = 5x$
   (b) $T(x,y) = 2x + 3y$
   (c) $T(x,y) = (5x + 2y, \; x - y)$
   (d) $T(x,y) = (4x - y, \; x + 2y, \; 3x + 4y, \; 6x - 7y)$
   (e) $T(x,y,z) = (4x - 2y, \; x + y - 2z)$
   (f) $T(x,y,z) = (2x + y - z, \; 3x + 2y + 3z, \; x + 4y - z)$
   (g) $T(x,y,z) = (6x + y - z, \; 4x - 2y + 3z)$

2. In Exercises 1(b), (c), and (d) use the matrix to compute $T(1,0)$, $T(0,1)$, $T(1,1)$, $T(2,5)$, $T(-6,7)$, and $T(-4,-2)$.

3. In Exercises 1(e), (f), and (g) use the matrix to compute $T(1,0,0)$, $T(0,1,0)$, $T(0,0,1)$, $T(1,1,0)$, $T(1,0,1)$, $T(1,-5,2)$, and $T(3,1,-4)$.

Selected Answers:

1(a) (5)

1(b) (2   3)

1(c) $\begin{pmatrix} 5 & 2 \\ 1 & -1 \end{pmatrix}$

1(d) $\begin{pmatrix} 4 & -1 \\ 1 & 2 \\ 3 & 4 \\ 6 & -7 \end{pmatrix}$

1(e) $\begin{pmatrix} 4 & -2 & 0 \\ 1 & 1 & -2 \end{pmatrix}$

1(f) $\begin{pmatrix} 2 & 1 & -1 \\ 3 & 2 & 3 \\ 1 & 4 & -1 \end{pmatrix}$

1(g) $\begin{pmatrix} 6 & 1 & -1 \\ 4 & -2 & 3 \end{pmatrix}$

2(b) $T(2,5) = 19$ **2(c)** $T(1,1) = (7,0)$ **2(d)** $T(-6,7) = (-31,8,10,-85)$
3(e) $T(1,0,1) = (4,-1)$ **3(f)** $T(1,1,0) = (3,5,5)$ **3(g)** $T(1,0,1) = (5,7)$

We have shown how each linear transformation $T: \mathbb{R}^n \to \mathbb{R}^m$ determines a unique matrix. On the other hand each $m \times n$ matrix determines a unique linear transformation $T: \mathbb{R}^n \to \mathbb{R}^m$.

Example Consider the matrix

$$\begin{pmatrix} 1 & 3 & 5 \\ 4 & 1 & 2 \end{pmatrix}.$$

This matrix determines a function $T: \mathbb{R}^3 \to \mathbb{R}^2$

$$\begin{pmatrix} 1 & 3 & 5 \\ 4 & 1 & 2 \end{pmatrix} \begin{pmatrix} x \\ y \\ z \end{pmatrix} = \begin{pmatrix} x + 3y + 5z \\ 4x + y + 2z \end{pmatrix}$$

Thus the linear transformation determined by the matrix is given by the equation $T(x,y,z) = (x + 3y + 5z, 4x + y + 2z)$.

Example Consider the 3×4 matrix

$$\begin{pmatrix} 5 & 1 & 4 & 2 \\ 3 & 1 & 2 & 1 \\ 0 & 5 & 6 & -2 \end{pmatrix}.$$

This determines the linear function $T: \mathbb{R}^4 \to \mathbb{R}^3$ with equation
$$T(x_1,x_2,x_3,x_4) = (5x_1 + x_2 + 4x_3 + 2x_4, \; 3x_1 + x_2 + 2x_3 + x_4, \; 5x_2 + 6x_3 - 2x_4)$$
since

$$\begin{pmatrix} 5 & 1 & 4 & 2 \\ 3 & 1 & 2 & 1 \\ 0 & 5 & 6 & -2 \end{pmatrix} \begin{pmatrix} x_1 \\ x_2 \\ x_3 \\ x_4 \end{pmatrix} = \begin{pmatrix} 5x_1 + x_2 + 4x_3 + 2x_4 \\ 3x_1 + x_2 + 2x_3 + x_4 \\ 5x_2 + 6x_3 - 2x_4 \end{pmatrix}$$

These examples show how, in general, an $m \times n$ matrix

$$\begin{pmatrix} a_{11} & a_{12} & \cdots & a_{1n} \\ a_{21} & a_{22} & \cdots & a_{2n} \\ \cdot & \cdot & & \cdot \\ \cdot & \cdot & & \cdot \\ \cdot & \cdot & & \cdot \\ a_{m1} & a_{m2} & \cdots & a_{mn} \end{pmatrix}$$

determines the linear transformation $T: \mathbb{R}^n \to \mathbb{R}^m$ with equation

$$\begin{aligned} T(x_1,x_2, \ldots ,x_n) = (&a_{11}x_1 + a_{12}x_2 + \cdots + a_{1n}x_n, \\ &a_{21}x_1 + a_{22}x_2 + \cdots + a_{2n}x_n, \ldots, \\ &a_{m1}x_1 + a_{m2}x_2 + \cdots + a_{mn}x_n) \end{aligned}$$

since

$$\begin{pmatrix} a_{11} & a_{12} & \cdots & a_{1n} \\ a_{21} & a_{22} & \cdots & a_{2n} \\ \cdot & \cdot & & \cdot \\ \cdot & \cdot & & \cdot \\ \cdot & \cdot & & \cdot \\ a_{m1} & a_{m2} & \cdots & a_{mn} \end{pmatrix} \begin{pmatrix} x_1 \\ x_2 \\ \cdot \\ \cdot \\ \cdot \\ x_n \end{pmatrix} = \begin{pmatrix} a_{11}x_1 + a_{12}x_2 + \cdots + a_{1n}x_n \\ a_{21}x_1 + a_{22}x_2 + \cdots + a_{2n}x_n \\ \cdot \\ \cdot \\ \cdot \\ a_{m1}x_1 + a_{m2}x_2 + \cdots + a_{mn}x_n \end{pmatrix} .$$

The function T determined by the matrix is clearly linear because of the form of the component functions.

Definition 4-11

The linear transformation $T: \mathbb{R}^n \to \mathbb{R}^m$ determined by the $m \times n$ matrix in the manner outlined above is called *the linear transformation of the matrix*.

Proposition 4-7

Each $m \times n$ matrix determines a unique linear transformation $T: \mathbb{R}^n \to \mathbb{R}^m$ as outlined above.

Brief Exercises

1. Write out the defining formulas for the linear transformations determined by the following matrices. Identify the domain and range spaces of the linear transformation; check this dimension against the sizes of the matrices.

(a) $\begin{pmatrix} 1 & 0 \\ 0 & 1 \end{pmatrix}$

(e) $\begin{pmatrix} 1 & 5 & 0 \\ 0 & -1 & 2 \\ 3 & 1 & 4 \end{pmatrix}$

(b) $\begin{pmatrix} 2 & 0 \\ 0 & 3 \end{pmatrix}$

(f) $\begin{pmatrix} 1 & 2 & 1 & 3 \\ 0 & 1 & 5 & 6 \\ -2 & 3 & 7 & 2 \end{pmatrix}$

(c) $\begin{pmatrix} 1 & 1 & 2 \\ 0 & -1 & 3 \end{pmatrix}$

(g) $\begin{pmatrix} 2 & 1 & 4 \\ 8 & 6 & 3 \\ 2 & 1 & 5 \\ 1 & 4 & 2 \end{pmatrix}$

(d) $\begin{pmatrix} 2 & 1 \\ -1 & 1 \\ 4 & 2 \end{pmatrix}$

Answers: **1(a)** $T(x,y) = (x,y)$ **1(b)** $T(x,y) = (2x,3y)$ **1(c)** $T(x,y,z) = (x + y + 2z, -y + 3z)$ **1(d)** $T(x,y) = (2x + y, -x + y, 4x + 2y)$ **1(e)** $T(x,y,z) = (x + 5y, -y + 2z, 3x + y + 4z)$ **1(f)** $T(x,y,z,w) = (x + 2y + z + 3w, y + 5z + 6w, -2x + 3y + 7z + 2w)$ **1(g)** $T(x,y,z) = (2x + y + 4z, 8x + 6y + 3z, 2x + y + 5z, x + 4y + 2z)$

SUMMARY *Every linear transformation* $T: \mathbb{R}^n \to \mathbb{R}^m$ *determines a unique* $m \times n$ *matrix which represents it in the sense that the components of* $T(x_1, \ldots, x_n)$ *can be computed using the scheme*

$$\begin{pmatrix} a_{11} & \cdots & a_{1n} \\ \cdot & & \cdot \\ \cdot & & \cdot \\ \cdot & & \cdot \\ a_{m1} & \cdots & a_{mn} \end{pmatrix} \begin{pmatrix} x_1 \\ \cdot \\ \cdot \\ \cdot \\ x_n \end{pmatrix} = \begin{pmatrix} a_{11}x_1 + \cdots + a_{1n}x_n \\ \cdot \\ \cdot \\ \cdot \\ a_{m1}x_1 + \cdots + a_{mn}x_n \end{pmatrix} .$$

Further, using the same scheme, every m × n matrix determines a unique linear transformation T: $\mathbb{R}^n \to \mathbb{R}^m$. *Therefore, linear transformations described in coordinate terms, matrices used in a certain fashion (as outlined in this section), and functions which are additive and homogeneous are three different ways of looking at the same thing. Consequently, we use whichever viewpoint is easiest in a particular problem.*

Exercises for Section 4-3

1. Find the matrix for the following linear transformations from $\mathbb{R}^2$ to $\mathbb{R}^2$ and, using the matrix, compute $T(5,1)$, $T(-3,7)$, and $T(17,-18)$.

   (a) $T(x,y) = (3x + y, 2x - 5y)$ (c) $T(x,y) = (4x + 2y, -5x + 3y)$
   (b) $T(x,y) = (2x + y, x - y)$

2. Find the matrix for the following linear transformations from $\mathbb{R}^3$ to $\mathbb{R}^2$.

   (a) $T(x,y,z) = (x - 2y + z, x + y - 3z)$
   (b) $T(x,y,z) = (2x + y + 3z, x - y + z)$
   (c) $T(x,y,z) = (5x + 2y + z, x + y)$

3. Find the matrix for the following linear transformations from $\mathbb{R}^2$ to $\mathbb{R}^4$.

   (a) $T(x,y) = (x + 2y, x - y, 3x + 5y, 4x - y)$
   (b) $T(x,y) = (x + y, x + y, 2x + y, 2x + y)$

4. Find the matrix for the linear transformation from $\mathbb{R}^3$ to $\mathbb{R}^3$ having the property that $T(1,0,1) = (5,1,2)$, $T(0,1,1) = (2,1,4)$, and $T(1,1,0) = (-4,0,1)$.

5. Find the matrix for the linear transformation from $\mathbb{R}^2$ to $\mathbb{R}^3$ having the property that $T(1,0) = (5,1,6)$ and $T(1,1) = (6,1,2)$.

6. Same exercise as 5 if $T(1,2) = (5,1,6)$ and $T(2,1) = (4,0,2)$.

7. If the matrix for $T: \mathbb{R}^2 \to \mathbb{R}^3$ is

$$\begin{pmatrix} 2 & 1 \\ 0 & 3 \\ -5 & 4 \end{pmatrix}$$

   find $T(2,1)$, $T(0,-3)$, $T(7,8)$, $T(16,2)$, and $T(x,y)$.

8. If the matrix for $T: \mathbb{R}^3 \to \mathbb{R}^3$ is

$$\begin{pmatrix} 1 & 1 & 2 \\ 0 & -5 & 3 \\ 8 & 1 & 6 \end{pmatrix}$$

find $T(3,1,1)$, $T(0,-1,2)$, $T(2,4,1)$, $T(3,1,-6)$, and $T(x,y,z)$.

9. If the matrix for $T: \mathbb{R}^2 \to \mathbb{R}^2$ is

$$\begin{pmatrix} 2 & 1 \\ -3 & 4 \end{pmatrix}$$

find $T(2,1)$, $T(3,4)$, and $T(x,y)$.

10. If the matrix for $T: \mathbb{R}^2 \to \mathbb{R}^2$ is

$$\begin{pmatrix} 5 & 1 \\ 6 & -2 \end{pmatrix}$$

what does an application of T do to the line

$$(5,2) + t(-3,1)?$$

11. More generally than 10, if $T: \mathbb{R}^2 \to \mathbb{R}^2$ is a linear transformation, what affect does applying T have on the line $X_0 + tX_1$? (You should conclude that T sends a line into a line or a point.)

12. If the matrix for $T: \mathbb{R}^2 \to \mathbb{R}^2$ is

$$\begin{pmatrix} 2 & 1 \\ -3 & 2 \end{pmatrix}$$

find the effect of applying T to the triangle with vertices at $(1,1)$, $(3,2)$, and $(-1,5)$.

13. If the matrix $T: \mathbb{R}^2 \to \mathbb{R}^2$ is

$$\begin{pmatrix} 5 & 1 \\ 4 & -2 \end{pmatrix}$$

find the effect of applying T to the square with vertices at $(0,0)$, $(1,0)$, $(1,1)$, and $(0,1)$.

14. If the matrix for $T: \mathbb{R}^3 \to \mathbb{R}^3$ is

$$\begin{pmatrix} 2 & 1 & 5 \\ 3 & 1 & 4 \\ 2 & 1 & 3 \end{pmatrix}$$

find the effect of applying T to the cube with vertices at $(0,0,0)$, $(1,0,0)$, $(0,1,0)$, $(0,0,1)$, $(1,1,0)$, $(1,0,1)$, $(0,1,1)$, and $(1,1,1)$.

SECTION 4-4
Arithmetic of Linear Transformations and Matrices

PREVIEW *We multiply a matrix by a scalar by multiplying each entry by the scalar. We add matrices having the same size by adding entries occupying the same position in the two matrices. If the number of columns in a matrix A equals the number of rows in B, the product AB is obtained by applying A to the columns of B; thus the (i,k) entry of AB is $a_{i1}b_{1k} + a_{i2}b_{2k} + \cdots + a_{in}b_{nk}$, computed by applying the ith row of A to the kth column of B. The correspondence between linear transformations and matrices is given by*

$$A_{S+T} = A_S + A_T \qquad A_{rT} = rA_T \qquad A_{T \circ S} = A_T A_S.$$

Multiplication of matrices is associative but not commutative. The matrices I_n and functions $I_{\mathbb{R}^n}$ play for matrices and linear transformations a role somewhat similar to that played by the number 1 in multiplication of real numbers.

We have discussed addition and scalar multiplication of vectors and of functions. Now we discuss addition and scalar multiplication of matrices. We can multiply any matrix by any scalar. However, we only add matrices which have the same size, i.e., which have the same number of rows and the same number of columns.

Definition 4-12

If A is a matrix and r is a scalar, then rA is the matrix obtained by multiplying every entry of A by the scalar r.

If A and B are matrices having the same size, then $A + B$ is the matrix obtained by adding entries of A and B which occupy the same position; i.e., the (i,j) entry of $A + B$ is obtained by adding the (i,j) entry of A and the (i,j) entry of B.

Example Suppose we are given the matrices

$$A = \begin{pmatrix} 1 & 2 & 1 \\ 3 & 1 & 4 \end{pmatrix} \qquad B = \begin{pmatrix} 3 & -1 & 2 \\ 4 & 5 & 1 \end{pmatrix}.$$

Then

$$A + B = \begin{pmatrix} 4 & 1 & 3 \\ 7 & 6 & 5 \end{pmatrix} \quad \text{and} \quad 5A = \begin{pmatrix} 5 & 10 & 5 \\ 15 & 5 & 20 \end{pmatrix}.$$

Example Suppose we are given the matrices

$$A = \begin{pmatrix} 1 & 2 \\ 3 & 1 \\ 1 & 1 \end{pmatrix} \qquad B = \begin{pmatrix} -4 & 5 \\ 3 & -2 \\ 5 & -2 \end{pmatrix} \qquad C = \begin{pmatrix} 2 & 1 \\ 1 & 3 \end{pmatrix}.$$

We can form $A + B$ and $B + A$ because A and B are the same size. However, we cannot form $A + C$ or $B + C$ because A and C are different sizes; the same reason applies to B and C.

$$A + B = \begin{pmatrix} -3 & 7 \\ 6 & -1 \\ 6 & -1 \end{pmatrix} \qquad 5A = \begin{pmatrix} 5 & 10 \\ 15 & 5 \\ 5 & 5 \end{pmatrix} \qquad 6B = \begin{pmatrix} -24 & 30 \\ 18 & -12 \\ 30 & -12 \end{pmatrix}$$

$$-3C = \begin{pmatrix} -6 & -3 \\ -3 & -9 \end{pmatrix}$$

Brief Exercises

Do the following matrix computations: $A + B + C$, $2A - 3B + 4C$, $5A + 6C$, $B + A$, and $4A - 3C$.

1.
$$A = \begin{pmatrix} 2 & 1 \\ 3 & 4 \end{pmatrix} \qquad B = \begin{pmatrix} -3 & 2 \\ 1 & -1 \end{pmatrix} \qquad C = \begin{pmatrix} 2 & 1 \\ 5 & -6 \end{pmatrix}$$

2.
$$A = \begin{pmatrix} 3 & 1 & 4 \\ -1 & 2 & 5 \end{pmatrix} \qquad B = \begin{pmatrix} 5 & 1 & 0 \\ 0 & 4 & 2 \end{pmatrix} \qquad C = \begin{pmatrix} -3 & 1 & 2 \\ 1 & -4 & 5 \end{pmatrix}$$

3.
$$A = \begin{pmatrix} 2 & 1 & 3 \\ 0 & 7 & 4 \\ 1 & 5 & 0 \end{pmatrix} \qquad B = \begin{pmatrix} 1 & 1 & -1 \\ 0 & 1 & 2 \\ -1 & 3 & 4 \end{pmatrix} \qquad C = \begin{pmatrix} 2 & 0 & 5 \\ 1 & 4 & 2 \\ 3 & 1 & 1 \end{pmatrix}$$

Selected Answers:

1
$$A + B + C = \begin{pmatrix} 1 & 4 \\ 9 & -3 \end{pmatrix} \qquad 2A - 3B + 4C = \begin{pmatrix} 21 & 0 \\ 23 & -13 \end{pmatrix}$$

2
$$2A - 3B + 4C = \begin{pmatrix} -21 & 3 & 16 \\ 2 & -24 & 24 \end{pmatrix} \qquad 5A + 6C = \begin{pmatrix} -3 & 11 & 32 \\ 1 & -14 & 55 \end{pmatrix}$$

3

$$B + A = \begin{pmatrix} 3 & 2 & 2 \\ 0 & 8 & 6 \\ 0 & 8 & 4 \end{pmatrix} \qquad 4A - 3C = \begin{pmatrix} 2 & 4 & -3 \\ -3 & 16 & 10 \\ -5 & 17 & -3 \end{pmatrix}$$

Matrices can also be multiplied as we now explain.

Example Suppose we have the matrices

$$A = \begin{pmatrix} 2 & 1 \\ 3 & -2 \\ 0 & 4 \end{pmatrix} \qquad \text{and} \qquad B = \begin{pmatrix} 2 & 3 \\ 1 & -1 \end{pmatrix}.$$

We illustrate computing AB.

$$\begin{pmatrix} 2 & 1 \\ 3 & -2 \\ 0 & 4 \end{pmatrix} \begin{pmatrix} 2 & 3 \\ 1 & -1 \end{pmatrix}$$

Think of the columns of B as vectors to which you apply the matrix A (as in Section 4-3). Applying A to the first column of B gives the first column of AB.

$$\begin{pmatrix} 2 & 1 \\ 3 & -2 \\ 0 & 4 \end{pmatrix} \begin{pmatrix} 2 \\ 1 \end{pmatrix} = \begin{pmatrix} 5 \\ 4 \\ 4 \end{pmatrix}$$

Applying A to the second column of B gives the second column of AB.

$$\begin{pmatrix} 2 & 1 \\ 3 & -2 \\ 0 & 4 \end{pmatrix} \begin{pmatrix} 3 \\ -1 \end{pmatrix} = \begin{pmatrix} 5 \\ 11 \\ -4 \end{pmatrix}$$

Thus

$$AB = \begin{pmatrix} 2 & 1 \\ 3 & -2 \\ 0 & 4 \end{pmatrix} \begin{pmatrix} 2 & 3 \\ 1 & -1 \end{pmatrix} = \begin{pmatrix} 5 & 5 \\ 4 & 11 \\ 4 & -4 \end{pmatrix}.$$

The product BA cannot be formed. One reason is that the matrix B cannot be applied to the columns of A; the columns of A are too long.

$$\begin{pmatrix} 2 & 3 \\ 1 & -2 \end{pmatrix} \begin{pmatrix} 2 & 1 \\ 3 & -2 \\ 0 & 4 \end{pmatrix}$$

Example   Suppose we have the matrices

$$A = \begin{pmatrix} 3 & 1 & 0 \\ -2 & 0 & 3 \\ 1 & 2 & 1 \end{pmatrix} \qquad B = \begin{pmatrix} -1 & 2 & 1 \\ 3 & 1 & -2 \\ 4 & 0 & 2 \end{pmatrix}.$$

Because A and B are both 3×3 matrices, both products AB and BA can be formed.

$$AB = \begin{pmatrix} 3 & 1 & 0 \\ -2 & 0 & 3 \\ 1 & 2 & 1 \end{pmatrix} \begin{pmatrix} -1 & 2 & 1 \\ 3 & 1 & -2 \\ 4 & 0 & 2 \end{pmatrix} = \begin{pmatrix} 0 & 7 & 1 \\ 14 & -4 & 4 \\ 9 & 4 & -1 \end{pmatrix}$$

$$BA = \begin{pmatrix} -1 & 2 & 1 \\ 3 & 1 & -2 \\ 4 & 0 & 2 \end{pmatrix} \begin{pmatrix} 3 & 1 & 0 \\ -2 & 0 & 3 \\ 1 & 2 & 1 \end{pmatrix} = \begin{pmatrix} -6 & 1 & 7 \\ 5 & -1 & 1 \\ 14 & 8 & 2 \end{pmatrix}$$

This example shows that even if the two products AB and BA can be formed, they may be different.

These two examples illustrate matrix multiplication. For a product of matrices to be formed, the number of columns in the left-hand matrix must equal the number of rows in the right-hand matrix. We have described the product AB as formed by applying A to the columns of B to obtain the columns of AB. There is another way to describe the individual entries of a product AB: the (i,k) entry of AB is obtained by applying the ith row of A to the kth column of B. We illustrate this in the next example.

Example   We compute the product of the matrices

$$A = \begin{pmatrix} 2 & 1 \\ 3 & 4 \\ -2 & 0 \end{pmatrix} \qquad \text{and} \qquad B = \begin{pmatrix} 1 & 2 \\ -4 & 3 \end{pmatrix}.$$

Now

$$AB = \begin{pmatrix} 2 & 1 \\ 3 & 4 \\ -2 & 0 \end{pmatrix} \begin{pmatrix} 1 & 2 \\ -4 & 3 \end{pmatrix} = \begin{pmatrix} -2 & 7 \\ -13 & 18 \\ -2 & -4 \end{pmatrix}.$$

Note that the (2,1) entry of AB is obtained with the second row of A and the first column of B.

$$\begin{pmatrix} 2 & 1 \\ \boxed{3 \;\; 4} \\ -2 & 0 \end{pmatrix} \begin{pmatrix} \boxed{1} & 2 \\ \boxed{-4} & 3 \end{pmatrix} = \begin{pmatrix} -2 & 7 \\ \boxed{-13} & 18 \\ -2 & -4 \end{pmatrix} \text{ and } \begin{pmatrix} 2 & 1 \\ 3 & 4 \\ \boxed{-2 \;\; 0} \end{pmatrix} \begin{pmatrix} 1 & \boxed{2} \\ -4 & \boxed{3} \end{pmatrix} = \begin{pmatrix} -2 & 7 \\ -13 & 18 \\ -2 & \boxed{-4} \end{pmatrix}$$

As another example, note that the (3,2) entry of AB is obtained with the 3rd row of A and the 2nd column of B.

Definition 4-13

If A is an $m \times n$ matrix and B is an $n \times p$ matrix, the product AB is the $m \times p$ matrix whose (i,k) entry is

$$a_{i1}b_{1k} + a_{i2}b_{2k} + \cdots + a_{in}b_{nk}$$

obtained by applying the ith row of A to the kth column of B.

$$\begin{pmatrix} & & \vdots & & \\ & & \vdots & & \\ a_{i1} & a_{i2} & \cdots & a_{in} \\ & & \vdots & & \\ & & \vdots & & \end{pmatrix} \begin{pmatrix} & & b_{1k} & & \\ & & b_{2k} & & \\ \cdots & & \vdots & & \cdots \\ & & \vdots & & \\ & & b_{nk} & & \end{pmatrix}$$

kth column

ith row

$$(a_{i1}b_{1k} + a_{i2}b_{2k} + \cdots + a_{in}b_{nk})$$

Brief Exercises

Perform the matrix computations AB, AC, BA, CA, BC, and CB in the following instances, if possible.

1.
$$A = \begin{pmatrix} 2 & 1 \\ 1 & 3 \end{pmatrix} \qquad B = \begin{pmatrix} 5 & 0 \\ 6 & 9 \end{pmatrix} \qquad C = \begin{pmatrix} 1 & 3 \\ 4 & 2 \end{pmatrix}$$

2.
$$A = \begin{pmatrix} 1 & 1 & 2 \\ 3 & 1 & 4 \\ 1 & -1 & 2 \end{pmatrix} \qquad B = \begin{pmatrix} 0 & 1 & 2 \\ 1 & 0 & 3 \\ 5 & 1 & 2 \end{pmatrix} \qquad C = \begin{pmatrix} 1 & 0 & 2 \\ -1 & 3 & 4 \\ 1 & 5 & 2 \end{pmatrix}$$

3.
$$A = \begin{pmatrix} 1 & 1 \\ -1 & 0 \end{pmatrix} \qquad B = \begin{pmatrix} 2 & 1 & 3 \\ 1 & 5 & 4 \end{pmatrix} \qquad C = \begin{pmatrix} 1 & 1 \\ 2 & 3 \end{pmatrix}$$

Answers:

1
$$AB = \begin{pmatrix} 16 & 9 \\ 23 & 27 \end{pmatrix} \qquad AC = \begin{pmatrix} 6 & 8 \\ 13 & 9 \end{pmatrix} \qquad BA = \begin{pmatrix} 10 & 5 \\ 21 & 33 \end{pmatrix}$$

$$CA = \begin{pmatrix} 5 & 10 \\ 10 & 10 \end{pmatrix} \qquad BC = \begin{pmatrix} 5 & 15 \\ 42 & 36 \end{pmatrix} \qquad CB = \begin{pmatrix} 23 & 27 \\ 32 & 18 \end{pmatrix}$$

2
$$AB = \begin{pmatrix} 11 & 3 & 9 \\ 21 & 7 & 17 \\ 9 & 3 & 3 \end{pmatrix} \quad AC = \begin{pmatrix} 2 & 13 & 10 \\ 6 & 23 & 18 \\ 4 & 7 & 2 \end{pmatrix} \quad BA = \begin{pmatrix} 5 & -1 & 8 \\ 4 & -2 & 8 \\ 10 & 4 & 18 \end{pmatrix}$$

$$CA = \begin{pmatrix} 3 & -1 & 6 \\ 12 & -2 & 18 \\ 18 & 4 & 26 \end{pmatrix} \quad BC = \begin{pmatrix} 1 & 13 & 8 \\ 4 & 15 & 8 \\ 6 & 13 & 18 \end{pmatrix} \quad CB = \begin{pmatrix} 10 & 3 & 6 \\ 23 & 3 & 15 \\ 15 & 3 & 21 \end{pmatrix}$$

3
$$AB = \begin{pmatrix} 3 & 6 & 7 \\ -2 & -1 & -3 \end{pmatrix} \qquad AC = \begin{pmatrix} 3 & 4 \\ -1 & -1 \end{pmatrix} \qquad CA = \begin{pmatrix} 0 & 1 \\ -1 & 2 \end{pmatrix}$$

$$CB = \begin{pmatrix} 3 & 6 & 7 \\ 7 & 17 & 18 \end{pmatrix} \qquad BA \text{ and } BC \text{ not defined}$$

In Section 4-3 correspondence between linear functions and matrices was established. This correspondence extends to addition and scalar multiplication of linear functions and matrices, and it extends to composition of linear functions and multiplication of matrices.

Example Suppose $S: \mathbb{R}^2 \to \mathbb{R}^3$ and $T: \mathbb{R}^2 \to \mathbb{R}^3$ are linear transformations defined by the equations

$$S(x,y) = (5x - y, 2x, x - 3y)$$
$$T(x,y) = (x + y, x - y, 3x - 4y).$$

Then $S + T$ and rT are the functions defined by the equations

$$(S + T)(x,y) = (6x, 3x - y, 4x - 7y)$$

and

$$(rT)(x,y) = (rx + ry, rx - ry, 3rx - 4ry).$$

From the defining equations you can see that $S + T$ and rT are linear. The matrices for S, T, $S + T$, and rT are

$$A_S = \begin{pmatrix} 5 & -1 \\ 2 & 0 \\ 1 & -3 \end{pmatrix} \quad A_T = \begin{pmatrix} 1 & 1 \\ 1 & -1 \\ 3 & -4 \end{pmatrix} \quad A_{S+T} = \begin{pmatrix} 6 & 0 \\ 3 & -1 \\ 4 & -7 \end{pmatrix}$$

$$A_{rT} = \begin{pmatrix} r & r \\ r & -r \\ 3r & -4r \end{pmatrix}.$$

Thus the matrix A_{S+T} for $S + T$ is the sum of the matrices for S and T, and the matrix for rT is r times the matrix A_T. In symbols, $A_{S+T} = A_S + A_T$ and $A_{rT} = rA_T$.

This example illustrates the general situation. If $S: \mathbb{R}^n \rightarrow \mathbb{R}^m$ and $T: \mathbb{R}^n \rightarrow \mathbb{R}^m$ are linear functions, then $S + T$ and rT are also linear. For suppose

$$S(x_1, \ldots ,x_n) = (a_{11}x_1 + \cdots + a_{1n}x_n, \ldots , a_{m1}x_1 + \cdots + a_{mn}x_n)$$

and

$$T(x_1, \ldots ,x_n) = (b_{11}x_1 + \cdots + b_{1n}x_n, \ldots , b_{m1}x_1 + \cdots + b_{mn}x_n)$$

then

$$(S + T)(x_1, \ldots ,x_n) = ((a_{11} + b_{11})x_1 + \cdots + (a_{1n} + b_{1n})x_n, \ldots ,$$
$$(a_{m1} + b_{m1})x_1 + \cdots + (a_{mn} + b_{mn})x_n)$$

and

$$(rT)(x, \ldots ,x) = (rb_{11}x_1 + \cdots + rb_{1n}x_n, \ldots ,$$
$$rb_{m1}x_1 + \cdots + rb_{mn}x_n).$$

The defining equations for $S + T$ and rT clearly show that $S + T$ and rT are linear. Furthermore, they show that the matrix A_{S+T} for $S + T$ is the sum of A_S and A_T; also the matrix A_{rT} for rT is the scalar multiple rA_T of the matrix for T.

Proposition 4-8

If $S: \mathbb{R}^n \to \mathbb{R}^m$ and $T: \mathbb{R}^n \to \mathbb{R}^m$ are linear transformations, and if r is a scalar, then $S + T$ and rT are linear transformations, too. Furthermore, the matrix for $S + T$ is the sum of the matrix for S and the matrix for T.

$$A_{S+T} = A_S + A_T$$

The matrix for rT is the scalar multiple of the matrix for T by the scalar r.

$$A_{rT} = rA_T$$

Brief Exercises

1. In the following, compute $S + T$ and rT to verify that they are linear.

   (a) $S(x,y) = 3x - y$, $T(x,y) = 6x - 4y$

   (b) $S(x,y) = (x + y, x - y)$, $T(x,y) = (7x - 4y, -6x + 2y)$

   (c) $S(x,y) = (5x + y, x + 2y, 3x - y)$, $T(x,y) = (4x - y, x + 2y, 3x + y)$

   (d) $S(x,y,z) = (2x + y - z, x + 2y + z, 3x + 4y + 4z)$,

   $T(x,y,z) = (5x - 3y + 4z, 2x + y + z, x - 2z)$

2. In each of the parts of Exercise 1, write out explicitly the matrices for S, T, $S + T$, and rT.

Selected Answers: **2(a)** $A_S = \begin{pmatrix} 3 & -1 \end{pmatrix}$, $A_T = \begin{pmatrix} 6 & -4 \end{pmatrix}$, $A_{S+T} = \begin{pmatrix} 9 & -5 \end{pmatrix}$,
$A_{rT} = \begin{pmatrix} 6r & -4r \end{pmatrix}$

2(d)

$$A_S = \begin{pmatrix} 2 & 1 & -1 \\ 1 & 2 & 1 \\ 3 & 4 & 4 \end{pmatrix} \qquad A_T = \begin{pmatrix} 5 & -3 & 4 \\ 2 & 1 & 1 \\ 1 & 0 & -2 \end{pmatrix}$$

$$A_{S+T} = \begin{pmatrix} 7 & -2 & 3 \\ 3 & 3 & 2 \\ 4 & 4 & 2 \end{pmatrix} \qquad A_{rT} = \begin{pmatrix} 5r & -3r & 4r \\ 2r & r & r \\ r & 0 & -2r \end{pmatrix}$$

Now we discuss the fact that the composition of linear functions is a linear function. We begin with two examples.

Example Suppose $S: \mathbb{R}^2 \to \mathbb{R}^3$ is the function defined by

$$S(x,y) = (3x - 2y, 5x + 6y)$$

and suppose that $T: \mathbb{R}^2 \to \mathbb{R}^2$ is the linear function defined by

$$T(u,v) = (u + v, u - v).$$

We verify that $T \circ S$ is a linear transformation. (Because we are dealing with composition of functions, we must distinguish between $\mathbb{R}^2$ viewed as the domain of S and $\mathbb{R}^2$ viewed as the domain of T. We therefore use (x,y) for vectors in the domain of S and (u,v) for vectors in the domain of T.) Now we calculate

$$(T \circ S)(x,y) = T(S(x,y)) = T(3x - 2y, 5x + 6y)$$
$$= ((3x - 2y) + (5x + 6y), (3x - 2y) - (5x + 6y))$$
$$= (8x + 4y, -2x - 8y).$$

This equation shows that $T \circ S$ is linear. The matrices for S, T, and $T \circ S$ are

$$A_S = \begin{pmatrix} 3 & -2 \\ 5 & 6 \end{pmatrix} \qquad A_T = \begin{pmatrix} 1 & 1 \\ 1 & -1 \end{pmatrix} \qquad A_{T \circ S} = \begin{pmatrix} 8 & 4 \\ -2 & -8 \end{pmatrix}.$$

Thus $A_{T \circ S} = A_T A_S$, as you readily verify.

Example   Suppose $S: \mathbb{R}^2 \to \mathbb{R}^4$ and $T: \mathbb{R}^4 \to \mathbb{R}^3$ are defined by the equations

$$S(x,y) = (x, y, x - y, 3x + 2y)$$
$$T(r,s,t,u) = (r + s + 2u, t - 3r + s, r + t).$$

In this example we verify that $T \circ S$ is linear, and we show how the matrix $A_{T \circ S}$ for $T \circ S$ can be calculated from the matrices A_S for S and A_T for T.

First, we verify that $T \circ S$ is linear.

$$(T \circ S)(x,y) = T(S(x,y))$$
$$= T(x, y, x - y, 3x + 2y)$$
$$= (x + y + 2(3x + 2y), (x - y) - 3x + y, x + (x - y))$$
$$= (7x + 5y, -2x, 2x - y)$$

This equation shows that $T \circ S$ is linear. The matrices for S, T, and $T \circ S$ are

$$A_S = \begin{pmatrix} 1 & 0 \\ 0 & 1 \\ 1 & -1 \\ 3 & 2 \end{pmatrix} \qquad A_T = \begin{pmatrix} 1 & 1 & 0 & 2 \\ -3 & 1 & 1 & 0 \\ 1 & 0 & 1 & 0 \end{pmatrix} \qquad A_{T \circ S} = \begin{pmatrix} 7 & 5 \\ -2 & 0 \\ 2 & -1 \end{pmatrix}.$$

Thus $A_{T \circ S} = A_T A_S$, as you can verify.

These two examples show that the composite of linear transformations is linear, and they show how to calculate the matrix for a composite of linear transformations. We can show that the composite $T \circ S$ of two linear transformations $S: \mathbb{R}^n \to \mathbb{R}^m$ and $T: \mathbb{R}^m \to \mathbb{R}^p$ is a linear transformation by examining the coordinates. Suppose

$$S(x_1, \ldots, x_n) = (a_{11}x_1 + \cdots + a_{1n}x_n, \ldots, a_{m1}x_1 + \cdots + a_{mn}x_n)$$

and

$$T(y_1, \ldots, y_m) = (b_{11}y_1 + \cdots + b_{1m}y_m, \ldots, b_{p1}y_1 + \cdots + b_{pm}y_m).$$

To calculate $(T \circ S)(x_1, \ldots, x_n)$ we need to substitute

$$a_{11}x_1 + \cdots + a_{1n}x_n \qquad \text{for} \qquad y_1$$

$$\vdots \qquad\qquad \vdots \qquad\qquad \vdots$$

$$a_{m1}x_1 + \cdots + a_{mn}x_n \qquad \text{for} \qquad y_m$$

in the coordinates of $T(y_1, \ldots, y_m)$. This is tedious, but the resulting coordinates would show that $T \circ S$ is linear.

However, it is easier to show that $T \circ S$ is additive and homogeneous.

$$
\begin{array}{ll}
(T \circ S)(\mathbf{X} + \mathbf{Y}) = T(S(\mathbf{X} + \mathbf{Y})) & \text{(Definition of } T \circ S) \\
\qquad = T(S(\mathbf{X}) + S(\mathbf{Y})) & (S \text{ is additive)} \\
\qquad = T(S(\mathbf{X})) + T(S(\mathbf{Y})) & (T \text{ is additive)} \\
\qquad = (T \circ S)(\mathbf{X}) + (T \circ S)(\mathbf{Y}) & \text{(Definition of } T \circ S)
\end{array}
$$

Also

$$
\begin{array}{ll}
(T \circ S)(r\mathbf{X}) = T(S(r\mathbf{S})) & \text{(Definition of } T \circ S) \\
\qquad = T(rS(\mathbf{X})) & (S \text{ is homogeneous)} \\
\qquad = r[T(S(\mathbf{X}))] & (T \text{ is homogeneous)} \\
\qquad = r[(T \circ S)(\mathbf{X})] & \text{(Definition of } T \circ S)
\end{array}
$$

Proposition 4-9_____

If $\mathscr{U}$, $\mathscr{V}$, and $\mathscr{W}$ are vector spaces over $\mathbb{R}$, and if $S: \mathscr{U} \to \mathscr{V}$ and $T: \mathscr{V} \to \mathscr{W}$ are linear transformations, then $(T \circ S): \mathscr{U} \to \mathscr{W}$ is also a linear transformation.

_____

The last two examples also show that the matrix $A_{T \circ S}$ for a composite of linear transformations is the product of the matrices A_T and A_S. This is true in general.

We obtain the columns of $A_{T\circ S}$ by computing $(T \circ S)(1, \dots, 0)$, $(T \circ S)(0, 1, 0, \dots,$ $(T \circ S)(0, \dots, 0, 1)$, i.e. by applying A_T to each column of A_S. This shows that $A_{T\circ S} = A_T A_S$.

Proposition 4-10

If $S: \mathbb{R}^p \to \mathbb{R}^n$ and $T: \mathbb{R}^n \to \mathbb{R}^m$ are linear transformations with matrices A_S and A_T, then $A_{T\circ S} = A_T A_S$, i.e., the matrix for a composite $T \circ S$ is the product of the matrices for T and for S.

In the previous section we showed how to compute $T(\mathbf{X})$ by using the matrix A for T and writing the vector $\mathbf{X}$ vertically.

$$\begin{pmatrix} a_{11} & a_{12} & \cdots & a_{1n} \\ a_{21} & a_{22} & \cdots & a_{2n} \\ \cdot & \cdot & & \cdot \\ \cdot & \cdot & & \cdot \\ \cdot & \cdot & & \cdot \\ a_{m1} & a_{m2} & \cdots & a_{mn} \end{pmatrix} \begin{pmatrix} x_1 \\ x_2 \\ \cdot \\ \cdot \\ \cdot \\ x_n \end{pmatrix} = \begin{pmatrix} a_{11}x_1 + a_{12}x_2 + \cdots + a_{1n}x_n \\ a_{21}x_1 + a_{22}x_2 + \cdots + a_{2n}x_n \\ \cdot \\ \cdot \\ \cdot \\ a_{m1}x_1 + a_{m2}x_2 + \cdots + a_{mn}x_n \end{pmatrix}$$

This computation can be interpreted as the product of an $m \times n$ matrix A and the $n \times 1$ matrix $\mathbf{X}$.

Let $\mathbf{Y}$ represent the vector in $\mathbb{R}^m$ (written vertically) or the $m \times 1$ matrix appearing on the right-hand side of the equation above. We abbreviate the equation by writing

$$A\mathbf{X} = \mathbf{Y}$$

and we interpret this equation in two ways. One way to interpret it is to think of matrix multiplication with the $m \times n$ matrix A, the $n \times 1$ matrix $\mathbf{X}$, and the $m \times 1$ matrix $\mathbf{Y}$. The other way to interpret it is to think of A as determining a linear transformation which operates on the vector $\mathbf{X}$ in $\mathbb{R}^n$ to produce a vector $\mathbf{Y}$ in $\mathbb{R}^m$. We use both points of view, ordinarily choosing the one which makes our work easier.

Brief Exercises

In the following cases, verify Propositions 4-9 and 4-10 by showing $T \circ S$ is linear and that $A_{T\circ S} = A_T A_S$.

1. $S(x,y) = (3x + y, x - y)$
   $T(u,v) = (2u + 4v, 3u - 2v)$

2. $S(x,y,z) = (x - y + z, 2x + 3y - z)$
   $T(u,v) = (3u + 4v, u - v)$

3. $S(x,y,z) = (x + 4y - 3z, x + y + 2z, 2x + y)$
   $T(u,v,w) = (2u + w, 5u - v + 2w, 6u + 3v - w)$

Since we can perform the operations of addition, scalar multiplication, and composition with functions, and we can perform the operations of addition, scalar multiplication, and multiplication with matrices, we can do arithmetic with them. In this part of this section we discuss some similarities and some differences between arithmetic for real numbers and for n-tuples and arithmetic for linear transformations from $\mathbb{R}^n$ to $\mathbb{R}^m$ and for $m \times n$ matrices.

Example   Suppose $T: \mathbb{R}^2 \to \mathbb{R}^3$ is defined by $T(x,y) = (2x + y, x - 3y, x + y)$. Its matrix is

$$A = \begin{pmatrix} 2 & 1 \\ 1 & -3 \\ 1 & 1 \end{pmatrix}.$$

Let $T_0: \mathbb{R}^2 \to \mathbb{R}^3$ be defined by $T_0(x,y) = (0,0,0)$. Its matrix is

$$0 = \begin{pmatrix} 0 & 0 \\ 0 & 0 \\ 0 & 0 \end{pmatrix}.$$

Then it is clear that $(T + T_0)(x,y) = T(x,y)$, so that $T + T_0 = T$; i.e., T_0 acts like a zero among the linear transformations. So we call it the zero linear transformation from $\mathbb{R}^2$ to $\mathbb{R}^3$. It is also clear that

$$A + 0 = A;$$

thus the 3×2 matrix 0 acts like a zero among the 3×2 matrices. So we call it the zero 3×2 matrix.

Furthermore, let $-T: \mathbb{R}^2 \to \mathbb{R}^3$ be the function defined by $(-T)(x,y) = (-2x - y, -x + 3y, -x - y)$. Then

$$T + (-T) = T_0$$

and so $-T$ acts like the negative of T among the linear functions from $\mathbb{R}^2$ to $\mathbb{R}^3$. Similarly the matrix

$$-A = \begin{pmatrix} -2 & -1 \\ -1 & 3 \\ -1 & -1 \end{pmatrix}$$

is the negative of the matrix A, since $A + (-A) = 0$.

Definition 4-14

The *zero linear transformation* $T_0: \mathbb{R}^n \to \mathbb{R}^m$ is defined by

$$T_0(\mathbf{X}) = 0 \qquad \text{or} \qquad T(x_1, \ldots, x_n) = (0, \ldots, 0).$$

The *negative of* T is defined by

$$(-T)(\mathbf{X}) = -T(\mathbf{X}) \qquad \text{or} \qquad (-T)(x_1, \ldots, x_n) = -T(x_1, \ldots, x_n).$$

The *zero* m × n *matrix 0* is the $m \times n$ matrix having every entry equal to 0. The *negative* of the matrix A is the matrix $-A$ obtained by multiplying each entry of A by -1. As with real numbers, $S - T$ is shorthand for $S + (-T)$, and $A - B$ is shorthand for $A + (-B)$.

Brief Exercises

1. Let R, S, and T be the linear transformations from $\mathbb{R}^2$ to $\mathbb{R}^2$ defined by $R(x,y) = (2x + 3y, x - y)$, $S(x,y) = (2x - y, x + y)$, and $T(x,y) = (3x - 4y, x + 2y)$. Find the defining equations for $3R + 5S - T$, $4T - 3S + 3R$, and $6S - 2R + T$.

2. Let

$$A = \begin{pmatrix} 1 & 2 & -1 \\ 3 & 0 & 4 \end{pmatrix} \qquad\qquad B = \begin{pmatrix} 2 & 1 & 3 \\ 5 & 0 & 2 \end{pmatrix}.$$

Compute $-A$, $-B$, $5A - B$, and $6B + 11A$.

3. With A and B as in Exercise 2 solve the following equations. C is the unknown matrix.

(a) $2C + A = B$ **(b)** $5C + 4A = 3B$ **(c)** $3C - 4A = 5B$

4. With A and B as in Exercise 2 solve for the unknown matrices C and D.

$$2C + 3D = 5A$$

$$C + 4D = 4B$$

5. With R, S, and T defined as in Exercise 1, solve for the unknown functions U and V.

$$5U + 2V = 3R$$

$$U + 4V = 6T$$

Selected Answers:

3(a) $C = \begin{pmatrix} \frac{1}{2} & -\frac{1}{2} & 2 \\ 1 & 0 & -1 \end{pmatrix}$

3(c) $C = (\frac{1}{3})\begin{pmatrix} 14 & 13 & 11 \\ 37 & 0 & 26 \end{pmatrix}$

3(b) $C = (\frac{1}{5})\begin{pmatrix} 2 & -5 & 13 \\ 3 & 0 & -10 \end{pmatrix}$

4 $C = \begin{pmatrix} -\frac{4}{5} & \frac{28}{5} & -\frac{56}{5} \\ 0 & 0 & \frac{56}{5} \end{pmatrix}$

$D = \begin{pmatrix} \frac{11}{5} & -\frac{2}{5} & \frac{29}{5} \\ 5 & 0 & -\frac{4}{5} \end{pmatrix}$

5 $U(x,y) = (-\frac{2}{3}x + \frac{14}{3}y, -2y)$, $V(x,y) = (\frac{14}{3}x - \frac{43}{6}y, \frac{3}{2}x + \frac{7}{2}y)$

Another similarity between arithmetic for real numbers and arithmetic for matrices is that multiplication is associative. To see this, suppose A, B, and C are three matrices having sizes so that the products $A(BC)$ and $(AB)C$ are defined. Then they are equal: $A(BC) = (AB)C$. One way to see this is to recall that A, B, and C determine some linear transformations, say R, S, and T, respectively (see Proposition 4-7). Because the matrices have the proper sizes so that $A(BC)$ and $(AB)C$ are defined, the composite functions $R \circ (S \circ T)$ and $(R \circ S) \circ T$ are defined. By Proposition 4-1, $R \circ (S \circ T) = (R \circ S) \circ T$. Therefore, their matrices, $A(BC)$ and $(AB)C$, are equal.

Proposition 4-11_____

Multiplication of matrices is associative, i.e., $A(BC) = (AB)C$ provided A, B, and C are the proper sizes so that all the products are defined.

_____

As important as the similarities are between arithmetic for linear transformations and matrices and arithmetic for real numbers, the differences are equally important. These differences occur when we draw parallels between multiplication of real numbers and either composition of functions or multiplication of matrices.

One important difference is that with real numbers we have $st = ts$, whereas with matrices A and B we do not necessarily have $AB = BA$, as we noted in an example earlier in this section.

Another point of difference between arithmetic with real numbers and arithmetic of matrices is this: the number 1 has the property that $t \cdot 1 = 1 \cdot t = t$ for *every* real number t. There are matrices and linear transformations which have properties like this, but the situation is more complicated than with real numbers.

Example   Suppose A is the 2×3 matrix

$$A = \begin{pmatrix} 1 & 2 & -3 \\ 4 & 5 & -2 \end{pmatrix}.$$

If some matrix B has the property that $AB = A$, then B must be a 3×3 matrix. You can check that

$$\begin{pmatrix} 1 & 2 & -3 \\ 4 & 5 & -2 \end{pmatrix} \begin{pmatrix} 1 & 0 & 0 \\ 0 & 1 & 0 \\ 0 & 0 & 1 \end{pmatrix} = \begin{pmatrix} 1 & 2 & -3 \\ 4 & 5 & -2 \end{pmatrix}.$$

However, if a matrix C has the property that $CA = A$, then C must be a 2×2 matrix. You can verify that

$$\begin{pmatrix} 1 & 0 \\ 0 & 1 \end{pmatrix} \begin{pmatrix} 1 & 2 & -3 \\ 4 & 5 & -2 \end{pmatrix} = \begin{pmatrix} 1 & 2 & -3 \\ 4 & 5 & -2 \end{pmatrix}.$$

This shows that it is impossible to find *one* matrix B so that $AB = BA = A$. One reason is that A is a 2×3 matrix, i.e. it is not a square matrix.

Example   Suppose A is the 3×3 matrix

$$A = \begin{pmatrix} 1 & -3 & 4 \\ 5 & 2 & 0 \\ -1 & -5 & -4 \end{pmatrix}.$$

You can check that

$$\begin{pmatrix} 1 & -3 & 4 \\ 5 & 2 & 0 \\ -1 & -5 & -4 \end{pmatrix} \begin{pmatrix} 1 & 0 & 0 \\ 0 & 1 & 0 \\ 0 & 0 & 1 \end{pmatrix} = \begin{pmatrix} 1 & -3 & 4 \\ 5 & 2 & 0 \\ -1 & -5 & -4 \end{pmatrix}$$

and

$$\begin{pmatrix} 1 & 0 & 0 \\ 0 & 1 & 0 \\ 0 & 0 & 1 \end{pmatrix} \begin{pmatrix} 1 & -3 & 4 \\ 5 & 2 & 0 \\ -1 & -5 & -4 \end{pmatrix} = \begin{pmatrix} 1 & -3 & 4 \\ 5 & 2 & 0 \\ -1 & -5 & -4 \end{pmatrix}.$$

From these two examples you can see that there is no one matrix which operates with matrices like the number 1 with real numbers. However, there are matrices which operate somewhat like the number 1.

Definition 4-15_____

The $n \times n$ *identity matrix* or the *identity matrix of order n* is the $n \times n$ matrix with each (i,i) entry equal to 1 and all other entries equal to 0. This matrix is denoted by I_n.

$$I_n = \begin{pmatrix} 1 & 0 & \cdots & 0 \\ 0 & 1 & \cdots & 0 \\ \cdot & \cdot & & \cdot \\ \cdot & \cdot & & \cdot \\ \cdot & \cdot & & \cdot \\ 0 & 0 & \cdots & 1 \end{pmatrix}$$

(As usual, n represents a positive integer.)

From the previous two examples, it is easy to see the next proposition.

Proposition 4-12_____

If A is an $m \times n$ matrix, then $I_m A = A$ and $A I_n = A$. So if A is a square matrix, i.e., is an $n \times n$ matrix, then $A I_n = I_n A = A$.

We have emphasized in Section 4-3 that there is a correspondence between matrices and linear transformations. From Proposition 4-7 we know that the matrix I_n defines a linear transformation from $\mathbb{R}^n$ to $\mathbb{R}^n$; the defining equation for this transformation is $T(x_1, \ldots, x_n) = (x_1, \ldots, x_n)$ or $T(\mathbf{X}) = \mathbf{X}$. We call this function the *identity function on* $\mathbb{R}^n$, and we denote it by $I_{\mathbb{R}^n}$. If the domain is clear from the context, then we denote this function simply by I.

Definition 4-16_____

The *identity function* from a vector space to itself is the function $I_{\mathscr{V}}: \mathscr{V} \to \mathscr{V}$ defined by requiring that, for each $\mathbf{X}$ in $\mathscr{V}$, $I_{\mathscr{V}}(\mathbf{X}) = \mathbf{X}$. If the domain of the function is clear from the context, the identity is denoted simply by I. Notice that I is a linear transformation.

Proposition 4-13_____

Suppose that $T: \mathbb{R}^n \to \mathbb{R}^m$ is a linear transformation; then $T \circ I_{\mathbb{R}^n} = T$ and $I_{\mathbb{R}^m} \circ T = T$.

SUMMARY *The sum, scalar multiple, and composite of linear transformations is again a linear transformation. Two matrices of the same size are added by adding entries which occupy the same position in the two matrices. The scalar multiple of a matrix is obtained by multiplying each entry of the matrix by the scalar. The* *product AB of an m × n matrix A and an n × p matrix B is obtained by applying A to each column of B; i.e., the (i,k) entry of AB is the sum of the products of corresponding entries of the ith row of A and the kth column of B.*

$$a_{i1}b_{1k} + a_{i2}b_{2k} + \cdots + a_{in}b_{nk}$$

The matrix for the sum of linear transformations is the sum of their matrices. The matrix for a scalar multiple of a linear transformation is that scalar multiple of the matrix. The matrix for a composite of linear transformations is the product of their matrices in the same order.

Matrix multiplication is associative but not commutative. The identity matrices I_n are introduced. They have the property that if A is an m × n matrix, then

$$I_m A = A I_n = A.$$

Exercises for Section 4-4

1. In each of the following, write formulas for $S - 4T$ and $3S + 2T$.

   (a) $S(x,y) = (2x + y, x - y, 3x + 2y)$
       $T(x,y) = (4x - y, x + y, x - y)$

   (b) $S(x,y,z) = (x - y - z, 3x + 2y - z)$
       $T(x,y,z) = (5x + y - 2z, x + 3y + 2z)$

   (c) $S(x,y,z) = (2x + 2y - 3z, x - y + z)$
       $T(x,y,z) = (4x + y - 4z, x - y + 2z)$

2. In each of the parts of Exercise 1, find the matrix for S and the matrix for T (call them A and B, respectively), and compute $A + B$ and $4A - 3B$.

3. Do the following matrix computations.

   (a) $2\begin{pmatrix} 5 & 0 & 2 \\ -1 & 3 & 5 \end{pmatrix} + 6\begin{pmatrix} 1 & -2 & 1 \\ 3 & 4 & 5 \end{pmatrix}$

   (b) $\begin{pmatrix} 2 & 3 \\ 1 & -1 \\ 2 & 1 \end{pmatrix} + \begin{pmatrix} 0 & 0 \\ 0 & 0 \\ 0 & 0 \end{pmatrix}$

   (c) $\begin{pmatrix} 2 & 3 \\ 1 & -1 \\ 2 & 1 \end{pmatrix} + \begin{pmatrix} -2 & -3 \\ -1 & 1 \\ -2 & -1 \end{pmatrix}$

4. Compute $T \circ S$ in the following cases without the use of matrices.

    (a) $S(x,y,z) = (3x - 4y + z, x - y + z)$

       $T(u,v) = (7u - 4v, 18u + 17v)$

    (b) $S(x,y,z,w) = (x + y - z + 2w, x - y + 2z - w, 4x - 3z)$

       $T(r,s,t) = (5r + 6s + t, -r + 3s + 2t, s + t)$

5. Compute $T \circ S$ in the cases listed in Exercise 4 by finding the matrix for S, the matrix for T, and then the matrix for $T \circ S$.

Do the following matrix multiplications.

6.
$$
\begin{pmatrix} 2 & 5 & 1 & 6 \\ 8 & 1 & -2 & 3 \\ 8 & 0 & 2 & 1 \end{pmatrix}
\begin{pmatrix} 4 & -5 \\ 1 & 2 \\ 8 & 1 \\ 6 & 0 \end{pmatrix}
$$

7.
$$
\begin{pmatrix} 5 & 2 & 1 & 3 \\ 8 & 6 & 1 & 2 \\ 0 & 1 & 5 & 8 \end{pmatrix}
\begin{pmatrix} 1 & 4 & 8 \\ 2 & 0 & 1 \\ -1 & 1 & 0 \\ 5 & 8 & 6 \end{pmatrix}
$$

8. (a)
$$
\begin{pmatrix} 1 & 0 & 0 \\ 0 & 1 & 0 \\ 0 & 0 & 1 \end{pmatrix}
\begin{pmatrix} 5 & 2 & 1 & 3 \\ 8 & 6 & 1 & 2 \\ 0 & 1 & 5 & 8 \end{pmatrix}
$$

    (b)
$$
\begin{pmatrix} 5 & 2 & 1 & 3 \\ 8 & 6 & 1 & 2 \\ 0 & 1 & 5 & 8 \end{pmatrix}
\begin{pmatrix} 1 & 0 & 0 & 0 \\ 0 & 1 & 0 & 0 \\ 0 & 0 & 1 & 0 \\ 0 & 0 & 0 & 1 \end{pmatrix}
$$

9. (a)
$$
\begin{pmatrix} 0 & 1 & 0 \\ 1 & 0 & 0 \\ 0 & 0 & 1 \end{pmatrix}
\begin{pmatrix} 5 & 2 & 1 & 3 \\ 8 & 6 & 1 & 2 \\ 0 & 1 & 5 & 8 \end{pmatrix}
$$

    (b)
$$
\begin{pmatrix} 1 & 0 & 0 \\ 0 & 1 & 0 \\ 0 & 0 & 5 \end{pmatrix}
\begin{pmatrix} 5 & 2 & 1 & 3 \\ 8 & 6 & 1 & 2 \\ 0 & 1 & 5 & 8 \end{pmatrix}
$$

10. (a) $\begin{pmatrix} 1 & 0 & -2 \\ 0 & 1 & 0 \\ 0 & 0 & 1 \end{pmatrix} \begin{pmatrix} 5 & 2 & 1 & 3 \\ 8 & 6 & 1 & 2 \\ 0 & 1 & 5 & 8 \end{pmatrix}$

(b) $\begin{pmatrix} 1 & 0 & 0 \\ -2 & 1 & 0 \\ 0 & 0 & 1 \end{pmatrix} \begin{pmatrix} 5 & 2 & 1 & 3 \\ 8 & 6 & 1 & 2 \\ 0 & 1 & 5 & 8 \end{pmatrix}$

11. Compute $A(BC)$ and $(AB)C$ where

$$A = \begin{pmatrix} 2 & 1 & 0 \\ 5 & -1 & 3 \\ 1 & 2 & 1 \end{pmatrix} \quad B = \begin{pmatrix} 2 & 1 & 5 & 6 \\ 0 & -1 & 2 & 1 \\ 0 & 0 & -1 & 0 \end{pmatrix} \quad C = \begin{pmatrix} 5 & 2 \\ 1 & 8 \\ 2 & 1 \\ 0 & 3 \end{pmatrix}.$$

12. Let

$$A = \begin{pmatrix} 5 & -1 & 2 \\ 3 & 4 & 1 \\ 6 & 8 & 0 \end{pmatrix} \quad B = \begin{pmatrix} 3 & 4 & 3 \\ 2 & 1 & 1 \\ 1 & -1 & 0 \end{pmatrix} \quad C = \begin{pmatrix} 1 & 0 & -1 \\ 2 & 1 & 3 \\ 0 & -2 & 4 \end{pmatrix}$$

$$D = \begin{pmatrix} 1 & 5 & 1 \\ 2 & -1 & 3 \end{pmatrix} \quad E = \begin{pmatrix} -1 & 2 & 4 \\ 6 & -5 & 2 \end{pmatrix}.$$

Perform the following calculations. AB, AC, BA, CA, BC, CB, $A(B + C)$, $C(B + 3A)$, and $(E + D)B$

13. If A is a square matrix, then the product AA is defined. We use A^2 for this product, A^3 for AAA, etc. Compute the following: $2A^2 + 3A$, $A^2 - 4A + 3I_2$, where

$$A = \begin{pmatrix} 1 & 2 \\ 0 & 3 \end{pmatrix}.$$

14. For the 3×3 matrix A below, compute $4A^3 - 3A^2 + 2A$, $A^3 - 7A^2 + 16A - 22I_3$, if

$$A = \begin{pmatrix} 1 & 2 & 0 \\ -1 & 2 & 1 \\ 3 & 0 & 4 \end{pmatrix}.$$

More Challenging Exercises

15. In the argument supporting Proposition 4-11 we made the statement, "Because the matrices have the proper sizes so that $A(BC)$ and $(AB)C$ are defined, the composite functions $R \circ (S \circ T)$ and $(R \circ S) \circ T$ are defined." Give a justification for this statement in the following cases.

(a) A is 2×3, B is 3×4, and C is 4×3
(b) A is 4×2, B is 2×5, and C is 5×5
(c) A is $m \times n$, B is $n \times p$, and C is $p \times q$

16. It is a fact that if A and B are matrices of the sizes so that AB is defined, then $(AB)^T = B^T A^T$. Verify this in the following particular cases.

(a)
$$A = \begin{pmatrix} 2 & 1 \\ 1 & 3 \\ 4 & 1 \end{pmatrix} \qquad B = \begin{pmatrix} 1 & 1 \\ 3 & 4 \end{pmatrix}$$

(b)
$$A = \begin{pmatrix} 1 & 4 & 1 \\ 2 & 1 & 3 \\ 1 & -1 & 2 \end{pmatrix} \qquad B = \begin{pmatrix} 1 & 5 & 0 \\ -1 & -3 & 1 \\ 4 & 2 & 3 \end{pmatrix}$$

17. Give an argument to support the statement that $(AB)^T = B^T A^T$ in general with, say, A being $m \times n$ and B being $n \times p$. (Hint: the (i,k) entry of AB is $a_{i1}b_{1k} + \cdots + a_{in}b_{nk}$. This is therefore the (k,i) entry of $(AB)^T$. Compare it with the (k,i) entry of $B^T A^T$.)

18. Verify that the set of 3×5 matrices satisfy all the properties listed in Definition 2-5. That is, show that the 3×5 matrices, with addition and scalar multiplication as defined in this section, form a vector space over $\mathbb{R}$. Show that the dimension of this vector space is 15. (Hint: consider the matrices $E_{i,j}$ having the number 1 in the (i,j) entry and 0 elsewhere.)

19. Formulate and justify a statement about the $m \times n$ matrices which is similar to the statement made about the 3×5 matrices in Exercise 18.

20. If A is an $m \times n$ matrix and $\mathbf{X}$ and $\mathbf{X}'$ are $n \times 1$ matrices (or vectors in $\mathbb{R}^n$), show that $A(\mathbf{X} + \mathbf{X}') = A\mathbf{X} + A\mathbf{X}'$ and $A(r\mathbf{X}) = rA\mathbf{X}$ if r is a scalar. Is this still true if $\mathbf{X}$ and $\mathbf{X}'$ are $n \times p$ matrices?

SECTION 4-5
Some Applications

PREVIEW *In this section we indicate how matrices can be used outside mathematics. It is impossible, in the space of a short book such as this one, to give an adequate survey of all the ways matrices can be used. Consequently, we have attempted to select some applications which give the general idea of how they can be used and which we ourselves have found interesting.*

Matrices are particularly useful when we must deal with large-scale problems, those in which there are many factors influencing what we are trying to analyze. However, the physical size of the page of this book and the time it takes to write out a large matrix have imposed limits on the scale of the problem we can present. Even though the applications are not, therefore, as realistic as we would like, we believe that considering small-scale examples will permit you to imagine how problems on a larger scale can be handled.

Finally, in some cases we do little more than describe a kind of use of matrices and then refer you to other sources for a fuller discussion. You should consult these other books to help you get further insight into the significance and power of matrix theory and linear algebra.

Example (Addition of matrices) An owner of two drugstores sells four lines of ball point pens in five colors. He can represent his inventory in each store as a 5×4 matrix as follows.

	Brand	1	2	3	4
	Red	5	3	16	2
	Blue	4	1	0	3
Store 1	Yellow	5	4	6	7
	Green	10	9	8	3
	Black	2	3	1	1
	Red	2	0	8	3
	Blue	4	1	2	1
Store 2	Yellow	2	1	0	3
	Green	4	3	2	2
	Black	6	2	2	4

The combined inventory is then the sum of these matrices.

As you can readily understand, it is not essential that inventory be handled with matrices. The important point is that some standardized pattern for storing the information can save space and time. Rather than recording that Store 2 has 4 blue Brand 1 pens, we can save time and space by writing 4 in a particular position. The position of the number then carries the information about the color and the brand. Matrices, therefore, provide a vehicle for recording information in an economical and systematic fashion.

Example Suppose we have five positions labeled 1 through 5 and suppose we have paths between these points as indicated in Figure 4-7. This information can

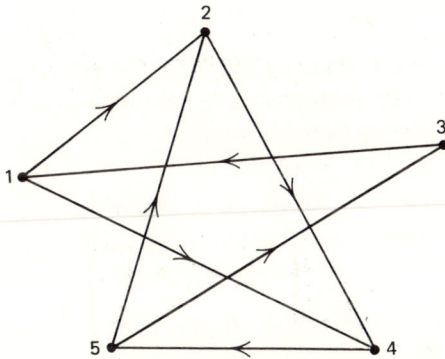

Figure 4-7

be recorded in the following matrix A.

$$
\begin{array}{c}
\\
\text{From 1} \\
\text{From 2} \\
\text{From 3} \\
\text{From 4} \\
\text{From 5}
\end{array}
\begin{array}{c}
\text{To 1 To 2 To 3 To 4 To 5} \\
\begin{pmatrix}
0 & 1 & 0 & 1 & 0 \\
0 & 0 & 0 & 1 & 0 \\
1 & 0 & 0 & 0 & 0 \\
0 & 0 & 0 & 0 & 1 \\
0 & 1 & 1 & 0 & 0
\end{pmatrix}
\end{array}
$$

If we can go directly from i to j, then we place the number 1 in the (i,j) position; if we cannot, then we place 0 in that position.

The matrix A^2 is

$$A^2 = \begin{pmatrix} 0 & 0 & 0 & 1 & 1 \\ 0 & 0 & 0 & 0 & 1 \\ 0 & 1 & 0 & 1 & 0 \\ 0 & 1 & 1 & 0 & 0 \\ 1 & 0 & 0 & 1 & 0 \end{pmatrix}.$$

This matrix shows the two-step paths in the diagram. To see this, consider the expression for the entries of A^2 in terms of the entries of A. Let a_{ij} represent the (i,j) entry of A. Then the (i,j) entry of A^2 is given by

$$a_{i1}a_{1j} + a_{i2}a_{2j} + \cdots + a_{i5}a_{5j}.$$

The term $a_{i2}a_{2j}$, for example, is nonzero if and only if $a_{i2} = 1$ and $a_{2j} = 1$, i.e. if and only if there is a path from i to 2 and a path from 2 to j. The term $a_{i2}a_{2j}$ tells whether there is a two-step path from i to j through 2; thus the (i,j) entry of A^2 gives the number of two-step paths from i to j.

The matrix A^3 shows all the three-step paths.

$$A^3 = \begin{pmatrix} 0 & 1 & 1 & 0 & 1 \\ 0 & 1 & 1 & 0 & 0 \\ 0 & 0 & 0 & 1 & 1 \\ 1 & 0 & 0 & 1 & 0 \\ 0 & 1 & 0 & 1 & 1 \end{pmatrix}$$

To see why A^3 shows the number of three-step paths in the diagram, let b_{ij} represent the (i,j) entry of A^2. The (i,j) entry of A^3 is therefore given by

$$a_{i1}b_{1j} + a_{i2}b_{2j} + \cdots + a_{i5}b_{5j}.$$

The term $a_{i1}b_{1j}$, for example, is nonzero provided $a_{i1} = 1$ and $b_{1j} \neq 0$, i.e., if there is one-step path from i to 1 and a two-step path from 1 to j. Thus the (i,j) entry of A^3 gives the number of three-step paths from i to j. A similar interpretation holds for A^4 and higher powers of A.

The matrix $I_5 + A$ shows the various possibilities of zero-step and one-step paths. The matrix $I_5 + A + A^2$ shows the total number of possible zero-step, one-step, and two-step paths.

$$I_5 + A + A^2 = \begin{pmatrix} 1 & 1 & 0 & 2 & 1 \\ 0 & 1 & 0 & 1 & 1 \\ 1 & 1 & 1 & 1 & 0 \\ 0 & 1 & 1 & 1 & 1 \\ 1 & 1 & 1 & 1 & 1 \end{pmatrix}$$

The (1,4) entry shows there are two paths from 1 to 4 using no more than two steps. The fact that every entry of the bottom row is nonzero indicates that from 5 you can get to any other point in two steps or less.

The same kind of analysis can be used to study networks of communications, for example, or routes of transportation.

Example Suppose we wish to study the pattern of friendship in a six-person group. We construct a matrix by placing the number 1 in the (i,j) position if the ith person likes the jth person. Suppose we obtain the matrix

$$A = \begin{pmatrix} 0 & 1 & 1 & 0 & 0 & 1 \\ 1 & 0 & 0 & 1 & 1 & 1 \\ 1 & 1 & 0 & 0 & 1 & 0 \\ 0 & 1 & 1 & 0 & 1 & 1 \\ 1 & 1 & 0 & 1 & 0 & 1 \\ 0 & 1 & 1 & 0 & 1 & 0 \end{pmatrix}.$$

In sociology it is sometimes of value to detect cliques within the group. For this purpose, a clique is understood to be a collection of at least three persons from the group, each reciprocating the relation with all the others, and no more persons from the group can be added and still have everyone reciprocating the relation with all the others. If there is such a clique, there must be at least three persons, say persons P_1, P_2, and P_3, each of whom reciprocate the relation with the other two. Thus we can think of a relationship path from P_1 to P_2 to P_3 and back to P_1. In other words, for each member of a clique it is possible to find a three-step path from himself back to himself. Suppose we form a new matrix B by setting the (i,j) entry of B equal to 1 if both the (i,j) and the (j,i) entry of A are equal to 1; and by setting the (i,j) entry of B equal to 0 in any other case. Thus the number 1 in the (i,j) entry of B indicates that persons P_i and P_j reciprocate the relation with each other.

$$B = \begin{pmatrix} 0 & 1 & 1 & 0 & 0 & 0 \\ 1 & 0 & 0 & 1 & 1 & 1 \\ 1 & 0 & 0 & 0 & 0 & 0 \\ 0 & 1 & 0 & 0 & 1 & 0 \\ 0 & 1 & 0 & 1 & 0 & 1 \\ 0 & 1 & 0 & 0 & 1 & 0 \end{pmatrix}$$

Consequently the matrix B records the various reciprocated relations. The matrix B^2 records the number of two-step reciprocated relations, and B^3 records the number of three-step reciprocated relations. In this case

$$
B^3 = \begin{pmatrix}
0 & 5 & 2 & 1 & 2 & 1 \\
5 & 4 & 0 & 6 & 6 & 6 \\
2 & 0 & 0 & 1 & 1 & 1 \\
1 & 6 & 1 & 2 & 5 & 2 \\
2 & 6 & 1 & 5 & 4 & 5 \\
1 & 6 & 1 & 2 & 5 & 2
\end{pmatrix}.
$$

The fact that the (2,2), (4,4), (5,5), and (6,6) entries are nonzero means that Persons 2, 4, 5, and 6 are each members of a clique. You can check the original matrix A to see that there are two three-person cliques, one consisting of Persons 2, 4, and 5, the other consisting of Persons 2, 5, and 6.

Example   Matrices can be used to analyze many kinds of electrical circuits. A simple example is that of a resistor ladder circuit or resistor circuits connected in cascade. Take, for example, the circuit illustrated in Figure 4-8. We can predict

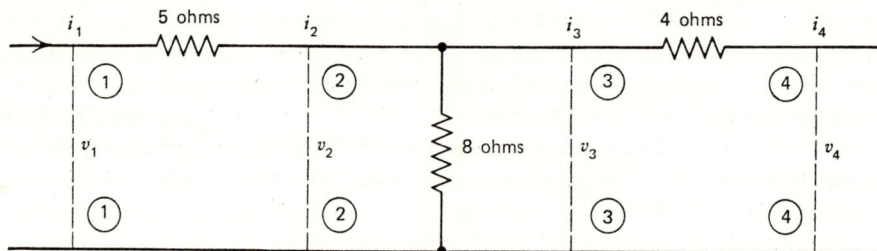

Figure 4-8

the voltage and current at the points labeled 4 if we know the current and voltage at the points labeled 1. We consider the circuit as being constructed from the two simple situations shown in Figures 4-9 and 4-10. In the first type, Ohm's law and Kirchhoff's laws give

$$
i_1 = i_2
$$

$$
v_2 = v_1 - i_1 r
$$

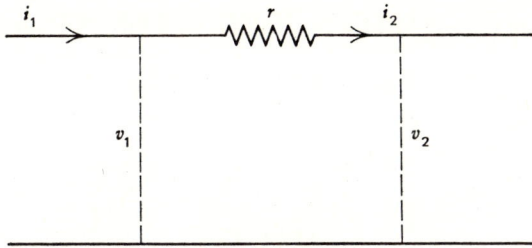

Figure 4-9

or, in matrix formulation,

$$\begin{pmatrix} v_2 \\ i_2 \end{pmatrix} = \begin{pmatrix} 1 & -r \\ 0 & 1 \end{pmatrix} \begin{pmatrix} v_1 \\ i_1 \end{pmatrix}.$$

In the second type we obtain

$$i_1 = i_2 + i$$

$$v_1 = ir = v_2$$

or, in matrix formulation,

$$\begin{pmatrix} v_2 \\ i_2 \end{pmatrix} = \begin{pmatrix} 1 & 0 \\ -1/r & 1 \end{pmatrix} \begin{pmatrix} v_1 \\ i_1 \end{pmatrix}.$$

In the case illustrated above, v_1 and i_1 are the measurements made at 1, v_2 and i_2 at 2, v_3 and i_3 at 3, and v_4 and i_4 at 4. We obtain

$$\begin{pmatrix} v_2 \\ i_2 \end{pmatrix} = \begin{pmatrix} 1 & -5 \\ 0 & 1 \end{pmatrix} \begin{pmatrix} v_1 \\ i_1 \end{pmatrix}$$

$$\begin{pmatrix} v_3 \\ i_3 \end{pmatrix} = \begin{pmatrix} 1 & 0 \\ -\frac{1}{8} & 1 \end{pmatrix} \begin{pmatrix} v_2 \\ i_2 \end{pmatrix}$$

$$\begin{pmatrix} v_4 \\ i_4 \end{pmatrix} = \begin{pmatrix} 1 & -4 \\ 0 & 1 \end{pmatrix} \begin{pmatrix} v_3 \\ i_3 \end{pmatrix}.$$

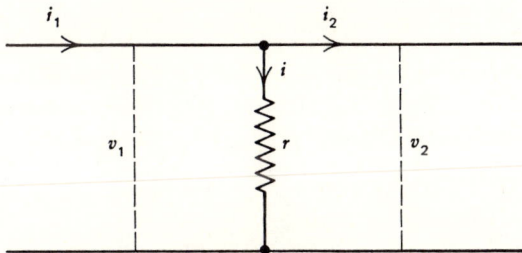

Figure 4-10

Then v_4 and i_4 can be computed from v_1 and i_1 with the aid of matrix multiplication.

$$\begin{pmatrix} v_4 \\ i_4 \end{pmatrix} = \begin{pmatrix} 1 & -4 \\ 0 & 1 \end{pmatrix}\begin{pmatrix} v_3 \\ i_3 \end{pmatrix} = \begin{pmatrix} 1 & -4 \\ 0 & 1 \end{pmatrix}\begin{pmatrix} 1 & 0 \\ -\frac{1}{8} & 1 \end{pmatrix}\begin{pmatrix} v_2 \\ i_2 \end{pmatrix}$$

$$= \begin{pmatrix} 1 & -4 \\ 0 & 1 \end{pmatrix}\begin{pmatrix} 1 & 0 \\ -\frac{1}{8} & 1 \end{pmatrix}\begin{pmatrix} 1 & -5 \\ 0 & 1 \end{pmatrix}\begin{pmatrix} v_1 \\ i_1 \end{pmatrix}$$

$$= \begin{pmatrix} \frac{3}{2} & -\frac{23}{2} \\ -\frac{1}{8} & \frac{13}{8} \end{pmatrix}\begin{pmatrix} v_1 \\ i_1 \end{pmatrix}$$

Example   Suppose that long years of observations show that the following matrix gives the probability of having a certain type of weather follow another from one day to the next.

	Rain	Snow	Clear
Rain	.3	.2	.5
Snow	.3	.3	.4
Clear	.4	.2	.4

Then the probabilities of having a certain type of weather on Wednesday given the type of weather on Monday is given by

$$A^2 = \begin{pmatrix} .35 & .22 & .43 \\ .34 & .23 & .43 \\ .34 & .22 & .44 \end{pmatrix}.$$

Similarly, the probabilities of having a certain type of weather on Friday given the type of weather on Monday is given by the matrix A^4.

Many references to books containing applications of matrices to other areas can be found in the exercises at the end of Chapter 2 of Noble's book [9] and throughout the book by Campbell [3]. See also the references at the end of the book.

5

Nonsingular Linear Transformations and Nonsingular Matrices

SECTION 5-1

Introduction to Nonsingular Linear Transformations and Matrices; A Method for Finding the Inverse of a Nonsingular Matrix

PREVIEW *An $n \times n$ matrix A is nonsingular if and only if there is an $n \times n$ matrix A^{-1} so that $A^{-1}A = AA^{-1} = I_n$. A linear transformation $T: \mathbb{R}^n \to \mathbb{R}^n$ is nonsingular if and only if there is a linear transformation $T^{-1}; \mathbb{R}^n \to \mathbb{R}^n$ so that $T^{-1} \circ T = T \circ T^{-1} = I_{\mathbb{R}^n}$. A linear transformation is nonsingular if and only if its matrix is nonsingular. A procedure is presented which tells whether a matrix is nonsingular and which yields the inverse of A if A is nonsingular.*

One of the most useful facts about real numbers is that every nonzero real number r has an inverse or reciprocal; i.e., if $r \neq 0$, there is a number r^{-1} having the property that $rr^{-1} = r^{-1}r = 1$. This enables us to solve equations such as $2x = 3$. We multiply both sides by 2^{-1} or $\frac{1}{2}$ to obtain $2^{-1}(2x) = \frac{3}{2}$. We use associativity of multiplication: $(2^{-1}2)x = \frac{3}{2}$. Then we have, by the definition of 2^{-1}, $1x = \frac{3}{2}$. Thus we can solve equations $ax = b$ as long as $a \neq 0$, or as long as there is another number a^{-1} so that $a^{-1}a = 1$. In a similar way we can solve matrix equations $AX = Y$ if A has an inverse as we illustrate in the next examples.

Example Consider the system of linear equations

$$3x + 2y = 1$$

$$x - y = 2.$$

We can view this as a matrix equation $AX = Y$

$$\begin{pmatrix} 3 & 2 \\ 1 & -1 \end{pmatrix} \begin{pmatrix} x \\ y \end{pmatrix} = \begin{pmatrix} 1 \\ 2 \end{pmatrix}$$

where $\mathbf{X} = \begin{pmatrix} x \\ y \end{pmatrix}$ and $\mathbf{Y} = \begin{pmatrix} 1 \\ 2 \end{pmatrix}$ are thought of as 2×1 matrices. If we can find a

204

matrix A^{-1} so that $AA^{-1} = A^{-1}A = I_2$, then we can solve this matrix equation by multiplying by A^{-1} on the left.

$$AX = Y$$

$$A^{-1}(AX) = A^{-1}Y$$

$$(A^{-1}A)X = A^{-1}Y$$

$$I_2X = A^{-1}Y$$

$$X = A^{-1}Y$$

In this particular case the matrix

$$\begin{pmatrix} \frac{1}{5} & \frac{2}{5} \\ \frac{1}{5} & -\frac{3}{5} \end{pmatrix}$$

is the one we want because

$$\begin{pmatrix} \frac{1}{5} & \frac{2}{5} \\ \frac{1}{5} & -\frac{3}{5} \end{pmatrix} \begin{pmatrix} 3 & 2 \\ 1 & -1 \end{pmatrix} = \begin{pmatrix} 1 & 0 \\ 0 & 1 \end{pmatrix}$$

and

$$\begin{pmatrix} 3 & 2 \\ 1 & -1 \end{pmatrix} \begin{pmatrix} \frac{1}{5} & \frac{2}{5} \\ \frac{1}{5} & -\frac{3}{5} \end{pmatrix} = \begin{pmatrix} 1 & 0 \\ 0 & 1 \end{pmatrix}.$$

The matrix

$$\begin{pmatrix} \frac{1}{5} & \frac{2}{5} \\ \frac{1}{5} & -\frac{3}{5} \end{pmatrix}$$

is called *the inverse of*

$$A = \begin{pmatrix} 3 & 2 \\ 1 & -1 \end{pmatrix}.$$

We denote it by A^{-1}; thus $A^{-1}A = AA^{-1} = I_2$. We solve the system of equations

$$AX = Y: \qquad \begin{pmatrix} 3 & 2 \\ 1 & -1 \end{pmatrix} \begin{pmatrix} x \\ y \end{pmatrix} = \begin{pmatrix} 1 \\ 2 \end{pmatrix}.$$

Multiply both sides by A^{-1} of the left.

$$A^{-1}(AX) = A^{-1}Y: \qquad \begin{pmatrix} \frac{1}{5} & \frac{2}{5} \\ \frac{1}{5} & -\frac{3}{5} \end{pmatrix} \left[\begin{pmatrix} 3 & 2 \\ 1 & -1 \end{pmatrix} \begin{pmatrix} x \\ y \end{pmatrix} \right] = \begin{pmatrix} \frac{1}{5} & \frac{2}{5} \\ \frac{1}{5} & -\frac{3}{5} \end{pmatrix} \begin{pmatrix} 1 \\ 2 \end{pmatrix}$$

Matrix multiplication is associative.

$$(A^{-1}A)X = A^{-1}Y: \qquad \left[\begin{pmatrix} \frac{1}{5} & \frac{2}{5} \\ \frac{1}{5} & -\frac{3}{5} \end{pmatrix} \begin{pmatrix} 3 & 2 \\ 1 & -1 \end{pmatrix} \right] \begin{pmatrix} x \\ y \end{pmatrix} = \begin{pmatrix} \frac{1}{5} & \frac{2}{5} \\ \frac{1}{5} & -\frac{3}{5} \end{pmatrix} \begin{pmatrix} 1 \\ 2 \end{pmatrix}$$

Simplify.

$$I_2X = A^{-1}Y: \qquad \begin{pmatrix} 1 & 0 \\ 0 & 1 \end{pmatrix}\begin{pmatrix} x \\ y \end{pmatrix} = \begin{pmatrix} 1 \\ -1 \end{pmatrix}$$

$$X = A^{-1}Y: \qquad \begin{pmatrix} x \\ y \end{pmatrix} = \begin{pmatrix} 1 \\ -1 \end{pmatrix}$$

Example   Consider the system of linear equations

$$4x - 2y + z = 2$$
$$3x + y - z = 1$$
$$5x + 2y + z = 3.$$

This system can be viewed as a matrix equation

$$\begin{pmatrix} 4 & -2 & 1 \\ 3 & 1 & -1 \\ 5 & 2 & 1 \end{pmatrix}\begin{pmatrix} x \\ y \\ z \end{pmatrix} = \begin{pmatrix} 2 \\ 1 \\ 3 \end{pmatrix}$$

where $\begin{pmatrix} x \\ y \\ z \end{pmatrix}$ and $\begin{pmatrix} 2 \\ 1 \\ 3 \end{pmatrix}$ are viewed as 3×1 matrices. If A is the coefficient matrix for this system, then its inverse is the matrix

$$\begin{pmatrix} \frac{3}{29} & \frac{4}{29} & \frac{1}{29} \\ -\frac{8}{29} & -\frac{1}{29} & \frac{7}{29} \\ \frac{1}{29} & -\frac{18}{29} & \frac{10}{29} \end{pmatrix}$$

since, as you can verify, $A^{-1}A = AA^{-1} = I_3$. Therefore we can solve the system of equations by multiplying both sides by A^{-1} on the left. The result of the computation is

$$\begin{pmatrix} x \\ y \\ z \end{pmatrix} = \begin{pmatrix} \frac{13}{29} \\ \frac{4}{29} \\ \frac{14}{29} \end{pmatrix}.$$

These examples motivate the following definition.

Definition 5-1_____

An $n \times n$ matrix A is called *nonsingular* or *invertible* if there is an $n \times n$ matrix A^{-1} so that $A^{-1}A = AA^{-1} = I_n$. The matrix A^{-1} is called the *inverse of A*. An $n \times n$ matrix A is *singular* or *noninvertible* if it does not have an inverse.

Note that we use the terms "nonsingular" and "invertible" only for square matrices.

An $n \times n$ matrix A cannot have two inverses, for suppose A has inverses A^{-1} and C, i.e., $A^{-1}A = AA^{-1} = I_n$ and $CA = AC = I_n$. Then $C = CI_n = C(AA^{-1}) = (CA)A^{-1} = I_nA^{-1} = A^{-1}$.

It is quite possible for an $n \times n$ matrix to be singular, however.

Example Consider the 2×2 matrix

$$A = \begin{pmatrix} 2 & 1 \\ 4 & 2 \end{pmatrix}.$$

This matrix is singular because no 2×2 matrix is its inverse. Suppose

$$\begin{pmatrix} a & b \\ c & d \end{pmatrix}$$

were the inverse of A. Then we would have

$$\begin{pmatrix} 2 & 1 \\ 4 & 2 \end{pmatrix}\begin{pmatrix} a & b \\ c & d \end{pmatrix} = \begin{pmatrix} 1 & 0 \\ 0 & 1 \end{pmatrix}$$

and

$$\begin{pmatrix} a & b \\ c & d \end{pmatrix}\begin{pmatrix} 2 & 1 \\ 4 & 2 \end{pmatrix} = \begin{pmatrix} 1 & 0 \\ 0 & 1 \end{pmatrix}$$

or

$$\begin{pmatrix} 2a + c & 2b + d \\ 4a + 2c & 4b + 2d \end{pmatrix} = \begin{pmatrix} 1 & 0 \\ 0 & 1 \end{pmatrix}$$

and

$$\begin{pmatrix} 2a + 4b & a + 2b \\ 2c + 4d & c + 2d \end{pmatrix} = \begin{pmatrix} 1 & 0 \\ 0 & 1 \end{pmatrix}.$$

Consequently, $2a + c = 1$ and $4a + 2c = 0$ from the first equality, and $2a + 4b = 1$ and $a + 2b = 0$ from the second equality; both are impossible. Thus A must be singular.

Nonsingular matrices are those having an inverse. There is a corresponding idea for linear transformations.

Definition 5-2

A linear transformation $T: \mathbb{R}^n \to \mathbb{R}^n$ is *nonsingular* or *invertible* if there is a linear function $T^{-1}: \mathbb{R}^n \to \mathbb{R}^n$ with the property that $T^{-1} \circ T = I$ and $T \circ T^{-1} = I$. The function T^{-1} is called the *inverse of* T. A linear transformation is *singular* or *noninvertible* if it does not have an inverse.

A linear transformation $T: \mathbb{R}^n \to \mathbb{R}^n$ cannot have two inverses, for if T^{-1} and S were inverses of T, i.e., $S \circ T = I$ and $T \circ S = I$, then $T^{-1} = T^{-1} \circ I = T^{-1} \circ (T \circ S) = (T^{-1} \circ T) \circ S = I \circ S = S$.

Furthermore, there is a close relation between nonsingular linear transformations and nonsingular matrices. If $T: \mathbb{R}^n \to \mathbb{R}^n$ is a nonsingular linear transformation, then it has an inverse $T^{-1}: \mathbb{R}^n \to \mathbb{R}^n$. Let A be the matrix for T and B the matrix for T^{-1}. Since $T^{-1} \circ T = I$, we have $BA = I_n$ by Proposition 4-10; since $T \circ T^{-1} = I$, we also have $AB = I_n$. Therefore the matrix for T^{-1} is A^{-1}. In other words, if T is nonsingular, the matrix for A is nonsingular, and A^{-1} is the matrix for T^{-1}.

On the other hand, suppose A is a nonsingular matrix. Let T be the linear function defined by A and S the linear function defined by A^{-1}. Again by Proposition 4-10 and the fact that $AA^{-1} = I_n$ and $A^{-1}A = I_n$, we have that $T \circ S = I$ and $S \circ T = I$. In other words, S is the inverse of T and hence T is nonsingular.

Proposition 5-1_____

A linear transformation $T: \mathbb{R}^n \to \mathbb{R}^n$ is nonsingular if and only if its matrix A is nonsingular. If T is nonsingular, then the matrix for T^{-1} is A^{-1}.

So the question then arises, how do we find the inverse of a square matrix A, provided A has an inverse? We present a method which will produce the inverse of A. In the next two sections we give some indication as to why the method works.

Example Let A be the 2×2 matrix

$$A = \begin{pmatrix} 3 & 2 \\ 1 & -1 \end{pmatrix}.$$

In this example we illustrate a method for finding A^{-1} if it exists.

(1) Construct a new matrix by placing I_2 to the right of A.

$$\left(\begin{array}{cc|cc} 3 & 2 & 1 & 0 \\ 1 & -1 & 0 & 1 \end{array} \right)$$

(2) Transform the left half of this matrix to row echelon form; apply each row operation to the whole row of the 2×4 matrix.

$$\left(\begin{array}{cc|cc} 3 & 2 & 1 & 0 \\ 1 & -1 & 0 & 1 \end{array} \right) \xrightarrow{R_{1,2}} \left(\begin{array}{cc|cc} 1 & -1 & 0 & 1 \\ 3 & 2 & 1 & 0 \end{array} \right)$$

$$\xrightarrow{R_{2,1}^{(-3)}} \left(\begin{array}{cc|cc} 1 & -1 & 0 & 1 \\ 0 & 5 & 1 & -3 \end{array} \right) \xrightarrow{R_3^{(\frac{1}{5})}} \left(\begin{array}{cc|cc} 1 & -1 & 0 & 1 \\ 0 & 1 & \frac{1}{5} & -\frac{3}{5} \end{array} \right)$$

$$\xrightarrow{R_{1,2}^{(1)}} \left(\begin{array}{cc|cc} 1 & 0 & \frac{1}{5} & \frac{2}{5} \\ 0 & 1 & \frac{1}{5} & -\frac{3}{5} \end{array} \right)$$

The matrix in the right half is the inverse of A, as you can readily check.

Example Let A be the 3×3 matrix

$$\begin{pmatrix} 1 & 1 & 0 \\ 0 & 1 & 1 \\ 1 & 0 & 1 \end{pmatrix}.$$

We illustrate a method for finding the inverse of A.

(1) Form a new matrix by placing I_3 to the right of A.

$$\left(\begin{array}{ccc|ccc} 1 & 1 & 0 & 1 & 0 & 0 \\ 0 & 1 & 1 & 0 & 1 & 0 \\ 1 & 0 & 1 & 0 & 0 & 1 \end{array}\right)$$

(2) Transform the left half of this matrix to row echelon form, applying each row operation to the whole row of the 3×6 matrix.

$$\left(\begin{array}{ccc|ccc} 1 & 1 & 0 & 1 & 0 & 0 \\ 0 & 1 & 1 & 0 & 1 & 0 \\ 1 & 0 & 1 & 0 & 0 & 1 \end{array}\right)$$

$$\xrightarrow{R_{3,1}{}^{(-1)}} \left(\begin{array}{ccc|ccc} 1 & 1 & 0 & 1 & 0 & 0 \\ 0 & 1 & 1 & 0 & 1 & 0 \\ 0 & -1 & 1 & -1 & 0 & 1 \end{array}\right)$$

$$\xrightarrow[R_{3,2}{}^{(1)}]{R_{1,2}{}^{(-1)}} \left(\begin{array}{ccc|ccc} 1 & 0 & -1 & 1 & -1 & 0 \\ 0 & 1 & 1 & 0 & 1 & 0 \\ 0 & 0 & 2 & -1 & 1 & 1 \end{array}\right)$$

$$\xrightarrow[\substack{R_{1,3}{}^{(1)} \\ R_{2,3}{}^{(-1)}}]{R_3{}^{(1/2)}} \left(\begin{array}{ccc|ccc} 1 & 0 & 0 & \frac{1}{2} & -\frac{1}{2} & \frac{1}{2} \\ 0 & 1 & 0 & \frac{1}{2} & \frac{1}{2} & -\frac{1}{2} \\ 0 & 0 & 1 & -\frac{1}{2} & \frac{1}{2} & \frac{1}{2} \end{array}\right)$$

The matrix in the right half is the inverse of A, as you can readily verify.

Example Let A be the 3×3 matrix

$$\begin{pmatrix} 1 & 1 & 0 \\ 0 & 1 & 1 \\ 1 & 2 & 1 \end{pmatrix}.$$

This matrix has no inverse. We illustrate how this can be discovered using the procedure for finding an inverse which we outlined in the previous two examples.

(1) Construct a new matrix by placing I_3 to the right of A.

$$\begin{pmatrix} 1 & 1 & 0 & \vdots & 1 & 0 & 0 \\ 0 & 1 & 1 & \vdots & 0 & 1 & 0 \\ 1 & 2 & 1 & \vdots & 0 & 0 & 1 \end{pmatrix}$$

(2) Transform the right half to row echelon form, applying each row operation to the whole row of the 3×6 matrix.

$$\begin{pmatrix} 1 & 1 & 0 & \vdots & 1 & 0 & 0 \\ 0 & 1 & 1 & \vdots & 0 & 1 & 0 \\ 1 & 2 & 1 & \vdots & 0 & 0 & 1 \end{pmatrix}$$

$$\xrightarrow{R_{3,1}^{(-1)}} \begin{pmatrix} 1 & 1 & 0 & \vdots & 1 & 0 & 0 \\ 0 & 1 & 1 & \vdots & 0 & 1 & 0 \\ 0 & 1 & 1 & \vdots & -1 & 0 & 1 \end{pmatrix}$$

$$\xrightarrow{R_{3,2}^{(-1)}} \begin{pmatrix} 1 & 1 & 0 & \vdots & 1 & 0 & 0 \\ 0 & 1 & 1 & \vdots & 0 & 1 & 0 \\ 0 & 0 & 0 & \vdots & -1 & -1 & 1 \end{pmatrix}$$

$$\xrightarrow{R_{1,2}^{(-1)}} \begin{pmatrix} 1 & 0 & -1 & \vdots & 1 & -1 & 0 \\ 0 & 1 & 1 & \vdots & 0 & 1 & 0 \\ 0 & 0 & 0 & \vdots & -1 & -1 & 1 \end{pmatrix}$$

The left-hand matrix is in row echelon form, but the row echelon form is *not* I_3. This is the fact which tells us that A has no inverse.

The three examples above illustrate a procedure for finding the inverse of an $n \times n$ matrix A:

(1) form the $n \times 2n$ matrix B

$$B = (A \mid I_n)$$

by placing I_n to the right of A; and

(2) transform A to row echelon form C, applying each row operation to the entire row of B; say the result look like $(C \quad D)$. If $C = I_n$, then $D = A^{-1}$.

If C is different from I_n (i.e., has at least one row of zeros), then A does not have an inverse.

The next two sections show why this procedure works.

Brief Exercises

1. Determine whether the following matrices have inverses by applying to them the procedure outlined above.

(a) $\begin{pmatrix} 2 & 1 \\ 1 & 2 \end{pmatrix}$

(c) $\begin{pmatrix} 4 & 1 & 3 \\ 1 & 0 & 1 \\ 2 & 1 & 2 \end{pmatrix}$

(b) $\begin{pmatrix} 2 & 1 \\ 6 & 3 \end{pmatrix}$

(d) $\begin{pmatrix} 4 & 2 & 3 \\ 1 & 2 & 1 \\ 3 & 0 & 2 \end{pmatrix}$

2. Determine whether the following linear transformations are nonsingular; if they are, find their inverse function.

(a) $T(x,y) = (3x - y, x + 2y)$
(b) $T(x,y) = (x + y, 2x + 2y)$
(c) $T(x,y,z) = (3x - z, x + y + z, 2y - z)$

Answers: $\mathbf{1(a)}$ $\begin{pmatrix} \frac{2}{3} & -\frac{1}{3} \\ -\frac{1}{3} & \frac{2}{3} \end{pmatrix}$ $\mathbf{1(b)}$ No inverse $\mathbf{1(c)}$ $\begin{pmatrix} 1 & -1 & 1 \\ 0 & -2 & -1 \\ -1 & 2 & 1 \end{pmatrix}$

$\mathbf{1(d)}$ No inverse $\mathbf{2(a)}$ $T^{-1}(u,v) = \frac{1}{7}(2u + v, -u + 3v)$ $\mathbf{2(b)}$ No inverse
$\mathbf{2(c)}$ $T^{-1}(r,s,t) = \frac{1}{11}(3r + 2s - t, -r + 3s + 4t, -2r + 6s - 3t)$

SUMMARY *An $n \times n$ matrix A is nonsingular if and only if there is an $n \times n$ matrix A^{-1} with the property that $A^{-1}A = AA^{-1} = I_n$. A linear transformation $T: \mathbb{R}^n \to \mathbb{R}^n$ is nonsingular if and only if there is a linear transformation $T^{-1}: \mathbb{R}^n \to \mathbb{R}^n$ so that $T^{-1} \circ T = I$ and $T \circ T^{-1} = I$. Then a linear transformation $T: \mathbb{R}^n \to \mathbb{R}^n$ is nonsingular if and only if its matrix A is nonsingular.*

We describe a method for determining whether an $n \times n$ matrix A has an inverse. Form the $n \times 2n$ matrix $B = (A \mid I_n)$ by placing I_n to the right of A. Using row operations on B, transform B so that the left half (i.e., the matrix A) is transformed to row echelon form. Say the result is $(C \mid D)$. The original matrix A is nonsingular if and only if $C = I_n$. If $C = I_n$, then $D = A^{-1}$.

Exercises for Section 5-1

1. Determine whether the following matrices are nonsingular; if they are, find their inverses.

(a)
$$\begin{pmatrix} 2 & 1 & 3 \\ 4 & 1 & 2 \\ 0 & -1 & 2 \end{pmatrix}$$

(d)
$$\begin{pmatrix} 1 & 1 & 0 & 2 \\ 2 & 1 & 3 & 0 \\ 1 & 0 & -1 & 0 \\ 1 & 1 & 0 & 0 \end{pmatrix}$$

(b)
$$\begin{pmatrix} 2 & 1 & 3 \\ 0 & 1 & 2 \\ -1 & 1 & 1 \end{pmatrix}$$

(e)
$$\begin{pmatrix} 1 & 1 & 0 & 1 \\ 0 & 1 & 1 & 0 \\ 1 & 0 & -1 & 2 \\ 0 & 1 & 0 & 1 \end{pmatrix}$$

(c)
$$\begin{pmatrix} 1 & 0 & 1 & 1 \\ 3 & 1 & 2 & 2 \\ -1 & 1 & 0 & 3 \\ 0 & 1 & 1 & 0 \end{pmatrix}$$

2. In the following systems of equations $A\mathbf{X} = \mathbf{Y}$, compute the inverse of the coefficient matrix A and find $\mathbf{X}$ by $\mathbf{X} = A^{-1}(A\mathbf{X}) = A^{-1}\mathbf{Y}$.

(a) $2x + y + z = 1$
$x - y + z = 3$
$x + 2y - z = 1$

(d) $3x + 2y - z = 4$
$x - y + z = 5$
$x + y - z = 1$

(b) $x + y = 1$
$2x - y = 3$

(e) $x - y + 2z = 2$
$x + y + z = 1$
$2x + y + 3z = 3$

(c) $2x + y = 2$
$x - y = 5$

3. Give an argument which shows that if A is a nonsingular matrix, then A^2 is nonsingular. What is the inverse of A^2? What about the matrix A^k, where k is a positive integer?

SECTION 5-2
Elementary Matrices

PREVIEW   *In Section 5-1 we discussed a procedure for finding the inverse of a square matrix; it involved applying row operations to the matrix. In this section we show how row operations can be interpreted through matrix multiplication with elementary matrices. We also show that elementary matrices are nonsingular and that the inverse of an elementary matrix is an elementary matrix.*

First we discuss elementary matrices. The $n \times n$ elementary matrices are those which can be obtained by applying one row operation to I_n.

Definition 5-3

The $n \times n$ elementary matrices are the matrices obtained by applying one elementary row operations to I_n.

Our notation for the elementary matrices is the following:

(1) $F_{i,j}$ denotes the matrix obtained from I_n using $R_{i,j}$,

(2) $F_i(x)$ denotes the matrix obtained from I_n using $R_i(x)$, and

(3) $F_{i,j}(x)$ denotes the matrix obtained from I_n using $R_{i,j}(x)$.

Example   The 2×2 elementary matrices are

$$F_{1,2} = \begin{pmatrix} 0 & 1 \\ 1 & 0 \end{pmatrix} \qquad F_1(x) = \begin{pmatrix} x & 0 \\ 0 & 1 \end{pmatrix} \qquad F_2(x) = \begin{pmatrix} 1 & 0 \\ 0 & x \end{pmatrix}$$

$$F_{1,2}(x) = \begin{pmatrix} 1 & x \\ 0 & 1 \end{pmatrix} \qquad F_{2,1}(x) = \begin{pmatrix} 1 & 0 \\ x & 1 \end{pmatrix}.$$

Usually we include I_2 as an elementary matrix, too. One can think of it as $F_1(1)$, for example.

Example   Some of the 3×3 elementary matrices are listed below.

$$F_{1,2} = \begin{pmatrix} 0 & 1 & 0 \\ 1 & 0 & 0 \\ 0 & 0 & 1 \end{pmatrix} \qquad F_{1,3} = \begin{pmatrix} 0 & 0 & 1 \\ 0 & 1 & 0 \\ 1 & 0 & 0 \end{pmatrix}$$

$$F_2(x) = \begin{pmatrix} 1 & 0 & 0 \\ 0 & x & 0 \\ 0 & 0 & 1 \end{pmatrix} \qquad F_{3,1}(x) = \begin{pmatrix} 1 & 0 & 0 \\ 0 & 1 & 0 \\ x & 0 & 1 \end{pmatrix}$$

At times it is convenient to picture the entries in matrices. For the elementary matrices we use the following device: for $F_{i,j}$ we use

$$
\begin{array}{c}
\quad\;\; i \qquad\;\; j \\
\begin{array}{c} i \\[20pt] j \end{array}
\left(
\begin{array}{ccccc}
\vdots & & \vdots & & \\
\cdots & 0 & \cdots & 1 & \cdots \\
& \vdots & & \vdots & \\
\cdots & 1 & \cdots & 0 & \cdots \\
& \vdots & & \vdots &
\end{array}
\right).
\end{array}
$$

For $F_i(x)$ we use

$$
\begin{array}{c}
i
\left(
\begin{array}{ccc}
\vdots & & \\
\cdots & x & \cdots \\
& \vdots &
\end{array}
\right).
\end{array}
$$

For $F_{i,j}(x)$ we use

$$
\begin{array}{c}
\quad\;\; i \qquad\;\; j \\
\begin{array}{c} i \\[20pt] j \end{array}
\left(
\begin{array}{ccccc}
\vdots & & \vdots & & \\
\cdots & 1 & \cdots & x & \cdots \\
& \vdots & & \vdots & \\
\cdots & 0 & \cdots & 1 & \cdots \\
& \vdots & & \vdots &
\end{array}
\right).
\end{array}
$$

You are to understand that the entries which are not explicitly written are the same as the corresponding entries in I_n.

One of the useful facts about elementary matrices is that you can perform a row operation on an $m \times n$ matrix A by multiplying A on the left with the appropriate $m \times m$ elementary matrix. (You can also perform column operations by multiplying A on the right with the appropriate $n \times n$ elementary matrix; see Exercise 4 at the end of this section.)

Example Let A be the 3×4 matrix

$$
A = \begin{pmatrix}
11 & 5 & 2 & -3 \\
0 & 4 & -2 & 7 \\
6 & 8 & -4 & 9
\end{pmatrix}.
$$

We perform some row operations on A by multiplying A on the left by an elementary matrix.

To perform $R_{1,3}$ on A, multiply A on the left with $F_{1,3}$.

$$\begin{pmatrix} 0 & 0 & 1 \\ 0 & 1 & 0 \\ 1 & 0 & 0 \end{pmatrix} \begin{pmatrix} 11 & 5 & 2 & -3 \\ 0 & 4 & -2 & 7 \\ 6 & 8 & -4 & 9 \end{pmatrix} = \begin{pmatrix} 6 & 8 & -4 & 9 \\ 0 & 4 & -2 & 7 \\ 11 & 5 & 2 & -3 \end{pmatrix}$$

To perform $R_3(x)$ on A, multiply A on the left with $F_3(x)$.

$$\begin{pmatrix} 1 & 0 & 0 \\ 0 & 1 & 0 \\ 0 & 0 & x \end{pmatrix} \begin{pmatrix} 11 & 5 & 2 & -3 \\ 0 & 4 & -2 & 7 \\ 6 & 8 & -4 & 9 \end{pmatrix} = \begin{pmatrix} 11 & 5 & 2 & -3 \\ 0 & 4 & -2 & 7 \\ 6x & 8x & -4x & 9x \end{pmatrix}$$

To perform $R_{3,1}(x)$ on A, multiply A on the left with $F_{3,1}(x)$.

$$\begin{pmatrix} 1 & 0 & 0 \\ 0 & 1 & 0 \\ x & 0 & 1 \end{pmatrix} \begin{pmatrix} 11 & 5 & 2 & -3 \\ 0 & 4 & -2 & 7 \\ 6 & 8 & -4 & 9 \end{pmatrix}$$
$$= \begin{pmatrix} 11 & 5 & 2 & -3 \\ 0 & 4 & -2 & 7 \\ 6+11x & 8+5x & -4+2x & 9-3x \end{pmatrix}$$

Proposition 5-2

Let A be an $m \times n$ matrix. The effect of applying $R_{i,j}$, $R_i(x)$, and $R_{i,j}(x)$ to A can be obtained by multiplying A on the left with $F_{i,j}$, $F_i(x)$, and $F_{i,j}(x)$, respectively (The elementary matrices referred to here are $m \times m$ matrices).

In Proposition 1-1 we saw that any matrix A can be transformed to row echelon form with row operations. Since row operations can be effected by multiplying on the left with elementary matrices, it follows that there are elementary matrices $G_1, \ldots, G_k$ so that $G_k \cdots G_1 A$ is in row echelon form.

Proposition 5-3

If A is an $m \times n$ matrix, then there are $m \times m$ elementary matrices $G_1, \ldots, G_k$ so that $G_k \cdots G_1 A$ is in row echelon form.

Brief Exercises

1. Perform the following operations on the matrix A by premultiplying with the appropriate elementary matrix.

$$A = \begin{pmatrix} 2 & 1 & 3 & 4 \\ 1 & 5 & 1 & 2 \\ 3 & 1 & 6 & 1 \\ 2 & 1 & 3 & 4 \end{pmatrix}$$

$$R_{1,2}, \ R_{2,3}(-1), \ R_{4,1}(-2), \ R_{3,4}(2), \ \text{and} \ R_1(2)$$

2. For the following matrices A, find elementary matrices $G_1, \ldots, G_k$ so that $G_k \cdots G_1 A$ is in row echelon form.

(a) $A = \begin{pmatrix} 2 & 1 \\ 1 & 3 \end{pmatrix}$ **(b)** $A = \begin{pmatrix} 1 & 2 & 1 \\ 0 & 1 & 1 \\ 1 & 0 & 1 \end{pmatrix}$ **(c)** $A = \begin{pmatrix} 2 & 1 & 1 \\ 1 & 0 & 2 \end{pmatrix}$

Selected Answers: **2(a)** $F_{1,2}(-3)F_2(-\frac{1}{5})F_{2,1}(-2)F_{1,2}A$ is in row echelon form
2(b) $F_{1,2}F_{3,2}F_{3,2}(-1)F_{2,1}(-1)F_1(\frac{1}{2})F_{1,3}(-1)A$ is in row echelon form
2(c) $F_{1,2}F_{1,2}(-2)A$ is in row echelon form

The next fact we need is that elementary matrices are nonsingular and that the inverse of an elementary matrix is also an elementary matrix.

Example   In this example we show that the 2×2 elementary matrices are nonsingular and that the inverse of an elementary matrix is also elementary.
The inverse of $F_{1,2}$ is $F_{1,2}$ since $F_{1,2}F_{1,2} = I_2$.

$$\begin{pmatrix} 0 & 1 \\ 1 & 0 \end{pmatrix}\begin{pmatrix} 0 & 1 \\ 1 & 0 \end{pmatrix} = \begin{pmatrix} 1 & 0 \\ 0 & 1 \end{pmatrix}$$

The inverse of $F_1(x)$ is $F_1(x^{-1})$ since $F_1(x^{-1})F_1(x) = I_2$ and $F_1(x)F_1(x^{-1}) = I_2$.

$$\begin{pmatrix} x^{-1} & 0 \\ 0 & 1 \end{pmatrix}\begin{pmatrix} x & 0 \\ 0 & 1 \end{pmatrix} = \begin{pmatrix} 1 & 0 \\ 0 & 1 \end{pmatrix}$$

$$\begin{pmatrix} x & 0 \\ 0 & 1 \end{pmatrix}\begin{pmatrix} x^{-1} & 0 \\ 0 & 1 \end{pmatrix} = \begin{pmatrix} 1 & 0 \\ 0 & 1 \end{pmatrix}$$

Similarly, the inverse of $F_2(x)$ is $F_2(x^{-1})$.

The inverse of $F_{2,1}(x)$ is $F_{2,1}(-x)$ since $F_{2,1}(-x)F_{2,1}(x) = I_2$ and $F_{2,1}(x)F_{2,1}(-x) = I_2$.

$$\begin{pmatrix} 1 & 0 \\ -x & 1 \end{pmatrix}\begin{pmatrix} 1 & 0 \\ x & 1 \end{pmatrix} = \begin{pmatrix} 1 & 0 \\ 0 & 1 \end{pmatrix}$$

$$\begin{pmatrix} 1 & 0 \\ x & 1 \end{pmatrix}\begin{pmatrix} 1 & 0 \\ -x & 1 \end{pmatrix} = \begin{pmatrix} 1 & 0 \\ 0 & 1 \end{pmatrix}$$

Example In this example we treat some 3×3 elementary matrices, showing that they are nonsingular and that their inverses are elementary.

The inverse of $F_{2,3}$ is $F_{2,3}$ since $F_{2,3}F_{2,3} = I_3$.

$$\begin{pmatrix} 1 & 0 & 0 \\ 0 & 0 & 1 \\ 0 & 1 & 0 \end{pmatrix}\begin{pmatrix} 1 & 0 & 0 \\ 0 & 0 & 1 \\ 0 & 1 & 0 \end{pmatrix} = \begin{pmatrix} 1 & 0 & 0 \\ 0 & 1 & 0 \\ 0 & 0 & 1 \end{pmatrix}$$

The inverse of $F_3(x)$ is $F_3(x^{-1})$ since $F_3(x^{-1})F_3(x) = F_3(x)F_3(x^{-1}) = I_3$.

$$\begin{pmatrix} 1 & 0 & 0 \\ 0 & 1 & 0 \\ 0 & 0 & x^{-1} \end{pmatrix}\begin{pmatrix} 1 & 0 & 0 \\ 0 & 1 & 0 \\ 0 & 0 & x \end{pmatrix} = \begin{pmatrix} 1 & 0 & 0 \\ 0 & 1 & 0 \\ 0 & 0 & 1 \end{pmatrix}$$

$$\begin{pmatrix} 1 & 0 & 0 \\ 0 & 1 & 0 \\ 0 & 0 & x \end{pmatrix}\begin{pmatrix} 1 & 0 & 0 \\ 0 & 1 & 0 \\ 0 & 0 & x^{-1} \end{pmatrix} = \begin{pmatrix} 1 & 0 & 0 \\ 0 & 1 & 0 \\ 0 & 0 & 1 \end{pmatrix}$$

The inverse of $F_{3,2}(x)$ is $F_{3,2}(-x)$ since $F_{3,2}(-x)F_{3,2}(x) = F_{3,2}(x)F_{3,2}(-x) = I_3$.

$$\begin{pmatrix} 1 & 0 & 0 \\ 0 & 1 & 0 \\ 0 & -x & 1 \end{pmatrix}\begin{pmatrix} 1 & 0 & 0 \\ 0 & 1 & 0 \\ 0 & x & 1 \end{pmatrix} = \begin{pmatrix} 1 & 0 & 0 \\ 0 & 1 & 0 \\ 0 & 0 & 1 \end{pmatrix}$$

$$\begin{pmatrix} 1 & 0 & 0 \\ 0 & 1 & 0 \\ 0 & x & 1 \end{pmatrix}\begin{pmatrix} 1 & 0 & 0 \\ 0 & 1 & 0 \\ 0 & -x & 1 \end{pmatrix} = \begin{pmatrix} 1 & 0 & 0 \\ 0 & 1 & 0 \\ 0 & 0 & 1 \end{pmatrix}$$

These examples illustrate the general situation for $n \times n$ elementary matrices. The inverse of $F_{i,j}$ is $F_{i,j}$, the inverse of $F_i(x)$ is $F_i(x^{-1})$, and the inverse of $F_{i,j}(x)$ is $F_{i,j}(-x)$ since

$$F_{i,j}F_{i,j} = I_n$$

$$F_i(x^{-1})F_i(x) = F_i(x)F_i(x^{-1}) = I_n$$

$$F_{i,j}(-x)F_{i,j}(x) = F_{i,j}(x)F_{i,j}(-x) = I_n.$$

You can also visualize the products

$$
\begin{array}{c}
\quad\quad i \quad\quad j \\
\begin{matrix} i \\ \\ j \end{matrix}
\begin{pmatrix}
\cdots & 0 & \cdots & 1 & \cdots \\
 & \vdots & & \vdots & \\
\cdots & 1 & \cdots & 0 & \cdots
\end{pmatrix}
\end{array}
\quad
\begin{array}{c}
\quad\quad i \quad\quad j \\
\begin{pmatrix}
\cdots & 0 & \cdots & 1 & \cdots \\
 & \vdots & & \vdots & \\
\cdots & 1 & \cdots & 0 & \cdots
\end{pmatrix}
\end{array}
$$

$$
=
\begin{array}{c}
\quad\quad i \quad\quad j \\
\begin{pmatrix}
\cdots & 1 & \cdots & 0 & \cdots \\
 & \vdots & & \vdots & \\
\cdots & 0 & \cdots & 1 & \cdots
\end{pmatrix}
\begin{matrix} i \\ \\ j \end{matrix}
\end{array}
$$

$$
\begin{array}{c}
\quad\quad i \\
i \begin{pmatrix} \cdots & x^{-1} & \cdots \\ & \vdots & \end{pmatrix}
\end{array}
\quad
\begin{array}{c}
\quad\quad i \\
\begin{pmatrix} \cdots & x & \cdots \\ & \vdots & \end{pmatrix}
\end{array}
$$

$$
=
\begin{pmatrix} \cdots & x & \cdots \\ & \vdots & \end{pmatrix}
\begin{pmatrix} \cdots & x^{-1} & \cdots \\ & \vdots & \end{pmatrix}
=
\begin{pmatrix} \cdots & 1 & \cdots \\ & \vdots & \end{pmatrix}
$$

$$
\begin{array}{cc}
\begin{matrix} & \;\;i & & \;\;j \end{matrix} & \begin{matrix} \;\;i & & \;\;j \end{matrix} \\
\begin{matrix} i \\ \\ j \end{matrix}
\begin{pmatrix}
\vdots & & \vdots & \\
\cdots & 1 & \cdots & -x & \cdots \\
\vdots & & \vdots & \\
\cdots & 0 & \cdots & 1 & \cdots \\
\vdots & & \vdots &
\end{pmatrix}
&
\begin{pmatrix}
\vdots & & \vdots & \\
\cdots & 1 & \cdots & x & \cdots \\
\vdots & & \vdots & \\
\cdots & 0 & \cdots & 1 & \cdots \\
\vdots & & \vdots &
\end{pmatrix}
\end{array}
$$

$$
=
\begin{pmatrix}
\vdots & & \vdots & \\
\cdots & 1 & \cdots & x & \cdots \\
\vdots & & \vdots & \\
\cdots & 0 & \cdots & 1 & \cdots \\
\vdots & & \vdots &
\end{pmatrix}
\begin{pmatrix}
\vdots & & \vdots & \\
\cdots & 1 & \cdots & -x & \cdots \\
\vdots & & \vdots & \\
\cdots & 0 & \cdots & 1 & \cdots \\
\vdots & & \vdots &
\end{pmatrix}
$$

$$
=
\begin{pmatrix}
\vdots & & \vdots & \\
\cdots & 1 & \cdots & 0 & \cdots \\
\vdots & & \vdots & \\
\cdots & 0 & \cdots & 1 & \cdots \\
\vdots & & \vdots &
\end{pmatrix}
$$

Proposition 5-4

The elementary matrices are nonsingular, and the inverse of an elementary matrix is an elementary matrix.

SUMMARY *An $n \times n$ elementary matrix is a square matrix obtained by applying exactly one row operation to I_n. We obtain $F_{i,j}$ using $R_{i,j}$, $F_i(x)$ using $R_i(x)$, and $F_{i,j}(x)$ using $R_{i,j}(x)$. The elementary matrices can be used to effect row operations:*

$R_{i,j}$ on A is effected by $F_{i,j}A$,
$R_i(x)$ on A is effected by $F_i(x)A$, and
$R_{i,j}(x)$ on A is effected by $F_{i,j}(x)A$.

The fact that any $m \times n$ matrix A can be transformed by row operations to row echelon form can be expressed by saying that there are $m \times m$ elementary matrices $G_1, \ldots, G_k$ so that $G_k \cdots G_1 A$ is in row echelon form. Furthermore, elementary *matrices are nonsingular, and the inverse of an elementary matrix is an elementary matrix.*

$$(F_{i,j})^{-1} = F_{i,j}$$
$$(F_i(x))^{-1} = F_i(x^{-1})$$
$$(F_{i,j}(x))^{-1} = F_{i,j}(-x).$$

Exercises for Section 5-2

1. Let A be the matrix

$$\begin{pmatrix} 5 & 1 & 8 & 2 \\ 3 & -1 & 4 & 1 \\ 0 & 2 & -1 & 4 \end{pmatrix}.$$

Perform the following row operations on A by multiplying A on the left with the appropriate elementary matrices: $R_{1,3}$, $R_{2,3}$, $R_2(5)$, $R_3(2)$, $R_{1,3}(2)$, and $R_{3,1}(-2)$.

2. Find the inverse for each of the following elementary 3×3 matrices, and express each inverse as an elementary matrix: $F_{1,2}$, $F_1(-2)$, $F_{3,2}(-2)$, $F_{2,1}(2)$, and $F_{3,1}$.

3. Same exercise as 2 with the following 4×4 elementary matrices: $F_{2,4}$, $F_2(-5)$, $F_{3,1}$, $F_{2,3}(-1)$, $F_{2,4}(-3)$, and $F_{1,2}(4)$.

4. Can column operations on a matrix A be effected by multiplying A on the right with elementary matrices? If so, state the fact in a way similar to Proposition 5-2.

5. Is there a version of Proposition 5-3 involving column echelon form? Give reasons.

6. In each of the following, let A be the given matrix and let B be the row echelon form for A. By keeping track of the row operations used to transform A to row echelon form, express B as a product $G_k \cdots G_1 A$, where $G_k, \ldots, G_1$ are elementary matrices.

(a) $\begin{pmatrix} 5 & 1 & 1 \\ 3 & 1 & 4 \end{pmatrix}$ (b) $\begin{pmatrix} 0 & 1 & 1 \\ 1 & 2 & 1 \end{pmatrix}$ (c) $\begin{pmatrix} -1 & 2 & 1 & 3 \\ 4 & 1 & 3 & 2 \\ 1 & 1 & 1 & 1 \end{pmatrix}$

7. For each matrix in Exercise 6 let A be the given matrix and let B be the row echelon form for A. Express A as a product $H_1 \cdots H_k B$, where $H_1, \ldots, H_k$ are elementary matrices.

SECTION 5-3
Nonsingular Matrices

PREVIEW *The product AB of nonsingular matrices A and B is nonsingular; the inverse is the product of the inverses in the reverse order: $(AB)^{-1} = B^{-1}A^{-1}$. Using this result we show that an $n \times n$ matrix is nonsingular if and only if its row echelon form is I_n. We justify the method described in Section 5-1 for finding the inverse of a nonsingular matrix. Also, we show that a matrix A is singular if and only if there is a nonzero vector $(x_1, \ldots, x_n)$ satisfying*

$$\begin{pmatrix} a_{11} & \cdots & a_{1n} \\ \cdot & & \cdot \\ \cdot & & \cdot \\ \cdot & & \cdot \\ a_{n1} & \cdots & a_{nn} \end{pmatrix} \begin{pmatrix} x_1 \\ \cdot \\ \cdot \\ \cdot \\ x_n \end{pmatrix} = \begin{pmatrix} 0 \\ \cdot \\ \cdot \\ \cdot \\ 0 \end{pmatrix}.$$

The corresponding statement for linear transformations is also true. Then we use this result for matrices to show that if A and B are $n \times n$ matrices and $AB = I_n$, then $A = B^{-1}$ and $B = A^{-1}$. The final result is that if A and B are $n \times n$ matrices and AB is nonsingular, then A and B are both nonsingular.

The first fact we need is that the product of two nonsingular matrices is nonsingular, and that the composite of nonsingular linear transformations is nonsingular. We illustrate this fact in the following example.

Example Suppose $S: \mathbb{R}^2 \to \mathbb{R}^2$ and $T: \mathbb{R}^2 \to \mathbb{R}^2$ are defined by the equations

$$S(x,y) = (x + 2y, 2x - y)$$
$$T(x,y) = (x + y, x - y).$$

These functions are nonsingular; their inverses are

$$S^{-1}(x,y) = (\tfrac{1}{5}x + \tfrac{2}{5}y, \tfrac{2}{5}x - \tfrac{1}{5}y)$$
$$T^{-1}(x,y) = (\tfrac{1}{2}x + \tfrac{1}{2}y, \tfrac{1}{2}x - \tfrac{1}{2}y).$$

You can check that

$$(T \circ S)(x,y) = T(x + 2y, 2x - y) = (3x + y, -x + 3y).$$

Also

$$(T \circ S)^{-1}(x,y) = (\tfrac{3}{10}x - \tfrac{1}{10}y, \tfrac{1}{10}x + \tfrac{3}{10}y).$$

We claim that the inverse of $T \circ S$ is the composite of the inverses in the reverse order, i.e., $(T \circ S)^{-1} = S^{-1} \circ T^{-1}$. You can check that

$$(S^{-1} \circ T^{-1})(x,y) = S^{-1}(\tfrac{1}{2}x + \tfrac{1}{2}y, \tfrac{1}{2}x - \tfrac{1}{2}y)$$

$$= (\tfrac{3}{10}x - \tfrac{1}{10}y, \tfrac{1}{10}x + \tfrac{3}{10}y).$$

The same kind of fact is true for the product of nonsingular matrices. In this case, the matrices A_S and A_T for S and T, respectively, are

$$A_S = \begin{pmatrix} 1 & 2 \\ 2 & -1 \end{pmatrix} \qquad A_T = \begin{pmatrix} 1 & 1 \\ 1 & -1 \end{pmatrix}.$$

Now

$$A_T A_S = \begin{pmatrix} 1 & 1 \\ 1 & -1 \end{pmatrix}\begin{pmatrix} 1 & 2 \\ 2 & -1 \end{pmatrix} = \begin{pmatrix} 3 & 1 \\ -1 & 3 \end{pmatrix}.$$

You can check the following.

$$A_S^{-1} = \begin{pmatrix} \tfrac{1}{5} & \tfrac{2}{5} \\ \tfrac{2}{5} & -\tfrac{1}{5} \end{pmatrix} \qquad A_T^{-1} = \begin{pmatrix} \tfrac{1}{2} & \tfrac{1}{2} \\ \tfrac{1}{2} & -\tfrac{1}{2} \end{pmatrix} \qquad (A_T A_S)^{-1} = \begin{pmatrix} \tfrac{3}{10} & -\tfrac{1}{10} \\ \tfrac{1}{10} & \tfrac{3}{10} \end{pmatrix}$$

This product can also be computed by

$$A_S^{-1} A_T^{-1} = \begin{pmatrix} \tfrac{1}{5} & \tfrac{2}{5} \\ \tfrac{2}{5} & -\tfrac{1}{5} \end{pmatrix}\begin{pmatrix} \tfrac{1}{2} & \tfrac{1}{2} \\ \tfrac{1}{2} & -\tfrac{1}{2} \end{pmatrix} = \begin{pmatrix} \tfrac{3}{10} & -\tfrac{1}{10} \\ \tfrac{1}{10} & \tfrac{3}{10} \end{pmatrix}.$$

The facts we illustrated in the last example are true in general. If A and B are nonsingular $n \times n$ matrices, then

$$(B^{-1}A^{-1})(AB) = B^{-1}[(A^{-1}A)B] = B^{-1}(I_n B) = B^{-1}B = I_n$$

$$(AB)(B^{-1}A^{-1}) = A[(BB^{-1})A^{-1}] = A(I_n A^{-1}) = AA^{-1} = I_n.$$

This shows that $(AB)^{-1} = B^{-1}A^{-1}$.

Because of the correspondence between matrices and linear transformations, a similar result holds; if S and T are nonsingular linear transformations from $\mathbb{R}^n$ to $\mathbb{R}^n$, then $(S \circ T)^{-1} = T^{-1} \circ S^{-1}$.

Proposition 5-5

If S and T are nonsingular linear transformations from $\mathbb{R}^n$ to $\mathbb{R}^n$, then $S \circ T$ is nonsingular and $(S \circ T)^{-1} = T^{-1} \circ S^{-1}$.

If A and B are $n \times n$ nonsingular matrices, then AB is nonsingular and $(AB)^{-1} = B^{-1}A^{-1}$.

This proposition also applies to products of more than two matrices or composites of more than two linear transformations. For matrices, for example, if A, B, and C are nonsingular $n \times n$ matrices, then $A(BC)$ is also nonsingular since it is the product of the nonsingular matrices A and BC. Thus

$$[A(BC)]^{-1} = (BC)^{-1}A^{-1} = C^{-1}B^{-1}A^{-1}.$$

Thus the product of nonsingular matrices is nonsingular, and the inverse of a product is the product of the inverses in the reverse order.

Brief Exercises

Verify Proposition 5-5 in the following instances.

1. $A = \begin{pmatrix} 2 & 1 \\ 1 & 3 \end{pmatrix}$ $\qquad B = \begin{pmatrix} 1 & 0 \\ 2 & 3 \end{pmatrix}$

2. $A = \begin{pmatrix} 1 & 1 & 0 \\ 0 & 1 & 1 \\ 1 & 0 & 1 \end{pmatrix}$ $\qquad B = \begin{pmatrix} 1 & 2 & 1 \\ 0 & 1 & 1 \\ 1 & 2 & 0 \end{pmatrix}$

3. In each of the following, find the matrices for the following linear transformations: $T \circ S$, $S \circ T$, $S^{-1} \circ T$, and $T^{-1} \circ S^{-1}$, where S and T are

(a) $S(x,y) = (x + y, 2x + y)$, $T(x,y) = (x - y, x + 2y)$.

(b) $S(x,y,z) = (x + z, x + y, y + z)$, $T(x,y, z) = (2x, y + 3z, x + z)$.

Selected Answers:

3(a)

$A_{T \circ S} = \begin{pmatrix} -1 & 0 \\ 5 & 3 \end{pmatrix}$ $\qquad A_{S \circ T} = \begin{pmatrix} 2 & 1 \\ 3 & 0 \end{pmatrix}$

$A_{S^{-1} \circ T} = \begin{pmatrix} 0 & 3 \\ 1 & -4 \end{pmatrix}$ $\qquad A_{T^{-1} \circ S^{-1}} = \begin{pmatrix} 0 & \frac{1}{3} \\ 1 & -\frac{2}{3} \end{pmatrix}$

3(b)

$A_{T \circ S} = \begin{pmatrix} 2 & 0 & 2 \\ 1 & 4 & 3 \\ 1 & 1 & 2 \end{pmatrix}$ $\qquad A_{S \circ T} = \begin{pmatrix} 3 & 0 & 1 \\ 2 & 1 & 3 \\ 1 & 1 & 4 \end{pmatrix}$

$A_{S^{-1} \circ T} = \begin{pmatrix} \frac{1}{2} & \frac{1}{2} & 1 \\ -\frac{1}{2} & \frac{1}{2} & 2 \\ \frac{3}{2} & -\frac{1}{2} & -1 \end{pmatrix}$ $\qquad A_{T^{-1} \circ S^{-1}} = \begin{pmatrix} \frac{1}{4} & \frac{1}{4} & -\frac{1}{4} \\ -\frac{5}{4} & \frac{11}{4} & -\frac{7}{4} \\ \frac{1}{4} & -\frac{3}{4} & \frac{3}{4} \end{pmatrix}$

The next point we discuss is how to determine whether an $n \times n$ matrix A is nonsingular. Suppose we transform A to its row echelon form C by row operations. By Proposition 5-3 this means that there are elementary matrices $G_1, \ldots, G_k$ with the property that $C = G_k \cdots G_1 A$. If A is nonsingular, then, because $G_1, \ldots, G_k$ are all nonsingular (Proposition 5-4) and because the product of nonsingular matrices is nonsingular (Proposition 5-5), it follows that C is nonsingular. So, if A is nonsingular, its row echelon form must be nonsingular.

Fortunately, there is only one $n \times n$ matrix in row echelon form which is nonsingular; that matrix is I_n. Any other $n \times n$ matrix in row echelon form has a row of zeros and therefore has no inverse; in fact, any matrix with a zero row cannot have an inverse as the following product shows.

$$\begin{pmatrix} 0 & 0 & \cdots & 0 \end{pmatrix} \begin{pmatrix} & & & \\ & & & \\ & & & \end{pmatrix} = \begin{pmatrix} 0 & 0 & \cdots & 0 \end{pmatrix} \neq I_n.$$

The result of this reasoning is this: if A is nonsingular, then its row echelon form is I_n.

The converse of this statement is also true. For suppose the row echelon form for A is I_n. That means (Proposition 5-3) that there are elementary matrices $G_1, \ldots, G_k$ so that

$$I_n = G_k \cdots G_1 A.$$

Multiplying on the left by G_k^{-1}, then by G_{k-1}^{-1}, and so on until finally by G_1^{-1}, we find that

$$G_1^{-1} \cdots G_{k-1}^{-1} G_k^{-1} = A.$$

All the matrices on the left-hand side are elementary matrices (Proposition 5-4) and therefore nonsingular; consequently, A is nonsingular (Proposition 5-5). Notice also that we have A expressed as a product of elementary matrices.

Proposition 5-6_____

Suppose A is an $n \times n$ matrix. The following statements are equivalent:
(1) A is nonsingular,
(2) the row echelon form for A is I_n,
(3) there are elementary matrices $G_1, \ldots, G_k$ so that $I_n = G_k \cdots G_1 A$, and,
(4) A is expressible as the product of elementary matrices.

_____

The results in Proposition 5-6 allow us to justify the method described in Section 5-1 for finding A^{-1}. Suppose that we transform A to I_n by row operations, i.e., there are elementary matrices $G_1, \ldots, G_k$ so that $I_n = G_k \cdots G_1 A$. Since A is nonsingular, multiply both sides of this equation by A^{-1} on the right to obtain $A^{-1} = G_k \cdots G_1 = G_k \cdots G_1 I_n$. This shows that if a sequence of row operations transforms A to I_n, that same sequence of row operations transforms I_n to A^{-1}.

Brief Exercises

1. Write the matrices below as products of elementary matrices. Check your work, since there are many correct answers.

(a) $\begin{pmatrix} 2 & 1 \\ -1 & 1 \end{pmatrix}$ (b) $\begin{pmatrix} 3 & 1 & 4 \\ 1 & -2 & 1 \\ 1 & 0 & 2 \end{pmatrix}$ (c) $\begin{pmatrix} 1 & 0 & 1 \\ 1 & 1 & 0 \\ 0 & 1 & 1 \end{pmatrix}$

Answers: Here are some possible answers

1(a) $A = F_{2,1}(1)F_2(-3)F_{1,2}(2)F_{1,2}$

1(b) $A = F_{1,3}(3)F_{2,3}(1)F_{2,1}(-2)F_2(-5)F_{1,2}(-2)F_{3,2}(2)F_{3,2}F_{1,2}$

1(c) $A = F_{2,1}(1)F_{3,2}(1)F_3(2)F_{1,3}(1)F_{2,3}(-1)$

We have defined matrices and linear transformations to be nonsingular if they have inverses. There is another way to decide whether a square matrix A or a linear transformation $T: \mathbb{R}^n \to \mathbb{R}^n$ is nonsingular. We consider the matrix situation.

In this discussion it is convenient if we think of $X = (x_1, \ldots, x_n)$ and $0 = (0, \ldots, 0)$ either as n-tuples written vertically or as $n \times 1$ matrices.

The criterion is this: A is nonsingular if and only if the only vector satisfying $AX = 0$ is the zero vector $X = 0$. The reason is easy to understand. If A is nonsingular, then A^{-1} exists and $X = 0$ is the only solution of $AX = 0$ because $X = A^{-1}(AX) = A^{-1}0 = 0$. On the other hand, if A is singular, then the row echelon form B for A has at least one zero row. This means that the system of equations represented by $BX = 0$ is a homogeneous system in n unknowns with fewer than n equations. By Proposition 3-7 there are nonzero vectors X satisfying $BX = 0$. But the solutions for $AX = 0$ are the same as those for $BX = 0$. So, if A is singular, there is a nonzero vector X satisfying the equations $AX = 0$.

Proposition 5-7

The $n \times n$ matrix A is nonsingular if and only if the only vector X satisfying $AX = 0$ is $X = 0$.

This proposition has a corresponding statement for linear transformations $T: \mathbb{R}^n \to \mathbb{R}^n$.

Proposition 5-8

The linear transformation $T: \mathbb{R}^n \to \mathbb{R}^n$ is nonsingular if and only if the only vector X for which $T(X) = 0$ is the vector $X = 0$.

Brief Exercises

1. Determine whether the following are nonsingular. For those matrices A which are singular, find a vector $\mathbf{X}$ different from $\mathbf{0}$ for which $A\mathbf{X} = \mathbf{0}$.

(a) $\begin{pmatrix} 2 & 1 \\ 4 & 2 \end{pmatrix}$

(c) $\begin{pmatrix} 2 & 1 & 5 \\ 1 & 0 & 2 \\ 0 & 1 & 1 \end{pmatrix}$

(b) $\begin{pmatrix} 1 & 0 \\ 2 & 1 \end{pmatrix}$

(d) $\begin{pmatrix} 1 & 1 & 0 \\ 0 & 1 & 1 \\ 2 & 3 & 1 \end{pmatrix}$

2. Determine whether the following are singular. For those transformations T which are singular, find a nonzero vector $\mathbf{X}$ for which $T(\mathbf{X}) = \mathbf{0}$.

(a) $T(x,y) = (2x + y, x - y)$
(b) $T(x,y) = (3x - y, 6x - 2y)$
(c) $T(x,y,z) = (x + z, y + z, x + z)$
(d) $T(x,y,z) = (2x + y - z, x - 2y + z, 3x - z)$

Answers: **1(a)** $(1,-2)$ **1(b)** Nonsingular **1(c)** $(-2,-1,1)$ **1(d)** $(1,-1,1)$
2(a) Nonsingular **2(b)** $(1,3)$ **2(c)** $(-1,-1,1)$ **2(d)** Nonsingular.

In Definition 5-1 the inverse of an $n \times n$ matrix A is an $n \times n$ matrix B having the property that $AB = BA = I_n$. We can use Proposition 5-7 to conclude that $B = A^{-1}$ just from knowing that $AB = I_n$. Thus it is sufficient to check that $AB = I_n$; we do not also need to compute $BA = I_n$ to conclude that B is the inverse of A.

To see this we first notice that B is nonsingular. Suppose $\mathbf{X}$ is a vector in $\mathbb{R}^n$ and, writing $\mathbf{X}$ and $\mathbf{0}$ as $n \times 1$ matrices, suppose $B\mathbf{X} = \mathbf{0}$. Then

$$\mathbf{0} = A\mathbf{0} = A(B\mathbf{X}) = (AB)\mathbf{X} = I_n\mathbf{X} = \mathbf{X}.$$

By Proposition 5-7 B is nonsingular. Since B has an inverse, multiply both sides of the equation $AB = I_n$ on the right with B^{-1} and we obtain $A = B^{-1}$. Therefore, A is also nonsingular, $A = B^{-1}$, and $B = A^{-1}$.

Consequently, if A and B are $n \times n$ matrices and $AB = I_n$, then A and B are nonsingular, $A = B^{-1}$, and $B = A^{-1}$.

A further consequence of Proposition 5-7 is that if C and D are $n \times n$ matrices and CD is nonsingular, then C and D must be nonsingular. The reasons for this are easy to see. Since CD is nonsingular, there is an inverse E for CD: $(CD)E = I_n$ or $C(DE) = I_n$. By what we have just shown, the matrices C and DE are nonsingular, and C is the inverse of DE. Therefore $(DE)C = I_n$ or $D(EC) = I_n$; again D is nonsingular by what we have just shown.

We summarize these facts for future reference in the next proposition.

Proposition 5-9_____

Suppose A and B are $n \times n$ matrices.

(1) If $AB = I_n$, then A and B are nonsingular with $A = B^{-1}$ and $B = A^{-1}$.

(2) If AB is nonsingular, then A and B are both nonsingular; equivalently, if either A or B is singular, then AB is singular, too.

_____

SUMMARY   *The first fact we show is that the product AB of nonsingular matrices is nonsingular and that $(AB)^{-1} = B^{-1}A^{-1}$. From this we show that an $n \times n$ matrix is nonsingular if and only if one of the following is true:*

(1) *the row echelon form for A is I_n,*

(2) *there are elementary matrices $G_1, \ldots , G_k$ so that $I_n = G_k \cdots G_1 A$, and*

(3) *A is expressible as the product of elementary matrices.*

A further criterion for nonsingularity is that the square matrix A is nonsingular if and only if the only vector $\mathbf{X}$ for which $A\mathbf{X} = \mathbf{0}$ is $\mathbf{X} = \mathbf{0}$ (in this equation $\mathbf{X}$ and $\mathbf{0}$ are written as $n \times 1$ matrices). Using this we obtain a shorter criterion for checking whether B is the inverse of A: if A and B are $n \times n$ matrices and $AB = I_n$, then A and B are nonsingular, $A = B^{-1}$, and $B = A^{-1}$. Furthermore, if A and B are $n \times n$ matrices and AB is nonsingular, then A and B are nonsingular.

Exercises for Section 5-3

1. In each of the following, let A be the given matrix. Express A in the form $G_k \cdots G_1 B$, where $G_k, \ldots , G_1$ are elementary matrices and B is the row echelon form of A.

   (a) $\begin{pmatrix} 2 & 1 & 3 \\ 1 & 0 & 1 \end{pmatrix}$ (b) $\begin{pmatrix} 2 & 1 \\ 0 & 3 \\ 1 & 1 \end{pmatrix}$ (c) $\begin{pmatrix} 1 & 2 & 1 \\ 0 & 1 & 1 \\ 1 & 1 & 3 \end{pmatrix}$

2. Suppose a square matrix has a column of zeros. Can it be nonsingular? Which square matrix in column echelon form is nonsingular? Is there some statement similar to Proposition 5-6 involving column operations? If so, state it and give supporting reasons.

More Challenging Exercises

3. Suppose $T: \mathbb{R}^n \to \mathbb{R}^n$ is a linear transformation and $S: \mathbb{R}^n \to \mathbb{R}^n$ is a function having the property that $S \circ T = I$ and $T \circ S = I$. Show that S must be linear.

4. Suppose $T: \mathbb{R}^n \to \mathbb{R}^n$ is a nonsingular linear transformation and $X_1, \ldots, X_n$ form a basis for $\mathbb{R}^n$. Show that $T(X_1), \ldots, T(X_n)$ form a basis for $\mathbb{R}^n$.

5. Give an argument to show that if $T: \mathbb{R}^n \to \mathbb{R}^n$ is a nonsingular linear transformation, then the only vector X satisfying $T(X) = 0$ is the zero vector.

6. Suppose $T: \mathbb{R}^n \to \mathbb{R}^n$ is a linear transformation having the property that the only vector X satisfying $T(X) = 0$ is the zero vector $X = 0$. Show that if $X_1, \ldots, X_n$ is a basis for $\mathbb{R}^n$, then $T(X_1), \ldots, T(X_n)$ is a basis for $\mathbb{R}^n$.

7. Suppose $T: \mathbb{R}^n \to \mathbb{R}^n$ is a linear transformation having the property that the only vector X satisfying $T(X) = 0$ is the zero vector $X = 0$. Show that T is nonsingular by an argument which does not use matrices.

8. Give an example of a 3×5 matrix A and a 5×3 matrix B for which $AB = I_3$ and $BA \neq I_5$.

9. Show that if A is an $m \times n$ matrix, then there are $n \times n$ elementary matrices $H_1, \ldots, H_t$ such that $AH_1 \cdots H_t$ is in column echelon form. (See Exercises 4 and 5 of Section 5-2.)

10. Give an argument to show that A is nonsingular if and only if its column echelon form is the identity matrix.

11. Give another argument for Proposition 5-7; transform A to column echelon form (see Exercise 9). If A is singular, then the last column in $H_1 \cdots H_t$ is a nonzero vector X satisfying $AX = 0$.

6

The Determinant Function

SECTION 6-1
Expansion of the Determinant Along a Row of the Matrix

PREVIEW *The determinant of the 2 × 2 matrix*

$$\begin{pmatrix} a & b \\ c & d \end{pmatrix}$$

is the number ad − bc. The concept of a submatrix is introduced, and the notation $A(i\,|\,j)$ is used to denote the matrix obtained from A by removing the ith row and the jth column. The formula for the determinant of an n × n matrix A (expansion along the ith row) is:

$$\det A = (-1)^{i+1}a_{i1} \det A(i\,|\,1) + (-1)^{i+2}a_{i2} \det A(i\,|\,2)$$
$$+ \cdots + (-1)^{i+n}a_{in} \det A(i\,|\,n).$$

Then we give examples of the use of this formula.

In Section 5-1 we discussed a method for determining whether a square matrix is nonsingular. This method also shows how to compute the inverse of a nonsingular matrix. The method depends on transforming the matrix to row echelon form. We apply this method to a general 2 × 2 matrix

$$A = \begin{pmatrix} a & b \\ c & d \end{pmatrix}$$

to find the inverse of A if it exists. Here are the calculations.

$$\left(\begin{array}{cc|cc} a & b & 1 & 0 \\ c & d & 0 & 1 \end{array}\right)$$

If $a = c = 0$, then the matrix is singular and does not have an inverse. So we suppose either $a \neq 0$ or $c \neq 0$. For our illustration we assume $a \neq 0$. (It really

230

does not matter; if $c \neq 0$ the answer comes out the same. We just have to do one or the other, so we choose $a \neq 0$ as the case to describe.)

$$\xrightarrow{R_1\left(\frac{1}{a}\right)} \left(\begin{array}{cc|cc} 1 & \dfrac{b}{a} & \dfrac{1}{a} & 0 \\ c & d & 0 & 1 \end{array} \right)$$

$$\xrightarrow{R_{2,1}(-c)} \left(\begin{array}{cc|cc} 1 & \dfrac{b}{a} & \dfrac{1}{a} & 0 \\ 0 & \dfrac{ad - bc}{a} & \dfrac{-c}{a} & 1 \end{array} \right)$$

The next step is the crucial one. If $ad - bc = 0$, the row echelon form for A is not I_2 and so A is singular. If $ad - bc \neq 0$, then the row echelon form of A is I_2 and A is nonsingular. To find A^{-1}, we assume $ad - bc \neq 0$ and continue.

$$\xrightarrow{R_2\left(\frac{a}{ad - bc}\right)} \left(\begin{array}{cc|cc} 1 & \dfrac{b}{a} & \dfrac{1}{a} & 0 \\ 0 & 1 & \dfrac{-c}{ad - bc} & \dfrac{a}{ad - bc} \end{array} \right)$$

$$\xrightarrow{R_{1,2}\left(\frac{-b}{a}\right)} \left(\begin{array}{cc|cc} 1 & 0 & \dfrac{d}{ad - bc} & \dfrac{-b}{ad - bc} \\ 0 & 1 & \dfrac{-c}{ad - bc} & \dfrac{a}{ad - bc} \end{array} \right)$$

Thus our calculations show that the matrix

$$\begin{pmatrix} a & b \\ c & d \end{pmatrix}$$

is nonsingular if and only if $ad - bc \neq 0$. If it is nonsingular, then its inverse is

$$\begin{pmatrix} \dfrac{d}{ad - bc} & \dfrac{-b}{ad - bc} \\ \dfrac{-c}{ad - bc} & \dfrac{a}{ad - bc} \end{pmatrix} = \dfrac{1}{ad - bc} \begin{pmatrix} d & -b \\ -c & a \end{pmatrix}.$$

Definition 6-1

The *determinant* of the 2 × 2 matrix

$$\begin{pmatrix} a & b \\ c & d \end{pmatrix}$$

is the number $ad - bc$.

Example With the formula above we can quickly determine whether a 2 × 2 matrix is nonsingular. For example, the matrix

$$\begin{pmatrix} 2 & 1 \\ 1 & 3 \end{pmatrix}$$

is nonsingular, since $ad - bc = 6 - 1 = 5 \neq 0$. Its inverse is

$$\begin{pmatrix} \frac{3}{5} & -\frac{1}{5} \\ -\frac{1}{5} & \frac{2}{5} \end{pmatrix}$$

as you can easily check.

On the other hand, the matrix

$$\begin{pmatrix} 2 & 4 \\ 1 & 2 \end{pmatrix}$$

is singular because $ad - bc = 4 - 4 = 0$. This matrix does not have an inverse.

Similarly the matrix

$$\begin{pmatrix} 5 & 1 \\ 3 & -4 \end{pmatrix}$$

is nonsingular, since its determinant is $ad - bc = (5)(-4) - (1)(3) = -23$; its inverse is

$$\begin{pmatrix} \frac{4}{23} & \frac{1}{23} \\ \frac{3}{23} & -\frac{5}{23} \end{pmatrix}.$$

If we try the same procedure on the 3 × 3 matrix

$$A = \begin{pmatrix} a & b & c \\ d & e & f \\ p & q & r \end{pmatrix}$$

we find, after considerable computations, that A is nonsingular if and only if $aer + bfp + cdq - cep - afq - bdr \neq 0$. Further, if A is nonsingular, then

$$A^{-1} = \frac{1}{aer + bfp + cdq - cep - afq - bdr} \begin{pmatrix} er - fq & cq - br & bf - ce \\ fp - dr & ar - cp & cd - af \\ dq - ep & bp - aq & ae - bd \end{pmatrix}.$$

Clearly these formulas are too complicated to remember without some device to aid the memory.

The formula for the determinant of a 3×3 matrix can be remembered in the following way. Write down the matrix and then repeat the first and second columns again as shown here.

$$\begin{array}{ccccc} a & b & c & a & b \\ d & e & f & d & e \\ p & q & r & p & q \end{array}$$

Compute the products along the arrows

giving a factor $+1$ to those products along the arrows going to the right and a factor of -1 to those products along the arrows going to the left.

Example We compute the determinant of the 3×3 matrix

$$A = \begin{pmatrix} 2 & 1 & 5 \\ 4 & -2 & 3 \\ 0 & 6 & -7 \end{pmatrix}.$$

Write the first and second columns again as below.

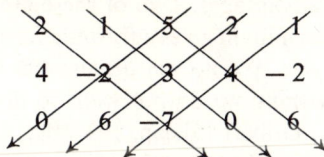

Using the arrow technique described above we get

$$\det A = 2(-2)(-7) + 1 \cdot 3 \cdot 0 + 5 \cdot 4 \cdot 6 - 5(-2) \cdot 0 - 2 \cdot 3 \cdot 6 - 1 \cdot 4(-7)$$

$$= 28 + 0 + 120 + 0 - 36 + 28 = 140.$$

Example   We compute the determinant of the 3 × 3 matrix

$$\begin{pmatrix} 8 & 10 & -2 \\ 5 & -3 & 1 \\ 2 & 0 & -4 \end{pmatrix}.$$

Write the first and second columns again as below.

Using the arrow technique described above we get

$$\det A = 8(-3)(-4) + 10 \cdot 1 \cdot 2 + (-2) \cdot 5 \cdot 0 - (-2)(-3) \cdot 2 - 8 \cdot 1 \cdot 0$$
$$- 10 \cdot 5(-4) = 96 + 20 + 0 - 12 - 0 + 200 = 304.$$

Definition 6-2

The determinant of the 3 × 3 matrix

$$\begin{pmatrix} a & b & c \\ d & e & f \\ p & q & r \end{pmatrix}$$

is

$$aer + bfp + cdq - cep - afq - bdr.$$

The formulas for the determinant and the inverse of a 3 × 3 matrix are complicated. But the formulas for the determinant of larger square matrices get much more complicated as the size of the matrix increases. Over the years three principal descriptions of the determinant of square matrices have been developed, each with its own advantages and disadvantages. Two of these are particularly useful for us.

One description of the determinant gives a formula for computing the determinant of a square matrix of any size. We do not discuss this because we would have to explain permutations, something we do not want to do. You can find treatments of this in other books, e.g., Gilbert [5], pp. 104–109.

The determinant can also be described by the way its value changes when row operations are applied to the matrix. We use this description in Section 6-2.

A third description shows how to compute the determinant of a square matrix of one size by computing the determinant of square matrices one size smaller. This is the method we discuss in this section.

To introduce this description of the determinant, we need the concept of a submatrix. It is a matrix obtained from another by removing some entire rows and/or some entire columns.

Example Suppose A is the 3×4 matrix

$$\begin{pmatrix} 1 & 2 & 0 & 1 \\ 3 & 1 & 1 & 2 \\ 1 & -1 & 3 & 4 \end{pmatrix}.$$

We can get many submatrices by removing various entire rows and/or entire columns. Here are some examples.

(1) Remove columns 3 and 4, row 3.

$$\begin{pmatrix} 1 & 2 \\ 3 & 1 \end{pmatrix}$$

(2) Remove row 1 and column 2.

$$\begin{pmatrix} 3 & 1 & 2 \\ 1 & 3 & 4 \end{pmatrix}$$

(3) Remove rows 1 and 2.

$$(1 \quad -1 \quad 3 \quad 4)$$

(4) The following is *not* a submatrix of A.

$$\begin{pmatrix} 1 & 2 & 0 \\ 3 & 1 & 2 \\ 1 & 3 & 4 \end{pmatrix}$$

Definition 6-3

A *submatrix* of the matrix A is a matrix which can be obtained from A by removing some entire rows and/or some entire columns. If we remove rows 1 and 2 and column 2, we indicate the resulting matrix by $A(1,2 \,|\, 2)$. The numbers of the rows removed are listed before the vertical bar, the numbers of the columns removed are listed after the vertical bar. If no row or column is removed we use a dash to fill the spaces: $A(- \,|\quad)$ or $A(\quad|\,-)$.

Example Suppose A is the 4×4 matrix

$$\begin{pmatrix} 1 & 2 & 1 & 3 \\ 0 & 5 & 1 & 2 \\ -1 & 2 & 4 & 1 \\ 3 & 1 & 3 & 5 \end{pmatrix}.$$

Here are some submatrices of A.

$$A(1 \mid 2,3) = \begin{pmatrix} 0 & 2 \\ -1 & 1 \\ 3 & 5 \end{pmatrix}$$

$$A(- \mid 1) = \begin{pmatrix} 2 & 1 & 3 \\ 5 & 1 & 2 \\ 2 & 4 & 1 \\ 1 & 3 & 5 \end{pmatrix}$$

$$A(2 \mid 3) = \begin{pmatrix} 1 & 2 & 3 \\ -1 & 2 & 1 \\ 3 & 1 & 5 \end{pmatrix}$$

Brief Exercises

1. If A is the following matrix, find $A(2 \mid 1)$, $A(1 \mid 5)$, $A(2,3 \mid 4,5)$, $A(5 \mid 5)$, $A(4,5 \mid 4,5)$, and $A(3,4,5 \mid 3,4,5)$.

$$A = \begin{pmatrix} 2 & 1 & 3 & 4 & 1 \\ 6 & 8 & 1 & 3 & 2 \\ 1 & 1 & 2 & 1 & 4 \\ 0 & 5 & -1 & 3 & 6 \\ 8 & -1 & 2 & 1 & 5 \end{pmatrix}$$

2. Compute the determinants of the following matrices.

(a) $\begin{pmatrix} 2 & 1 \\ 1 & 5 \end{pmatrix}$

(b) $\begin{pmatrix} 4 & 0 \\ 3 & 2 \end{pmatrix}$

(c) $\begin{pmatrix} 2 & 0 \\ 0 & -4 \end{pmatrix}$

(d) $\begin{pmatrix} 0 & 3 \\ 1 & 0 \end{pmatrix}$

(e) $\begin{pmatrix} 1 & 3 & 0 \\ 1 & -1 & 2 \\ 0 & 1 & 4 \end{pmatrix}$

(f) $\begin{pmatrix} 2 & 1 & 5 \\ 4 & 1 & 2 \\ 3 & -1 & 5 \end{pmatrix}$

Answers:

1

$$A(2\mid 1) = \begin{pmatrix} 1 & 3 & 4 & 1 \\ 1 & 2 & 1 & 4 \\ 5 & -1 & 3 & 6 \\ -1 & 2 & 1 & 5 \end{pmatrix}$$

$$A(1\mid 5) = \begin{pmatrix} 6 & 8 & 1 & 3 \\ 1 & 1 & 2 & 1 \\ 0 & 5 & -1 & 3 \\ 8 & -1 & 2 & 1 \end{pmatrix}$$

$$A(2,3\mid 4,5) = \begin{pmatrix} 2 & 1 & 3 \\ 0 & 5 & -1 \\ 8 & -1 & 2 \end{pmatrix}$$

$$A(5\mid 5) = \begin{pmatrix} 2 & 1 & 3 & 4 \\ 6 & 8 & 1 & 3 \\ 1 & 1 & 2 & 1 \\ 0 & 5 & -1 & 3 \end{pmatrix}$$

$$A(4,5\mid 4,5) = \begin{pmatrix} 2 & 1 & 3 \\ 6 & 8 & 1 \\ 1 & 1 & 2 \end{pmatrix}$$

$$A(3,4,5\mid 3,4,5) = \begin{pmatrix} 2 & 1 \\ 6 & 8 \end{pmatrix}$$

2(a) 9   **2(b)** 8   **2(c)** -8 **2(d)** -3 **2(e)** -18 **2(f)** -35

Now we show how to compute the determinant of a square matrix of a particular size by computing the determinant of square matrices which are one size smaller. For example, we know how to compute the determinant of a 2×2 matrix from Definition 6-1. This method shows how to compute the determinant of a 3×3 matrix. Once that is known, the method shows how to compute the determinant of a 4×4 matrix. Then we use the method again to compute the determinant of a 5×5 matrix, and so on.

For completeness we include the definition of the determinant of a 1×1 matrix.

Definition 6-4

The determinant of a 1×1 matrix (a) is defined to be the number a. We write $\det (a) = a$.

Suppose that A is an $n \times n$ matrix ($n \geq 2$). Pick any row of A; say you pick the ith row of A. Then the *expansion for the determinant of A along the ith row* is

$$\det A = (-1)^{i+1}a_{i1} \det A(i\mid 1) + (-1)^{i+2}a_{i2} \det A(i\mid 2)$$
$$+ \cdots + (-1)^{i+n}a_{in} \det A(i\mid n).$$

Notice that $A(i\mid 1)$, for example, is an $(n-1) \times (n-1)$ matrix.

The formula can be described in words, too. Take the first entry a_{i1} in the ith row. Remove the row and column containing a_{i1} and compute the determinant of the resulting submatrix $A(i \mid 1)$. Multiply these numbers by $(-1)^{i+1}$, the power being the sum of the row and column in which a_{i1} appears. This gives the first term

$$(-1)^{i+1} a_{i1} \det A(i \mid 1).$$

The other terms are computed in a similar fashion; for example, the second term is

$$(-1)^{i+2} a_{i2} \det A(i \mid 2),$$

and the third term is

$$(-1)^{i+3} a_{i3} \det A(i \mid 3).$$

Notice that we actually have something of a proposition included in this definition in that we are saying that the expansion along any row always gives the same value. This is not obvious, but justifying it involves technicalities which we do not pursue. We illustrate this definition and the fact that expansion along any row yields the same value.

Example   In this example we compute the determinant of the 2×2 matrix

$$A = \begin{pmatrix} 1 & 2 \\ 3 & 5 \end{pmatrix}$$

using Definition 6-4.

The expansion along the first row is

$$(-1)^{1+1} \cdot 1 \cdot \det A(1 \mid 1) + (-1)^{1+2} \cdot 2 \cdot \det A(1 \mid 2)$$
$$= 1 \det (5) - 2 \det (3)$$
$$= 5 - 6 = -1.$$

The expansion along the second row is

$$(-1)^{2+1} \cdot 3 \cdot \det A(2 \mid 1) + (-1)^{2+2} \cdot 5 \cdot \det A(2 \mid 2)$$
$$= -3 \det (2) + 5 \det (1)$$
$$= -6 + 5 = -1.$$

Example   In this example we compute the determinant of the 3×3 matrix

$$A = \begin{pmatrix} 2 & 1 & -1 \\ 3 & 0 & 4 \\ -1 & 2 & 1 \end{pmatrix}$$

using Definition 6-4. The expansion along the first row is

$$(-1)^{1+1} \cdot 2 \cdot \det A(1 \mid 1) + (-1)^{1+2} \cdot 1 \cdot \det A(1 \mid 2) + (-1)^{1+3}(-1) \det A(1 \mid 3)$$

$$= (-1)^{1+1} \cdot 2 \cdot \det \begin{pmatrix} 0 & 4 \\ 2 & 1 \end{pmatrix} + (-1)^{1+2} \cdot 1 \cdot \det \begin{pmatrix} 3 & 4 \\ -1 & 1 \end{pmatrix}$$

$$+ (-1)^{1+3}(-1) \det \begin{pmatrix} 3 & 0 \\ -1 & 2 \end{pmatrix}$$

$$= -16 - 7 - 6 = -29.$$

The expansion along the third row is

$$(-1)^{3+1}(-1) \det A(3 \mid 1) + (-1)^{3+2} \cdot 2 \cdot \det A(3 \mid 2) + (-1)^{3+3} \cdot 1 \cdot \det A(3 \mid 3)$$

$$= -\det \begin{pmatrix} 1 & -1 \\ 0 & 4 \end{pmatrix} - 2 \det \begin{pmatrix} 2 & -1 \\ 3 & 4 \end{pmatrix} + \det \begin{pmatrix} 2 & 1 \\ 3 & 0 \end{pmatrix}$$

$$= -4 - 22 - 3 = -29.$$

Example In this example we compute the determinant of the 4 × 4 matrix

$$A = \begin{pmatrix} 2 & 1 & 0 & 1 \\ 3 & -1 & 2 & 4 \\ 1 & 3 & 5 & 1 \\ 2 & 1 & 6 & -3 \end{pmatrix}$$

using the expansion along the 3rd row.

$$\det A = (-1)^{3+1} \cdot 1 \cdot \det A(3 \mid 1) + (-1)^{3+2} \cdot 3 \cdot \det A(3 \mid 2)$$
$$\qquad + (-1)^{3+3} \cdot 5 \cdot \det A(3 \mid 3) + (-1)^{3+4} \cdot 1 \cdot \det A(3 \mid 4)$$

$$= \det \begin{pmatrix} 1 & 0 & 1 \\ -1 & 2 & 4 \\ 1 & 6 & -3 \end{pmatrix} - 3 \det \begin{pmatrix} 2 & 0 & 1 \\ 3 & 2 & 4 \\ 2 & 6 & -3 \end{pmatrix}$$

$$+ 5 \det \begin{pmatrix} 2 & 1 & 1 \\ 3 & -1 & 4 \\ 2 & 1 & -3 \end{pmatrix} - \det \begin{pmatrix} 2 & 1 & 0 \\ 3 & -1 & 2 \\ 2 & 1 & 6 \end{pmatrix}$$

$$= (-6 + 0 - 6 - 2 - 24 - 0) - 3(-12 + 0 + 18 - 4 - 48 - 0)$$
$$+ 5(6 + 8 + 3 + 2 - 8 + 9) - (-12 + 4 + 0 - 0 - 4 - 18)$$
$$= -38 - 3(-46) + 5(20) - (-30) = 230$$

Notice that the process of computing a determinant by expanding along a row changes the problem from that of computing the determinant of a 4 × 4 matrix to one of computing the determinants of four 3 × 3 matrices. Thus, if necessary, one can compute the determinant of a large square matrix by repeated use of Definition 6-4 until one has to compute the determinant of a relatively small matrix.

SUMMARY   *The determinant of the 2 × 2 matrix*

$$\begin{pmatrix} a & b \\ c & d \end{pmatrix}$$

is ad − bc. The determinant of the 3 × 3 matrix

$$\begin{pmatrix} a & b & c \\ d & e & f \\ p & q & r \end{pmatrix}$$

is figured by computing the products along the arrows indicated

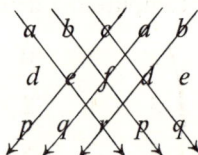

giving +1 to the products along arrows pointing to the right and −1 to those along arrows pointing to the left.

   The determinant for larger matrices A, say n × n, can be computed by the formula

$$\det A = (-1)^{i+1}a_{i1} \det A(i\,|\,1) + (-1)^{i+2}a_{i2} \det A(i\,|\,2)$$
$$+ \cdots + (-1)^{i+n}a_{in} \det A(i\,|\,n),$$

where $A(i\,|\,j)$ is the $(n-1) \times (n-1)$ matrix obtained from A by removing the ith row and the jth column. This is called the expansion of the determinant along the ith row.

Exercises for Section 6-1

1. Compute the determinants of the following matrices by expanding each one along every one of its rows.

   (a) $\begin{pmatrix} 2 & 1 & 3 \\ 1 & -1 & 2 \\ 1 & 5 & 4 \end{pmatrix}$ **(b)** $\begin{pmatrix} 1 & 0 & 0 \\ 0 & 1 & 0 \\ 0 & 0 & 1 \end{pmatrix}$

(c) $\begin{pmatrix} 1 & 0 & 0 \\ 1 & 1 & 0 \\ 1 & 1 & 1 \end{pmatrix}$ (d) $\begin{pmatrix} 3 & 4 & 1 \\ 0 & 2 & 1 \\ 0 & 0 & 3 \end{pmatrix}$

(e) $\begin{pmatrix} 5 & 1 & 0 & 2 \\ 3 & -1 & 1 & 2 \\ 1 & 4 & 1 & 2 \\ 9 & 4 & 2 & 6 \end{pmatrix}$ (f) $\begin{pmatrix} 2 & 5 & -1 & 4 \\ 3 & 2 & 1 & 1 \\ -1 & 3 & 2 & 4 \\ 5 & 1 & 6 & 7 \end{pmatrix}$

2. Compute the determinant of the following matrices three ways by expanding along each row.

(a) $\begin{pmatrix} 2 & 1 & 5 \\ 3 & 1 & 0 \\ 2 & 0 & 3 \end{pmatrix}$ (b) $\begin{pmatrix} 1 & 7 & 0 \\ -2 & 5 & 1 \\ 3 & 4 & -2 \end{pmatrix}$ (c) $\begin{pmatrix} 2 & 5 & 2 \\ -1 & 3 & 0 \\ 6 & 4 & 4 \end{pmatrix}$

3. Compute the determinant of the following 4 × 4 matrix by expanding along each of its rows.

$$\begin{pmatrix} 1 & 2 & 0 & 5 \\ -1 & 3 & 4 & 1 \\ 2 & -1 & 4 & 1 \\ 6 & 8 & 2 & 0 \end{pmatrix}$$

4. Compute the determinant of the following 5 × 5 matrix along its rows.

$$\begin{pmatrix} 1 & 1 & 2 & 1 & 5 \\ 0 & -2 & 3 & 4 & 1 \\ 2 & 1 & -3 & 8 & 1 \\ 2 & 1 & 4 & 0 & 3 \\ 1 & 0 & 2 & 1 & 3 \end{pmatrix}$$

SECTION 6-2
Techniques for Computing the Determinant

PREVIEW *This section begins with a description of the determinant in terms of the way its value changes when row operations are applied to the matrix. Then* *using the fact that* $\det A = \det A^T$, *we show that the determinant of a matrix can be computed by expanding along any column of the matrix. Further, the column operations on a matrix produce the same changes in the value of the determinant as the corresponding row operations. Then we conclude the section by observing that* $\det A = 0$ *if A has a row or a column of zeros, or, more generally, if a row (column) is a linear combination of the other rows (columns).*

As we mentioned in Section 6-1, we present two descriptions of the determinant function. The second description is concerned with the changes in the value of the determinant when row operations are used on the matrix. It is possible to show that these properties of the determinant follow from the definition used in Section 6-1; we do not present this argument because the technicalities seem to distract from the ideas which we are pursuing.

In the next two examples we illustrate how the value of the determinant of a matrix changes as row operations are applied.

Example   In this example we show the effect of row operations on the value of the determinant of 2×2 matrices. Let A be the matrix

$$\begin{pmatrix} 1 & 2 \\ 3 & 4 \end{pmatrix}.$$

Note that $\det A = -2$. Apply $R_{1,2}$ by multiplying with $F_{1,2}$ on the left (see Proposition 5-2).

$$F_{1,2}A = \begin{pmatrix} 3 & 4 \\ 1 & 2 \end{pmatrix}$$

Then $\det F_{1,2}A = 2$, so $\det F_{1,2}A = -\det A$. The value of the determinant is multiplied by a factor of -1 if two rows are interchanged.

Suppose we apply $R_2(x)$ to A by premultiplying with $F_2(x)$.

$$F_2(x)A = \begin{pmatrix} 1 & 2 \\ 3x & 4x \end{pmatrix}$$

Then $\det F_2(x)A = -2x = x \det A$. The value of the determinant is multiplied by the factor x if a row is multiplied by x.

Finally, suppose we apply $R_{2,1}(x)$ to A by premultiplying with $F_{2,1}(x)$.

$$F_{2,1}(x)A = \begin{pmatrix} 1 & 2 \\ 3+x & 4+2x \end{pmatrix}$$

Then $\det F_{2,1}(x)A = (4 + 2x) - 2(3 + x) = -2 = \det A$. The value of the determinant is unchanged by application of row operations of the type $F_{i,j}(x)$, provided we require that $i \neq j$.

These results are not just accidents because of the particular entries in the matrix A. We can check the same results on the matrix

$$B = \begin{pmatrix} a & b \\ c & d \end{pmatrix}.$$

We leave the verifications to you, but you should carry out the calculations for the following equalities.

$$\det F_{1,2}B = -\det B$$

$$\det F_1(x)B = x \det B$$

$$\det F_2(x)B = x \det B$$

$$\det F_{1,2}(x)B = \det B$$

$$\det F_{2,1}(x)B = \det B$$

$$\det I_2 = 1$$

Example   In this example we illustrate that the facts we just obtained for 2×2 matrices are also true for 3×3 matrices. We take a numerical example first. Let A be the 3×3 matrix

$$\begin{pmatrix} 1 & 0 & 1 \\ 2 & 3 & 1 \\ -2 & 1 & 4 \end{pmatrix}.$$

We compute the determinant of the following matrices as an indication of the general situation.

$$F_{2,3}A = \begin{pmatrix} 1 & 0 & 1 \\ -2 & 1 & 4 \\ 2 & 3 & 1 \end{pmatrix}$$

$$F_{1,3}A = \begin{pmatrix} -2 & 1 & 4 \\ 2 & 3 & 1 \\ 1 & 0 & 1 \end{pmatrix}$$

$$F_1(x)A = \begin{pmatrix} x & 0 & x \\ 2 & 3 & 1 \\ -2 & 1 & 4 \end{pmatrix}$$

$$F_2(x)A = \begin{pmatrix} 1 & 0 & 1 \\ 2x & 3x & x \\ -2 & 1 & 4 \end{pmatrix}$$

$$F_{1,3}(x)A = \begin{pmatrix} 1 - 2x & x & 1 + 4x \\ 2 & 3 & 1 \\ -2 & 1 & 4 \end{pmatrix}$$

$$F_{2,1}(x)A = \begin{pmatrix} 1 & 0 & 1 \\ 2 + x & 3 & 1 + x \\ -2 & 1 & 4 \end{pmatrix}$$

By inspecting these matrices you can easily check that

$$\det F_{2,3}A = \det F_{1,3}A = -\det A,$$

$$\det F_1(x)A = \det F_2(x)A = x \det A,$$

$$\det F_{1,3}(x)A = \det F_{2,1}(x)A = \det A,$$

$$\det I_3 = 1.$$

These examples illustrate facts which are true for square matrices of any size. We state these in the next proposition.

Proposition 6-1

Suppose A is an $n \times n$ matrix.

(1) Interchanging two rows of A changes the value of the determinant by a factor of -1: $\det F_{i,j}A = -\det A$, $i \neq j$.
(2) Multiplying a row by the scalar x changes the value of the determinant by a factor of x: $\det F_i(x)A = x \det A$.
(3) Adding a multiple of one row to another leaves the value of the determinant unchanged: $\det F_{i,j}(x)A = \det A$, $i \neq j$.
(4) The determinant of the identity is 1: $\det I_n = 1$.

Our aim in presenting the second description of the determinant is to combine it with the first description to simplify our numerical calculations. So, in the spirit of this aim, in this section we use a fact about determinants whose discussion we delay until Section 6-4. It is this: if A is an $n \times n$ matrix, then $\det A = \det A^T$.

The transpose A^T of a matrix A was introduced in Definition 1-7 in the exercises for Section 1-2. It is formed by using the columns of A for the rows of A^T (or by using the rows of A for the columns of A^T). There are two principal consequences which help us in computing determinants. The first is that the determinant of a matrix A can be computed by expanding along any column of A as well as along any row. The reason is that expanding det A^T along a row of A^T corresponds to expanding det A along a column of A.

Example   Let A be the matrix

$$A = \begin{pmatrix} 1 & 2 & -1 \\ 3 & 5 & 0 \\ -4 & -2 & 6 \end{pmatrix}.$$

Then

$$A^T = \begin{pmatrix} 1 & 3 & -4 \\ 2 & 5 & -2 \\ -1 & 0 & 6 \end{pmatrix}.$$

If we expand det A^T along the first row we obtain

$$\det A^T = (-1)^{1+1} \cdot 1 \cdot \det \begin{pmatrix} 5 & -2 \\ 0 & 6 \end{pmatrix} + (-1)^{1+2} \cdot 3 \cdot \det \begin{pmatrix} 2 & -2 \\ -1 & 6 \end{pmatrix}$$

$$+ (-1)^{1+3}(-4) \det \begin{pmatrix} 2 & 5 \\ -1 & 0 \end{pmatrix}.$$

This corresponds to the expansion of det A along the first column of A.

$$\det A = (-1)^{1+1} \cdot 1 \cdot \det \begin{pmatrix} 5 & 0 \\ -2 & 6 \end{pmatrix} + (-1)^{2+1} \cdot 3 \cdot \det \begin{pmatrix} 2 & -1 \\ -2 & 6 \end{pmatrix}$$

$$+ (-1)^{3+1}(-4) \det \begin{pmatrix} 2 & -1 \\ 5 & 0 \end{pmatrix}$$

The 2×2 matrices appearing in the expansion of det A are transposes of those appearing in the expansion of det A^T.

Example   In this example we compute the determinant of the following matrix by expanding along its columns.

$$A = \begin{pmatrix} 2 & 1 & 0 \\ 5 & -3 & 4 \\ -7 & 8 & -6 \end{pmatrix}$$

Expanding along the first column we get

$$\det A = (-1)^{1+1} \cdot 2 \cdot \det \begin{pmatrix} -3 & 4 \\ 8 & -6 \end{pmatrix} + (-1)^{2+1} \cdot 5 \cdot \det \begin{pmatrix} 1 & 0 \\ 8 & -6 \end{pmatrix}$$

$$+ (-1)^{3+1}(-7) \det \begin{pmatrix} 1 & 0 \\ -3 & 4 \end{pmatrix} = -26.$$

Expanding along the second column we get

$$\det A = (-1)^{1+2} \cdot 1 \cdot \det \begin{pmatrix} 5 & 4 \\ -7 & -6 \end{pmatrix} + (-1)^{2+2}(-3) \det \begin{pmatrix} 2 & 0 \\ -7 & -6 \end{pmatrix}$$

$$+ (-1)^{3+2} \cdot 8 \cdot \det \begin{pmatrix} 2 & 0 \\ 5 & 4 \end{pmatrix} = -26.$$

Finally, expanding along the third column, we obtain, by omitting zero terms,

$$\det A = (-1)^{2+3} \cdot 4 \cdot \det \begin{pmatrix} 2 & 1 \\ -7 & 8 \end{pmatrix} + (-1)^{3+3}(-6) \det \begin{pmatrix} 2 & 1 \\ 5 & -3 \end{pmatrix} = -26.$$

These examples illustrate the fact we have already mentioned: the determinant of A can be computed by an expansion along any column as well as along any row.

Proposition 6-2
Suppose A is an $n \times n$ matrix. Choose a column of A, say the jth column. Then the determinant of A can be computed by the *expansion along the jth column of A*:

$$\det A = (-1)^{1+j}a_{1j} \det A(1\,|\,j) + (-1)^{2+j}a_{2j} \det A(2\,|\,j)$$

$$+ \cdots + (-1)^{n+j}a_{nj} \det A(n\,|\,j).$$

Brief Exercises

1. Compute the determinant of the following matrices by expanding along each column.

(a) $\begin{pmatrix} 2 & 1 \\ 1 & 3 \end{pmatrix}$

(c) $\begin{pmatrix} 1 & 5 & 1 \\ -2 & 1 & 3 \\ 5 & 4 & 2 \end{pmatrix}$

(b) $\begin{pmatrix} 1 & 0 & 2 \\ -1 & 3 & 1 \\ 4 & 1 & -2 \end{pmatrix}$

(d) $\begin{pmatrix} 1 & 1 & 2 \\ -1 & 0 & 1 \\ 3 & 2 & 1 \end{pmatrix}$

Answers: 1(a) 5 1(b) −33 1(c) 72 1(d) −2

The second consequence of the fact that $\det A = \det A^T$ for a square matrix A involves column operations on a matrix. We discussed column operations on a matrix earlier in Definition 1-5. This was in the Exercises for Section 1-2. As you recall, the column operations are:

(1) interchange of the ith and jth columns, denoted by $C_{i,j}$,
(2) multiplication of the ith column by a nonzero scalar x, denoted by $C_i(x)$, and
(3) addition of x times the jth column to the ith column, denoted by $C_{i,j}(x)$, $i \neq j$.

We say that the row operations and column operations correspond to each other in the following way: $R_{i,j}$ and $C_{i,j}$ correspond to each other, $R_i(x)$ and $C_i(x)$ correspond to each other, and $R_{i,j}(x)$ and $C_{i,j}(x)$ correspond to each other. The fact which interests us is that when column operations are applied to a square matrix, the value of the determinant changes in the same way as when the corresponding row operations are applied.

Example Let A be the matrix

$$A = \begin{pmatrix} 1 & 3 & 4 \\ -2 & 0 & 5 \\ 6 & -4 & 2 \end{pmatrix}.$$

Then

$$A^T = \begin{pmatrix} 1 & -2 & 6 \\ 3 & 0 & -4 \\ 4 & 5 & 2 \end{pmatrix}.$$

Suppose we interchange the first and third columns of A to obtain the matrix

$$B = \begin{pmatrix} 4 & 3 & 1 \\ 5 & 0 & -2 \\ 2 & -4 & 6 \end{pmatrix}.$$

Then B^T can be obtained by interchanging the first and third rows of A^T. Therefore, $\det B = \det B^T = -\det A^T = -\det A$.

Similarly, if we multiply the second column of A by the nonzero scalar x and obtain the matrix

$$C = \begin{pmatrix} 1 & 3x & 4 \\ -2 & 0 & 5 \\ 6 & -4x & 2 \end{pmatrix},$$

then C^T can be obtained by multiplying the second row of A^T by x. Therefore $\det C = \det C^T = x \det A^T = x \det A$.

In a similar fashion you can readily see that if we add a multiple of one column of A to another and obtain the matrix D, then D^T can be obtained from A^T by the corresponding row operation. Then $\det D = \det D^T = \det A^T = \det A$.

This example illustrates the fact that the determinant changes value in the same way when a column operation is applied as when the corresponding row operation is applied. The reason for this is easy to understand: if B is obtained from A by a column operation, then B^T can be obtained from A^T by the corresponding row operation. Thus if the column operation is $C_{i,j}$, $\det B = \det B^T = -\det A^T = -\det A$. If the column operation is $C_i(x)$, $\det B = \det B^T = x \det A^T = x \det A$. If the column operation is $C_{i,j}(x)$, $\det B = \det B^T = \det A^T = \det A$.

Proposition 6-3_____

Suppose A is a square matrix.
(1) Interchanging two columns of A changes the value of the determinant by a factor of -1.
(2) Multiplying a column of A by a nonzero scalar x changes the value of the determinant by a factor of x.
(3) The value of the determinant is unchanged if a multiple of one column is added to another.

Now we give some examples to show how we can simplify our calculations by using row and column operations.

Example Suppose we wish to compute the determinant of the matrix

$$A = \begin{pmatrix} 1 & 2 & 1 & -1 \\ 2 & 1 & 3 & 2 \\ 1 & 4 & 1 & 2 \\ -1 & 0 & 1 & 3 \end{pmatrix}.$$

We can expand $\det A$ along any row, as indicated by Definition 6-4. You can readily see that it would be advantageous to compute $\det A$ by expanding along the fourth row; because there is a zero in the (4,2) position, we only need compute the determinant of three 3×3 matrices rather than four if we expand along a different row.

We can simplify our calculations even more, however, by using column opera-tions on A to produce more zeros in the fourth row. Use $C_{3,1}(1)$ to obtain

$$A_1 = \begin{pmatrix} 1 & 2 & 2 & -1 \\ 2 & 1 & 5 & 2 \\ 1 & 4 & 2 & 2 \\ -1 & 0 & 0 & 3 \end{pmatrix},$$

and then use $C_{4,1}(3)$ to obtain

$$A_2 = \begin{pmatrix} 1 & 2 & 2 & 2 \\ 2 & 1 & 5 & 8 \\ 1 & 4 & 2 & 5 \\ -1 & 0 & 0 & 0 \end{pmatrix}.$$

By Proposition 6-3, $\det A = \det A_1 = \det A_2$. If we expand $\det A_2$ along the fourth row we obtain

$$(-1)^{4+1}(-1) \det \begin{pmatrix} 2 & 2 & 2 \\ 1 & 5 & 8 \\ 4 & 2 & 5 \end{pmatrix} = 36.$$

The point of this example is that if we apply row and/or column operations along with the formula for expanding the determinant, we can sometimes simplify our calculations.

Example In this example we use Propositions 6-1, 6-2, and 6-3 along with the definition of the determinant to compute the determinant of the 5×5 matrix

$$A = \begin{pmatrix} 2 & 1 & 2 & 3 & 1 \\ 3 & 0 & 2 & -1 & 2 \\ 1 & 2 & 1 & 3 & 0 \\ 0 & -1 & 3 & 1 & 2 \\ 2 & 0 & 4 & 1 & -1 \end{pmatrix}.$$

Use $R_{4,1}(1)$ and $R_{3,1}(-2)$ to obtain

$$A_1 = \begin{pmatrix} 2 & 1 & 2 & 3 & 1 \\ 3 & 0 & 2 & -1 & 2 \\ -3 & 0 & -3 & -3 & -2 \\ 2 & 0 & 5 & 4 & 3 \\ 2 & 0 & 4 & 1 & -1 \end{pmatrix}.$$

By Proposition 6-1, $\det A_1 = \det A$. Expanding $\det A_1$ along the second column we get

$$\det A_1 = (-1)^{1+2} \cdot 1 \cdot \det \begin{pmatrix} 3 & 2 & -1 & 2 \\ -3 & -3 & -3 & -2 \\ 2 & 5 & 4 & 3 \\ 2 & 4 & 1 & -1 \end{pmatrix}.$$

Let A_2 be the 4×4 matrix above. Then $\det A = -\det A_2$. Apply the column operations $C_{1,3}(3)$, $C_{2,3}(2)$, and $C_{4,3}(2)$ to A_2 to obtain

$$A_3 = \begin{pmatrix} 0 & 0 & -1 & 0 \\ -12 & -9 & -3 & -8 \\ 14 & 13 & 4 & 11 \\ 5 & 6 & 1 & 1 \end{pmatrix}.$$

By Proposition 6-3, $\det A_2 = \det A_3$, so $\det A = -\det A_3$. Expanding $\det A_3$ along the first row we obtain

$$\det A_3 = (-1)^{1+3}(-1) \det \begin{pmatrix} -12 & -9 & -8 \\ 14 & 13 & 11 \\ 5 & 6 & 1 \end{pmatrix}.$$

Let A_4 be the 3×3 matrix above. Since $\det A_3 = -\det A_4$ we have that $\det A = \det A_4$. Use the row operations $R_{2,1}(1)$, $R_{1,3}(8)$, and $R_{2,3}(-3)$ on A_4 to obtain

$$A_5 = \begin{pmatrix} 28 & 39 & 0 \\ -13 & -14 & 0 \\ 5 & 6 & 1 \end{pmatrix}.$$

By Proposition 6-1, det $A_5 = $ det A_4, so det $A = $ det A_5. Expanding det A_5 along the third column we obtain

$$(-1)^{3+3} \cdot 1 \cdot \det \begin{pmatrix} 28 & 39 \\ -13 & -14 \end{pmatrix} = -392 + 507 = 115.$$

These examples indicate how you can simplify some of your calculations by using row operations and column operations before expanding the determinant.

Brief Exercises

1. Compute the determinant of each of the following matrices. Use row and column operations to simplify your numerical calculations.

(a) $\begin{pmatrix} 2 & 4 & 5 \\ 6 & -6 & 7 \\ -4 & -2 & 8 \end{pmatrix}$

(c) $\begin{pmatrix} 2 & 1 & 3 & 4 \\ -1 & 0 & -2 & 1 \\ 5 & 1 & 4 & 1 \\ 2 & -1 & 3 & 5 \end{pmatrix}$

(b) $\begin{pmatrix} 6 & 8 & 1 & 3 \\ 2 & 0 & 4 & -2 \\ 3 & -3 & 6 & -6 \\ 4 & 5 & -6 & -8 \end{pmatrix}$

Answers: **1(a)** -552 **1(b)** -678 **1(c)** -65

We have presented techniques for computing the determinant of a square matrix. We end this section with a few observations which are sometimes helpful.

If the matrix A has a zero row (or a zero column), then det $A = 0$; expand det A along that row (or column).

There are other situations in which we can conclude that det $A = 0$. We illustrate two of them in the next two examples.

Example Consider the matrix

$$A = \begin{pmatrix} 2 & 1 & 3 & 4 \\ 5 & 1 & 2 & 0 \\ -1 & 3 & 1 & 2 \\ 5 & 1 & 2 & 0 \end{pmatrix}.$$

Notice that the second and fourth rows are equal. If we apply $R_{4,2}(-1)$ to A and obtain the matrix

$$A_1 = \begin{pmatrix} 2 & 1 & 3 & 4 \\ 5 & 1 & 2 & 0 \\ -1 & 3 & 1 & 2 \\ 0 & 0 & 0 & 0 \end{pmatrix},$$

then, by Proposition 6-2, $\det A_1 = \det A$. But $\det A_1 = 0$; expand along the fourth row. This shows why $\det A = 0$ if two different rows of A are equal.

Example   Consider the matrix

$$A = \begin{pmatrix} 2 & 1 & 3 & 4 \\ 5 & 1 & 2 & 0 \\ -1 & 3 & 1 & 2 \\ -2 & 4 & 5 & 10 \end{pmatrix}.$$

We illustrate that if one row of A is a linear combination of the other rows, then $\det A = 0$. Label the rows of A by $\mathbf{X}_1$, $\mathbf{X}_2$, $\mathbf{X}_3$, and $\mathbf{X}_4$.

$$\mathbf{X}_1 = (2,1,3,4),$$
$$\mathbf{X}_2 = (5,1,2,0),$$
$$\mathbf{X}_3 = (-1,3,1,2),$$
$$\mathbf{X}_4 = (-2,4,5,10)$$

You can check that $\mathbf{X}_4 = 2\mathbf{X}_1 - \mathbf{X}_2 + \mathbf{X}_3$. Let A_1 be the matrix obtained by applying the row operations $R_{4,1}(-2)$, $R_{4,2}(1)$, and $R_{4,3}(-1)$. Then, by Proposition 6-1, $\det A_1 = \det A$. But

$$A_1 = \begin{pmatrix} 2 & 1 & 3 & 4 \\ 5 & 1 & 2 & 0 \\ -1 & 3 & 1 & 2 \\ 0 & 0 & 0 & 0 \end{pmatrix},$$

and so $0 = \det A_1 = \det A$.

   This shows why $\det A = 0$ if one row of A is a linear combination of the other rows; using row operations of the type $R_{i,j}(x)$, we can obtain a matrix A_1 having a row of zeros. Thus $0 = \det A_1 = \det A$.

SUMMARY *In this section we discuss the main techniques for computing the determinant of a square matrix. First, we discuss the changes in the value of the determinant caused by use of a row or a column operation:*

(1) *interchanging two rows or two columns changes the value of the determinant by a factor of −1,*
(2) *multiplying a row or a column by the scalar x changes the value of the determinant by a factor of x, and*
(3) *adding a multiple of one row to another or of one column to another leaves the value of the determinant unchanged.*

The determinant can also be computed by expanding along a column of A, say the jth.

$$\det A = (-1)^{1+j}a_{1j} \det A(1 \mid j) + \cdots + (-1)^{n+j}a_{nj} \det A(n \mid j)$$

Finally, some miscellaneous useful facts:

(1) *if A has a zero row or column, then $\det A = 0$;*
(2) *if A has two rows or two columns that are equal, then $\det A = 0$;*
(3) *if some row (column) of A is a linear combination of the other rows (columns) of A, then $\det A = 0$.*

Exercises for Section 6-2

1. Compute the determinant of the following matrices by expanding along some column.

(a) $\begin{pmatrix} 2 & 1 & 5 \\ 3 & 1 & 0 \\ 2 & 0 & 3 \end{pmatrix}$

(b) $\begin{pmatrix} 1 & 7 & 0 \\ -2 & 5 & 1 \\ 3 & 4 & -2 \end{pmatrix}$

(c) $\begin{pmatrix} 2 & 5 & 2 \\ -1 & 3 & 0 \\ 6 & 4 & 4 \end{pmatrix}$

(d) $\begin{pmatrix} 1 & 2 & 0 & 5 \\ -1 & 3 & 4 & 1 \\ 2 & -1 & 4 & 1 \\ 6 & 8 & 2 & 0 \end{pmatrix}$

(e) $\begin{pmatrix} 1 & 1 & 2 & 1 & 5 \\ 0 & -2 & 3 & 4 & 1 \\ 2 & 1 & -3 & 8 & 1 \\ 2 & 1 & 4 & 0 & 3 \\ 1 & 0 & 2 & 1 & 3 \end{pmatrix}$

2. Verify that det A = det A^T in the following cases.

(a) $\begin{pmatrix} 2 & 1 \\ 0 & 3 \end{pmatrix}$ **(b)** $\begin{pmatrix} 1 & 5 \\ 9 & 8 \end{pmatrix}$ **(c)** $\begin{pmatrix} 1 & 0 & 5 \\ 2 & 3 & 1 \\ 6 & -1 & 2 \end{pmatrix}$ **(d)** $\begin{pmatrix} 8 & 1 & 3 \\ 1 & 2 & 1 \\ 3 & 0 & 5 \end{pmatrix}$

3. Compute the determinant of the following matrix A and its transpose A^T by expanding A along its second column and A^T along its second row.

$$\begin{pmatrix} 2 & 1 & 3 & 4 \\ 0 & 5 & -6 & 7 \\ 1 & 2 & 0 & 1 \\ 3 & 1 & -2 & 0 \end{pmatrix}$$

4. Use any of the techniques in this chapter to compute the determinant of the following matrices.

(a) $\begin{pmatrix} 1 & 2 & 1 \\ 3 & 0 & 1 \\ 1 & 2 & 1 \end{pmatrix}$

(b) $\begin{pmatrix} 2 & 1 & 5 \\ -3 & 1 & 6 \\ -1 & 2 & 11 \end{pmatrix}$

(c) $\begin{pmatrix} 5 & 1 & 8 \\ 6 & 2 & 11 \\ 0 & 1 & 2 \end{pmatrix}$

(d) $\begin{pmatrix} 1 & 2 & 0 & 1 \\ 3 & 0 & 1 & 0 \\ 5 & 2 & 1 & 0 \\ 6 & 8 & 7 & 2 \end{pmatrix}$

(e) $\begin{pmatrix} 2 & 1 & 8 & 3 \\ 0 & 2 & 4 & 6 \\ -1 & 5 & 1 & 2 \\ 7 & 0 & 1 & 2 \end{pmatrix}$

(f) $\begin{pmatrix} 1 & 5 & 1 & 0 & 2 \\ 3 & 4 & -2 & 1 & 0 \\ 5 & -1 & 0 & 2 & 1 \\ 1 & 2 & 1 & 0 & -3 \\ 1 & 1 & 0 & 1 & 2 \end{pmatrix}$

SECTION 6-3
Matrix of Cofactors, Another Method for Finding the Inverse of a Nonsingular Matrix

PREVIEW   *For every n × n matrix A we construct a matrix A' called the matrix of cofactors of A. It has the property that $AA' = (\det A)I_n$. Consequently, if $\det A \neq 0$, then A is nonsingular and*

$$A^{-1} = (\tfrac{1}{\det A})A'.$$

Using this expression for the inverse of a nonsingular matrix, we obtain Cramer's Rule for solving a system of n equations in n unknowns which has a nonsingular coefficient matrix.

The formula for the expansion of $\det A$ along a column of A,

$$\det A = (-1)^{1+j}a_{1j} \det A(1 \mid j) + \cdots + (-1)^{n+j}a_{nj} \det A(n \mid j),$$

gives us another way to find the inverse of a nonsingular matrix. The term $(-1)^{i+j} \det A(i \mid j)$ is called the (i,j)-cofactor of A.

Example   Consider the 3×3 matrix

$$A = \begin{pmatrix} 2 & 1 & 0 \\ 3 & 4 & -1 \\ -2 & 5 & -4 \end{pmatrix}.$$

We form another matrix A' by placing the (i,j)-cofactor of A in the (j,i) position (*not* the (i,j) position).

$$\begin{pmatrix} \det \begin{pmatrix} 4 & -1 \\ 5 & -4 \end{pmatrix} & -\det \begin{pmatrix} 1 & 0 \\ 5 & -4 \end{pmatrix} & \det \begin{pmatrix} 1 & 0 \\ 4 & -1 \end{pmatrix} \\ -\det \begin{pmatrix} 3 & -1 \\ -2 & -4 \end{pmatrix} & \det \begin{pmatrix} 2 & 0 \\ -2 & -4 \end{pmatrix} & -\det \begin{pmatrix} 2 & 0 \\ 3 & -1 \end{pmatrix} \\ \det \begin{pmatrix} 3 & 4 \\ -2 & 5 \end{pmatrix} & -\det \begin{pmatrix} 2 & 1 \\ -2 & 5 \end{pmatrix} & \det \begin{pmatrix} 2 & 1 \\ 3 & 4 \end{pmatrix} \end{pmatrix}$$

$$= \begin{pmatrix} -11 & 4 & -1 \\ 14 & -8 & 2 \\ 23 & -12 & 5 \end{pmatrix}$$

You can check that

$$AA' = A'A = \begin{pmatrix} -8 & 0 & 0 \\ 0 & -8 & 0 \\ 0 & 0 & -8 \end{pmatrix} = (-8)I_3 = (\det A)I_3.$$

Thus

$$A \cdot (\tfrac{1}{\det A} \cdot A') = I_3.$$

Since A and $(\tfrac{1}{\det A})A'$ are both 3×3 matrices, Proposition 5-9 shows us that $A^{-1} = (\tfrac{1}{\det A})A'$. We show why the equality $AA' = (\det A)I_3$ is true. First consider the computation of the $(1,1)$ entry of AA'; we use the first row of A and the first column of A'.

$$2 \det \begin{pmatrix} 4 & -1 \\ 5 & -4 \end{pmatrix} - 1 \det \begin{pmatrix} 3 & -1 \\ -2 & -4 \end{pmatrix} + 0 \det \begin{pmatrix} 3 & 4 \\ -2 & 5 \end{pmatrix}$$

This is the expansion of $\det A$ along the first row of A. The first column of A' was chosen precisely so that its combination with the first row of A yields $\det A$.

Next consider the computation of the $(2,1)$ entry of AA'; we use the second row of A and the first column of A'.

$$3 \det \begin{pmatrix} 4 & -1 \\ 5 & -4 \end{pmatrix} - 4 \det \begin{pmatrix} 3 & -1 \\ -2 & -4 \end{pmatrix} - \det \begin{pmatrix} 3 & 4 \\ -2 & 5 \end{pmatrix}$$

This is the expansion for the determinant of a matrix having $(3 \quad 4 \quad -1)$ as first row and having second and third rows the same as those of A; i.e., this is the expansion along the first row for the determinant of

$$\begin{pmatrix} 3 & 4 & -1 \\ 3 & 4 & -1 \\ -2 & 5 & -4 \end{pmatrix}.$$

Since the first and second rows are equal, the determinant of this matrix is 0.

The essential point is this: to compute the first column of AA', we compute the determinants of matrices having the form

$$\begin{pmatrix} - & - & - \\ 3 & 4 & -1 \\ -2 & 5 & -4 \end{pmatrix}$$

in which the blank row is replaced in turn by the first, second, and third rows of A. Thus the first column of AA' is

$$\begin{pmatrix} \det A \\ 0 \\ 0 \end{pmatrix}.$$

Similarly, the second column of AA' is found by taking the determinants of matrices having the form

$$\begin{pmatrix} 2 & 1 & 0 \\ - & - & - \\ -2 & 5 & -4 \end{pmatrix}$$

in which the blank row is replaced in turn by the first, second, and third rows of A. Thus the second column of AA' is

$$\begin{pmatrix} 0 \\ \det A \\ 0 \end{pmatrix}.$$

Similarly, the third column of AA' consists of the determinants of the matrices

$$\begin{pmatrix} 2 & 1 & 0 \\ 3 & 4 & -1 \\ - & - & - \end{pmatrix}$$

obtained by substituting the various rows of A in the row of blanks. Therefore, the third column of AA' is

$$\begin{pmatrix} 0 \\ 0 \\ \det A \end{pmatrix}.$$

The situation we have just described for 3×3 matrices in the example above occurs in general. We need these general facts, so we record the necessary definitions.

Definition 6-5

Suppose A is an $n \times n$ matrix. The (i,j)-*cofactor of A* is $(-1)^{i+j} \det A(i \,|\, j)$. The *matrix of cofactors of A* (adjoint of A, adjugate of A) is the $n \times n$ matrix A' whose (i,j) entry is $(-1)^{j+i} \det A(j \,|\, i)$. Thus the (j,i)-cofactor is the (i,j)-entry of A'.

Suppose A is an $n \times n$ matrix. The jth column of AA' consists of the determinants of the matrices

$$j\text{th row} \begin{pmatrix} a_{11} & a_{12} & \cdots & a_{1n} \\ \cdot & \cdot & & \cdot \\ \cdot & \cdot & & \cdot \\ \cdot & \cdot & & \cdot \\ - & - & & - \\ \cdot & \cdot & & \cdot \\ \cdot & \cdot & & \cdot \\ \cdot & \cdot & & \cdot \\ a_{n1} & a_{n2} & \cdots & a_{nn} \end{pmatrix}$$

obtained by substituting successively the rows of A in the blanks. Therefore the jth column of AA' is

$$\begin{pmatrix} 0 \\ \cdot \\ \cdot \\ \cdot \\ 0 \\ \det A \\ 0 \\ \cdot \\ \cdot \\ \cdot \\ 0 \end{pmatrix} \quad j\text{th entry}$$

This shows that $AA' = (\det A)I_n$.

Proposition 6-4

If A is an $n \times n$ matrix and A' is the matrix of cofactors, then $AA' = (\det A)I_n$. Therefore, if $\det A \neq 0$, $A[\frac{1}{\det A} A'] = I_n$ and $A^{-1} = (\frac{1}{\det A})A'$.

Brief Exercises

1. Compute the matrix of cofactors of the following matrices. For those which are nonsingular, find the inverse using Proposition 6-4.

(a) $\begin{pmatrix} 2 & 1 \\ 1 & 2 \end{pmatrix}$

(d) $\begin{pmatrix} 1 & 0 & 0 \\ 1 & 1 & 0 \\ 1 & 1 & 1 \end{pmatrix}$

(b) $\begin{pmatrix} 3 & 1 \\ 6 & 2 \end{pmatrix}$

(e) $\begin{pmatrix} 1 & 3 & 2 \\ 1 & 0 & 1 \\ 0 & 5 & 2 \end{pmatrix}$

(c) $\begin{pmatrix} 1 & 0 & 7 \\ 1 & 1 & 0 \\ 0 & 1 & 1 \end{pmatrix}$

Answers: The following are the matrices of cofactors.

1(a) $\begin{pmatrix} 2 & -1 \\ -1 & 2 \end{pmatrix}$ **1(d)** $\begin{pmatrix} 1 & 0 & 0 \\ -1 & 1 & 0 \\ 0 & -1 & 1 \end{pmatrix}$

1(b) $\begin{pmatrix} 2 & -1 \\ -6 & 3 \end{pmatrix}$ **1(e)** $\begin{pmatrix} -5 & 4 & 3 \\ -2 & 2 & 1 \\ 5 & -5 & -3 \end{pmatrix}$

1(c) $\begin{pmatrix} 1 & 7 & -7 \\ -1 & 1 & 7 \\ 1 & -1 & 1 \end{pmatrix}$

We can use the expression for A^{-1} obtained in Proposition 6-4 to find the solution of a system of n equations in n unknowns when the coefficient matrix is nonsingular. The formulas we obtain for the solutions are called *Cramer's Rule*.

Example Consider the system of linear equations

$$2x - y + 3z = 7$$
$$x + 2y - z = 1$$
$$x + y + z = 3.$$

The coefficient matrix is

$$A = \begin{pmatrix} 2 & -1 & 3 \\ 1 & 2 & -1 \\ 1 & 1 & 1 \end{pmatrix}.$$

It is nonsingular since $\det A = 5$. The matrix of cofactors for A is

$$A' = \begin{pmatrix} 3 & 4 & -5 \\ -2 & -1 & 5 \\ -1 & -3 & 5 \end{pmatrix}.$$

Thus the inverse of A is

$$A^{-1} = (\tfrac{1}{5}) \begin{pmatrix} 3 & 4 & -5 \\ -2 & -1 & 5 \\ -1 & -3 & 5 \end{pmatrix}.$$

Consequently

$$
\begin{pmatrix} x \\ y \\ z \end{pmatrix} = (\tfrac{1}{5}) \begin{pmatrix} 3 & 4 & -5 \\ -2 & -1 & 5 \\ -1 & -3 & 5 \end{pmatrix} \begin{pmatrix} 7 \\ 1 \\ 3 \end{pmatrix}.
$$

To complete the calculation we apply each row of A' to $(7,1,3)$.

Recall how A' is formed. The first row of A' consists of the cofactors corresponding to the first column of A. Applying the first row of A' to $(7,1,3)$ is therefore the same as computing the determinant of the matrix

$$
\begin{pmatrix} 7 & -1 & 3 \\ 1 & 2 & -1 \\ 3 & 1 & 1 \end{pmatrix}.
$$

The second row of A' is obtained as cofactors of the second column of A. Therefore, applying the second row of A' to $(7,1,3)$ is the same as computing the determinant of the matrix

$$
\begin{pmatrix} 2 & 7 & 3 \\ 1 & 1 & -1 \\ 1 & 3 & 1 \end{pmatrix}.
$$

Finally, the third row of A' is obtained as cofactors of the third column of A. Consequently, applying the third row of A' to $(7,1,3)$ is the same as computing the determinant of the matrix

$$
\begin{pmatrix} 2 & -1 & 7 \\ 1 & 2 & 1 \\ 1 & 1 & 3 \end{pmatrix}.
$$

These observations give the following formulas, sometimes called Cramer's Rule.

$$
x = \frac{\det \begin{pmatrix} 7 & -1 & 3 \\ 1 & 2 & -1 \\ 3 & 1 & 1 \end{pmatrix}}{\det \begin{pmatrix} 2 & -1 & 3 \\ 1 & 2 & -1 \\ 1 & 1 & 1 \end{pmatrix}}
$$

$$y = \frac{\det \begin{pmatrix} 2 & 7 & 3 \\ 1 & 1 & -1 \\ 1 & 3 & 1 \end{pmatrix}}{\det \begin{pmatrix} 2 & -1 & 3 \\ 1 & 2 & -1 \\ 1 & 1 & 1 \end{pmatrix}}$$

$$z = \frac{\det \begin{pmatrix} 2 & -1 & 7 \\ 1 & 2 & 1 \\ 1 & 1 & 3 \end{pmatrix}}{\det \begin{pmatrix} 2 & -1 & 3 \\ 1 & 2 & -1 \\ 1 & 1 & 1 \end{pmatrix}}$$

In other words, the first unknown is computed as follows: replace the first column of A by the constants, compute the determinant of this matrix, and divide by det A. The second unknown is computed by replacing the second column of A by the constants, computing the determinant of this matrix, and dividing by det A. A similar procedure is followed for the other unknowns.

Example Consider the system of equations

$$x - y + z = 1$$
$$x + y - z = 2$$
$$x + y + z = 5.$$

We solve this system using the procedure outlined in the last example.

$$x = \frac{\det \begin{pmatrix} 1 & -1 & 1 \\ 2 & 1 & -1 \\ 5 & 1 & 1 \end{pmatrix}}{\det \begin{pmatrix} 1 & -1 & 1 \\ 1 & 1 & -1 \\ 1 & 1 & 1 \end{pmatrix}} = \frac{6}{4}$$

$$y = \frac{\det \begin{pmatrix} 1 & 1 & 1 \\ 1 & 2 & -1 \\ 1 & 5 & 1 \end{pmatrix}}{\det \begin{pmatrix} 1 & -1 & 1 \\ 1 & 1 & -1 \\ 1 & 1 & 1 \end{pmatrix}} = \frac{8}{4}$$

$$z = \frac{\det \begin{pmatrix} 1 & -1 & 1 \\ 1 & 1 & 2 \\ 1 & 1 & 5 \end{pmatrix}}{\det \begin{pmatrix} 1 & -1 & 1 \\ 1 & 1 & -1 \\ 1 & 1 & 1 \end{pmatrix}} = \frac{6}{4}$$

The procedure outlined in the examples can be used to solve a system of linear equations in which the number of equations equals the number of unknowns and in which the determinant of the coefficient matrix is nonzero. If A is the coefficient matrix, then the first unknown is computed by substituting the constants in the first column of A, computing the determinant of this matrix, and dividing by $\det A$. For the second unknown the procedure is the same, except that the constants are substituted into the second column of A. For the third unknown, the constants are substituted in the third column, and so on.

Proposition 6-5

(Cramer's Rule) Let A be the coefficient matrix for the system of n linear equations in n unknowns.

$$a_{11}x_1 + a_{12}x_2 + \cdots + a_{1n}x_n = y_1$$

$$a_{21}x_1 + a_{22}x_2 + \cdots + a_{2n}x_n = y_2$$

$$\vdots \qquad \vdots \qquad \qquad \vdots \qquad \vdots$$

$$a_{n1}x_1 + a_{n2}x_2 + \cdots + a_{nn}x_n = y_n$$

Suppose det $A \neq 0$. Then

$$x_1 = \frac{\det \begin{pmatrix} y_1 & a_{12} & \cdots & a_{1n} \\ y_2 & a_{22} & \cdots & a_{2n} \\ \cdot & \cdot & & \cdot \\ \cdot & \cdot & & \cdot \\ \cdot & \cdot & & \cdot \\ y_n & a_{n2} & \cdots & a_{nn} \end{pmatrix}}{\det A}$$

$$x_2 = \frac{\det \begin{pmatrix} a_{11} & y_1 & \cdots & a_{1n} \\ a_{21} & y_2 & \cdots & a_{2n} \\ \cdot & \cdot & & \cdot \\ \cdot & \cdot & & \cdot \\ \cdot & \cdot & & \cdot \\ a_{n1} & y_n & \cdots & a_{nn} \end{pmatrix}}{\det A}$$

$$\vdots$$

$$x_n = \frac{\det \begin{pmatrix} a_{11} & a_{12} & \cdots & y_1 \\ a_{21} & a_{22} & \cdots & y_2 \\ \cdot & \cdot & & \cdot \\ \cdot & \cdot & & \cdot \\ a_{n1} & a_{n2} & \cdots & y_n \end{pmatrix}}{\det A} \, .$$

The reasons why these formulas work are precisely the ones we mentioned in the examples. Write the system of equations with the unknowns and constants written as $n \times 1$ matrices.

$$AX = Y$$

Since A is nonsingular, $X = A^{-1}Y$. Use the form of A^{-1} obtained in Proposition 6-4.

$$X = \left(\tfrac{1}{\det A}\right)A'Y$$

Thus we obtain x_1 by applying the first row of A' to Y and then dividing by det A. Since the first row of A' consists of the cofactors of the first column of A, applying the first row of A' to **Y** is the same as computing the determinant of the matrix obtained by replacing the first column of A with **Y**. Similarly, the value for x_2 is obtained by applying the second row of A' to **Y** which is the same as computing the determinant of the matrix obtained by replacing the second column of A with **Y**, and so on for the other unknowns.

Brief Exercises

Use Cramer's Rule to solve the following systems of equations, if possible.

1. $3x - y = 2$ **2.** $x + 2y = 2$ **3.** $x + y - z = 1$
   $x + y = 3$ $x + y = 1$ $x - y + 2z = 3$
                                   $2x - y + z = 5$

Answers:

1 $x = \dfrac{\det \begin{pmatrix} 2 & -1 \\ 3 & 1 \end{pmatrix}}{\begin{pmatrix} 3 & -1 \\ 1 & 1 \end{pmatrix}} = \dfrac{5}{4}$
$\qquad$
$y = \dfrac{\det \begin{pmatrix} 3 & 2 \\ 1 & 3 \end{pmatrix}}{\det \begin{pmatrix} 3 & -1 \\ 1 & 1 \end{pmatrix}} = \dfrac{7}{4}$

2 $x = \dfrac{\det \begin{pmatrix} 2 & 2 \\ 1 & 1 \end{pmatrix}}{\det \begin{pmatrix} 1 & 2 \\ 1 & 1 \end{pmatrix}} = \dfrac{0}{-1} = 0$
$\qquad$
$y = \dfrac{\det \begin{pmatrix} 1 & 2 \\ 1 & 1 \end{pmatrix}}{\det \begin{pmatrix} 1 & 2 \\ 1 & 1 \end{pmatrix}} = \dfrac{-1}{-1} = 1$

3 $x = \dfrac{\det \begin{pmatrix} 1 & 1 & -1 \\ 3 & -1 & 2 \\ 5 & -1 & 1 \end{pmatrix}}{\det \begin{pmatrix} 1 & 1 & -1 \\ 1 & -1 & 2 \\ 2 & -1 & 1 \end{pmatrix}} = \dfrac{6}{3} = 2$
$\qquad$
$y = \dfrac{\det \begin{pmatrix} 1 & 1 & -1 \\ 1 & 3 & 2 \\ 2 & 5 & 1 \end{pmatrix}}{\det \begin{pmatrix} 1 & 1 & -1 \\ 1 & -1 & 2 \\ 2 & -1 & 1 \end{pmatrix}} = \dfrac{-3}{3} = -1$

$z = \dfrac{\det \begin{pmatrix} 1 & 1 & 1 \\ 1 & -1 & 3 \\ 2 & -1 & 5 \end{pmatrix}}{\det \begin{pmatrix} 1 & 1 & -1 \\ 1 & -1 & 2 \\ 2 & -1 & 1 \end{pmatrix}} = \dfrac{0}{3} = 0$

SUMMARY   *In this section formulas for the determinant are used to obtain the inverse of a nonsingular matrix. The (i,j)-cofactor of the $n \times n$ matrix A is the number $(-1)^{i+j} \det A(i \mid j)$. The matrix A' cofactors of A is the $n \times n$ matrix having the (j,i)-cofactor $(-1)^{j+i} \det A(j \mid i)$ as its (i,j) entry. Because of the way A' is constructed, we obtain*

$$AA' = (\det A)I_n.$$

Consequently, if $\det A \neq 0$,

$$A^{-1} = \left(\tfrac{1}{\det A}\right)A'.$$

With this formula for A^{-1} we obtain Cramer's Rule for solving the system of equations $A\mathbf{X} = \mathbf{Y}$ if A is nonsingular. The ith unknown x_i is given by

$$x_i = \frac{\det \begin{pmatrix} a_{11} & \cdots & y_1 & \cdots & a_{1n} \\ a_{21} & \cdots & y_2 & \cdots & a_{2n} \\ \cdot & & \cdot & & \cdot \\ \cdot & & \cdot & & \cdot \\ \cdot & & \cdot & & \cdot \\ a_{n1} & \cdots & y_n & \cdots & a_{nn} \end{pmatrix}}{\det A}$$

ith col.

i.e., replace the ith column of A with the constants $\mathbf{Y}$, compute the determinant of the resulting matrix, and divide by $\det A$.

Exercises for Section 6-3

1. Compute the matrix of cofactors of the following matrices.

(a) $\begin{pmatrix} 2 & 1 \\ 0 & 2 \end{pmatrix}$
   (c) $\begin{pmatrix} 1 & 2 & 1 \\ 3 & 1 & 0 \\ 2 & 2 & 1 \end{pmatrix}$

(b) $\begin{pmatrix} 3 & 1 \\ -4 & 2 \end{pmatrix}$
   (d) $\begin{pmatrix} 1 & 1 & 2 \\ -1 & 0 & 3 \\ 4 & 2 & 0 \end{pmatrix}$

2. For the nonsingular matrices in Exercise 1, compute the inverses by dividing the matrix of cofactors by the determinant of the matrix.

Solve the following systems of equations using Cramer's Rule.

3. $2x + y = 5$
    $x - y = 6$

4. $x + y - 2z = 7$
    $x - y + 2z = 3$
    $2x + y + z = 5$

5. $2x - y + 3z = 8$
    $x + 4y - 5z = 0$
    $6x - y + z = 2$

SECTION 6-4
Some Facts About the Determinant

PREVIEW   *In this section we give some arguments for the following important facts about the determinant:*

(1) $\det A \neq 0$ *if and only if* A *is nonsingular,*
(2) $\det AB = \det A \det B$, *and*
(3) $\det A = \det A^T$.

In Section 6-1 we observed that for a 2×2 matrix A, $\det A \neq 0$ if and only if A is nonsingular. We now show that this is true for all square matrices, no matter what their size. Furthermore, in Section 6-2 we used the fact that if A is a square matrix, then $\det A = \det A^T$. In this section we give some explanation of this fact, and we give some explanation of the fact that $\det AB = (\det A)(\det B)$. To establish these facts, we consider elementary matrices first.

Refer to Proposition 6-1 and use $A = I_n$ in the formulas. We therefore have that $\det F_{i,j} = -1$, $\det F_i(x) = x$, and $\det F_{i,j}(x) = 1$. These three equations show that if G is an elementary matrix, then $\det G \neq 0$.

Second, observe that parts (1), (2), and (3) of Proposition 6-1 can be rephrased by saying that if G is an elementary matrix, then

$$\det GA = (\det G)(\det A).$$

Third, from the definition of the three kinds of elementary matrices in Definition 5-3 we have that $F_{i,j}^T = F_{i,j}$, $F_i(x)^T = F_i(x)$, and $F_{i,j}(x)^T = F_{j,i}(x)$. Consequently, if G is an elementary matrix, then $\det G^T = \det G$.

Proposition 6-6
If A is an $n \times n$ matrix and G is an $n \times n$ elementary matrix, then
(1) $\det G \neq 0$,
(2) $\det GA = (\det G)(\det A)$, and
(3) G^T is elementary and $\det G^T = \det G$.

Now suppose that A is a square matrix. By Proposition 5-3 elementary matrices $G_1, \ldots, G_k$ can be found so that $B = G_k \cdots G_1 A$ is in row echelon form. Using part (2) of Proposition 6-6 again and again we get

$$\det B = \det (G_k \cdots G_1 A) = (\det G_k) \det (G_{k-1} \cdots G_1 A) = \cdots$$

$$= (\det G_k)(\det G_{k-1}) \cdots (\det G_1)(\det A).$$

By part (1) of Proposition 6-6, all the factors $(\det G_k), \ldots, (\det G_1)$ are nonzero, so their product is also nonzero. Thus $\det B = r \det A$, where $r \neq 0$. If A is nonsingular, then $B = I_n$ (Proposition 5-6) and $\det I_n \neq 0$. So if A is nonsingular, then $\det A \neq 0$. However, if A is singular, then B has a row of zeros; thus $\det B = 0$ and so $\det A = 0$.

Proposition 6-7

A square matrix A is nonsingular if and only if $\det A \neq 0$.

To see why the determinant of a product equals the product of the determinants, let us use A and B to represent any $n \times n$ matrices. If A is nonsingular, then A is expressible as a product of elementary matrices $H_1, \ldots, H_t$ (Proposition 5-6): $A = H_1 \cdots H_t$. Using part (2) of Proposition 6-6 again and again, we get

$$\det AB = \det (H_1 \cdots H_t B)$$

$$= (\det H_1) \cdots (\det H_t)(\det B)$$

$$= (\det A)(\det B).$$

On the other hand, if A is singular, then $\det A = 0$. But, by Proposition 5-9, AB is also singular and so $\det AB = 0$. So if A is singular, then $\det AB = 0 = (\det A)(\det B)$.

Proposition 6-8

If A and B are $n \times n$ matrices, then

$$\det AB = (\det A)(\det B).$$

Finally we explain why $\det A = \det A^T$. To do this we show that A is nonsingular if and only if A^T is nonsingular. Suppose A is nonsingular. Then $A^{-1}A = I_n$. Taking the transpose of both sides of this equation and recalling from Exercise 17 of Section 4-4 that the transpose of a product is the product of the transposes in the reverse order, we have that $A^T(A^{-1})^T = I_n$. By Proposition 5-9 it follows that A^T is nonsingular. On the other hand, if A^T is nonsingular, then its inverse exists; call it B. Then $BA^T = I_n$. Again taking the transpose of both sides of this equation, and recalling that $(A^T)^T = A$, we have $AB^T = I_n$, and so, again by Proposition 5-9, A is nonsingular.

Consequently, A is nonsingular if and only if A^T is nonsingular. So if A is singular, then so is A^T; in that case, $\det A = 0 = \det A^T$.

However, if A is nonsingular, then A is expressible as a product of elementary matrices, call them $G_1, \ldots, G_k$, say: $A = G_1 \cdots G_k$. Taking transposes of both sides we have $A^T = G_k^T \cdots G_1^T$. Since $G_1^T, \ldots, G_k^T$ are elementary matrices, we have that

$$\det A^T = (\det G_k^T) \cdots (\det G_1^T)$$
$$= (\det G_1) \cdots (\det G_k)$$
$$= \det A.$$

(We have used parts (2) and (3) of Proposition 6-6.)

Proposition 6-9_____

If A is an $n \times n$ matrix, then $\det A = \det A^T$.

_____

In Section 6-1 we introduced the determinant by the formula

$$\det A = (-1)^{i+1}a_{i1} \det A(i \mid 1) + \cdots + (-1)^{i+n}a_{in} \det A(i \mid n).$$

In this section, however, we have relied upon the description of the determinant given in Proposition 6-1.

$$\det F_{i,j}A = -\det A \qquad (i \neq j)$$
$$\det F_i(x)A = x \det A$$
$$\det F_{i,j}(x)A = \det \qquad A \; (i \neq j)$$
$$\det I_n = 1$$

We have not explained the connection between these two descriptions. The properties listed in Proposition 6-1 can be derived from the formula given in Definition 6-4; see the treatment in Staib [10], pp. 168–182. But it is also possible to derive the formula in Definition 6-4 from the properties in Proposition 6-1; see the discussion in Curtis [4], pp. 114–124. Consequently, we have relied on two equivalent descriptions of the determinant function.

SUMMARY   *Using the properties listed in Proposition 6-1, we show that the following facts are true about determinants:*

(1) *$\det A \neq 0$ if and only if A is nonsingular,*
(2) *$\det AB = \det A \det B$ (A and B both square and the same size), and*
(3) *$\det A = \det A^T$.*

Exercises for Section 6-4

Definition 6-6_____

A square matrix is a *diagonal matrix* if the nonzero entries occur only in the (i,i) positions. The positions $(1,1)$, $(2,2)$, ..., (n,n) constitute the *main diagonal* of a square matrix. If the entry in the (i,i) position of a diagonal matrix is a_i or a_{ii}, then the matrix is often denoted by diag $(a_1, ..., a_n)$ or by diag $(a_{11}, ..., a_{nn})$.

_____

1. Show that the determinant of the diagonal matrix diag $(a_1, ..., a_n)$ is $a_1 \cdots a_n$. That is, the determinant of a diagonal matrix equals the product of the entries along the main diagonal.

Definition 6-7_____

A square matrix is *upper* (*lower*) *triangular* if every entry below (above) the main diagonal is zero.

_____

2. Show that the determinant of a triangular matrix is the product of the entries along the main diagonal.

3. Show that if A is a square matrix and if $AB = BA$ for every square matrix B (of the same size as A), then A is a scalar multiple of the identity matrix.

7

Kernel, Image, Nullity, and Rank of Linear Transformations and Matrices

SECTION 7-1
Row Rank and Column Rank of a Matrix

PREVIEW *Each row of an m × n matrix A can be considered as a vector in $\mathbb{R}^n$, and each column can be viewed as a vector in $\mathbb{R}^m$. The rows of A generate a subspace of $\mathbb{R}^n$ called the row space of A, denoted by $\mathscr{R}(A)$. The columns of A generate a subspace of $\mathbb{R}^m$ called the column space of A, denoted by $\mathscr{C}(A)$. We show in this section that these two subspaces have the same dimension.*

If A is an $m \times n$ matrix over $\mathbb{R}$, then we consider the rows of A as vectors in $\mathbb{R}^n$. These rows generate a subspace of $\mathbb{R}^n$ (Proposition 3-2) called the row space of A. We denote it by $\mathscr{R}(A)$. This subspace consists of all linear combinations of the rows of A. Similarly the columns of A can be viewed as vectors in $\mathbb{R}^m$; all the linear combinations of the columns of A constitute a subspace of $\mathbb{R}^m$ called the column space of A. We denote the column space of A by $\mathscr{C}(A)$.

Example Let A be the 3×3 matrix

$$\begin{pmatrix} 1 & 2 & 0 \\ -1 & 0 & 2 \\ 0 & 2 & 2 \end{pmatrix}.$$

The row space $\mathscr{R}(A)$ of A is the subspace of $\mathbb{R}^3$ generated by $(1,2,0)$, $(-1,0,2)$, and $(0,2,2)$. The column space $\mathscr{C}(A)$ of A is the subspace of $\mathbb{R}^3$ generated by $(1,-1,0)$, $(2,0,2)$, and $(0,2,2)$. Since the third row equals the sum of the first and second rows, dim $\mathscr{R}(A) = 2$. Since the third column equals the second minus two times the first, dim $\mathscr{C}(A) = 2$.

Example Let A be the 3×4 matrix

$$\begin{pmatrix} 2 & 1 & 0 & 3 \\ -1 & 2 & 1 & 0 \\ 1 & 0 & 2 & 1 \end{pmatrix}.$$

The row space $\mathscr{R}(A)$ of A is the subspace of $\mathbb{R}^4$ generated by $(2,1,0,3)$, $(-1,2,1,0)$, and $(1,0,2,1)$. The column space $\mathscr{C}(A)$ of A is the subspace of $\mathbb{R}^3$ generated by $(2,-1,1)$, $(1,2,0)$, $(0,1,2)$, and $(3,0,1)$. It is easy to check that the three rows of A form a linearly independent set and so dim $\mathscr{R}(A) = 3$. You can verify that

$$(2,-1,1) = (-\tfrac{7}{13})(1,2,0) + (\tfrac{1}{13})(0,1,2) + (\tfrac{11}{13})(3,0,1)$$

272

and that the second, third, and fourth columns form a linearly independent set. Thus dim $\mathscr{C}(A) = 3$.

These examples illustrate the concepts of row space and column space of a matrix. They also indicate that dim $\mathscr{R}(A) = $ dim $\mathscr{C}(A)$.

Definition 7-1_____

If A is an $m \times n$ matrix, then $\mathscr{R}(A)$ denotes the subspace of $\mathbb{R}^n$ generated by the rows of A; it is called the *row space of A*. The *row rank of A* is the dimension of $\mathscr{R}(A)$. Similarly $\mathscr{C}(A)$ denotes the subspace of $\mathbb{R}^m$ generated by the columns of A; it is called the *column space of A*. The *column rank of A* is the dimension of $\mathscr{C}(A)$.

_____

Our goal in this section is to show that the row rank of a matrix A equals the column rank of A. Our strategy is this: let B represent the row echelon form of A. We show:

(1) row rank $B = $ column rank B,
(2) row rank $A = $ row rank B, and
(3) column rank $A = $ column rank B.

In the next two examples we illustrate the fact that if B is an $m \times n$ matrix in row echelon form, then the row rank of B equals the column rank of B.

Example Let B be the 4×7 matrix

$$\begin{pmatrix} 1 & 2 & 0 & 3 & 5 & 0 & -1 \\ 0 & 0 & 1 & 2 & -1 & 0 & 3 \\ 0 & 0 & 0 & 0 & 0 & 1 & 2 \\ 0 & 0 & 0 & 0 & 0 & 0 & 0 \end{pmatrix}.$$

The row rank of B is three, because the nonzero rows form a basis for $\mathscr{R}(B)$. This is easy to see because of the leading 1's in the first, third, and sixth columns; each leading 1 is the only nonzero entry in its column. Furthermore, the column rank of B is three because the columns containing the leading 1's form a basis for $\mathscr{C}(B)$.

$$\begin{pmatrix} \mathbf{1} & 2 & \mathbf{0} & 3 & 5 & \mathbf{0} & -1 \\ \mathbf{0} & 0 & \mathbf{1} & 2 & -1 & \mathbf{0} & 3 \\ \mathbf{0} & 0 & \mathbf{0} & 0 & 0 & \mathbf{1} & 2 \\ \mathbf{0} & 0 & \mathbf{0} & 0 & 0 & \mathbf{0} & 0 \end{pmatrix}$$

Example   Let B be the 5×8 matrix

$$\begin{pmatrix} 0 & 1 & 0 & 2 & 7 & 0 & 5 & 0 \\ 0 & 0 & 1 & 3 & 8 & 0 & -3 & 0 \\ 0 & 0 & 0 & 0 & 0 & 1 & 2 & 0 \\ 0 & 0 & 0 & 0 & 0 & 0 & 0 & 1 \\ 0 & 0 & 0 & 0 & 0 & 0 & 0 & 0 \end{pmatrix}$$

The row rank of B is four because the nonzero rows form a basis for $\mathscr{R}(B)$; again the leading 1's are the key to seeing this. Also, the column rank of B is four because the columns containing the leading 1's form a basis for $\mathscr{C}(B)$.

These examples give an insight as to why the row rank of B equals the column rank of B if B is in row echelon form. The row rank of B equals the number of nonzero rows; the leading 1's in the nonzero rows show that the nonzero rows form a basis for $\mathscr{R}(B)$. Furthermore, the columns containing the leading 1's form a basis for $\mathscr{C}(B)$; there are as many of these as there are nonzero rows.

Proposition 7-1

If B is a matrix in row echelon form, then the row rank of B equals the column rank of B; both equal the number of nonzero rows in B.

Brief Exercises

1. Determine the row rank and the column rank of the following matrices; give reasons.

   (a) $\begin{pmatrix} 1 & 1 \\ 0 & 0 \end{pmatrix}$ (b) $\begin{pmatrix} 1 & 1 & 0 \\ 0 & 0 & 1 \\ 0 & 0 & 0 \end{pmatrix}$ (c) $\begin{pmatrix} 1 & 0 & 1 & 0 & 2 \\ 0 & 1 & 1 & 0 & 1 \\ 0 & 0 & 0 & 1 & 3 \\ 0 & 0 & 0 & 0 & 0 \end{pmatrix}$

Answers: 1(a) 1   1(b) 2   1(c) 3

The next step in showing that the row rank of a matrix A equals the column rank of A is this: if B is the row echelon form for A, then the row rank of B equals the row rank of A. The way we do this is by showing that the row rank of a matrix is unchanged if we apply row operations; as a matter of fact, we actually show the row space is unchanged, and therefore the row rank is unchanged, too.

Example Consider the 3 × 4 matrix

$$A = \begin{pmatrix} 2 & 1 & 1 & 3 \\ 4 & 0 & 1 & 2 \\ 1 & 1 & 2 & 1 \end{pmatrix}.$$

If we apply a row operation of the type $F_{i,j}$, say $F_{1,3}$, then the rows of $F_{1,3}A$ and the rows of A are the same although listed in a different order. Thus $\mathscr{R}(F_{i,j}A) = \mathscr{R}(A)$.

$$F_{1,3}A = \begin{pmatrix} 1 & 1 & 2 & 1 \\ 4 & 0 & 1 & 2 \\ 2 & 1 & 1 & 3 \end{pmatrix}$$

If we apply a row operation of the type $F_i(x)$, $(x \neq 0)$, say $F_2(3)$, then the rows of $F_2(3)A$ are linear combinations of the rows of A.

$$F_2(3)A = \begin{pmatrix} 2 & 1 & 1 & 3 \\ 12 & 0 & 3 & 6 \\ 1 & 1 & 2 & 1 \end{pmatrix}$$

But the rows of A are also linear combinations of the rows of $F_2(3)A$. Consequently $\mathscr{R}(F_i(x)A) = \mathscr{R}(A)$.

Finally, if we apply a row operation of the type $F_{i,j}(x)$, $(i \neq j)$, say $F_{2,3}(-1)$, then the rows of $F_{2,3}(-1)A$ are linear combinations of the rows of A.

$$F_{2,3}(-1)A = \begin{pmatrix} 2 & 1 & 1 & 3 \\ 3 & -1 & -1 & 1 \\ 1 & 1 & 2 & 1 \end{pmatrix}$$

But it is also clear that the rows of A are linear combinations of the rows of $F_{2,3}(-1)A$. (The second row of A equals the sum of the second and third rows of $F_{2,3}(-1)A$.) Thus $\mathscr{R}(F_{2,3}(-1)A) = \mathscr{R}(A)$.

This example illustrates the general situation. The rows of $F_{i,j}A$ and A are the same although appearing in a different order; so $\mathscr{R}(F_{i,j}A) = \mathscr{R}(A)$. The rows of $F_i(x)A$ are linear combinations of the rows of A and vice versa; just consider the ith rows of the two matrices; the other rows are not changed. Thus $\mathscr{R}(F_i(x)A) = \mathscr{R}(A)$. Similarly, the rows of $F_{i,j}(x)A$ are linear combinations of the rows of A and vice versa. Again, just consider the ith rows of the two matrices; the other rows are not changed. So $\mathscr{R}(F_{i,j}(x)A) = \mathscr{R}(A)$.

These arguments show that if G is an elementary matrix, then $\mathscr{R}(GA) = \mathscr{R}(A)$. We know from Proposition 5-3 that there are elementary matrices $G_1, \ldots, G_k$ so that $B = G_k \cdots G_1 A$ is in row echelon form. Using the fact we have just established again and again, we obtain the following.

$$\mathscr{R}(B) = \mathscr{R}(G_k \cdots G_1 A) = \mathscr{R}(G_{k-1} \cdots G_1 A)$$

$$= \cdots = \mathscr{R}(G_1 A) = \mathscr{R}(A)$$

Therefore the row rank of B equals the row rank of A.

Proposition 7-2

If the matrix B is obtained from the matrix A by row operations, then $\mathscr{R}(A) = \mathscr{R}(B)$, and the row rank of A equals the row rank of B. In particular, if B is the row echelon form for A, then the row rank of A equals the row rank of B.

Brief Exercises

1. Determine the row rank of the following matrices.

(a) $\begin{pmatrix} 2 & 1 \\ 1 & 0 \end{pmatrix}$ (b) $\begin{pmatrix} 2 & 1 \\ 4 & 2 \end{pmatrix}$ (c) $\begin{pmatrix} 1 & 2 & 1 \\ 3 & 1 & 0 \\ 0 & 0 & 1 \end{pmatrix}$

Answers: **1(a)** 2   **1(b)** 1   **1(c)** 3

The third step in showing that the row rank of A equals the column rank of A is to show that if B is the row echelon form for A, then the column rank of B equals the column rank of A. Probably the easiest way to see this is to recall that there are elementary matrices $G_1, \ldots, G_k$ so that $B = G_k \cdots G_1 A$ (Proposition 5-3). The product $G_k \cdots G_1$ is nonsingular since elementary matrices are nonsingular (Proposition 5-4) and the product of nonsingular matrices is nonsingular (Proposition 5-5). So we know that $B = GA$, where G is the nonsingular matrix $G_k \cdots G_1$. What we do, then, is to show that the column rank of A equals the column rank of GA, where G is any nonsingular matrix.

Example   Let A be the 3×4 matrix

$$\begin{pmatrix} 2 & 1 & 0 & 4 \\ 3 & -2 & 1 & -2 \\ 1 & 0 & 2 & -1 \end{pmatrix}$$

and let G be the nonsingular 3×3 matrix

$$\begin{pmatrix} 1 & 2 & -1 \\ 3 & 0 & 5 \\ -4 & 2 & 0 \end{pmatrix}.$$

Label the columns of A by $\mathbf{A}_1$, $\mathbf{A}_2$, $\mathbf{A}_3$, and $\mathbf{A}_4$. Think of these as vectors from $\mathbb{R}^3$ written vertically. It is easy to check that $\mathbf{A}_1$, $\mathbf{A}_2$, and $\mathbf{A}_3$ form a basis for $\mathscr{C}(A)$, since $\mathbf{A}_4 = \mathbf{A}_1 + 2\mathbf{A}_2 - \mathbf{A}_3$, and since $\mathbf{A}_1$, $\mathbf{A}_2$, and $\mathbf{A}_3$ form a linearly independent set.

Notice that the columns of GA are $G\mathbf{A}_1$, $G\mathbf{A}_2$, $G\mathbf{A}_3$, and $G\mathbf{A}_4$. Since G defines a linear transformation (Proposition 4-7), we have $G\mathbf{A}_4 = G\mathbf{A}_1 + 2G\mathbf{A}_2 - G\mathbf{A}_3$. Thus $G\mathbf{A}_1$, $G\mathbf{A}_2$, and $G\mathbf{A}_3$ generate $\mathscr{C}(GA)$. Furthermore $G\mathbf{A}_1$, $G\mathbf{A}_2$, and $G\mathbf{A}_3$ from a linearly independent set; if a, b, and c are scalars so that

$$aG\mathbf{A}_1 + bG\mathbf{A}_2 + cG\mathbf{A}_3 = \mathbf{0}$$

then

$$G(a\mathbf{A}_1 + b\mathbf{A}_2 + c\mathbf{A}_3) = \mathbf{0}.$$

Since G is nonsingular, from Proposition 5-8 we have

$$a\mathbf{A}_1 + b\mathbf{A}_2 + c\mathbf{A}_3 = \mathbf{0}.$$

Since $\mathbf{A}_1$, $\mathbf{A}_2$, and $\mathbf{A}_3$ form a linearly independent set, $a = b = c = 0$. Consequently $G\mathbf{A}_1$, $G\mathbf{A}_2$, and $G\mathbf{A}_3$ form a basis for $\mathscr{C}(GA)$. Therefore, the column rank of $GA =$ column rank $A = 3$.

Example Let A be the 3×5 matrix

$$\begin{pmatrix} 1 & 0 & 1 & 1 & 2 \\ -1 & 2 & 1 & 1 & -4 \\ 4 & 1 & 5 & 0 & 2 \end{pmatrix}$$

and let G be any nonsingular 3×3 matrix. Denote the columns of A by $\mathbf{A}_1$, $\mathbf{A}_2$, $\mathbf{A}_3$, $\mathbf{A}_4$, and $\mathbf{A}_5$. You can check that $\mathbf{A}_1$, $\mathbf{A}_2$, and $\mathbf{A}_4$ form a basis for $\mathscr{C}(A)$ since they form a linearly independent set and since $\mathbf{A}_3 = \mathbf{A}_1 + \mathbf{A}_2$ and $\mathbf{A}_5 = \mathbf{A}_1 - 2\mathbf{A}_2 + \mathbf{A}_4$. The columns of GA are the vectors $G\mathbf{A}_1$, $G\mathbf{A}_2$, $G\mathbf{A}_3$, $G\mathbf{A}_4$, and $G\mathbf{A}_5$. Since G defines a linear transformation, we have $G\mathbf{A}_3 = G\mathbf{A}_1 + G\mathbf{A}_2$ and $G\mathbf{A}_5 = G\mathbf{A}_1 - 2G\mathbf{A}_2 + G\mathbf{A}_4$. Furthermore $G\mathbf{A}_1$, $G\mathbf{A}_2$, and $G\mathbf{A}_4$ form a linearly independent set; if a, b, and c are scalars so that

$$aG\mathbf{A}_1 + bG\mathbf{A}_2 + cG\mathbf{A}_4 = \mathbf{0}$$

then

$$G(a\mathbf{A}_1 + b\mathbf{A}_2 + c\mathbf{A}_4) = \mathbf{0}.$$

Since G is nonsingular we have, from Proposition 5-8, that

$$a\mathbf{A}_1 + b\mathbf{A}_2 + c\mathbf{A}_4 = \mathbf{0}.$$

Since $\mathbf{A}_1$, $\mathbf{A}_2$, and $\mathbf{A}_4$ form a linearly independent set, $a = b = c = 0$.

Consequently if $\mathbf{A}_1$, $\mathbf{A}_2$, and $\mathbf{A}_4$ form a basis for $\mathscr{C}(A)$, then $G\mathbf{A}_1$, $G\mathbf{A}_2$, and $G\mathbf{A}_4$ form a basis for $\mathscr{C}(GA)$.

These examples illustrate the general situation. Suppose A is an $m \times n$ matrix. Label its columns by $\mathbf{A}_1, \ldots, \mathbf{A}_n$. Then the columns of GA are $G\mathbf{A}_1, \ldots, G\mathbf{A}_n$. Furthermore, if a basis for $\mathscr{C}(A)$ is chosen from the columns of A, say $\mathbf{X}_1, \ldots, \mathbf{X}_r$ are columns of A which form a basis for $\mathscr{C}(A)$, then, provided G is nonsingular, $G\mathbf{X}_1, \ldots, G\mathbf{X}_r$ are columns of GA forming a basis for $\mathscr{C}(GA)$.

Proposition 7-3_____

If A is an $m \times n$ matrix and G is an $m \times m$ nonsingular matrix, then the column rank of A equals the column rank of GA. In particular, if B is the row echelon form for A, then the column rank of A equals the column rank of B.

If we combine the results of Propositions 7-1, 7-2, and 7-3, we have the following proposition.

Proposition 7-4_____

If A is an $m \times n$ matrix, then the row rank of A equals the column rank of A.

Definition 7-2_____

If A is an $m \times n$ matrix, then the *rank of A* is the row rank or the column rank of A.

Brief Exercises

1. Determine the rank of the following matrices.

(a) $\begin{pmatrix} 1 & 0 \\ 1 & 1 \end{pmatrix}$ (c) $\begin{pmatrix} 2 & 1 & 0 \\ 0 & 1 & 2 \end{pmatrix}$

(b) $\begin{pmatrix} 2 & 1 & 3 \\ 1 & 0 & 0 \end{pmatrix}$ (d) $\begin{pmatrix} 4 & 1 & 8 & 2 \\ 3 & 4 & 1 & 1 \\ 0 & 2 & 1 & 0 \end{pmatrix}$

Answers: **1(a)** 2   **1(b)** 2   **1(c)** 2   **1(d)** 3

SUMMARY *The rows of an $m \times n$ matrix A generate a subspace of $\mathbb{R}^n$ called the row space of A; its dimension is the row rank of A. The columns of A generate* *a subspace of $\mathbb{R}^m$ called the column space of A; its dimension is the column rank of A. The main result of this section is that the row rank of any matrix A equals the column rank of A. The rank of A is the same as the row rank and the column rank of A.*

Exercises for Section 7-1

1. Calculate the rank of the following matrices.

(a) $\begin{pmatrix} 1 & 1 \\ 2 & 2 \end{pmatrix}$

(c) $\begin{pmatrix} 5 & 1 & 8 & 6 \\ 1 & 2 & 1 & 3 \\ 0 & 1 & 1 & 1 \end{pmatrix}$

(b) $\begin{pmatrix} -1 & 2 & 1 \\ 1 & 3 & 2 \end{pmatrix}$

(d) $\begin{pmatrix} 1 & 1 & 0 & 1 \\ 1 & 0 & 1 & 1 \\ 1 & 1 & 1 & 0 \\ 0 & 1 & 1 & 0 \end{pmatrix}$

2. What is the relation between the rank of a matrix and the rank of a submatrix?

3. Explain why column operations do not change the rank of a matrix.

4. Explain the fact that a matrix has rank r if and only if it can be transformed to a matrix of the form

$$\begin{pmatrix} I_r & 0 \\ 0 & 0 \end{pmatrix}$$

by a combination of row and column transformations.

5. Suppose A is an $m \times n$ matrix having rank r. Explain why there is a nonsingular $m \times n$ matrix P and a nonsingular $n \times n$ matrix Q such that

$$PAQ = \begin{pmatrix} I_r & 0 \\ 0 & 0 \end{pmatrix}.$$

More Challenging Exercises

6. Explain the following: an $n \times n$ matrix is nonsingular if and only if its rank equals n.

7. Suppose $\mathbf{X}_1, \ldots, \mathbf{X}_k$ are vectors in $\mathbb{R}^n$, and suppose we form the vector $\mathbf{Y}_1$ by deleting certain coordinates from $\mathbf{X}_1$, we form $\mathbf{Y}_2$ by deleting the same coordinates from $\mathbf{X}_2$, and so on. If the vectors $\mathbf{Y}_1, \ldots, \mathbf{Y}_k$ are linearly independent, is the same true of the vectors $\mathbf{X}_1, \ldots, \mathbf{X}_k$?

8. Suppose A is a matrix having rank r. Show that there must be an $r \times r$ submatrix of A which is nonsingular. Also show that if $s > r$, then there cannot be an $s \times s$ submatrix of A which is nonsingular.

9. Suppose A and B are matrices of the same size. Explain why A and B have the same rank if and only if there are nonsingular matrices P and Q of the proper sizes so that $B = PAQ$.

10. Give an argument to show that the rank of AB is at most equal to the smaller of the rank of A and the rank of B.

SECTION 7-2
Kernel and Image of Linear Transformations, Dimension Theorem

PREVIEW *We introduce the kernel and image of a linear transformation $T: \mathscr{V} \to \mathscr{W}$ and show they are subspaces of $\mathscr{V}$ and $\mathscr{W}$, respectively. Then we show that if $\dim \mathscr{V} = n$, then $n = \dim \operatorname{Ker} T + \dim \operatorname{Im} T$. We also apply the term kernel and image to $m \times n$ matrices. Finally, we get a relation between the solution set for a nonhomogeneous linear problem $T(\mathbf{X}) = \mathbf{Y}_0$ and the solution set for its associated linear problem $T(\mathbf{X}) = \mathbf{0}$.*

There are two subspaces associated with a linear transformation T, the kernel or nullspace of T, and the image of T. First we describe the kernel of T.

Example Consider the linear transformation $T: \mathbb{R}^3 \to \mathbb{R}^2$ defined by the equation

$$T(x,y,z) = (4x - y + 3z, \, x + 2y + z).$$

The matrix for T is

$$A_T = \begin{pmatrix} 4 & -1 & 3 \\ 1 & 2 & 1 \end{pmatrix}.$$

The homogeneous system of linear equations corresponding to T is

$$4x - y + 3z = 0$$
$$x + 2y + z = 0.$$

The *kernel* or *nullspace of T* is the same as the solution set of the homogeneous system. We use Ker T as an abbreviation for Kernel T. So Ker T is the set of vectors (x,y,z) satisfying

$$4x - y + 3z = 0$$
$$x + 2y + z = 0$$

or, if we use $\mathbf{X} = (x,y,z)$, we can describe Ker T in coordinate-free form as the set of vectors $\mathbf{X}$ for which $T(\mathbf{X}) = \mathbf{0}$. The same set is described in matrix language. The *kernel of* A_T or *nullspace of* A_T is denoted by Ker A_T; it is the set of all vectors (x,y,z) satisfying the matrix equation

$$\begin{pmatrix} 4 & -1 & 3 \\ 1 & 2 & 1 \end{pmatrix} \begin{pmatrix} x \\ y \\ z \end{pmatrix} = \begin{pmatrix} 0 \\ 0 \end{pmatrix}.$$

If we want to find the dimension of the kernel of T, we solve the homogeneous system by transforming A_T to row echelon form.

$$\begin{pmatrix} 1 & 0 & \frac{7}{9} \\ 0 & 1 & \frac{1}{9} \end{pmatrix}$$

Therefore (x,y,z) is in Ker T if and only if $x = -\frac{7}{9}z$ and $y = -\frac{1}{9}z$, or if and only if

$$(x,y,z) = (-\tfrac{7}{9}z, -\tfrac{1}{9}z, z) = z(-\tfrac{7}{9}, -\tfrac{1}{9}, 1).$$

Thus Ker T is the line through the origin determined by the vector $(-\tfrac{7}{9}, -\tfrac{1}{9}, 1)$; it can also be described as the subspace of $\mathbb{R}^3$ generated by the vector $(-\tfrac{7}{9}, -\tfrac{1}{9}, 1)$.

Example If $T: \mathbb{R}^3 \rightarrow \mathbb{R}^3$ is defined by the equation

$$T(x,y,z) = (x + z, 3x + 2y - z, x + 4y + 2z)$$

then the kernel or nullspace of T is the set of vectors $\mathbf{X}$ which satisfy $T(\mathbf{X}) = \mathbf{0}$. It is the same as the solution set for the associated homogeneous system of linear equations

$$x \qquad\quad + z = 0$$
$$3x + 2y - z = 0$$
$$x + 4y + 2z = 0.$$

This set is also the kernel or nullspace of the matrix A_T, i.e., the set of (x,y,z) for which

$$\begin{pmatrix} 1 & 0 & 1 \\ 3 & 2 & -1 \\ 1 & 4 & 2 \end{pmatrix} \begin{pmatrix} x \\ y \\ z \end{pmatrix} = \begin{pmatrix} 0 \\ 0 \\ 0 \end{pmatrix}.$$

To find the kernel of T we solve the system of equations by transforming the matrix A_T to row echelon form

$$\begin{pmatrix} 1 & 0 & 0 \\ 0 & 1 & 0 \\ 0 & 0 & 1 \end{pmatrix}.$$

Thus (x,y,z) is in Ker T if and only if $x = 0$, $y = 0$, and $z = 0$. Thus Ker T consists of only the zero vector $\mathbf{0}$. As in the previous example, Ker T is a subspace of the domain of T.

The second subspace related to T is the image of T. This set is the same as the image of a function as introduced in Definition 4-1. It happens that this set is a vector space if the function is linear.

Example Consider again the linear transformation introduced in the first example. The image of T is the set of vectors $T(x,y,z) = (4x - y + 3z, x + 2y + z)$, where x, y, and z are real numbers. It can also be described as all vectors of the form

$$\begin{pmatrix} 4 & -1 & 3 \\ 1 & 2 & 1 \end{pmatrix} \begin{pmatrix} x \\ y \\ z \end{pmatrix}$$

where (x,y,z) is in $\mathbb{R}^3$. We use Im T as an abbreviation for the image of T. It can be described as all the vectors we can obtain by applying T to every vector $\mathbf{X}$ in $\mathbb{R}^3$.

We can be more explicit in describing Im T in this particular case since it consists of all vectors having the form

$$(4x - y + 3z, x + 2y + z) = x(4,1) + y(-1,2) + z(3,1).$$

Thus Im T is the subspace of $\mathbb{R}^2$ generated by the vectors $(4,1)$, $(-1,2)$, and $(3,1)$. This can also be described as the subspace of $\mathbb{R}^2$ generated by the columns of A_T, i.e., the column space of A_T.

Example The image of the linear transformation $T: \mathbb{R}^3 \to \mathbb{R}^3$ given in the second example, which was defined by the equation

$$T(x,y,z) = (x + z, 3x + 2y - z, x + 4y + 2z),$$

is all vectors having the form

$$(x + z, 3x + 2y - z, x + 4y + 2z) = x(1,3,1) + y(0,2,4) + z(1,-1,2).$$

Thus Im T is the subspace of $\mathbb{R}^3$ generated by the three vectors $(1,3,1)$, $(0,2,4)$, and $(1,-1,2)$. You can check that these three vectors are linearly independent and so Im $T = \mathbb{R}^3$ in this case. Notice also that Im T is the subspace of $\mathbb{R}^3$ generated by the columns of A_T, the matrix for T.

$$\begin{pmatrix} 1 & 0 & 1 \\ 3 & 2 & -1 \\ 1 & 4 & 2 \end{pmatrix}$$

Thus the image of T equals the column space of its matrix A_T.

These examples give motivation for our definitions of the kernel and image of T. Our definitions are given in the coordinate-free notation.

Definition 7-3_____

Suppose $\mathscr{V}$ and $\mathscr{W}$ are vector spaces over $\mathbb{R}$ and $T: \mathscr{V} \to \mathscr{W}$ is a linear transformation. Then the *kernel* or *nullspace of T* is the set of vectors $\mathbf{X}$ in $\mathscr{V}$ for which $T(\mathbf{X}) = \mathbf{0}$. The *image of T* is the set of vectors in $\mathscr{W}$ which are equal to $T(\mathbf{X})$ for some $\mathbf{X}$ in $\mathscr{V}$. We denote the kernel by Ker T and the image by Im T.

If T is a linear transformation from $\mathbb{R}^n$ to $\mathbb{R}^m$ given by the equation

$$T(x_1, \ldots, x_n) = (a_{11}x_1 + \cdots + a_{1n}x_n, \ldots, a_{m1}x_1 + \cdots + a_{mn}x_n)$$

then the kernel of T can also be described as the solution set for the homogeneous system of linear equations

$$a_{11}x_1 + \cdots + a_{1n}x_n = 0$$
$$\vdots \qquad \qquad \vdots$$
$$a_{m1}x_1 + \cdots + a_{mn}x_n = 0.$$

The image of T can be described as the vectors of the form

$$(a_{11}x_1 + \cdots + a_{1n}x_n, \ldots, a_{m1}x_1 + \cdots + a_{mn}x_n)$$
$$= x_1(a_{11}, \ldots, a_{m1}) + x_2(a_{12}, \ldots, a_{m2}) + \cdots + x_n(a_{1n}, \ldots, a_{mn}).$$

In other words, the image of T is the column space of the matrix for T.

$$A_T = \begin{pmatrix} a_{11} & a_{12} & \cdots & a_{1n} \\ a_{21} & a_{22} & \cdots & a_{2n} \\ \vdots & \vdots & & \vdots \\ a_{m1} & a_{m2} & \cdots & a_{mn} \end{pmatrix}$$

Because Im T is the set generated by the columns of A_T, we know from Proposition 3-2 that Im T is a subspace of $\mathbb{R}^m$. We can also verify that Ker T is a subspace of $\mathbb{R}^n$ by showing that if the n-tuples $(b_1, \ldots, b_n)$ and $(c_1, \ldots, c_n)$ are solutions for the homogeneous system, then so are $(b_1 + c_1, \ldots, b_n + c_n)$ and $(rb_1, \ldots, rb_n)$. However, we can show that Ker T and Im T are subspaces of $\mathscr{V}$ and $\mathscr{W}$ in a co-ordinate-free argument which is extremely simple.

Suppose $\mathbf{X}_1$ and $\mathbf{X}_2$ are in Ker T, i.e., $T(\mathbf{X}_1) = \mathbf{0}$ and $T(\mathbf{X}_2) = \mathbf{0}$. Then $T(\mathbf{X}_1 + \mathbf{X}_2) = T(\mathbf{X}_1) + T(\mathbf{X}_2) = \mathbf{0}$ and $T(r\mathbf{X}_1) = rT(\mathbf{X}_1) = r\mathbf{0} = \mathbf{0}$. Thus Ker T is a subspace of $\mathscr{V}$.

Suppose $\mathbf{Y}_1$ and $\mathbf{Y}_2$ are in Im T; i.e., there are vectors $\mathbf{X}_1$ and $\mathbf{X}_2$ in $\mathscr{V}$ so that $T(\mathbf{X}_1) = \mathbf{Y}_1$ and $T(\mathbf{X}_2) = \mathbf{Y}_2$. Then $T(\mathbf{X}_1 + \mathbf{X}_2) = T(\mathbf{X}_1) + T(\mathbf{X}_2) = \mathbf{Y}_1 + \mathbf{Y}_2$ and $T(r\mathbf{X}_1) = rT(\mathbf{X}_1) = r\mathbf{Y}_1$. This shows that if $\mathbf{Y}_1$ and $\mathbf{Y}_2$ are in Im T, then so are $\mathbf{Y}_1 + \mathbf{Y}_2$ and $r\mathbf{Y}_1$, i.e., Im T is a subspace of $\mathscr{W}$.

Proposition 7-5

If $\mathscr{V}$ and $\mathscr{W}$ are vector spaces over $\mathbb{R}$ and $T: \mathscr{V} \to \mathscr{W}$ is a linear transformation, then Ker T is a subspace of $\mathscr{V}$ and Im T is a subspace of $\mathscr{W}$.

The concepts of kernel or nullspace and of image of a linear transformation are applied to matrices, too, particularly when we think of matrices as representing linear transformations. The same kinds of facts are true for matrices.

Definition 7-4

If A is the $m \times n$ matrix

$$A = \begin{pmatrix} a_{11} & a_{12} & \cdots & a_{1n} \\ a_{21} & a_{22} & \cdots & a_{2n} \\ \cdot & \cdot & & \cdot \\ \cdot & \cdot & & \cdot \\ \cdot & \cdot & & \cdot \\ a_{m1} & a_{m2} & \cdots & a_{mn} \end{pmatrix}$$

then the *kernel* or *nullspace of A* is the collection of vectors $(x_1, \ldots, x_n)$ in $\mathbb{R}^n$ satisfying

$$\begin{pmatrix} a_{11} & a_{12} & \cdots & a_{1n} \\ a_{21} & a_{22} & \cdots & a_{2n} \\ \cdot & \cdot & & \cdot \\ \cdot & \cdot & & \cdot \\ \cdot & \cdot & & \cdot \\ a_{m1} & a_{m2} & \cdots & a_{mn} \end{pmatrix} \begin{pmatrix} x_1 \\ x_2 \\ \cdot \\ \cdot \\ \cdot \\ x_n \end{pmatrix} = \begin{pmatrix} 0 \\ 0 \\ \cdot \\ \cdot \\ \cdot \\ 0 \end{pmatrix}$$

The *image of A* is the collection of vectors expressible as

$$\begin{pmatrix} a_{11} & a_{12} & \cdots & a_{1n} \\ a_{21} & a_{22} & \cdots & a_{2n} \\ \cdot & \cdot & & \cdot \\ \cdot & \cdot & & \cdot \\ \cdot & \cdot & & \cdot \\ a_{m1} & a_{m2} & \cdots & a_{mn} \end{pmatrix} \begin{pmatrix} x_1 \\ x_2 \\ \cdot \\ \cdot \\ \cdot \\ x_n \end{pmatrix}$$

for any $(x_1,\ldots,x_n)$ in $\mathbb{R}^n$. Thus, the image of A equals the column space of A.

Proposition 7-6

If A is an $m \times n$ matrix, then the kernel of A is a subspace of $\mathbb{R}^n$ and the image of A is a subspace of $\mathbb{R}^m$.

Brief Exercises

1. For the linear transformations defined by the following equations, find a basis for Ker T and a basis for Im T.

   (a) $T(x,y) = (2x + y, 4x + 2y)$
   (b) $T(x,y) = (x - y, 5x + 6y, 6x + 5y)$
   (c) $T(x,y,z) = (x + y + 2z, 3x + y - z)$
   (d) $T(x,y,z) = (4x + y + z, 3x - y + 2z, x - y + z, 5x + y + 2z)$
   $Ker\,T = \{0\}$ $Im\,T = (4,3,15), (1,-1,-1,1), (1,2,1,2)$

2. Find a basis for the kernel and a basis for the image of the following matrices

   (a) $\begin{pmatrix} 1 & 1 & 2 \\ 3 & 0 & 1 \end{pmatrix}$ (b) $\begin{pmatrix} 2 & 1 \\ 3 & 2 \end{pmatrix}$ (c) $\begin{pmatrix} 5 & 1 & 0 & 2 \\ 3 & 4 & 1 & 2 \\ 0 & 1 & 0 & 1 \end{pmatrix}$

Answers: Some of the exercises have many correct answers. Here are some. **1(a)** Ker T: $(1,-2)$; Im T: $(1,2)$ **1(b)** Ker T is the zero subspace; Im T: $(1,5,6)$, $(-1,6,5)$ **1(c)** Ker T: $(\frac{3}{2},-\frac{7}{2},1)$; Im T: $(1,3)$, $(1,1)$ **1(d)** Ker T is the zero subspace; Im T: $(4,3,1,5)$, $(1,-1,-1,1)$, $(1,2,1,2)$ **2(a)** Ker A: $(-1,-5,3)$; Im A: $(1,0)$, $(0,1)$ **2(b)** Ker A is the zero subspace; Im T: $(1,0)$, $(0,1)$ **2(c)** Ker A: $(5,-1,-13,1)$; Im A: $(1,0,0)$, $(0,1,0)$, $(0,0,1)$

There is a relationship between the dimensions of the kernel and the image of a linear transformation or of a matrix. We investigate this relationship now. There are some standard terms used in this connection which we introduce in the next definition.

Definition 7-5 ─────────────────────────────

If $T: \mathbb{R}^n \to \mathbb{R}^m$ is a linear transformation, then the *rank of T* is the dimension of the image of T. The *nullity of T* is the dimension of the kernel of T.

   If A is an $m \times n$ matrix, the *rank of A* (as defined in Definition 7-2) is the dimension of the column or row space of A. The *nullity of A* is the dimension of the kernel of A.

───

If A is the matrix for T, then the kernel of A is the same as the kernel of T, and the image of T is the same as the column space of A. Consequently, the nullity of T equals the nullity of A, and the rank of T equals the rank of A.

   We discuss these concepts for an $m \times n$ matrix in the next two examples.

Example   Consider the 3×4 matrix

$$A = \begin{pmatrix} 1 & 0 & 2 & 1 \\ 1 & 1 & 3 & 2 \\ -1 & 2 & 1 & 0 \end{pmatrix}.$$

We investigate the relation between the dimension of the nullspace of A (i.e., the nullity of A) and the dimension of the image of A (the rank of A).

   Transform A to row echelon form.

$$\begin{pmatrix} 1 & 0 & 0 & 3 \\ 0 & 1 & 0 & 2 \\ 0 & 0 & 1 & -1 \end{pmatrix}$$

It is clear that the rank of A is 3.

   Now we consider the nullspace of A. It is the set of vectors (x_1, x_2, x_3, x_4) which are solutions of

$$\begin{pmatrix} 1 & 0 & 2 & 1 \\ 1 & 1 & 3 & 2 \\ -1 & 2 & 1 & 0 \end{pmatrix} \begin{pmatrix} x_1 \\ x_2 \\ x_3 \\ x_4 \end{pmatrix} = \begin{pmatrix} 0 \\ 0 \\ 0 \end{pmatrix},$$

i.e., the solution set for a homogeneous system of equations. Therefore, the nullspace of A is the same as the nullspace of its row echelon form. Thus (x_1, x_2, x_3, x_4)

is in the nullspace of A if and only if

$$x_1 \qquad\qquad + 3x_4 = 0$$
$$x_2 \qquad + 2x_4 = 0$$
$$x_3 - x_4 = 0$$

or, equivalently,

$$x_1 = -3x_4$$
$$x_2 = -2x_4$$
$$x_3 = x_4.$$

Therefore (x_1, x_2, x_3, x_4) is in the nullspace of A if and only if $(x_1, x_2, x_3, x_4) = (-3x_4, -2x_4, x_4, x_4) = x_4(-3, -2, 1, 1)$. Therefore the nullity of A is 1. The rank of A is 3. So

$$\text{nullity } A + \text{rank } A = \text{number of columns in } A.$$

Let us view A as the coefficient matrix for a homogeneous system of linear equations. Here the rank of A is 3, and we can solve for three unknowns as linear functions of the remaining unknowns. So, the description of the solution set is made in terms of one unknown; this shows that the dimension of the nullspace of A is 1.

Thus the rank of A corresponds to the number of unknowns solved for in the echelon form, and the nullity of A corresponds to the number of unknowns not solved for in the echelon form. Thus the sum of these two numbers equals the total number of unknowns.

Example Let A be the matrix

$$\begin{pmatrix} 2 & 0 & 1 & 3 & 4 \\ 1 & 1 & 0 & -2 & 1 \\ 3 & 1 & 1 & 2 & 5 \\ 5 & -1 & 3 & 12 & 11 \end{pmatrix}.$$

Transform A to row echelon form.

$$\begin{pmatrix} 1 & 0 & \frac{1}{2} & 0 & 2 \\ 0 & 1 & -\frac{1}{2} & 0 & -1 \\ 0 & 0 & 0 & 1 & 0 \\ 0 & 0 & 0 & 0 & 0 \end{pmatrix}$$

The rank of A is 3. If we view A as the coefficient matrix for a homogeneous system of linear equations, we can solve for three unknowns x_1, x_2, and x_4 in terms of x_3 and x_5.

$$x_1 = -\tfrac{1}{2}x_3 - 2x_5$$
$$x_2 = \tfrac{1}{2}x_3 + x_5$$
$$x_4 = 0$$

The description of a vector (x_1,x_2,x_3,x_4,x_5) in the nullspace of A uses the two unknowns x_3 and x_5.

$$(x_1,x_2,x_3,x_4,x_5) = (-\tfrac{1}{2}x_3 - 2x_5, \tfrac{1}{2}x_3 + x_5, x_3, 0, x_5)$$
$$= x_3(-\tfrac{1}{2},\tfrac{1}{2},1,0,0) + x_5(-2,1,0,0,1)$$

So again rank A + nullity A = number of columns in A = number of unknowns.

These examples illustrate the fact which interests us. Suppose A is an $m \times n$ matrix. The rank r of A is the number of nonzero rows in the echelon form for A. To describe the vectors in the kernel of A, we solve for r unknowns as linear functions of the remaining $n - r$ unknowns. Then, in describing vectors $(x_1, \ldots, x_n)$ in the nullspace, we replace these r unknowns by the r linear functions of $n - r$ unknowns and find that the nullity of A is $n - r$.

Proposition 7-7_____

(Dimension Theorem) If A is an $m \times n$ matrix, then rank A + nullity $A = n$. If $T: \mathbb{R}^n \to \mathbb{R}^m$ is a linear transformation, then dim Ker T + dim Im $T = n$ or nullity T + rank $T = n$.

_____

There is a way to view the equality in Proposition 7-7 in a more geometric fashion when dealing with a linear transformation $T: \mathbb{R}^n \to \mathbb{R}^m$. We illustrate this in the next two examples. We restrict the examples to $\mathbb{R}^3$ to be able to draw pictures.

Example   Consider the linear function $T: \mathbb{R}^3 \to \mathbb{R}^3$ defined by

$$T(x,y,z) = (2x + y, x + 3z, 3x + y + 3z).$$

The kernel of T consists of vectors (x,y,z) of the form

$$(x,y,z) = (-3z,6z,z) = z(-3,6,1).$$

So pick a basis containing the vector $(-3,6,1)$. For example, $(-3,6,1)$, $(2,1,0)$, and $(1,0,3)$ form a basis for $\mathbb{R}^3$. Also $T(-3,6,1) = (0,0,0)$, $T(2,1,0) = (5,2,7)$, and $T(1,0,3) = (2,10,12)$. The function T maps $\mathbb{R}^3$ onto the plane determined by $(5,2,7)$ and $(2,10,12)$, so it must collapse one dimension of $\mathbb{R}^3$ to map $\mathbb{R}^3$ onto a plane. It collapses $\mathbb{R}^3$ in the direction of the line determined by $(-3,6,1)$. We can think of T as operating in two steps: (1) Collapse or project $\mathbb{R}^3$ onto the plane determined by $(2,1,0)$ and $(1,0,3)$; make the projection parallel to the line determined by $(-3,6,1)$. (2) Map the plane determined by $(2,1,0)$ and $(1,0,3)$ onto the plane determined by $(5,2,7)$ and $(2,10,12)$ (see Figure 7-1).

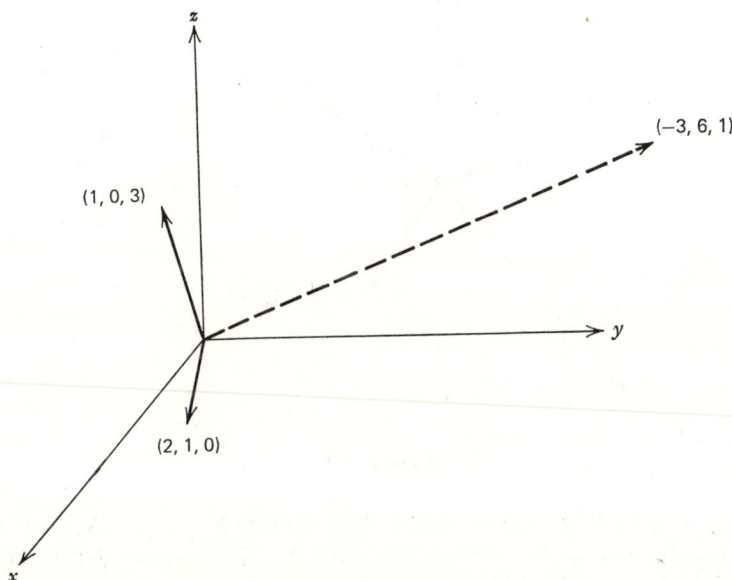

Figure 7-1

Example Let $T: \mathbb{R}^3 \to \mathbb{R}^3$ be defined by

$$T(x,y,z) = (x + 2y - z, 3x + 6y - 3z, 2x + 4y - 2z).$$

The kernel of T is the set of vectors (x,y,z) of the form

$$(x,y,z) = (-2y + z, y, z) = y(-2,1,0) + z(1,0,1).$$

Thus the kernel of T is the two-dimensional subspace of $\mathbb{R}^3$ with $(-2,1,0)$ and $(1,0,1)$ constituting its basis. Pick another vector, e.g., $(1,2,-1)$, so that $(-2,1,0)$, $(1,0,1)$, and $(1,2,-1)$ form a basis for $\mathbb{R}^3$. Now $T(-2,1,0) = (0,0,0)$, $T(1,0,1) = (0,0,0)$, and $T(1,2,-1) = (6,18,12)$. The effect of T can be thought of in two steps: (1) Project $\mathbb{R}^3$ onto the line determined by $(1,2,-1)$; do this projection parallel to the plane determined by $(-2,1,0)$ and $(1,0,1)$. (2) Map the line determined by $(1,2,-1)$ onto the line determined by $(6,18,12)$ (see Figure 7-2).

The geometric flavor of the previous two examples can be preserved somewhat in the general situation of a linear transformation $T: \mathscr{V} \to \mathscr{W}$ if the dimension of $\mathscr{V}$

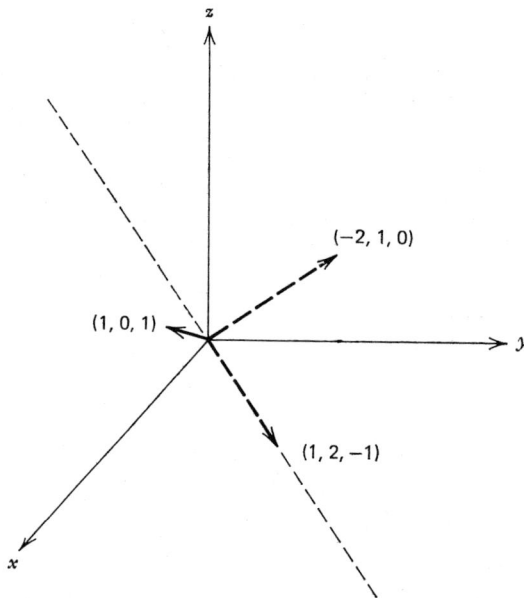

Figure 7-2

is n. Let $\mathbf{X}_1, \ldots, \mathbf{X}_m$ be a basis for Ker T. Pick vectors $\mathbf{X}_{m+1}, \ldots, \mathbf{X}_n$ in $\mathscr{V}$ so that $\mathbf{X}_1, \ldots, \mathbf{X}_m, \mathbf{X}_{m+1}, \ldots, \mathbf{X}_n$ form a basis for $\mathscr{V}$ (Proposition 3-12). Then the vectors $T(\mathbf{X}_{m+1}), \ldots, T(\mathbf{X}_n)$ form a basis for Im T (Exercise 8). We can therefore think of T as operating in two steps: (1) project $\mathscr{V}$ onto the subspace generated by $\mathbf{X}_{m+1}, \ldots, \mathbf{X}_n$; do this by collapsing $\mathscr{V}$ along Ker T, and (2) map the subspace generated by $\mathbf{X}_{m+1}, \ldots, \mathbf{X}_n$ onto the subspace of $\mathscr{W}$ whose basis is $T(\mathbf{X}_{m+1}), \ldots, T(\mathbf{X}_n)$.

SUMMARY *If $T: \mathscr{V} \to \mathscr{W}$ is a linear transformation, the kernel or nullspace of T is the set of vectors $\mathbf{X}$ in $\mathscr{V}$ satisfying $T(\mathbf{X}) = \mathbf{0}$, and the image of T is the set of vectors in $\mathscr{W}$ obtained by applying T to every vector in $\mathscr{V}$. Similar terms apply to an $m \times n$ matrix A. The kernel or nullspace of A is the set of vectors $\mathbf{X}$ which satisfy $A\mathbf{X} = \mathbf{0}$. The image of A is the set of all vectors of the form $A\mathbf{X}$. The nullity of T (of A) is the dimension of the nullspace or kernel of T (of A). The rank of T is the dimension of the image of T. If $T: \mathscr{V} \to \mathscr{W}$ is a linear transformation and dim $\mathscr{V} = n$, then*

$$n = \text{nullity } T + \text{rank } T.$$

Similarly, if A is an $m \times n$ matrix, then

$$n = \text{nullity } A + \text{rank } A.$$

Exercises for Section 7-2

1. The matrix for a linear transformation T is

$$\begin{pmatrix} 2 & 1 & 3 \\ 4 & 8 & 1 \end{pmatrix}.$$

Find a basis for the kernel of T and a basis for the image of T.

2. The matrix for a linear transformation T is

$$\begin{pmatrix} 5 & 0 & 3 & 1 \\ 2 & 0 & 1 & 4 \\ 1 & 0 & -1 & 2 \end{pmatrix}.$$

Find a basis for the kernel of T and a basis for the image of T.

3. In each of the parts of this exercise, find a basis for the nullspace of the matrix and a basis for the image of the matrix.

(a) $\begin{pmatrix} 2 & 1 & 3 \\ 0 & 1 & 1 \end{pmatrix}$

(c) $\begin{pmatrix} 1 & 2 & 1 \\ 3 & 1 & 0 \\ 1 & -1 & 2 \\ 1 & 1 & 1 \end{pmatrix}$

(b) $\begin{pmatrix} 1 & 1 & 2 & 1 \\ 4 & -1 & 2 & 3 \\ 1 & 1 & 2 & 1 \end{pmatrix}$

(d) $\begin{pmatrix} 1 & 1 & 2 & 1 & 3 \\ 5 & -1 & 2 & 4 & 1 \\ 6 & 1 & 0 & 2 & 0 \end{pmatrix}$

4. Consider the linear transformation $T: \mathbb{R}^3 \to \mathbb{R}^3$ defined by the matrix

$$\begin{pmatrix} 1 & 0 & 1 \\ 1 & 1 & 0 \\ 2 & 1 & 1 \end{pmatrix}.$$

Show dim Ker $T = 1$ and dim Im $T = 2$. Find a generating set for Im T Pick a nonzero vector X_1 in Ker T. Pick any two vectors X_2 and X_3 so that X_1, X_2, and X_3 form a basis for $\mathbb{R}^3$. (Make specific numerical choices.) Show by actual computation that $T(X_2)$ and $T(X_3)$ form a basis for Im T.

5. Consider the linear transformation $T: \mathbb{R}^2 \to \mathbb{R}^2$ defined by the matrix

$$\begin{pmatrix} 2 & 1 \\ 0 & -2 \end{pmatrix}.$$

Show dim Ker $T = 0$ and dim Im $T = 2$. Pick any two vectors $\mathbf{X}_1$ and $\mathbf{X}_2$ which form a basis for $\mathbb{R}^2$. (Make specific numerical choices.) Show by computation that $T(\mathbf{X}_1)$ and $T(\mathbf{X}_2)$ form a basis for $\mathbb{R}^2$.

6. Consider the linear transformation $T: \mathbb{R}^4 \to \mathbb{R}^3$ defined by the matrix

$$\begin{pmatrix} 2 & 1 & 0 & 3 \\ -1 & 1 & -3 & 0 \\ 3 & 0 & 3 & 3 \end{pmatrix}.$$

Show dim Ker $T = 2$. Pick a basis for Ker T consisting of $\mathbf{X}_1$ and $\mathbf{X}_2$ (make specific numerical choices). Pick two vectors $\mathbf{X}_3$ and $\mathbf{X}_4$ so that $\mathbf{X}_1$, $\mathbf{X}_2$, $\mathbf{X}_3$, and $\mathbf{X}_4$ form a basis for $\mathbb{R}^4$. Show by actual computation that $T(\mathbf{X}_3)$ and $T(\mathbf{X}_4)$ form a basis for Im T.

More Challenging Exercises

7. If $T: \mathcal{V} \to \mathcal{W}$ is a linear transformations and dim $\mathcal{V} = n$, then let $\mathbf{X}_1, \ldots, \mathbf{X}_m$ be a basis for Ker T and $\mathbf{X}_1, \ldots, \mathbf{X}_m, \mathbf{X}_{m+1}, \ldots, \mathbf{X}_n$ form a basis for $\mathcal{V}$. Show that $T(\mathbf{X}_{m+1}), \ldots, T(\mathbf{X}_n)$ form a basis for Im T.

8. In the exercises for Section 3-4 we introduced the concept of complementary subspaces for a vector space. Using the notation of Exercise 7, show that Ker T and the subspace generated by $\mathbf{X}_{m+1}, \ldots, \mathbf{X}_n$ are complementary subspaces.

SECTION 7-3
Systems of Equations

PREVIEW   *The solution set for a system of linear equations can be represented as one particular vector added to the solution set for the associated homogeneous system. After giving that representation, we discuss the possibility of solving for some unknowns in terms of the remaining unknowns in a system of equations.*

The first topic we discuss in this section is the description of the solution set for a system of linear equations in terms of the concepts of rank and nullity of the coefficient matrix.

Example   Consider the system of equations

$$3x + y + z = 1$$

$$x - y + z = 2$$

$$5x + 3y + z = 0.$$

Transform the augmented matrix for this system to row echelon form.

$$\begin{pmatrix} 1 & 0 & \frac{1}{2} & \frac{3}{4} \\ 0 & 1 & -\frac{1}{2} & -\frac{5}{4} \\ 0 & 0 & 0 & 0 \end{pmatrix}$$

The solution set is the set of vectors (x,y,z) satisfying

$$(x,y,z) = (\tfrac{3}{4} - \tfrac{1}{2}z, -\tfrac{5}{4} + \tfrac{1}{2}z, z)$$

$$= (\tfrac{3}{4}, -\tfrac{5}{4}, 0) + z(-\tfrac{1}{2}, \tfrac{1}{2}, 1).$$

The solution set for the homogeneous system having the same coefficient matrix is the set of vectors of the form $z(-\tfrac{1}{2},\tfrac{1}{2},1)$; this is also the kernel or the nullspace of the coefficient matrix. The solution set for the nonhomogeneous system can therefore be described as all vectors which can be expressed as $(\tfrac{3}{4},-\tfrac{5}{4},0)$ added to some vector in the kernel of the coefficient matrix. Geometrically, the kernel of the coefficient matrix is the line determined by $(-\tfrac{1}{2},\tfrac{1}{2},1)$, whereas the solution set for the nonhomogeneous system is the line through $(\tfrac{3}{4},-\tfrac{5}{4},0)$ parallel to the line determined by $(-\tfrac{1}{2},\tfrac{1}{2},1)$. In other words, the solution set for the nonhomogeneous system can be obtained from the solution set for the homogeneous system by translation through the vector $(\tfrac{3}{4},-\tfrac{5}{4},0)$ (see Figure 7-3).

Suppose we let $T: \mathbb{R}^3 \to \mathbb{R}^3$ be the linear transformation whose matrix is the coefficient matrix, i.e.,

$$T(x,y,z) = (3x + y + z, x - y + z, 5x + 3y + z).$$

The problem of finding all solutions to the system of equations can then be phrased as follows: find all vectors (x,y,z) for which $T(x,y,z) = (1,2,0)$. Or, if we let $\mathbf{X} = (x,y,z)$ and $\mathbf{Y}_0 = (1,2,0)$, we can phrase the problem this way: find all vectors $\mathbf{X}$ for which $T(\mathbf{X}) = \mathbf{Y}_0$. The solution to this problem is the set of all vectors expressible as

$$(\tfrac{3}{4}, -\tfrac{5}{4}, 0) + (a,b,c)$$

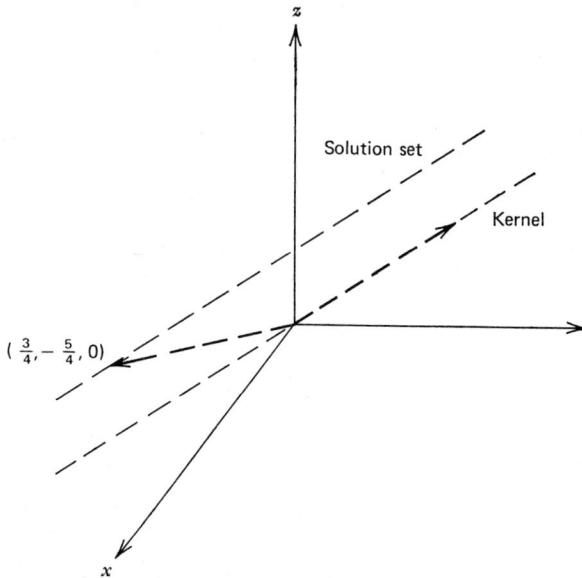

Figure 7-3

where (a,b,c) is in Ker T. Or, if we let $\mathbf{X_0} = (\frac{3}{4},-\frac{5}{4},0)$, the solution is the set of vectors expressible as $\mathbf{X_0} + \mathbf{X'}$, where $\mathbf{X'}$ is in Ker T.

Example   Consider the system of equations

$$2x - y + z - w = 3$$
$$x + 2y + 3z + 2w = 4$$
$$6x + 2y + 8z + 2w = 14.$$

Transform the augmented matrix to row echelon form

$$\begin{pmatrix} 1 & 0 & 1 & 0 & 2 \\ 0 & 1 & 1 & 1 & 1 \\ 0 & 0 & 0 & 0 & 0 \end{pmatrix}$$

The solution set consists of all vectors (x,y,z,w) satisfying

$$(x,y,z,w) = (2 - z, 1 - z - w, z, w)$$
$$= (2,1,0,0) + z(-1,-1,1,0) + w(0,-1,0,1).$$

The solution set for the homogeneous system having the same coefficient matrix is the set of vectors $z(-1,-1,1,0) + w(0,-1,0,1)$. Thus the solution set for the non-homogeneous system can be obtained by adding one particular solution $(2,1,0,0)$ to the various vectors in the kernel of the coefficient matrix.

The same thing can be stated using the linear transformation $T: \mathbb{R}^4 \to \mathbb{R}^3$ whose matrix is the coefficient matrix for the system above. Let $\mathbf{X} = (x,y,z,w)$, $\mathbf{Y}_0 = (3,4,14)$, and $\mathbf{X}_0 = (2,1,0,0)$. Then the original problem can be stated as follows: find all vectors $\mathbf{X}$ for which $T(\mathbf{X}) = \mathbf{Y}_0$. The solution set can be described as those vectors expressible as $\mathbf{X}_0 + \mathbf{X}'$, where $\mathbf{X}'$ is in Ker T.

These examples indicate the relationship which exists between the solution set for a homogeneous linear system $T(\mathbf{X}) = \mathbf{0}$ and the nonhomogeneous linear system $T(\mathbf{X}) = \mathbf{Y}_0$.

Proposition 7-8_____

If $T: \mathbb{R}^n \to \mathbb{R}^m$ is a linear transformation and $T(\mathbf{X}_0) = \mathbf{Y}_0$, then the vectors $\mathbf{X}$ which satisfy $T(\mathbf{X}) = \mathbf{Y}_0$ are exactly those which can be expressed as $\mathbf{X}_0 + \mathbf{X}'$ for some vector $\mathbf{X}'$ satisfying $T(\mathbf{X}') = \mathbf{0}$.

_____

Brief Exercises

1. Write the solution set for the following system of equations in the form used in Proposition 7-8.

   (a) $2x + y = 2$ (b) $5x + y + z = 1$ (c) $x + 2y + 3z = 1$

        $x - y = 2$ $2x - y - z = 3$ $x + 5y \quad\;\; = 2$

                                                      $-x + y - 6z = 0$

2. Write the solution set for the following linear problem in the form used in Proposition 7-8.

$$T(x,y,z) = (2,1,3) \quad \text{where}$$
$$T(x,y,z) = (x + y, x - 2y + z, 2x - y + z).$$

Answers: **1(a)** $(\tfrac{4}{3}, -\tfrac{2}{3}) + (0,0)$ **1(b)** $(\tfrac{4}{7}, -\tfrac{13}{7}, 0) + z(0,-1,1)$ **1(c)** $(\tfrac{1}{3}, \tfrac{1}{3}, 0) + z(-5,1,1)$ **2** $(\tfrac{5}{3}, \tfrac{1}{3}, 0) + z(-\tfrac{1}{3}, \tfrac{1}{3}, 1)$

The concept of the rank of a matrix has another interpretation in systems of linear equations. In solving a system of linear equations we solve for the first unknown in one equation and eliminate it from the remaining equations. Then we solve for the second unknown in another equation and eliminate it from the remaining equations, and so forth. At first glance, then, we might expect to be able to solve for as many unknowns as there are equations. However, we have seen many examples where this is not possible.

An intuitive way of understanding what is taking place is to think of each equation as providing one piece of information about the unknowns. We use one piece of information to solve for the first unknown, a different bit of information to solve for the second unknown, and so on. If we are to solve for every unknown, then we must have as many "different" bits of information as there are unknowns.

This intuition about different bits of information is put onto a computational basis with the introduction of the concepts of linear combination and rank. Suppose we have a system of equations which is consistent, i.e., has a solution. If the last row of the augmented matrix, for example, is a linear combination of the other rows, then the information provided by the last equation is already present in the earlier equations. The rank of the augmented matrix therefore tells us the number of "independent bits of information" provided by the equations. That is why the rank equals the number of unknowns for which we can solve.

Our procedure for solving a system of linear equations is to transform the augmented matrix to row echelon form. By doing this we do not decide in advance which unknowns to solve for; the selection is made by the procedure. The question we consider now is this: can we specify in advance which unknowns we want to solve for?

Example   Consider the system of equations
$$5x + 2y + z = 1$$
$$4x - 4y - 2z = 0.$$

It is easy to see that the rank of the coefficient matrix is 2, and so transforming the augmented matrix to row echelon form will allow us to solve for two unknowns. The row echelon form is
$$\begin{pmatrix} 1 & 0 & 0 & \frac{1}{7} \\ 0 & 1 & \frac{1}{2} & \frac{1}{7} \end{pmatrix}.$$

Thus we have solved for x and y in terms of z.

Can we solve for x and z in terms of y? If we can, then we should be able to transform the augmented matrix with row operations to a matrix having the form
$$\begin{pmatrix} 1 & - & 0 & - \\ 0 & - & 1 & - \end{pmatrix}.$$

This matrix has the leading 1's in the columns for x and z. You can check that the augmented matrix can be transformed to the matrix
$$\begin{pmatrix} 1 & 0 & 0 & \frac{1}{7} \\ 0 & 2 & 1 & \frac{2}{7} \end{pmatrix}.$$

This gives the equations

$$x = \tfrac{1}{7}$$
$$z = \tfrac{2}{7} - 2y.$$

If we want to solve for y and z in terms of x, then we try to transform the augmented matrix to a matrix having leading 1's in the second and third columns.

$$\begin{pmatrix} - & 1 & 0 & - \\ - & 0 & 1 & - \end{pmatrix}$$

You can check that it is not possible to solve for y and z from this system of equations. The reason is that the submatrix consisting of the columns corresponding to the unknowns y and z

$$\begin{pmatrix} 2 & 1 \\ -4 & -2 \end{pmatrix}$$

has rank equal to 1. It cannot be transformed by row operations to

$$\begin{pmatrix} 1 & 0 \\ 0 & 1 \end{pmatrix}$$

because row operations do not change the rank of a matrix (Proposition 7-2). On the other hand, the submatrix consisting of the first two columns

$$\begin{pmatrix} 5 & 2 \\ 4 & -4 \end{pmatrix}$$

has rank equal to 2 as does the submatrix consisting of the first and third columns

$$\begin{pmatrix} 5 & 1 \\ 4 & -2 \end{pmatrix}.$$

Therefore, they can be transformed to

$$\begin{pmatrix} 1 & 0 \\ 0 & 1 \end{pmatrix}.$$

For that reason, we can solve for x and y, for x and z, but not for y and z.

Example Consider the system of equations

$$x + y + z + w + v = 0$$
$$x + 3y - z \qquad - 2v = 0$$
$$x + 2y \qquad + w - v = 0$$
$$x + y + z - w + 3v = 0.$$

You can check that the rank of this matrix is equal to 3. Consequently, we cannot solve for more than three unknowns at one time. We are interested in determining which triples of unknowns we can solve for.

Suppose we want to solve for x, y, and z. We can do so if and only if the submatrix consisting of the first three columns has rank equal to 3.

$$\begin{pmatrix} 1 & 1 & 1 \\ 1 & 3 & -1 \\ 1 & 2 & 0 \\ 1 & 1 & 1 \end{pmatrix}$$

However, you can check that this matrix has rank equal to 2. Therefore we cannot solve for x, y, and z in terms of w and v.

We can solve for x, y, and w if and only if 3 is the rank of the submatrix consisting of the columns corresponding to x, y, and w.

$$\begin{pmatrix} 1 & 1 & 1 \\ 1 & 3 & 0 \\ 1 & 2 & 1 \\ 1 & 1 & -1 \end{pmatrix}$$

This matrix has rank equal to 3, so we can solve for x, y, and w.

Continuing in this fashion we find that we can solve for the following triples of unknowns.

$$x, y, w$$
$$x, y, v$$
$$y, z, w$$
$$y, z, v$$

However, we cannot solve for x, y, z or for z, w, v.

These examples illustrate the general situation.

Proposition 7-9

We can solve for k designated unknowns in a system of linear equations if and only if the submatrix of the coefficient matrix consisting of columns corresponding to those unknowns has rank equal to k.

Brief Exercises

1. Consider the system of linear equations

$$5x + y - z = 0$$
$$x - 2y + 2z = 0.$$

Determine which pair of unknowns can be solved in terms of the remaining unknown.

2. In the system of linear equations

$$x + y + z - w = 0$$
$$x - y + 2z + 5w = 0$$
$$x + 2y \qquad - 4w = 0$$
$$2x + y + 3z + 2w = 0$$

determine which triples of unknowns can be solved for in terms of the remaining one.

Answers: **1** It is possible to solve for x and y, for x and z, but not for y and z
2 It is possible to solve for any three unknowns

SUMMARY *First we describe the solution set for a linear system. If $T: \mathbb{R}^n \to \mathbb{R}^m$ is a linear transformation and if $T(X_0) = Y_0$, then the vectors X satisfying $T(X) = Y_0$ are those expressible as $X_0 + X'$ for some vector X' in Ker T. Then we show that we can solve for a given set of k unknowns in a system of linear equations if and only if k is the rank of the submatrix of the coefficient matrix consisting of the columns corresponding to the unknowns in question.*

Exercises for Section 7-3

1. Solve the system of linear equations

$$x + y - 3z = a$$
$$x - y + 2z = b.$$

(a) Draw a diagram of the solution set for the homogeneous system, i.e., for $a = b = 0$.
(b) Draw diagrams of the solution sets for the nonhomogeneous systems obtained by using the following values for a and b.
(i) $a = 2, b = 0$ (ii) $a = 1, b = 1$
(c) What is the geometric relation between the different solution sets?

2. Solve the system of linear equations

$$4x - 6y + z = a$$

$$2x - 3y + \tfrac{1}{2}z = b.$$

(a) Are there values for a and b for which there is no solution?

(b) Draw a diagram of the solution set for the homogeneous system in which $a = b = 0$.

(c) Draw diagrams of the solution sets for the nonhomogeneous systems obtained by using the following values for a and b.
   (i) $a = 4, b = 2$ (ii) $a = 2, b = 1$ (iii) $a = -4, b = -2$

(d) What is the geometric relation between the different solution sets?

3. Solve the system of linear equations

$$x + y - z = a$$

$$x - y + z = b$$

$$3x + y + z = c.$$

(a) Are there values for a, b, and c for which there is no solution?

(b) Draw a diagram of the solution set for the homogeneous system $a = b = c = 0$.

(c) Draw diagrams of the solution sets for the nonhomogeneous systems obtained by using the following values for a, b, and c.
   (i) $a = 2, b = 0, c = 4$ (ii) $a = 0, b = 2, c = 2$

(d) What is the geometric relation between the different solution sets?

4. Suppose $T: \mathbb{R}^3 \to \mathbb{R}^2$ is the linear transformation defined by the matrix

$$\begin{pmatrix} 5 & 6 & -7 \\ 1 & -1 & 2 \end{pmatrix}.$$

(a) What is the dimension of the image of T?

(b) How does this relate to the possibility of solving the system of equations in Exercise 1?

(c) What is the dimension of the solution set for the homogeneous system?

5. Same exercise as 4 for T whose matrix is

$$\begin{pmatrix} 4 & -6 & 1 \\ 2 & -3 & \tfrac{1}{2} \end{pmatrix}.$$

Refer to Exercise 2.

6. Same exercise as 4 for T whose matrix is

$$\begin{pmatrix} 1 & 1 & -1 \\ 1 & -1 & 1 \\ 3 & 1 & 1 \end{pmatrix}.$$

Refer to Exercise 3.

7. Express the solution set of the following systems of equations in the form $X_0 + X'$, where X' is in Ker A and A is the coefficient matrix of the system.

(a) $x + y = 1$ **(b)** $3x + 2y + z = 1$

$\quad\ x - y = 2$ $\ x - y + 2z = 2$

8. Consider the following homogeneous system of linear equations.

$$x - y + z + w - v = 0$$
$$3x + y + 2z - w + v = 0$$
$$2x \quad\ - z + 2w - v = 0$$

(a) Show that the rank of the coefficient matrix equals 3.
(b) Determine which triples of variables are defined as functions of the remaining two variables.

9. Show that the coefficient matrix for the system

$$x + y + z - 2w = 0$$
$$3x - y + 2z + 3w = 0$$
$$x + y - z + w = 0$$

has rank equal to 3. Then determine which triples of unknowns are defined as linear functions of the remaining unknown.

8

Distance
in $\mathbb{R}^n$

SECTION 8-1
Introductory Ideas on Distance, Dot Product

PREVIEW *In this section we discuss distance between points and length of vectors in $\mathbb{R}^n$. We use our experience in $\mathbb{R}^2$ and $\mathbb{R}^3$ to guide us in the higher dimensions. We then introduce the operation of dot product and determine the relation between it and lengths of vectors.*

There are situations in which it is very useful to compute distance between points or lengths of vectors. We discuss one kind of distance called Euclidean distance.

The distance between the points labeled -2 and 4 on the real line can be computed by $|4 - (-2)| = 6$ or $|-2 - 4| = 6$. In general, the distance between x_1 and x_2 is given by $|x_1 - x_2|$ or by $|x_2 - x_1|$ (see Figure 8-1). We use the absolute value to take care of the two situations: x_1 to the left of x_2, x_1 to the right of x_2. (It also takes care of the case $x_1 = x_2$.) We also speak of the length of the line segment from x_1 to x_2 or from x_2 to x_1; it is the same as the distance between x_1 and x_2.

The distance between the two points $(1,2)$ and $(3,2)$ in $\mathbb{R}^2$ is $|3 - 1| = |1 - 3| = 2$ (see Figure 8-2). The distance between $(3,2)$ and $(3,5)$ is $|5 - 2| = |2 - 5| = 3$. Then the distance between $(1,2)$ and $(3,5)$ is computed using these horizontal and vertical distances along with the Pythagorean theorem.

$$\sqrt{2^2 + 3^2} = \sqrt{13}.$$

Notice that the vector $(2,3)$ begins at $(1,2)$ and ends at $(3,5)$. We use the same computation for the length of the vector $(2,3)$.

In general, the distance between (x_1,x_2) and (y_1,y_2), the length of the vector from (x_1,x_2) to (y_1,y_2), and the length of the vector from (y_1,y_2) to (x_1,x_2) are all given by

$$\sqrt{(x_1 - y_1)^2 + (x_2 - y_2)^2}$$

(see Figure 8-3).

We use the same kind of formula to determine the distance between points in $\mathbb{R}^3$, $(2,1,-3)$ and $(1,4,3)$, for example. It is

$$\sqrt{(2 - 1)^2 + (1 - 4)^2 + (-3 - 3)^2} = \sqrt{46}.$$

This also gives the length of the vector $(-1,3,6)$ which begins at $(2,1,-3)$ and ends at $(1,4,3)$, and it also gives the length of the vector $(1,-3,-6)$ which begins at

304

Figure 8-1

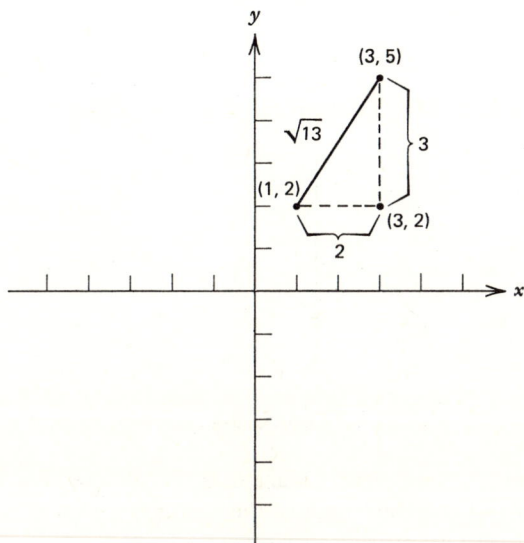

Figure 8-2

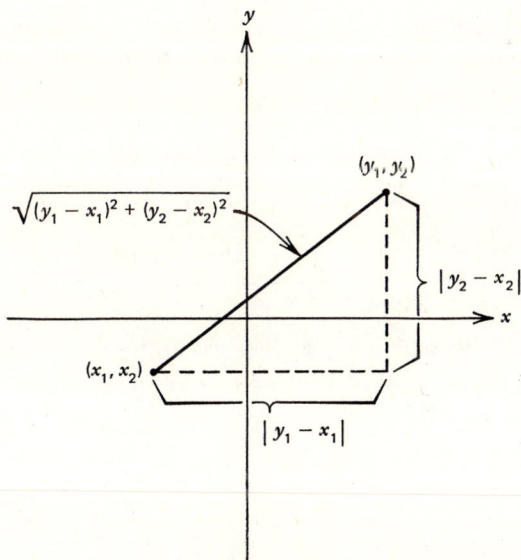

Figure 8-3

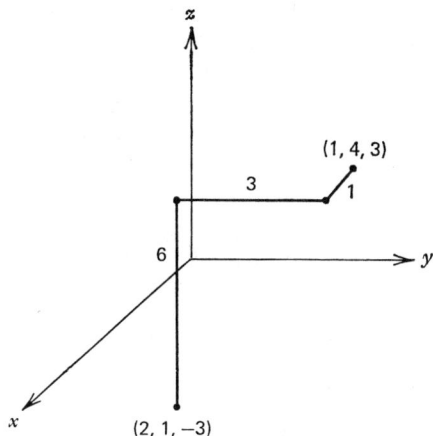

Figure 8-4

(1,4,3) and ends at $(2,1,-3)$ (see Figure 8-4). As before, the distance between (x_1,x_2,x_3) and (y_1,y_2,y_3) and the length of the vector from (y_1,y_2,y_3) to (x_1,x_2,x_3) or from (x_1,x_2,x_3) to (y_1,y_2,y_3) are given by

$$\sqrt{(x_1 - y_1)^2 + (x_2 - y_2)^2 + (x_3 - y_3)^2}.$$

Although our geometric intuition does not extend beyond three-dimensional space, we use the ideas of distance between points and the ideas of lengths of vectors for higher-dimensional spaces by relying on the arithmetical formulas.

Definition 8-1_____

The distance between the points $(x_1,x_2, \ldots ,x_n)$ and $(y_1,y_2, \ldots ,y_n)$ in $\mathbb{R}^n$ is

$$\sqrt{(x_1 - y_1)^2 + (x_2 - y_2)^2 + \cdots + (x_n - y_n)^2}.$$

The length of the vector $\mathbf{X} = (x_1,x_2, \ldots ,x_n)$ in $\mathbb{R}^n$, denoted by $|\mathbf{X}|$, is

$$|\mathbf{X}| = \sqrt{x_1^2 + x_2^2 + \cdots + x_n^2}.$$

(This may also be viewed as the distance from the point $(0,0, \ldots ,0)$ to $(x_1,x_2, \ldots ,x_n)$.) A vector is called a *unit vector* of its length equals 1, i.e., $\mathbf{X}$ is a unit vector if $|\mathbf{X}| = 1$.

Brief Exercises

1. Compute the distance between the following pairs of points. Draw diagrams.

    (a) $(1,2)$ and $(3,-5)$ **(c)** $(1,1)$ and $(3,-6)$

    (b) $(1,5)$ and $(-4,2)$ **(d)** $(2,3)$ and $(4,-3)$

2. Compute the distance between the following points.

(a) $(1,2,1)$ and $(3,-1,4)$ (c) $(1,1,-1,2)$ and $(4,1,8,6)$

(b) $(2,-1,3)$ and $(5,6,2)$ (d) $(1,-1,2,1,3)$ and $(1,2,6,1,2)$

3. Find the length of the following vectors.

(a) $(1,2,1)$ (c) $(1,6,-1,2,3)$

(b) $(-3,5,1)$ (d) $(5,3,2,1,4)$

Answers: **1(a)** $\sqrt{53}$ **1(b)** $\sqrt{34}$ **1(c)** $\sqrt{53}$ **1(d)** $\sqrt{40}$ **2(a)** $\sqrt{22}$
2(b) $\sqrt{59}$ **2(c)** $\sqrt{106}$ **2(d)** $\sqrt{26}$ **3(a)** $\sqrt{6}$ **3(b)** $\sqrt{35}$ **3(c)** $\sqrt{51}$
3(d) $\sqrt{55}$

There is an arithmetic operation between vectors in $\mathbb{R}^n$ which is related to length and which has some very useful geometric interpretations. Here we concentrate on the arithmetic of the dot or inner product, leaving a discussion of the geometric interpretations for the next section.

The dot product of two vectors (x_1,x_2) and (y_1,y_2) in $\mathbb{R}^2$, denoted by $(x_1,x_2) \cdot (y_1,y_2)$, is $x_1y_1 + x_2y_2$. Thus $(2,3) \cdot (-1,4) = (2)(-1) + (3)(4) = 10$.

The dot product of two vectors (x_1,x_2,x_3) and (y_1,y_2,y_3) in $\mathbb{R}^3$ is $(x_1,x_2,x_3) \cdot (y_1,y_2,y_3) = x_1y_1 + x_2y_2 + x_3y_3$. So $(2,1,4) \cdot (3,0,-2) = (2)(3) + (1)(0) + (4)(-2) = -2$.

Definition 8-2_____

If $X = (x_1, \ldots ,x_n)$ and $Y = (y_1, \ldots ,y_n)$ are vectors in $\mathbb{R}^n$, then the *dot product* or inner product of X and Y is $X \cdot Y = x_1y_1 + \cdots + x_ny_n$.

We can use the notation of dot product to express the length of vectors. For example, the length of $X = (2,1)$ is $|(2,1)| = \sqrt{2^2 + 1^2} = \sqrt{5}$. Also $(2,1) \cdot (2,1) = 2^2 + 1^2$. Thus $X \cdot X = |X|^2$ or $|X| = \sqrt{X \cdot X}$.

Similarly if $X = (2,-1,5)$, then $|X| = \sqrt{4 + 1 + 25} = \sqrt{30}$ and $X \cdot X = 4 + 1 + 25 = 30$. So $X \cdot X = |X|^2$ or $|X| = \sqrt{X \cdot X}$.

In general, if $X = (x_1, \ldots ,x_n)$, then $|X| = \sqrt{x_1^2 + \cdots + x_n^2}$, whereas $X \cdot X = x_1^2 + \cdots + x_n^2$. Thus $X \cdot X = |X|^2$ and $|X| = \sqrt{X \cdot X}$.

Proposition 8-1_____

If X is a vector in $\mathbb{R}^n$, then

$$X \cdot X = |X|^2 \quad \text{and} \quad |X| = \sqrt{X \cdot X}.$$

There is a formula for computing $(\mathbf{X} - \mathbf{Y}) \cdot (\mathbf{X} - \mathbf{Y})$ which we need in the next section.

$$
\begin{aligned}
(\mathbf{X} - \mathbf{Y}) \cdot (\mathbf{X} - \mathbf{Y}) &= (x_1 - y_1)^2 + (x_2 - y_2)^2 + \cdots + (x_n - y_n)^2 \\
&= (x_1^2 - 2x_1y_1 + y_1^2) + (x_2^2 - 2x_2y_2 + y_2^2) + \cdots + (x_n^2 - 2x_ny_n + y_n^2) \\
&= (x_1^2 + x_2^2 + \cdots + x_n^2) - 2(x_1y_1 + x_2y_2 + \cdots + x_ny_n) \\
&\quad + (y_1^2 + y_2^2 + \cdots + y_n^2) \\
&= \mathbf{X} \cdot \mathbf{X} - 2\mathbf{X} \cdot \mathbf{Y} + \mathbf{Y} \cdot \mathbf{Y}
\end{aligned}
$$

Proposition 8-2

If $\mathbf{X}$ and $\mathbf{Y}$ are vectors in $\mathbb{R}^n$, then

$$(\mathbf{X} - \mathbf{Y}) \cdot (\mathbf{X} - \mathbf{Y}) = \mathbf{X} \cdot \mathbf{X} - 2\mathbf{X} \cdot \mathbf{Y} + \mathbf{Y} \cdot \mathbf{Y}.$$

Brief Exercises

Refer to Section 3-2 for notation.

1. If $\mathbf{i} = (1,0)$ and $\mathbf{j} = (0,1)$, compute the following.

   (a) $(3\mathbf{i} + 2\mathbf{j}) \cdot (6\mathbf{i} - 7\mathbf{j})$ (b) $(-4\mathbf{i} + 7\mathbf{j}) \cdot (3\mathbf{i} - 2\mathbf{j})$

2. If $\mathbf{i} = (1,0,0)$, $\mathbf{j} = (0,1,0)$, and $\mathbf{k} = (0,0,1)$, compute the following.

   (a) $(2\mathbf{i} + 3\mathbf{j} - \mathbf{k}) \cdot (4\mathbf{i} - \mathbf{j})$ (c) $(\mathbf{i} - \mathbf{j}) \cdot (\mathbf{j} + \mathbf{k})$
   (b) $(\mathbf{i} + 2\mathbf{j} + \mathbf{k}) \cdot (\mathbf{i} + 2\mathbf{j} - \mathbf{k})$

3. Using the notation of Exercise 2, find the lengths of the following.

   (a) $2\mathbf{i} - 3\mathbf{j} + 4\mathbf{k}$ (c) $5\mathbf{j} - 2\mathbf{i} + 2\mathbf{k}$
   (b) $\mathbf{i} + 3\mathbf{j}$ (d) $6\mathbf{i} + 2\mathbf{j} - \mathbf{k}$

Answers: 1(a) 4   1(b) -26 2(a) 5   2(b) 4   2(c) -1 3(a) $\sqrt{29}$
3(b) $\sqrt{10}$ 3(c) $\sqrt{33}$ 3(d) $\sqrt{41}$

The computations for dot product can also be done as products of matrices. For example, if $\mathbf{X} = (2,1,3)$ and $\mathbf{Y} = (-1,2,4)$, the dot product $\mathbf{X} \cdot \mathbf{Y}$ can be computed by taking the matrix product of the 1×3 matrix

$$(2 \quad 1 \quad 3)$$

and the 3×1 matrix

$$\begin{pmatrix} -1 \\ 2 \\ 4 \end{pmatrix}.$$

Thus

$$\mathbf{X} \cdot \mathbf{Y} = \begin{pmatrix} 2 & 1 & 3 \end{pmatrix} \begin{pmatrix} -1 \\ 2 \\ 4 \end{pmatrix} = 12.$$

To do this, notice that the first vector $\mathbf{X}$ is considered as a 1×3 matrix, whereas the second vector is considered as a 3×1 matrix.

It is reasonable for us to think of the dot product as an operation defined for pairs of vectors $\mathbf{X}$ and $\mathbf{Y}$ in $\mathbb{R}^n$ without thinking of $\mathbf{X}$ and $\mathbf{Y}$ as matrices. It is also legitimate to think of the dot product as being computed as a product of matrices. However, it is not legitimate to have the vector $\mathbf{X}$ viewed as a $1 \times n$ matrix and the vector $\mathbf{Y}$ viewed as an $n \times 1$ matrix in the same context. We should choose which kind of matrix we want to represent the vectors. It does not make any essential difference which choice is made, but once a choice is made, it should be used consistently so that no confusion will arise.

When we think of computations with vectors in $\mathbb{R}^n$ as done with matrices, we will use $n \times 1$ matrices to represent vectors in $\mathbb{R}^n$ unless some comment calling off this agreement is made at the time. Thus, for example, we feel free to write $\mathbf{X} \cdot \mathbf{Y} = \mathbf{X}^T \mathbf{Y}$. The symbols $\mathbf{X}$ and $\mathbf{Y}$ on the left-hand side represent vectors in $\mathbb{R}^n$, whereas $\mathbf{X}$ and $\mathbf{Y}$ on the right-hand side represent $n \times 1$ matrices (and so $\mathbf{X}^T$ must be used to obtain a $1 \times n$ matrix).

An argument might be made that we need to make a distinction in our notation between $\mathbf{X}$ representing a vector in $\mathbb{R}^n$ and $\mathbf{X}$ representing an $n \times 1$ matrix. However, me do not do so because the traditional usage does not make such a distinction and because it is the notation you are most likely to see in situations outside mathematics.

SUMMARY *The distance between the points $(x_1, \ldots, x_n)$ and $(y_1, \ldots, y_n)$ in $\mathbb{R}^n$ is defined as*

$$\sqrt{(x_1 - y_1)^2 + \cdots + (x_n - y_n)^2}.$$

The length of a vector $\mathbf{X} = (x_1, \ldots, x_n)$ in $\mathbb{R}^n$ is

$$|\mathbf{X}| = \sqrt{x_1^2 + \cdots + x_n^2}.$$

If $|\mathbf{X}| = 1$, then $\mathbf{X}$ is called a unit vector. The dot product of $\mathbf{X} = (x_1, \ldots, x_n)$ and $\mathbf{Y} = (y_1, \ldots, y_n)$ in $\mathbb{R}^n$ is defined by $\mathbf{X} \cdot \mathbf{Y} = x_1 y_1 + \cdots + x_n y_n$. The dot product of vectors is related to lengths of vectors by the equations $\mathbf{X} \cdot \mathbf{X} = |\mathbf{X}|^2$ and $|\mathbf{X}| = \sqrt{\mathbf{X} \cdot \mathbf{X}}$. The computation for the dot product $\mathbf{X} \cdot \mathbf{Y}$ can also be done by thinking of $\mathbf{X}$ and $\mathbf{Y}$ as $n \times 1$ matrices and performing the matrix multiplication indicated on the right-hand side of the equation $\mathbf{X} \cdot \mathbf{Y} = \mathbf{X}^T \mathbf{Y}$.

Exercises for Section 8-1

1. Compute the distance between the following pairs of points. Draw diagrams.
   (a) (1,2) and (3,−5)
   (b) (1,5) and (−4,2)
   (c) (1,1) and (3,−6)
   (d) (1,2,1) and (3,−1,4)
   (e) (2,−1,3) and (5,6,2)
   (f) (1,1,−2,2) and (4,1,8,6)

2. "Normalize" the following vectors in this sense: in each case, find a scalar r so that $r\mathbf{X}$ is a unit vector if $\mathbf{X}$ is the given vector. Then obtain the unit vector.
   (a) (1,1,2)
   (b) (5,1,6,2)
   (c) (−1,−1,2,5)
   (d) (1,1,−1,2,6)

3. Find the length of the following vectors.
   (a) (1,2,1)
   (b) (−3,5,1)
   (c) (1,6,−1,2,3)
   (d) (5,3,2,1,4)

4. Compute the following dot products.
   (a) $\mathbf{i}\cdot\mathbf{i}$, $\mathbf{i}\cdot\mathbf{j}$, $\mathbf{i}\cdot\mathbf{k}$, $\mathbf{j}\cdot\mathbf{j}$, $\mathbf{j}\cdot\mathbf{k}$, $\mathbf{k}\cdot\mathbf{k}$
   (b) $(2\mathbf{i} + 3\mathbf{j} − 4\mathbf{k})\cdot(6\mathbf{i} − 3\mathbf{j} − 4\mathbf{k})$
   (c) $(\mathbf{i} − 5\mathbf{j} + 6\mathbf{k})\cdot(4\mathbf{i} + 7\mathbf{j} − 2\mathbf{k})$

5. Compute the lengths of the following vectors.
   (a) $2\mathbf{i} − 3\mathbf{j} + 4\mathbf{k}$
   (b) $\mathbf{i} + 3\mathbf{j}$
   (c) $5\mathbf{j} − 2\mathbf{i} + 2\mathbf{k}$
   (d) $6\mathbf{i} + 2\mathbf{j} − \mathbf{k}$

6. In $\mathbb{R}^5$, if $\mathbf{X} = (2,1,3,−1,4)$, $\mathbf{Y} = (0,1,−2,3,1)$, and $\mathbf{Z} = (1,1,2,−4,1)$, compute the following.
   (a) $\mathbf{X}\cdot(3\mathbf{Y} + 2\mathbf{Z})$
   (b) $(2\mathbf{X} + 3\mathbf{Y})\cdot(5\mathbf{X} − 6\mathbf{Y} + 3\mathbf{Z})$
   (c) $(4\mathbf{X} − 2\mathbf{Y} + 3\mathbf{Z})\cdot(4\mathbf{X} + 6\mathbf{Y} − 2\mathbf{Z})$

7. Compute the dot product of each of the following vectors with the unit vectors of the standard basis.
   (a) $5\mathbf{i} + 2\mathbf{j} + 3\mathbf{k}$
   (b) $8\mathbf{i} − 6\mathbf{j} + 17\mathbf{k}$
   (c) $x\mathbf{i} + y\mathbf{j} + z\mathbf{k}$

8. Find the coordinates of the following points. Draw diagrams if possible.
   (a) the midpoint of the line segment joining (1,2) and (3,−5)
   (b) the point $\frac{1}{3}$ of the way from (2,1) to (−3,4)
   (c) the point $\frac{2}{3}$ of the way from (2,1) to (−3,4)
   (d) the point $\frac{1}{4}$ of the way from (2,1) to (−3,4)
   (e) the point equidistant from (1,4) and (3,−3) and having first coordinate equal to 5
   (f) the midpoint of the line segment joining (1,1,2,1) and (5,−2,3,1)
   (g) the point $\frac{2}{3}$ of the way from (1,2,1,3) to (−1,6,8,2)

9. Find a point (x,y) in $\mathbb{R}^2$ so that $(2,1)$, $(1,3)$, and (x,y) are the vertices of an equilateral triangle. (Draw a diagram; the diagram should convince you that there are two answers to this problem.)

10. Verify the following properties of the dot product of vectors in $\mathbb{R}^n$.
    (a) $\mathbf{X} \cdot \mathbf{Y} = \mathbf{Y} \cdot \mathbf{X}$ (e) $|\mathbf{X} + \mathbf{Y}|^2 = |\mathbf{X}|^2 + 2\mathbf{X} \cdot \mathbf{Y} + |\mathbf{Y}|^2$
    (b) $\mathbf{X} \cdot \mathbf{X} = 0$ if and only if $\mathbf{X} = \mathbf{0}$ (f) $|r\mathbf{X}| = |r|\,|\mathbf{X}|$
    (c) $\mathbf{X} \cdot (\mathbf{Y} + \mathbf{Z}) = \mathbf{X} \cdot \mathbf{Y} + \mathbf{X} \cdot \mathbf{Z}$ (g) $(r\mathbf{X}) \cdot \mathbf{Y} = r(\mathbf{X} \cdot \mathbf{Y}) = \mathbf{X} \cdot (r\mathbf{Y})$
    (d) $(\mathbf{X} + \mathbf{Y}) \cdot \mathbf{Z} = \mathbf{X} \cdot \mathbf{Z} + \mathbf{Y} \cdot \mathbf{Z}$

11. Let A be an $n \times n$ matrix, and let $\mathbf{X}$ and $\mathbf{Y}$ be $n \times 1$ matrices (or elements of $\mathbb{R}^n$ written vertically). Give an argument to support the equality $(A\mathbf{X}) \cdot \mathbf{Y} = \mathbf{X} \cdot (A^T\mathbf{Y})$. (See Exercises 16 and 17 of Section 4-4.)

12. Verify the Pythagorean theorem in $\mathbb{R}^n$; i.e., if $\mathbf{X} \cdot \mathbf{Y} = 0$, then $|\mathbf{Y} - \mathbf{X}|^2 = |\mathbf{X}|^2 + |\mathbf{Y}|^2$.

13. Argue that if $\mathbf{X}$ and $\mathbf{Y}$ are vectors in $\mathbb{R}^n$, then $|\mathbf{X} + \mathbf{Y}|^2 + |\mathbf{X} - \mathbf{Y}|^2 = 2|\mathbf{X}|^2 + 2|\mathbf{Y}|^2$. This is called the "parallelogram law" in $\mathbb{R}^n$.

14. Verify that if $\mathbf{X}$ and $\mathbf{Y}$ are vectors in $\mathbb{R}^n$, then $\mathbf{X} \cdot \mathbf{Y} = (\frac{1}{4})[|\mathbf{X} + \mathbf{Y}|^2 - |\mathbf{X} - \mathbf{Y}|^2]$.

SECTION 8-2
Geometric Interpretation of Dot Products

PREVIEW *Geometrical facts about angles between vectors or lines and lengths of projections of vectors along other vectors are discussed in the geometrical setting of $\mathbb{R}^2$ and $\mathbb{R}^3$. We relate these facts to properties of the dot product. Since the dot product is defined in $\mathbb{R}^n$, we then use it to generalize the geometrical notions from $\mathbb{R}^2$ and $\mathbb{R}^3$ to $\mathbb{R}^n$ for $n > 3$.*

In trigonometry, a formula is developed which relates the lengths of the sides of a triangle with the cosine of an angle. This formula is called the "law of cosines." In Figure 8-5 we have drawn a triangle with vertices A, B, and C, and with the lengths

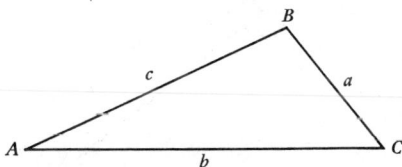

Figure 8-5

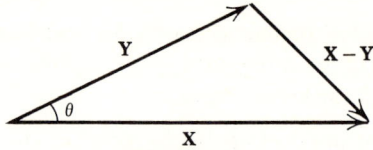

Figure 8-6

of the sides opposite A, B, and C denoted by a, b, and c, respectively. One version of the law of cosines is

$$a^2 = b^2 + c^2 - 2bc \cos A.$$

In Figure 8-6 we use the triangle of Figure 8-5 with different labeling; the side AC is $\mathbf{X}$, the side AB is $\mathbf{Y}$, the side BC is then $\mathbf{X} - \mathbf{Y}$, and the angle between $\mathbf{X}$ and $\mathbf{Y}$ is θ. The law of cosines becomes

$$|\mathbf{X} - \mathbf{Y}|^2 = |\mathbf{X}|^2 + |\mathbf{Y}|^2 - 2\,|\mathbf{X}|\,|\mathbf{Y}| \cos \theta.$$

We substitute expressions for $|\mathbf{X}|^2$, $|\mathbf{Y}|^2$, and $|\mathbf{X} - \mathbf{Y}|^2$ from Proposition 8-1 in this equation, and we obtain

$$(\mathbf{X} - \mathbf{Y}) \cdot (\mathbf{X} - \mathbf{Y}) = \mathbf{X} \cdot \mathbf{X} + \mathbf{Y} \cdot \mathbf{Y} - 2\,|\mathbf{X}|\,|\mathbf{Y}| \cos \theta.$$

From Proposition 8-2 we have

$$(\mathbf{X} - \mathbf{Y}) \cdot (\mathbf{X} - \mathbf{Y}) = \mathbf{X} \cdot \mathbf{X} + \mathbf{Y} \cdot \mathbf{Y} - 2\mathbf{X} \cdot \mathbf{Y}.$$

Thus

$$\mathbf{X} \cdot \mathbf{X} - 2\mathbf{X} \cdot \mathbf{Y} + \mathbf{Y} \cdot \mathbf{Y} = \mathbf{X} \cdot \mathbf{X} + \mathbf{Y} \cdot \mathbf{Y} - 2\,|\mathbf{X}|\,|\mathbf{Y}| \cos \theta.$$

From this equation we get

$$\mathbf{X} \cdot \mathbf{Y} = |\mathbf{X}|\,|\mathbf{Y}| \cos \theta. \tag{1}$$

The first geometric interpretation we obtain concerns the "angle between $\mathbf{X}$ and $\mathbf{Y}$." There are two angles determined by two nonzero vectors; they are θ and $2\pi - \theta$, as illustrated in Figure 8-7. Since $\cos(2\pi - \theta) = \cos \theta$, we agree to use

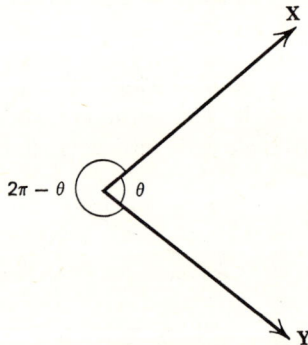

Figure 8-7

the angle between 0 and π as the angle between two vectors. In our calculations we can determine $\cos \theta$.

If $|\mathbf{X}| \neq 0$ and $|\mathbf{Y}| \neq 0$, then we can solve for $\cos \theta$ in equation (1) to obtain

$$\cos \theta = \frac{\mathbf{X} \cdot \mathbf{Y}}{|\mathbf{X}|\,|\mathbf{Y}|}. \tag{2}$$

Example In $\mathbb{R}^2$ let $\mathbf{X} = (2,1)$ and $\mathbf{Y} = (-3,2)$. Then $\cos \theta = [2(-3) + 1 \cdot 2]/(\sqrt{5})(\sqrt{13}) = -4/\sqrt{65}$ (see Figure 8-8). You can check that this information shows that the angle θ is approximately 120°.

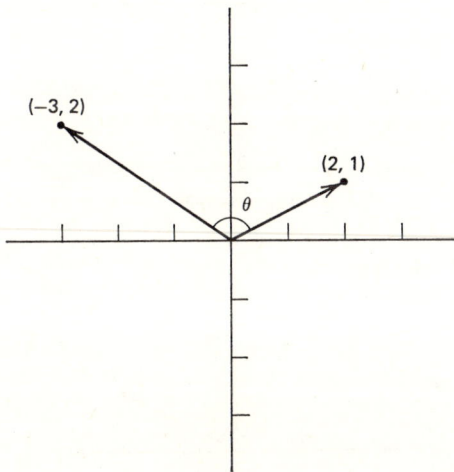

Figure 8-8

Example In $\mathbb{R}^2$ let $\mathbf{X} = (1,5)$ and $\mathbf{Y} = (-1,-2)$. Then $\cos \theta = [1(-1) + 5(-2)]/(\sqrt{26})(\sqrt{5}) = -11/\sqrt{130}$. This shows that the angle θ is approximately 165° (see Figure 8-9). If φ is the angle between $\mathbf{X}$ and the positive part of the x-axis, we can determine $\cos \varphi$ by finding the cosine of the angle between $\mathbf{X}$ and the vector $(1,0)$.

$$\cos \varphi = [1 \cdot 1 + 5 \cdot 0]/(\sqrt{26})(\sqrt{1}) = 1/\sqrt{26}$$

This shows that the angle φ is approximately 79°.

Example In $\mathbb{R}^2$ let $\mathbf{X} = (2,1)$ and $\mathbf{Y} = (-1,2)$. Then if θ is the angle between $\mathbf{X}$ and $\mathbf{Y}$, $\cos \theta = [2(-1) + 1 \cdot 2]/(\sqrt{5})(\sqrt{5}) = 0$. Since $\cos \theta = 0$, and since we agree that θ must lie between 0 and π, we see that $\theta = \frac{1}{2}\pi$. Consequently $\mathbf{X}$ and $\mathbf{Y}$ are perpendicular to each other (see Figure 8-10).

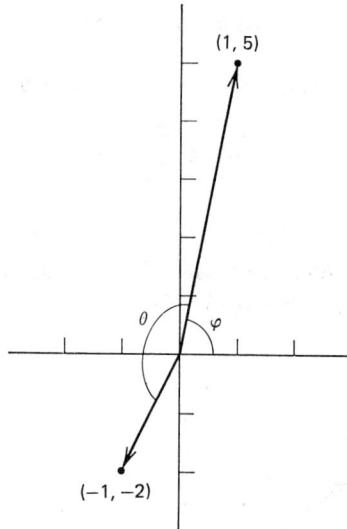

Figure 8-9

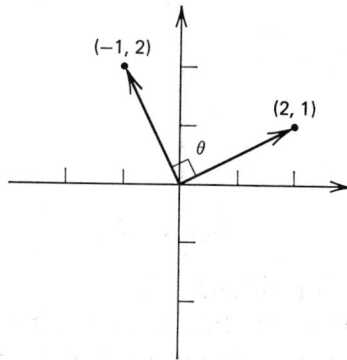

Figure 8-10

Example In $\mathbb{R}^3$ let $\mathbf{X} = (1,1,2)$ and $\mathbf{Y} = (3,-1,1)$. If $\mathbf{X}$ and $\mathbf{Y}$ are pictured as having a common beginning point, then they determine a plane passing through this common point. If θ is the angle between $\mathbf{X}$ and $\mathbf{Y}$ in that plane, then

$$\cos \theta = (3 - 1 + 2)/((\sqrt{6})(\sqrt{11}) = 4/\sqrt{66}$$

(see Figure 8-11). This means that θ is approximately $60.5°$. Since $\cos \theta \neq 0$ the vectors $\mathbf{X}$ and $\mathbf{Y}$ are not perpendicular to each other. If $\mathbf{Z} = (3,5,-4)$, then the

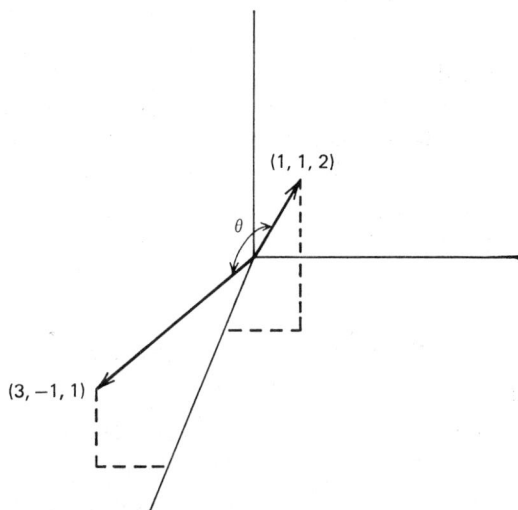

Figure 8-11

cosine of the angle between **X** and **Z** is $(3 + 5 - 8)/(\sqrt{6})(\sqrt{50}) = 0$, and the cosine of the angle between **Y** and **Z** is $(9 - 5 - 4)/(\sqrt{11})(\sqrt{50}) = 0$. Consequently **Z** is perpendicular to both **X** and **Y**.

These examples should give some feeling for the use of the dot product in $\mathbb{R}^2$ and $\mathbb{R}^3$ to measure the cosine of the angle between two vectors. Since we have the dot product in $\mathbb{R}^n$, even for $n > 3$, we can use it to introduce geometrical notions into $\mathbb{R}^n$ for $n > 3$. If we consider two vectors **X** and **Y** positioned at **0** in $\mathbb{R}^n$, they can be viewed as contained in some two-dimensional space which may be thought of as a plane, or very similar to $\mathbb{R}^2$. Therefore, it seems reasonable to introduce the geometrical concept of the cosine of the angle between these vectors in this "plane." We can do it by means of the dot product.

Definition 8-3

If **X** and **Y** are vectors in $\mathbb{R}^n$, we define the cosine of the angle between them by the equation

$$\mathbf{X} \cdot \mathbf{Y} = |\mathbf{X}|\,|\mathbf{Y}|\cos\theta.$$

If **X** and **Y** are both nonzero vectors, then

$$\cos\theta = \frac{\mathbf{X} \cdot \mathbf{Y}}{|\mathbf{X}|\,|\mathbf{Y}|}.$$

We say that **X** and **Y** are *perpendicular* or *orthogonal* to each other if and only if $\mathbf{X} \cdot \mathbf{Y} = 0$.

Example In $\mathbb{R}^5$ let $\mathbf{X} = (1,2,1,-1,3)$ and $\mathbf{Y} = (-1,0,4,-2,1)$. These vectors determine a plane in $\mathbb{R}^5$, i.e., if $\mathbf{X}$ and $\mathbf{Y}$ are "pictured" as beginning at the origin, then this plane is the two-dimensional subspace generated by $\mathbf{X}$ and $\mathbf{Y}$. Then the angle between $\mathbf{X}$ and $\mathbf{Y}$ is the angle between $\mathbf{X}$ and $\mathbf{Y}$ in this two-dimensional subspace. If θ is the angle between $\mathbf{X}$ and $\mathbf{Y}$, then $\cos\theta = (-1 + 0 + 4 + 2 + 3)/$ $(\sqrt{16})(\sqrt{22}) = 2/\sqrt{22}$. You can check that this shows that θ is approximately $65°$. The point of this example is that we can generalize the geometric ideas of $\mathbb{R}^2$ and $\mathbb{R}^3$ about the angle between vectors to spaces with dimension more than 2 or 3.

Brief Exercises

1. Compute the cosine of the angle between the following vectors.

  (a) $(2,1)$ and $(3,-4)$
  (b) $(1,1)$ and $(-1,1)$
  (c) $(1,2,3)$ and $(-1,2,-3)$
  (d) $(3,1,3)$ and $(-1,2,1)$
  (e) $(1,1,0,2)$ and $(3,1,2,0)$
  (f) $(5,0,1,2,3)$ and $(1,1,-1,4,2)$

2. Determine whether the pairs of vectors in Exercise 1 are orthogonal.

Answers: **1(a)** $2/5\sqrt{5}$ **1(b)** 0 **1(c)** $-6/14$ **1(d)** $2/\sqrt{114}$ **1(e)** $2/\sqrt{21}$ **1(f)** $18/\sqrt{897}$ **2** Only the pair in 1(b)

The second geometric fact we discuss concerns the orthogonal projection of a vector $\mathbf{X}$ upon a nonzero vector $\mathbf{Y}$. We illustrate this in Figures 8-12 and 8-13. Picture $\mathbf{X}$ and $\mathbf{Y}$ as having the same starting point. The orthogonal projection of $\mathbf{X}$ on $\mathbf{Y}$ is determined geometrically by dropping a perpendicular from the endpoint of $\mathbf{X}$ to the line determined by $\mathbf{Y}$. In both figures the orthogonal projection is the vector $\mathbf{Z}$.

Furthermore, we want to express $\mathbf{X}$ as the sum of a vector along a given vector $\mathbf{Y}$ and a vector perpendicular to $\mathbf{Y}$. Both figures indicate that if we can obtain the

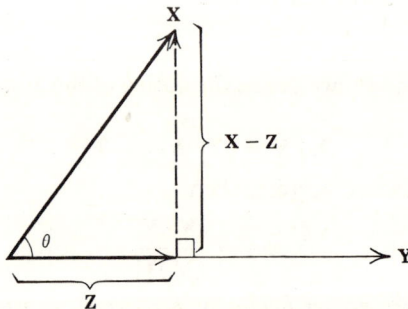

Figure 8-12

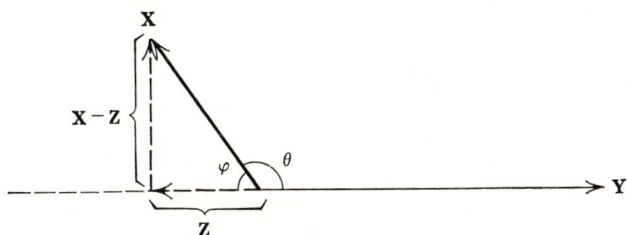

Figure 8-13

orthogonal projection of $\mathbf{X}$ on $\mathbf{Y}$ (the vector $\mathbf{Z}$ in both figures), then we use the vector $\mathbf{X} - \mathbf{Z}$ from the endpoint of $\mathbf{Z}$ to the endpoint of $\mathbf{X}$ for the vector orthogonal to $\mathbf{Y}$.

To determine the orthogonal projection of $\mathbf{X}$ on $\mathbf{Y}$, we note that in Figure 8-12 the length of $\mathbf{Z}$ is $|\mathbf{X}| \cos \theta$ (use trigonometry), whereas in Figure 8-13 the length of $\mathbf{Z}$ is $|\mathbf{X}| \cos \varphi = -(|\mathbf{X}| \cos \theta)$. In the first case, $\mathbf{Z}$ is a positive multiple of $\mathbf{Y}$, so

$$\mathbf{Z} = (|\mathbf{X}| \cos \theta) \frac{\mathbf{Y}}{|\mathbf{Y}|}.$$

In the second case, $\mathbf{Z}$ is a negative multiple of $\mathbf{Y}$, so

$$\mathbf{Z} = -(-|\mathbf{X}| \cos \varphi) \frac{\mathbf{Y}}{|\mathbf{Y}|} = (|\mathbf{X}| \cos \theta) \frac{\mathbf{Y}}{|\mathbf{Y}|}.$$

But $\mathbf{X} \cdot \mathbf{Y} = |\mathbf{X}| \, |\mathbf{Y}| \cos \theta$. Therefore,

$$\mathbf{Z} = \frac{\mathbf{X} \cdot \mathbf{Y}}{|\mathbf{Y}|^2} \mathbf{Y}.$$

Example   Let $\mathbf{X} = (2,1)$ and $\mathbf{Y} = (-3,2)$ in $\mathbb{R}^2$. In this example we write $\mathbf{X}$ as the sum of a vector along $\mathbf{Y}$ and a vector perpendicular to $\mathbf{Y}$ (see Figure 8-14). Then we write $\mathbf{Y}$ as the sum of a vector along $\mathbf{X}$ and a vector perpendicular to $\mathbf{X}$ (see Figure 8-15).

The perpendicular projection of $\mathbf{X}$ on $\mathbf{Y}$ is given by

$$\frac{\mathbf{X} \cdot \mathbf{Y}}{|\mathbf{Y}|^2} \mathbf{Y} = \left(-\frac{4}{13}\right)(-3,2) = \left(\frac{12}{13}, -\frac{8}{13}\right).$$

Thus the vector $\mathbf{X} = (2,1)$ can be expressed as

$$(2,1) = (\tfrac{12}{13}, -\tfrac{8}{13}) + [(2,1) - (\tfrac{12}{13}, -\tfrac{8}{13})]$$
$$= (\tfrac{12}{13}, -\tfrac{8}{13}) + (\tfrac{14}{13}, \tfrac{21}{13}).$$

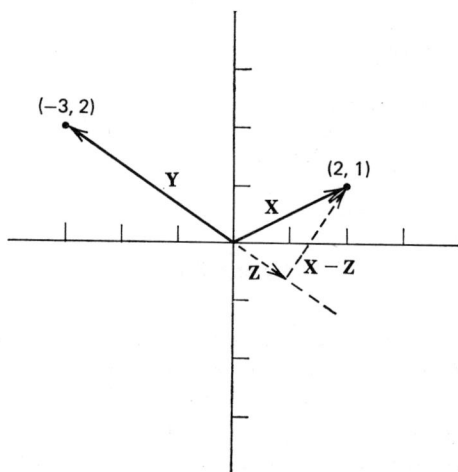

Figure 8-14

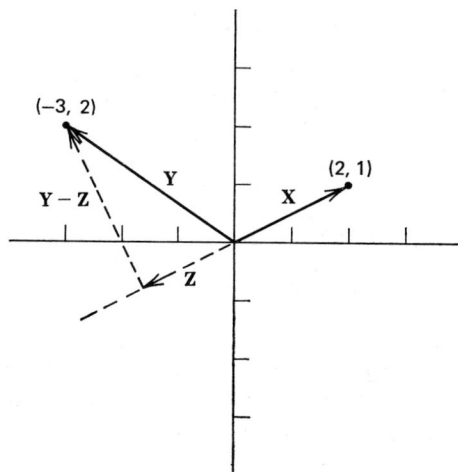

Figure 8-15

The vector $(\frac{12}{13}, -\frac{8}{13})$ is the orthogonal projection of $(2,1)$ on $(-3,2)$, and the vector $(\frac{14}{13}, \frac{21}{13})$ is perpendicular to $(-3,2)$. In Figure 8-14, the vector $\mathbf{Z} = (\frac{12}{13}, -\frac{8}{13})$ and $\mathbf{X} - \mathbf{Z} = (\frac{14}{13}, \frac{21}{13})$.

The perpendicular projection of $\mathbf{Y}$ on $\mathbf{X}$ is given by

$$\frac{(\mathbf{Y} \cdot \mathbf{X})}{|\mathbf{X}|^2} \mathbf{X} = \left(-\frac{4}{5}\right)(2,1) = \left(-\frac{8}{5}, -\frac{4}{5}\right).$$

Thus the vector **Y** can be expressed as the sum of two mutually perpendicular vectors, one parallel to **X** and the other perpendicular to **X**.

$$(-3,2) = (-\tfrac{8}{5},-\tfrac{4}{5}) + (-\tfrac{7}{5},\tfrac{14}{5}).$$

The vectors $(-\tfrac{8}{5},-\tfrac{4}{5})$ and $(-\tfrac{7}{5},\tfrac{14}{5})$ are perpendicular to each other as you can easily verify. In Figure 8-15 the vector $\mathbf{Z} = (-\tfrac{8}{5},-\tfrac{4}{5})$ and $\mathbf{Y} - \mathbf{Z} = (-\tfrac{7}{5},\tfrac{14}{5})$.

Example In $\mathbb{R}^3$ let $\mathbf{X} = (3,1,2)$ and $\mathbf{Y} = (-1,1,2)$. In this example we write **X** as the sum of a vector along **Y** and a vector perpendicular to **Y**. The orthogonal projection of **X** on **Y** is

$$\frac{(\mathbf{X}\cdot\mathbf{Y})}{|\mathbf{Y}|^2}\,\mathbf{Y} = \left(\frac{2}{6}\right)(-1,1,2) = \left(-\frac{1}{3},\frac{1}{3},\frac{2}{3}\right).$$

Thus we can express **X** as the sum

$$(3,1,2) = (-\tfrac{1}{3},\tfrac{1}{3},\tfrac{2}{3}) + (\tfrac{10}{3},\tfrac{2}{3},\tfrac{4}{3})$$

You can check easily that the vectors $(-\tfrac{1}{3},\tfrac{1}{3},\tfrac{2}{3})$ and $(\tfrac{10}{3},\tfrac{2}{3},\tfrac{4}{3})$ are perpendicular to each other. We have illustrated this in Figure 8-16.

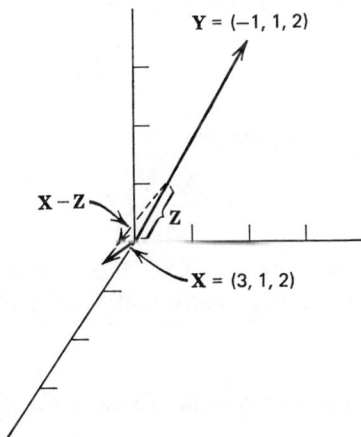

Figure 8-16

Since the orthogonal projection of **X** on **Y** for vectors in $\mathbb{R}^2$ and $\mathbb{R}^3$ can be expressed in terms of the dot product, we are led to introduce this terminology into $\mathbb{R}^n$. If **X** and **Y** are vectors in $\mathbb{R}^n$ and if **Y** is nonzero, let

$$\mathbf{Z} = \left(\frac{\mathbf{X}\cdot\mathbf{Y}}{|\mathbf{Y}|^2}\right)\mathbf{Y}.$$

Then **Z** is parallel to or lies along **Y**, and $\mathbf{X} - \mathbf{Z}$ is orthogonal to **Z**, as we verify.

$$(\mathbf{X} - \mathbf{Z}) \cdot \mathbf{Z} = \mathbf{X} \cdot \mathbf{Z} - \mathbf{Z} \cdot \mathbf{Z}$$

$$= \frac{(\mathbf{X} \cdot \mathbf{Y})^2}{|\mathbf{Y}|^2} - \frac{(\mathbf{X} \cdot \mathbf{Y})^2 (\mathbf{Y} \cdot \mathbf{Y})}{|\mathbf{Y}|^4} = 0$$

Example In $\mathbb{R}^5$ let $\mathbf{X} = (3,1,0,2,4)$ and $\mathbf{Y} = (-1,0,7,-2,3)$. In this example we express **X** as the sum of a vector parallel to **Y** and a vector perpendicular to **Y**. The orthogonal projection of **X** on **Y** is

$$\frac{(\mathbf{X} \cdot \mathbf{Y})}{|\mathbf{Y}|^2} \mathbf{Y} = \left(\frac{5}{63}\right)(-1,0,7,-2,3)$$

$$= \left(-\frac{5}{63}, 0, \frac{35}{63}, -\frac{10}{63}, \frac{15}{63}\right).$$

Consequently,

$$\mathbf{X} = (3,1,0,2,4) = \left(-\tfrac{5}{63}, 0, \tfrac{35}{63}, -\tfrac{10}{63}, \tfrac{15}{63}\right) + \left(\tfrac{194}{63}, 1, -\tfrac{35}{63}, \tfrac{136}{63}, \tfrac{237}{63}\right).$$

Proposition 8-3_____

If **X** and **Y** are vectors in $\mathbb{R}^n$, and if **Y** is nonzero, then the vector

$$\mathbf{X} - \left(\frac{\mathbf{X} \cdot \mathbf{Y}}{|\mathbf{Y}|^2}\right)\mathbf{Y}$$

is orthogonal to **Y**, and so the vector

$$\left(\frac{\mathbf{X} \cdot \mathbf{Y}}{|\mathbf{Y}|^2}\right)\mathbf{Y}$$

is called the orthogonal projection of **X** on **Y**. If the vector **Y** is a unit vector, i.e., $\mathbf{Y} \cdot \mathbf{Y} = 1$, then the orthogonal projection of **X** on **Y** is given by $(\mathbf{X} \cdot \mathbf{Y})\mathbf{Y}$.

_____

SUMMARY *Two geometric interpretations of the dot product are given in this section. First,*

$$\mathbf{X} \cdot \mathbf{Y} = |\mathbf{X}| \, |\mathbf{Y}| \cos \theta,$$

*where θ is the angle between **X** and **Y**. Or if **X** and **Y** are nonzero vectors,*

$$\cos \theta = \frac{\mathbf{X} \cdot \mathbf{Y}}{|\mathbf{X}| \, |\mathbf{Y}|}.$$

*From this, the vectors **X** and **Y** are orthogonal or perpendicular if and only if* $\mathbf{X} \cdot \mathbf{Y} = 0.$

The second geometric interpretation involves the orthogonal projection of a vector **X** on a nonzero vector **Y**. It is given by

$$\left(\frac{\mathbf{X} \cdot \mathbf{Y}}{|\mathbf{Y}|^2}\right)\mathbf{Y}.$$

If **Y** is a unit vector, then the orthogonal projection of **X** on **Y** is given by $(\mathbf{X} \cdot \mathbf{Y})\mathbf{Y}$.

Exercises for Section 8-2

1. Find the cosine of the angle between the pairs of vectors given below. Draw diagrams.

   (a) $(2,1)$ and $(-1,1)$ (c) $(5,-4)$ and $(2,3)$
   (b) $(3,2)$ and $(1,4)$ (d) $(2,5)$ and $(-5,2)$

2. Find the cosine of the angle between the given pairs of vectors.

   (a) $(2,1,3)$ and $(0,1,-\frac{1}{3})$ (c) $(3,1,5,1)$ and $(-1,1,2,1)$
   (b) $(1,5,1,3)$ and $(2,1,4,0)$ (d) $(4,1,2,1,3)$ and $(-1,2,1,2,4)$

3. Determine whether the vectors in each part are mutually orthogonal.

   (a) $(1,2)$ and $(-2,1)$ (c) $(2,5)$ and $(4,-2)$
   (b) $(3,4)$ and $(8,-6)$ (d) $(2,1,3)$, $(-1,2,0)$, $(-6,-3,5)$

4. Find the orthogonal projection of **X** upon **Y**.

   (a) $\mathbf{X} = (1,2)$ and $\mathbf{Y} = (3,1)$; draw diagram
   (b) $\mathbf{X} = (2,-1)$ and $\mathbf{Y} = (4,3)$; draw diagram
   (c) $\mathbf{X} = (1,1,2)$ and $\mathbf{Y} = (3,4,2)$
   (d) $\mathbf{X} = (2,1,1,3)$ and $\mathbf{Y} = (2,-3,2,4)$

5. In each of the following, express **X** as the sum of a vector along **Y** and a vector orthogonal to **Y**.

   (a) $\mathbf{X} = (1,2)$ and $\mathbf{Y} = (3,-2)$; draw diagram
   (b) $\mathbf{X} = (3,1)$ and $\mathbf{Y} = (-4,5)$; draw diagram
   (c) $\mathbf{X} = (-6,2)$ and $\mathbf{Y} = (2,1)$; draw diagram
   (d) $\mathbf{X} = (2,1,3)$ and $\mathbf{Y} = (1,-1,2)$
   (e) $\mathbf{X} = (5,0,6)$ and $\mathbf{Y} = (-1,2,-6)$
   (f) $\mathbf{X} = (3,1,4,2,3)$ and $\mathbf{Y} = (-1,2,1,4,-5)$

6. Find the distance from the point to the line in each part in the following way:
    (1) find a point on the line,
    (2) let **X** be the vector from the point on the line to the given point, and let **Y** be a vector along the line,
    (3) compute the orthogonal projection of **X** on **Y**; call it **Z**, and
    (4) find the length of **X** − **Z**.

    (a) $(-1,3)$ and $x + y = 7$
    (b) $(3,1)$ and $x - y = 1$
    (c) $(1,1,2)$ and the solution set for the equations $x + 2y + z = 1$ and $x - y + 2z = 1$

7. The solution set for the homogeneous system

$$2x + 3y = 0$$

consists of all vectors (x,y) which are orthogonal to $(2,3)$. Draw a diagram of the solution set.

8. The solution set for the homogeneous system

$$2x + 3y + z = 0$$
$$x - y + 2z = 0$$

consists of all vectors (x,y,z) which are orthogonal to the vectors $(2,3,1)$ and $(1,-1,2)$. (In this case, the solution set is the line through the origin which is perpendicular to the plane generated by $(2,3,1)$ and $(1,-1,2)$.) Confirm this by finding the solution set explicitly and verifying that every vector in the solution set is orthogonal to $(2,3,1)$ and $(1,-1,2)$.

9. Using the ideas of Exercises 7 and 8, argue that the solution set for a homogeneous system of equations $A\mathbf{X} = \mathbf{0}$ consists of all vectors which are orthogonal to every row of A. Relate the dimension of the solution set with the rank of A.

SECTION 8-3
Orthogonal and Orthonormal Bases

PREVIEW   *A particularly useful kind of basis for* $\mathbb{R}^n$ *is one in which each vector is orthogonal to all the others; this kind of basis is called an orthogonal basis. If, in addition, every vector in the basis is a unit vector, then the basis is called an orthonormal basis. We explain how to express a vector as a linear combination of* *vectors in an orthogonal or in an orthonormal basis. Then we show that every subspace of* $\mathbb{R}^n$ *has an orthonormal basis.*

Every vector in the standard basis for $\mathbb{R}^n$ is orthogonal to every other vector in the basis; furthermore, each vector in the standard basis is a unit vector. But the standard bases are not the only bases with these properties. Bases having these properties are particularly easy to deal with, so they have been given special names.

Definition 8-4_____

Suppose $\mathbf{X}_1, \ldots, \mathbf{X}_n$ form a basis for $\mathbb{R}^n$. They form an *orthogonal basis* if and only if each of the vectors in the basis is orthogonal to all the others; i.e., if $\mathbf{X}_i \cdot \mathbf{X}_j = 0$ for $i \neq j$. These vectors form an *orthonormal basis* if and only if each vector in the basis is orthogonal to all the others and each vector in the basis has length 1; i.e., $\mathbf{X}_i \cdot \mathbf{X}_j = 0$ if $i \neq j$ and $\mathbf{X}_i \cdot \mathbf{X}_i = 1$.

_____

The next few examples illustrate some orthonormal and some orthogonal bases.

Example The vectors $(\frac{1}{2}\sqrt{2}, \frac{1}{2}\sqrt{2})$ and $(-\frac{1}{2}\sqrt{2}, \frac{1}{2}\sqrt{2})$ form an orthonormal basis for $\mathbb{R}^2$ (see Figure 8-17). The vectors $(\frac{1}{2}\sqrt{3}, -1/2)$ and $(1/2, \frac{1}{2}\sqrt{3})$ also form an orthonormal basis for $\mathbb{R}^2$ (see Figure 8-18). In fact, if (a,b) is a nonzero vector in $\mathbb{R}^2$, then $(-b,a)$ and $(b,-a)$ are orthogonal to (a,b). Thus the vectors (a,b) and $(-b,a)$ form an orthogonal basis for $\mathbb{R}^2$ as do the vectors (a,b) and $(b,-a)$ (see Figure 8-19).

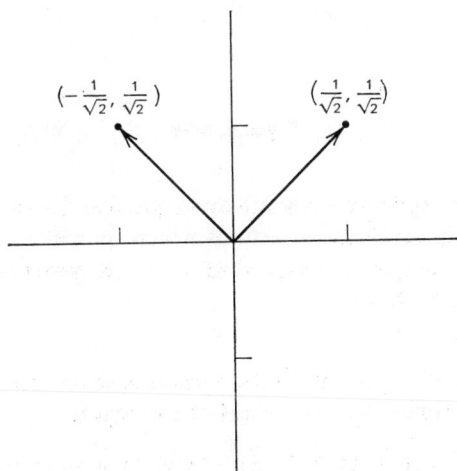

Figure 8-17

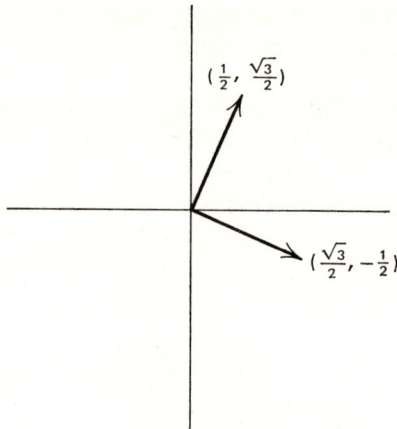

Figure 8-18

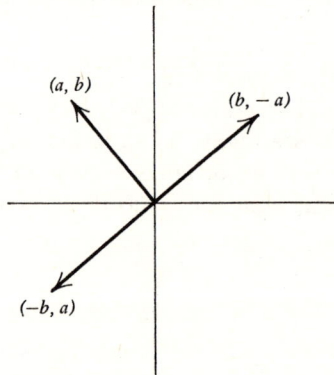

Figure 8-19

Example In this example we list some orthogonal and some orthonormal bases for $\mathbb{R}^3$. You can easily verify the assertions which we make.

The following are examples of bases which are orthogonal but not orthonormal.

(1) $(1,0,1)$, $(-1,1,1)$, $(1,2,-1)$
(2) $(1,0,0)$, $(0,1,1)$, $(0,-1,1)$

These bases are easily changed into orthonormal bases by multiplying each vector by a suitable scalar, namely the reciprocal of its length.

(1′) $(\frac{1}{2}\sqrt{2}, 0, \frac{1}{2}\sqrt{2})$, $(-\frac{1}{3}\sqrt{3}, \frac{1}{3}\sqrt{3}, \frac{1}{3}\sqrt{3})$, $(\frac{1}{6}\sqrt{6}, \frac{1}{3}\sqrt{6}, -\frac{1}{6}\sqrt{6})$
(2′) $(1,0,0)$, $(0, \frac{1}{2}\sqrt{2}, \frac{1}{2}\sqrt{2})$, $(0, -\frac{1}{2}\sqrt{2}, \frac{1}{2}\sqrt{2})$

Example   Two examples of orthogonal bases for $\mathbb{R}^4$ are
(1) $(1,0,1,1)$, $(1,0,-1,0)$, $(1,0,1,-2)$, $(0,1,0,0)$
(2) $(1,0,0,0)$, $(0,1,1,0)$, $(0,-1,1,0)$, $(0,0,0,1)$.
You can easily see what changes need to made to these vectors to obtain orthonormal bases from these orthogonal bases.

One reason for using orthonormal bases is that it is extremely easy to express a vector as a linear combination of vectors in an orthonormal basis. We illustrate this in the next two examples.

Example   Suppose we are given the orthonormal basis consisting of $(\frac{1}{5}\sqrt{5}, -\frac{2}{5}\sqrt{5})$ and $(\frac{2}{5}\sqrt{5}, \frac{1}{5}\sqrt{5})$ in $\mathbb{R}^2$, and suppose we want to express $(2,3)$ as a linear combination of the basis vectors; i.e., we want to find the scalars a and b so that

$$(2,3) = a(\tfrac{1}{5}\sqrt{5}, -\tfrac{2}{5}\sqrt{5}) + b(\tfrac{2}{5}\sqrt{5}, \tfrac{1}{5}\sqrt{5}).$$

This is quite easy to do if we use two properties of the dot product: $(\mathbf{X} + \mathbf{Y}) \cdot \mathbf{Z} = \mathbf{X} \cdot \mathbf{Z} + \mathbf{Y} \cdot \mathbf{Z}$ and $(a\mathbf{X}) \cdot \mathbf{Z} = a(\mathbf{X} \cdot \mathbf{Z})$.

$$
\begin{aligned}
(2,3) \cdot (\tfrac{1}{5}\sqrt{5}, -\tfrac{2}{5}\sqrt{5}) &= [a(\tfrac{1}{5}\sqrt{5}, -\tfrac{2}{5}\sqrt{5}) + b(\tfrac{2}{5}\sqrt{5}, \tfrac{1}{5}\sqrt{5})] \cdot (\tfrac{1}{5}\sqrt{5}, -\tfrac{2}{5}\sqrt{5}) \\
&= a(\tfrac{1}{5}\sqrt{5}, -\tfrac{2}{5}\sqrt{5}) \cdot (\tfrac{1}{5}\sqrt{5}, -\tfrac{2}{5}\sqrt{5}) \\
&\quad + b(\tfrac{2}{5}\sqrt{5}, \tfrac{1}{5}\sqrt{5}) \cdot (\tfrac{1}{5}\sqrt{5}, -\tfrac{2}{5}\sqrt{5}) \\
&= (a)(1) + (b)(0) = a
\end{aligned}
$$

Therefore, $a = -\frac{4}{5}\sqrt{5}$. Similarly,

$$
\begin{aligned}
(2,3) \cdot (\tfrac{2}{5}\sqrt{5}, \tfrac{1}{5}\sqrt{5}) &= [a(\tfrac{1}{5}\sqrt{5}, -\tfrac{2}{5}\sqrt{5}) + b(\tfrac{2}{5}\sqrt{5}, \tfrac{1}{5}\sqrt{5})] \cdot (\tfrac{2}{5}\sqrt{5}, \tfrac{1}{5}\sqrt{5}) \\
&= a(\tfrac{1}{5}\sqrt{5}, -\tfrac{2}{5}\sqrt{5}) \cdot (\tfrac{2}{5}\sqrt{5}, \tfrac{1}{5}\sqrt{5}) + b(\tfrac{2}{5}\sqrt{5}, \tfrac{1}{5}\sqrt{5}) \cdot (\tfrac{2}{5}\sqrt{5}, \tfrac{1}{5}\sqrt{5}) \\
&= (a)(0) + (b)(1) = b.
\end{aligned}
$$

Therefore, $b = \frac{7}{5}\sqrt{5}$. So

$$(2,3) = (-\tfrac{4}{5}\sqrt{5})(\tfrac{1}{5}\sqrt{5}, -\tfrac{2}{5}\sqrt{5}) + (\tfrac{7}{5}\sqrt{5})(\tfrac{2}{5}\sqrt{5}, \tfrac{1}{5}\sqrt{5}).$$

Notice that in each of these computations only one term results. The reason is that the vectors in the basis are orthogonal to each other.

Example   The three vectors $\mathbf{X}_1 = (1/\sqrt{2}, 0, 1/\sqrt{2})$, $\mathbf{X}_2 = (-1/\sqrt{3}, 1/\sqrt{3}, 1/\sqrt{3})$, and $\mathbf{X}_3 = (1/\sqrt{6}, 2/\sqrt{6}, -1/\sqrt{6})$ form an orthonormal basis for $\mathbb{R}^3$. We express $\mathbf{X} = (2,1,5)$ as a linear combination of these. To do this we need to find scalars x_1, x_2, and x_3 so that

$$\mathbf{X} = x_1\mathbf{X}_1 + x_2\mathbf{X}_2 + x_3\mathbf{X}_3.$$

We proceed as in the previous example.

$$\mathbf{X} \cdot \mathbf{X}_1 = x_1 \mathbf{X}_1 \cdot \mathbf{X}_1 + x_2 \mathbf{X}_2 \cdot \mathbf{X}_1 + x_3 \mathbf{X}_3 \cdot \mathbf{X}_1$$

Since we are dealing with an orthonormal basis, $\mathbf{X}_1 \cdot \mathbf{X}_1 = 1$, $\mathbf{X}_2 \cdot \mathbf{X}_1 = 0$, and $\mathbf{X}_3 \cdot \mathbf{X}_1 = 0$. Therefore,

$$\mathbf{X} \cdot \mathbf{X}_1 = x_1.$$

So, in this case,

$$x_1 = (2,1,5) \cdot (\tfrac{1}{2}\sqrt{2}, 0, \tfrac{1}{2}\sqrt{2}) = \tfrac{7}{2}\sqrt{2}.$$

Similarly,

$$\mathbf{X} \cdot \mathbf{X}_2 = x_1 \mathbf{X}_1 \cdot \mathbf{X}_2 + x_2 \mathbf{X}_2 \cdot \mathbf{X}_2 + x_3 \mathbf{X}_3 \cdot \mathbf{X}_2 = x_2$$

and

$$\mathbf{X} \cdot \mathbf{X}_3 = x_1 \mathbf{X}_1 \cdot \mathbf{X}_3 + x_2 \mathbf{X}_2 \cdot \mathbf{X}_3 + x_3 \mathbf{X}_3 \cdot \mathbf{X}_3 = x_3.$$

Therefore,

$$x_2 = (2,1,5) \cdot (-\tfrac{1}{3}\sqrt{3}, \tfrac{1}{3}\sqrt{3}, \tfrac{1}{3}\sqrt{3}) = \tfrac{4}{3}\sqrt{3}$$

and

$$x_3 = (2,1,5) \cdot (\tfrac{1}{6}\sqrt{6}, \tfrac{1}{3}\sqrt{6}, -\tfrac{1}{6}\sqrt{6}) = -\tfrac{1}{6}\sqrt{6}.$$

Consequently,

$$(2,1,5) = \tfrac{7}{2}\sqrt{2}\,(\tfrac{1}{2}\sqrt{2}, 0, \tfrac{1}{2}\sqrt{2}) + \tfrac{4}{3}\sqrt{3}\,(-\tfrac{1}{3}\sqrt{3}, \tfrac{1}{3}\sqrt{3}, \tfrac{1}{3}\sqrt{3})$$
$$- \tfrac{1}{6}\sqrt{6}\,(\tfrac{1}{6}\sqrt{6}, \tfrac{1}{3}\sqrt{6}, -\tfrac{1}{6}\sqrt{6}).$$

These two examples show that it is relatively easy to express any vector $\mathbf{X}$ in $\mathbb{R}^n$ as a linear combination of vectors in an orthonormal basis consisting of $\mathbf{X}_1, \ldots, \mathbf{X}_n$. If

$$\mathbf{X} = x_1 \mathbf{X}_1 + \cdots + x_n \mathbf{X}_n,$$

then

$$\mathbf{X} \cdot \mathbf{X}_1 = x_1 \mathbf{X}_1 \mathbf{X}_1 + \cdots + x_n \mathbf{X}_n \mathbf{X}_1 = x_1$$

This equation follows from the fact that $\mathbf{X}_1, \ldots, \mathbf{X}_n$ form an orthonormal basis. The vector $\mathbf{X}_1$ is orthogonal to all other vectors in the basis, i.e., $\mathbf{X}_i \cdot \mathbf{X}_1 = 0$ if $i \neq 1$, and $\mathbf{X}_1$ is a unit vector, i.e., $\mathbf{X}_1 \cdot \mathbf{X}_1 = 1$. Similarly,

$$\mathbf{X} \cdot \mathbf{X}_2 = x_1 \mathbf{X}_1 \cdot \mathbf{X}_2 + \cdots + x_n \mathbf{X}_n \cdot \mathbf{X}_2 = x_2,$$

and so on. Therefore we can express $\mathbf{X}$ as follows.

$$\mathbf{X} = (\mathbf{X} \cdot \mathbf{X}_1)\mathbf{X}_1 + \cdots + (\mathbf{X} \cdot \mathbf{X}_n)\mathbf{X}_n$$

Since $\mathbf{X}_1, \ldots, \mathbf{X}_n$ are unit vectors, the terms $(\mathbf{X} \cdot \mathbf{X}_i)\mathbf{X}_i$ represent the orthogonal projection of $\mathbf{X}$ upon $\mathbf{X}_i$ (see Proposition 8-3).

Proposition 8-4

If $X_1, \ldots, X_n$ form an orthonormal basis for $\mathbb{R}^n$, and if X is a vector in $\mathbb{R}^n$, then

$$X = (X \cdot X_1)X_1 + \cdots + (X \cdot X_n)X_n.$$

This expresses X as the sum of its orthogonal projections along each vector in the orthonormal basis.

Brief Exercises

1. The vectors given form an orthogonal basis for $\mathbb{R}^2$. By dividing each vector by its length, find an orthonormal basis for $\mathbb{R}^2$.

   (a) $(1,1)$, $(-1,1)$ (b) $(3,5)$, $(5,-3)$

2. Write the following vectors as linear combinations of each of the orthonormal bases in Exercise 1.
   $$(1,2), \quad (-1,1)$$

3. The vectors given form an orthogonal basis for $\mathbb{R}^3$. By dividing each vector by its length, find an orthonormal basis for $\mathbb{R}^3$.

   (a) $(2,0,0)$, $(0,3,0)$, $(0,0,2)$ (b) $(1,1,0)$, $(-1,1,0)$, $(0,0,2)$

4. Write the following vectors as linear combinations of each of the orthonormal bases in Exercise 3.
   $$(1,2,1), \quad (5,1,3)$$

Selected Answers: **1(a)** $(1/\sqrt{2}, 1/\sqrt{2})$, $(-1/\sqrt{2}, 1/\sqrt{2})$ **1(b)** $(3/\sqrt{34}, 5/\sqrt{34})$, $(5/\sqrt{34}, -3/\sqrt{34})$ **2** $(1,2) = (3/\sqrt{2})(1/\sqrt{2}, 1/\sqrt{2}) + (1/\sqrt{2})(-1/\sqrt{2}, 1/\sqrt{2})$. $(-1,1) = (2/\sqrt{34})(3/\sqrt{34}, 5/\sqrt{34}) - (8/\sqrt{34})(5/\sqrt{34}, -3/\sqrt{34})$ **3(a)** $(1,0,0)$, $(0,1,0)$, $(0,0,1)$ **3(b)** $(1/\sqrt{2}, 1/\sqrt{2}, 0)$, $(-1/\sqrt{2}, 1/\sqrt{2}, 0)$, $(0,0,1)$ **4** $(1,2,1) = (3/\sqrt{2})(1/\sqrt{2}, 1/\sqrt{2}, 0) + (1/\sqrt{2})(-1/\sqrt{2}, 1/\sqrt{2}, 0) + 1(0,0,1)$. $(5,1,3) = (6/\sqrt{2})(1/\sqrt{2}, 1/\sqrt{2}, 0) - (4/\sqrt{2})(-1/\sqrt{2}, 1/\sqrt{2}, 0) + 3(0,0,1)$

The final topic for this section is an explanation of how to construct an orthonormal basis for a subspace of $\mathbb{R}^n$. The first procedure we illustrate is particularly useful if the subspace is described by a system of homogeneous equations or if we want the orthogonal basis to contain some given mutually orthogonal vectors. It is based on the observation that the vectors $(x_1, \ldots, x_n)$ which are orthogonal to a given vector $(a_1, \ldots, a_n)$ must satisfy the equation

$$a_1 x_1 + \cdots + a_n x_n = 0,$$

i.e., the dot product of $(x_1, \ldots, x_n)$ and $(a_1, \ldots, a_n)$ is zero. This procedure gives an orthogonal basis; after that is constructed we divide each vector by its length to obtain an orthonormal basis.

Example Suppose we want to construct an orthogonal basis for $\mathbb{R}^3$. We can start with any vector, $(1,2,1)$, for example. The vectors orthogonal to $(1,2,1)$ are the vectors (x,y,z) satisfying

$$x + 2y + z = 0.$$

That is, the dot product of $(1,2,1)$ and (x,y,z) must be zero since they are to be orthogonal. Pick any nonzero vector which is a solution of this equation, say $(1,0,-1)$.

A third vector (x,y,z) in an orthogonal basis containing $(1,2,1)$ and $(1,0,-1)$ must satisfy the equations

$$x + 2y + z = 0$$
$$x \quad\quad - z = 0,$$

i.e., the dot product of (x,y,z) with $(1,2,1)$ and with $(1,0,-1)$ must be zero since they are to be orthogonal. The row echelon form for the coefficient matrix is

$$\begin{pmatrix} 1 & 0 & -1 \\ 0 & 1 & 1 \end{pmatrix}.$$

A vector orthogonal to $(1,2,1)$ and $(1,0,-1)$ must therefore satisfy the equations

$$x = z$$
$$y = -z.$$

Pick any nonzero vector which satisfies the equations, e.g., $(1,-1,1)$. Thus the vectors $(1,2,1)$, $(1,0,-1)$, and $(1,-1,1)$ form an orthogonal basis for $\mathbb{R}^3$. We can obtain an orthonormal basis by normalizing each vector in the orthogonal basis.

$$(\tfrac{1}{6}\sqrt{6}, \tfrac{1}{3}\sqrt{6}, \tfrac{1}{6}\sqrt{6}), (\tfrac{1}{2}\sqrt{2}, 0, -\tfrac{1}{2}\sqrt{2}), (\tfrac{1}{3}\sqrt{3}, -\tfrac{1}{3}\sqrt{3}, \tfrac{1}{3}\sqrt{3})$$

In this context it is customary to use the word "normalize" to mean "divide a vector by its length to obtain a unit vector." We use this terminology. Consequently, the procedure we describe here yields an orthogonal basis; we can obtain an orthonormal basis by normalizing the vectors in the orthogonal basis.

Example In this example we construct an orthogonal basis for $\mathbb{R}^4$ starting with $(3,1,0,2)$. Any vector (x,y,z,w) which is orthogonal to $(3,1,0,2)$ must satisfy the equation

$$3x + y + 2w = 0.$$

That is, the dot product of $(3,1,0,2)$ and (x,y,z,w) must be zero. Pick any nonzero vector which is a solution for this equation; say we choose $(0,-2,0,1)$.

The next step is to find a vector (x,y,z,w) which is orthogonal to $(3,1,0,2)$ and to $(0,-2,0,1)$; such a vector must satisfy the equations

$$3x + y \qquad + 2w = 0$$
$$- 2y \qquad + w = 0.$$

Transform the coefficient matrix to row echelon form.

$$\begin{pmatrix} 1 & 0 & 0 & \frac{5}{6} \\ 0 & 1 & 0 & -\frac{1}{2} \end{pmatrix}$$

It is easy to read off a nonzero solution, for example, $(-5,3,0,6)$.

To find a fourth vector in the orthogonal basis, we need a vector (x,y,z,w) which is orthogonal to $(1,0,0,\frac{5}{6})$, $(0,1,0,-\frac{1}{2})$, and $(-5,3,0,6)$.—Remember: row operations do not change the row space of a matrix (Proposition 7-2). Therefore, we can either seek a vector orthogonal to $(3,1,0,2)$ and $(0,-2,0,1)$ or we can seek one which is orthogonal to $(1,0,0,\frac{5}{6})$ and $(0,1,0,-\frac{1}{2})$; the computations are easier if we use the vectors which give a matrix in row echelon form.—A vector (x,y,z,w) which is orthogonal to $(1,0,0,\frac{5}{6})$, $(0,1,0,-\frac{1}{2})$, and $(-5,3,0,6)$ must satisfy the equations

$$x \qquad + \tfrac{5}{6}w = 0$$
$$y - \tfrac{1}{2}w = 0$$
$$-5x + 3y + 6w = 0.$$

The row echelon form for the coefficient matrix is

$$\begin{pmatrix} 1 & 0 & 0 & 0 \\ 0 & 1 & 0 & 0 \\ 0 & 0 & 0 & 1 \end{pmatrix}.$$

From this matrix it is easy to see that $(0,0,1,0)$ is a solution.

Therefore, we have constructed the orthogonal basis consisting of $(3,1,0,2)$, $(0,-2,0,1)$, $(-5,3,0,6)$, and $(0,0,1,0)$. If we want an orthonormal basis, we can normalize each of these vectors.

Example In this example we show how to construct an orthogonal basis for the subspace $\mathscr{W}$ of $\mathbb{R}^4$ consisting of the vectors (x,y,z,w) which satisfy $x - y + z - w = 0$. This subspace is the set of all vectors (x,y,z,w) which are orthogonal to $(1,-1,1,-1)$.

Pick any vector in $\mathscr{W}$, e.g., $(1,0,0,1)$. To find a second vector in the orthogonal basis for $\mathscr{W}$, we must determine a vector (x,y,z,w) in $\mathscr{W}$ which is orthogonal to

$(1,0,0,1)$; i.e., a vector (x,y,z,w) which satisfies the equations

$$x - y + z - w = 0$$
$$x \qquad\qquad + w = 0.$$

The row echelon form for the coefficient matrix is

$$\begin{pmatrix} 1 & 0 & 0 & 1 \\ 0 & 1 & -1 & 2 \end{pmatrix}.$$

A nonzero solution can be read off from this matrix, e.g., $(-1,-1,1,1)$. We now have two vectors in an orthogonal basis for $\mathcal{W}$: $(1,0,0,1)$ and $(-1,-1,1,1)$.

To find a third vector in an orthogonal basis, we must determine a vector (x,y,z,w) in $\mathcal{W}$ which is orthogonal to $(1,0,0,1)$ and $(-1,-1,1,1)$, i.e., a vector satisfying the equations

$$x - y + z - w = 0$$
$$x \qquad\qquad + w = 0$$
$$-x - y + z + w = 0.$$

Transform the coefficient matrix to row echelon form.

$$\begin{pmatrix} 1 & 0 & 0 & 0 \\ 0 & 1 & -1 & 0 \\ 0 & 0 & 0 & 1 \end{pmatrix}$$

A solution can be read off from the matrix, e.g., $(0,1,1,0)$.

Therefore, we have determined an orthogonal basis for consisting of $(1,0,0,1)$, $(-1,-1,1,1)$, and $(0,1,1,0)$. If an orthonormal basis is desired, we need only normalize each of these vectors and obtain

$$(\tfrac{1}{2}\sqrt{2}, 0, 0, \tfrac{1}{2}\sqrt{2}), \; (-\tfrac{1}{2}, -\tfrac{1}{2}, \tfrac{1}{2}, \tfrac{1}{2}), \; (0, \tfrac{1}{2}\sqrt{2}, \tfrac{1}{2}\sqrt{2}, 0).$$

These examples illustrate the general procedures we can follow. Suppose we wish to find an orthonormal basis for a subspace $\mathcal{W}$ in $\mathbb{R}^n$ which is described as the solution set for a homogeneous system of linear equations

$$a_{11}x_1 + \cdots + a_{1n}x_n = 0$$
$$\cdot \qquad\qquad \cdot \qquad \cdot$$
$$\cdot \qquad\qquad \cdot \qquad \cdot$$
$$\cdot \qquad\qquad \cdot \qquad \cdot$$
$$a_{m1}x_1 + \cdots + a_{mn}x_n = 0.$$

If we let $A_1, \ldots, A_m$ denote the rows of the coefficient matrix A for this system of linear equations, then we can write the system of equations as $A_1 \cdot X = 0, \ldots,$ $A_m \cdot X = 0$, where $X = (x_1, \ldots, x_n)$. Geometrically, $\mathscr{W}$ is the set of all vectors X which are orthogonal to the vectors $A_1, \ldots, A_m$.

Transform the coefficient matrix A to its row echelon form B, and label the nonzero rows of B by $B_1, \ldots, B_r$. Then $\mathscr{W}$ is still the set of vectors X for which $B_1 \cdot X = 0, \ldots, B_r \cdot X = 0$, because the rows of A and the rows of B generate the same subspace (Proposition 7-2). Find a nonzero vector X_1 which satisfies these equations. This is the first vector in the orthogonal basis for $\mathscr{W}$.

The second vector in the orthogonal basis for $\mathscr{W}$ must satisfy the equations

$$A_1 \cdot X = 0, \ldots, A_m \cdot X = 0$$

or equivalently the equations

$$B_1 \cdot X = 0, \ldots, B_r \cdot X = 0,$$

and it must be orthogonal to X_1, i.e., $X_1 \cdot X = 0$. So we find a nonzero solution for the system of equations whose coefficient matrix has rows $B_1, \ldots, B_r$, and X_1. Label this vector by X_2.

The third vector in the orthogonal basis for $\mathscr{W}$ must be orthogonal to $B_1, \ldots, B_r$, X_1, and X_2. Therefore we must find a nonzero solution to the homogeneous system of equations whose coefficient matrix has $B_1, \ldots, B_r$, X_1, and X_2 for its rows.

We continue until there are no further nonzero solutions, i.e., until the B's and the X's form a basis for $\mathbb{R}^n$. The important thing to note about the procedure is this: as we construct the orthogonal basis, each vector becomes a new row in the coefficient matrix used to determine the next vector in the basis. The reason is that the vectors in the orthogonal basis yet to be determined at any stage must be orthogonal to those already found.

Brief Exercises

1. Find an orthogonal basis for $\mathbb{R}^3$ containing $(1,0,1)$.

2. Find an orthogonal basis for $\mathbb{R}^3$ containing $(-1,2,1)$.

3. Find an orthogonal basis for $\mathbb{R}^3$ containing $(1,0,1)$ and $(1,1,-1)$.

4. Find an orthogonal basis for $\mathbb{R}^3$ containing $(1,1,1)$ and $(2,-1,-1)$.

5. Find an orthogonal basis for $\mathbb{R}^4$ containing $(1,1,0,1)$.

6. Find an orthogonal basis for the subspace $\mathscr{W}$ of $\mathbb{R}^4$ consisting of vectors (x,y,z,w) which satisfy the equations

$$2x - y + z - w = 0$$
$$x + y \quad\ + w = 0.$$

The second procedure we present for finding an orthogonal basis for a subspace $\mathcal{W}$ is useful if $\mathcal{W}$ is described as the subspace generated by some vectors. Since a generating set for a vector space always contains a basis (Proposition 3-5), we assume that $\mathcal{W}$ is described as the subspace having $X_1, \ldots, X_m$ as a basis. The process we describe is the Gram-Schmidt process. It depends upon the idea of the orthogonal projection of a vector on a subspace; this idea is closely related to the idea of the orthogonal projection of one vector on another. In the next example we illustrate this idea in three-dimensional space.

Example Suppose Y is a vector in three-dimensional space and X_1 and X_2 are mutually orthogonal unit vectors. We picture these vectors as all beginning at the origin (see Figure 8-20). The orthogonal projection of Y on X_1 is $(Y \cdot X_1)X_1$, and

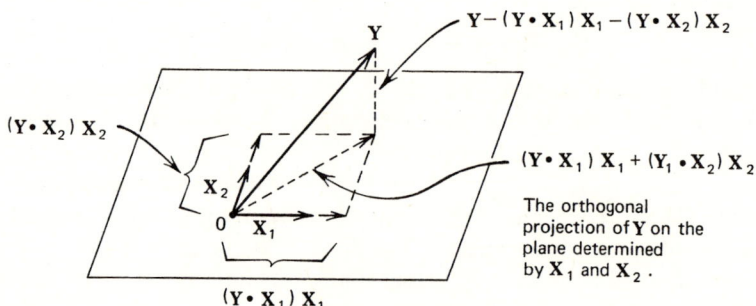

Figure 8-20

the orthogonal projection of Y on X_2 is $(Y \cdot X_2)X_2$. Then the vector $(Y \cdot X_1)X_1 + (Y \cdot X_2)X_2$ is the orthogonal projection of Y on the plane determined by X_1 and X_2 because

$$Y - [(Y \cdot X_1)X_1 + (Y \cdot X_2)X_2]$$

is orthogonal to X_1 and to X_2 and consequently to every vector in the plane determined by X_1 and X_2. This vector is orthogonal to X_1:

$$[Y - (Y \cdot X_1)X_1 - (Y \cdot X_2)X_2] \cdot X_1$$
$$= Y \cdot X_1 - (Y \cdot X_1)(X_1 \cdot X_1) - (Y \cdot X_2)(X_2 \cdot X_1)$$
$$= Y \cdot X_1 - Y \cdot X_1 = 0$$

because X_1 is a unit vector, i.e., $X_1 \cdot X_1 = 1$, and because X_1 and X_2 are mutually orthogonal, i.e., $X_2 \cdot X_1 = 0$. The vector is also orthogonal to X_2:

$$[Y - (Y \cdot X_1)X_1 - (Y \cdot X_2)X_2] \cdot X_2$$
$$= Y \cdot X_2 - (Y \cdot X_1)(X_1 \cdot X_2) - (Y \cdot X_2)(X_2 \cdot X_2)$$
$$= Y \cdot X_2 - Y \cdot X_2 = 0$$

because $\mathbf{X}_1$ and $\mathbf{X}_2$ are mutually orthogonal, i.e., $\mathbf{X}_1 \cdot \mathbf{X}_2 = 0$, and because $\mathbf{X}_2$ is a unit vector, i.e., $\mathbf{X}_2 \cdot \mathbf{X}_2 = 1$.

Because of these calculations we call the vector

$$(\mathbf{Y} \cdot \mathbf{X}_1)\mathbf{X}_1 + (\mathbf{Y} \cdot \mathbf{X}_2)\mathbf{X}_2$$

the orthogonal projection of $\mathbf{Y}$ on the plane determined by $\mathbf{X}_1$ and $\mathbf{X}_2$. Notice that this formula works because $\mathbf{X}_1$ and $\mathbf{X}_2$ are mutually orthogonal unit vectors.

This example gives an indication of the general situation. Suppose $\mathbf{X}_1, \ldots, \mathbf{X}_k$ are mutually orthogonal unit vectors in $\mathbb{R}^n$ and that $\mathbf{Y}$ is a vector in $\mathbb{R}^n$. Then the vector

$$\mathbf{Y} - (\mathbf{Y} \cdot \mathbf{X}_1)\mathbf{X}_1 - \cdots - (\mathbf{Y} \cdot \mathbf{X}_k)\mathbf{X}_k$$

is orthogonal to each one of the vectors $\mathbf{X}_1, \ldots, \mathbf{X}_k$. To see this, suppose $\mathbf{X}_i$ is one of the vectors $\mathbf{X}_1, \ldots, \mathbf{X}_k$. Then

$$[\mathbf{Y} - (\mathbf{Y} \cdot \mathbf{X}_1)\mathbf{X}_1 - \cdots - (\mathbf{Y} \cdot \mathbf{X}_k)\mathbf{X}_k] \cdot \mathbf{X}_i$$

$$= \mathbf{Y} \cdot \mathbf{X}_i - (\mathbf{Y} \cdot \mathbf{X}_1)(\mathbf{X}_1 \cdot \mathbf{X}_i) - \cdots - (\mathbf{Y} \cdot \mathbf{X}_k)(\mathbf{X}_k \cdot \mathbf{X}_i)$$

$$= \mathbf{Y} \cdot \mathbf{X}_i - \mathbf{Y} \cdot \mathbf{X}_i = 0$$

because $\mathbf{X}_i$ is a unit vector, i.e., $\mathbf{X}_i \cdot \mathbf{X}_i = 1$, and because $\mathbf{X}_i$ is orthogonal to every other $\mathbf{X}$, i.e., $\mathbf{X}_j \cdot \mathbf{X}_i = 0$ if $j \neq i$.

Proposition 8-5_____

If $\mathbf{X}_1, \ldots, \mathbf{X}_k$ are k mutually orthogonal unit vectors in $\mathbb{R}^n$ and $\mathbf{Y}$ is a vector in $\mathbb{R}^n$, then

$$\mathbf{Y} - (\mathbf{Y} \cdot \mathbf{X}_1)\mathbf{X}_1 - \cdots - (\mathbf{Y} \cdot \mathbf{X}_k)\mathbf{X}_k$$

is orthogonal to each of the vectors $\mathbf{X}_1, \ldots, \mathbf{X}_k$.

_____

Because of this result, we make the following definition.

Definition 8-5_____

If $\mathbf{X}_1, \ldots, \mathbf{X}_k$ are k mutually orthogonal unit vectors in $\mathbb{R}^n$ and $\mathbf{Y}$ is a vector in $\mathbb{R}^n$, then

$$(\mathbf{Y} \cdot \mathbf{X}_1)\mathbf{X}_1 + \cdots + (\mathbf{Y} \cdot \mathbf{X}_k)\mathbf{X}_k$$

is called the *orthogonal projection of $\mathbf{Y}$ on the subspace generated by* $\mathbf{X}_1, \ldots, \mathbf{X}_k$. The vector obtained by subtracting from $\mathbf{Y}$ its orthogonal projection on the subspace generated by $\mathbf{X}_1, \ldots, \mathbf{X}_k$ is orthogonal to each of the vectors $\mathbf{X}_1, \ldots, \mathbf{X}_k$.

_____

Another convenient fact about nonzero orthogonal vectors is that they form a linearly independent set. To see this, suppose that $\mathbf{X}_1, \ldots, \mathbf{X}_k$ are nonzero orthogonal vectors, and suppose

$$a_1\mathbf{X}_1 + \cdots + a_k\mathbf{X}_k = \mathbf{0}.$$

If $\mathbf{X}_i$ is any of the vectors $\mathbf{X}_1, \ldots, \mathbf{X}_k$, then

$$(a_1\mathbf{X}_1 + \cdots + a_k\mathbf{X}_k) \cdot \mathbf{X}_i = a_i(\mathbf{X}_i \cdot \mathbf{X}_i) = \mathbf{0} \cdot \mathbf{X}_i = 0.$$

Since $\mathbf{X}_i$ is a nonzero vector, $\mathbf{X}_i \cdot \mathbf{X}_i \neq 0$; consequently $a_i = 0$.

Proposition 8-6

If $\mathbf{X}_1, \ldots, \mathbf{X}_k$ are k nonzero mutually orthogonal vectors in $\mathbb{R}^n$, then they form a linearly independent set.

Now that we have the facts stated in Propositions 8-5 and 8-6, we can show that any subspace of $\mathbb{R}^n$ has an orthonormal basis. We do this by starting with a basis for the subspace (Proposition 3-11) and use a process called the Gram-Schmidt process to produce an orthonormal basis from the given basis. We illustrate the process in the next two examples.

Example The vectors $(1,2)$ and $(-2,3)$ form a basis for $\mathbb{R}^2$. Pick one of the vectors to begin the process; say we pick $(1,2)$. Normalize this vector to obtain the first vector in the orthonormal basis.

$$\mathbf{X}_1 = (\tfrac{1}{5}\sqrt{5}, \tfrac{2}{5}\sqrt{5})$$

The second step is to subtract from $(-2,3)$ its orthogonal projection on $\mathbf{X}_1$.

$$(-2,3) - [(-2,3) \cdot (\tfrac{1}{5}\sqrt{5}, \tfrac{2}{5}\sqrt{5})](\tfrac{1}{5}\sqrt{5}, \tfrac{2}{5}\sqrt{5}) = (-\tfrac{14}{5}, \tfrac{7}{5})$$

Normalize this vector to obtain the second vector in the orthonormal basis.

$$\mathbf{X}_2 = (-\tfrac{2}{5}\sqrt{5}, \tfrac{1}{5}\sqrt{5})$$

Although in the case of a basis in $\mathbb{R}^2$ you might have been able to figure out what $\mathbf{X}_2$ should be geometrically, we have gone through the arithmetic process because it is the process we can use in higher-dimensional spaces.

Example The vectors $(1,1,0)$, $(1,0,1)$, and $(0,1,1)$ form a basis for $\mathbb{R}^3$. We illustrate the Gram-Schmidt process for producing an orthonormal basis from these vectors.

Pick one vector, say $(1,1,0)$, and normalize it to obtain the first vector in the orthonormal basis.

$$\mathbf{X}_1 = (\tfrac{1}{2}\sqrt{2}, \tfrac{1}{2}\sqrt{2}, 0)$$

For the second step, subtract from the second vector, say $(1,0,1)$, its orthogonal projection on $\mathbf{X}_1$.

$$(1,0,1) - (\tfrac{1}{2}\sqrt{2})(\tfrac{1}{2}\sqrt{2}, \tfrac{1}{2}\sqrt{2}, 0) = (\tfrac{1}{2}, -\tfrac{1}{2}, 1)$$

Normalize this vector to obtain the second vector in the orthonormal basis.

$$\mathbf{X}_2 = (\tfrac{1}{6}\sqrt{6}, -\tfrac{1}{6}\sqrt{6}, \tfrac{1}{3}\sqrt{6})$$

For the third step, subtract from the third vector $(0,1,1)$ its orthogonal projection on the subspace determined by $\mathbf{X}_1$ and $\mathbf{X}_2$.

$$(0,1,1) - (\tfrac{1}{2}\sqrt{2})(\tfrac{1}{2}\sqrt{2}, \tfrac{1}{2}\sqrt{2}, 0) - (\tfrac{1}{6}\sqrt{6})(\tfrac{1}{6}\sqrt{6}, -\tfrac{1}{6}\sqrt{6}, \tfrac{1}{3}\sqrt{6}) = (-\tfrac{2}{3}, \tfrac{2}{3}, \tfrac{2}{3})$$

Normalize this vector to obtain the third vector in the orthonormal basis.

$$\mathbf{X}_3 = (-\tfrac{1}{3}\sqrt{3}, \tfrac{1}{3}\sqrt{3}, \tfrac{1}{3}\sqrt{3})$$

These two examples illustrate the general procedure for producing an orthonormal basis from a given basis. In the general discussion, let $\mathbf{Y}_1, \ldots, \mathbf{Y}_m$ denote the vectors forming the basis for a subspace $\mathscr{W}$ of $\mathbb{R}^n$, and we use $\mathbf{X}_1, \ldots, \mathbf{X}_m$ for the vectors in the orthonormal basis we construct from the $\mathbf{Y}$'s.

The first step is to normalize $\mathbf{Y}_1$.

$$\mathbf{X}_1 = \frac{\mathbf{Y}_1}{|\mathbf{Y}_1|}$$

The second step is to find the orthogonal projection of $\mathbf{Y}_2$ on $\mathbf{X}_1$, subtract it from $\mathbf{Y}_2$, and then normalize (see Figure 8-21).

$$\mathbf{X}_2 = \frac{\mathbf{Y}_2 - (\mathbf{Y}_2 \cdot \mathbf{X}_1)\mathbf{X}_1}{|\mathbf{Y}_2 - (\mathbf{Y}_2 \cdot \mathbf{X}_1)\mathbf{X}_1|}$$

The third step is to find the orthogonal projection of $\mathbf{Y}_3$ on the subspace generated by $\mathbf{X}_1$ and $\mathbf{X}_2$, subtract it from $\mathbf{Y}_3$, and normalize (see Figure 8-22).

$$\mathbf{X}_3 = \frac{\mathbf{Y}_3 - (\mathbf{Y}_3 \cdot \mathbf{X}_1)\mathbf{X}_1 - (\mathbf{Y}_3 \cdot \mathbf{X}_2)\mathbf{X}_2}{|\mathbf{Y}_3 - (\mathbf{Y}_3 \cdot \mathbf{X}_1)\mathbf{X}_1 - (\mathbf{Y}_3 \cdot \mathbf{X}_2)\mathbf{X}_2|}$$

Continue in this fashion until all the $\mathbf{Y}$'s have been used.

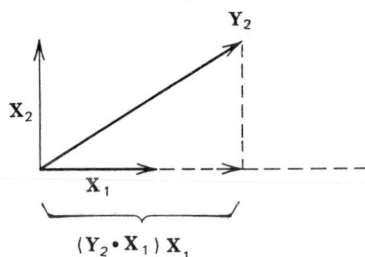

Figure 8-21

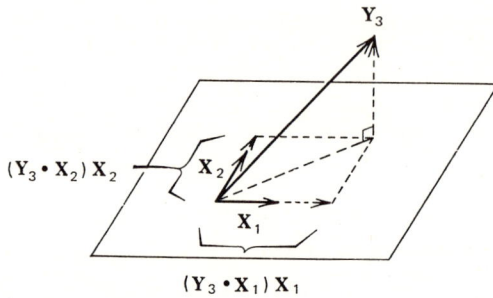

Figure 8-22

We can describe every step after the first as follows: we have some mutually orthogonal unit vectors $\mathbf{X}_1, \ldots, \mathbf{X}_k$. The next step is to find the orthogonal projection of $\mathbf{Y}_{k+1}$ on the subspace generated by $\mathbf{X}_1, \ldots, \mathbf{X}_k$, subtract it from $\mathbf{Y}_{k+1}$, and then normalize.

$$\mathbf{X}_{k+1} = \frac{\mathbf{Y}_{k+1} - (\mathbf{Y}_{k+1} \cdot \mathbf{X}_1)\mathbf{X}_1 - \cdots - (\mathbf{Y}_{k+1} \cdot \mathbf{X}_k)\mathbf{X}_k}{|\mathbf{Y}_{k+1} - (\mathbf{Y}_{k+1} \cdot \mathbf{X}_1)\mathbf{X}_1 - \cdots - (\mathbf{Y}_{k+1} \cdot \mathbf{X}_k)\mathbf{X}_k|}$$

One thing to notice about this process is that at each stage the $\mathbf{X}$'s that have been constructed generate the same subspace as the $\mathbf{Y}$'s that have been used. This is clearly the case with $\mathbf{X}_1$ and $\mathbf{Y}_1$ since they are scalar multiples of each other. For the steps past the first, suppose that we already know that $\mathbf{X}_1, \ldots, \mathbf{X}_k$ and $\mathbf{Y}_1, \ldots, \mathbf{Y}_k$ generate the same subspace. Then, from the equation above, it is easy to see that $\mathbf{X}_{k+1}$ is a linear combination of $\mathbf{X}_1, \ldots, \mathbf{X}_k, \mathbf{Y}_{k+1}$. Consequently $\mathbf{X}_1, \ldots, \mathbf{X}_k, \mathbf{X}_{k+1}$ can all be expressed as linear combinations of $\mathbf{Y}_1, \ldots, \mathbf{Y}_k,$ $\mathbf{Y}_{k+1}$. On the other hand, by solving for $\mathbf{Y}_{k+1}$ in the equation above, you can see that $\mathbf{Y}_{k+1}$ is a linear combination of $\mathbf{X}_1, \ldots, \mathbf{X}_k, \mathbf{X}_{k+1}$. Therefore $\mathbf{X}_1, \ldots, \mathbf{X}_{k+1}$ and $\mathbf{Y}_1, \ldots, \mathbf{Y}_{k+1}$ generate the same subspace.

Because $\mathbf{X}_1, \ldots, \mathbf{X}_k$ and $\mathbf{Y}_1, \ldots, \mathbf{Y}_k$ generate the same subspace, $\mathbf{Y}_{k+1}$ is not a linear combination of $\mathbf{X}_1, \ldots, \mathbf{X}_k$.—It cannot be a linear combination of $\mathbf{Y}_1, \ldots, \mathbf{Y}_k$ since the $\mathbf{Y}$'s form a linearly independent set.—Therefore, the vector

$$\mathbf{Y}_{k+1} - (\mathbf{Y}_{k+1} \cdot \mathbf{X}_1)\mathbf{X}_1 - \cdots - (\mathbf{Y}_{k+1} \cdot \mathbf{X}_k)\mathbf{X}_k$$

is not zero and so we can always normalize.

Proposition 8-7

Every subspace of $\mathbb{R}^n$ has an orthonormal basis.

SUMMARY *In this section we deal with orthogonal and orthonormal bases for subspaces of* $\mathbb{R}^n$. *A basis is an orthogonal basis if the vectors in it are mutually orthogonal. The basis is an orthonormal basis if the vectors are mutually orthonormal unit vectors.*

If $X_1, \ldots, X_n$ *form an orthonormal basis for* $\mathbb{R}^n$ *and if* **X** *is a vector in* $\mathbb{R}^n$, *then* **X** *can be expressed by*

$$X = (X \cdot X_1)X_1 + \cdots + (X \cdot X_n)X_n.$$

Mutually orthogonal unit vectors have pleasant properties. If $X_1, \ldots, X_k$ *are k mutually orthogonal unit vectors and* **Y** *is in* $\mathbb{R}^n$, *then the vector*

$$Y - (Y \cdot X_1)X_1 - \cdots - (Y \cdot X_k)X_k$$

is orthogonal to the vectors $X_1, \ldots, X_k$, *and therefore*

$$(Y \cdot X_1)X_1 + \cdots + (Y \cdot X_k)X_k$$

is the orthogonal projection of **Y** *on the subspace generated by* $X_1, \ldots, X_k$. *Furthermore, a set of mutually orthogonal vectors is a linearly independent set.*

Finally we show that every subspace $\mathscr{W}$ *of* $\mathbb{R}^n$ *contains an orthonormal basis. We give two procedures. One is particularly useful if* $\mathscr{W}$ *is expressed as the solution set for a homogeneous system of linear equations. The second, the Gram-Schmidt process, is useful if a basis for* $\mathscr{W}$ *is given.*

Exercises for Section 8-3

1. Verify that the following are orthonormal bases for the spaces indicated.

   (a) $\mathbb{R}^2$: $(\frac{1}{2}\sqrt{2}, \frac{1}{2}\sqrt{2})$, $(-\frac{1}{2}\sqrt{2}, \frac{1}{2}\sqrt{2})$

   (b) $\mathbb{R}^2$: $(\frac{1}{2}, \frac{1}{2}\sqrt{3})$, $(\frac{1}{2}\sqrt{3}, -\frac{1}{2})$

   (c) $\mathbb{R}^3$: $(\frac{1}{2}\sqrt{2}, \frac{1}{2}\sqrt{2}, 0)$, $(-\frac{1}{3}\sqrt{3}, \frac{1}{3}\sqrt{3}, \frac{1}{3}\sqrt{3})$, $(\frac{1}{6}\sqrt{6}, -\frac{1}{6}\sqrt{6}, \frac{1}{3}\sqrt{6})$

   (d) $\mathbb{R}^3$: $(\frac{1}{2}\sqrt{2}, \frac{1}{2}\sqrt{2}, 0)$, $(-\frac{1}{2}\sqrt{2}, \frac{1}{2}\sqrt{2}, 0)$, $(0,0,1)$

   (e) $\mathbb{R}^4$: $(\frac{1}{3}\sqrt{3}, 0, \frac{1}{3}\sqrt{3}, \frac{1}{3}\sqrt{3})$, $(\frac{1}{3}\sqrt{3}, \frac{1}{3}\sqrt{3}, 0, -\frac{1}{3}\sqrt{3})$,
   $(0, \frac{1}{3}\sqrt{3}, -\frac{1}{3}\sqrt{3}, \frac{1}{3}\sqrt{3})$, $(\frac{1}{3}\sqrt{3}, -\frac{1}{3}\sqrt{3}, -\frac{1}{3}\sqrt{3}, 0)$

2. Express the vectors listed below as linear combinations of the orthonormal bases in Exercises 1(a) and (b).

$$(3,1), (2,-1)$$

3. Express the vectors listed below as linear combinations of the orthonormal bases in Exercises 1(c) and (d).

$$(2,1,6), (3,-4,2)$$

4. Express the vectors $(1,2,-1,6)$ and $(7,0,-1,6)$ as linear combinations of the orthonormal bases listed in Exercise 1(e).

5. Find an orthogonal basis for $\mathbb{R}^3$ which contains the vector $(2,1,3)$. (There are many solutions to this.)

6. Find an orthogonal basis for the solution set of the homogeneous system of linear equations
$$3x - 2y + 4z + \ w = 0$$
$$5x - 6y - \ z - 3w = 0.$$

7. Same exercise as 6 for
$$x - \ y + z + 3w + 2v = 0$$
$$3x + 2y \quad\quad - \ w + \ v = 0.$$

8. If $\mathscr{W}$ is the subspace of $\mathbb{R}^3$ generated by $(2,0,-1)$ and $(3,1,1)$, find the orthogonal projection of $(1,0,2)$ on $\mathscr{W}$.

9. If $\mathscr{W}$ is the subspace of $\mathbb{R}^4$ generated by $(2,1,3,0)$ and $(1,1,2,1)$, find the orthogonal projection of the vector $(1,0,1,0)$ onto $\mathscr{W}$.

10. Find the point where the perpendicular from the given point to the given plane intersects the plane.

(a) $(1,2,1)$, $3x - y + z = 5$ (b) $(3,1,4)$, $x - 2y + 3z = 10$

More Challenging Exercises

11. Suppose $\mathscr{W}_1$ is a subspace of $\mathbb{R}^n$. Let $\mathscr{W}_2$ be the collection of all vectors $\mathbf{X}$ which are orthogonal to every $\mathbf{Y}$ in $\mathscr{W}_1$.

   (a) Show that $\mathscr{W}_2$ is a subspace of $\mathbb{R}^n$.
   (b) Show that $\mathscr{W}_2$ is a subspace of $\mathbb{R}^n$ complementary to $\mathscr{W}_1$. The subspace $\mathscr{W}_2$ is denoted by $\mathscr{W}_1^\perp$ and is called *the orthogonal complement* of $\mathscr{W}_1$. You have just proved that every subspace of $\mathbb{R}^n$ possesses an orthogonal complement. Can a subspace have two different orthogonal complements?

12. Give an argument that the solution set for a homogeneous system of linear equations is the orthogonal complement of the row space of the coefficient matrix.

13. Suppose $\mathscr{W}$ is the subspace of $\mathbb{R}^4$ generated by $(2,0,4,4)$ and $(-1,3,1,7)$. Find a system of two equations in four unknowns for which $\mathscr{W}$ is the solution set. (Hint: find a generating set for the orthogonal complement of $\mathscr{W}$.)

14. Suppose $\mathscr{W}$ is the subspace of $\mathbb{R}^5$ generated by the vectors $(2,1,3,1,4)$ and $(1,1,2,1,3)$. Find a system of three equations in five unknowns for which $\mathscr{W}$ is the solution set.

15. A square matrix is *orthogonal* if and only if $AA^T = I_n$. (Refer to Proposition 5-9 to see that $A^T = A^{-1}$.) Give an argument that the rows of A constitute an orthonormal basis for $\mathbb{R}^n$. Conversely, if $\mathbf{X}_1, \ldots, \mathbf{X}_n$ form an orthonormal basis for $\mathbb{R}^n$, then the $n \times n$ matrix A having $\mathbf{X}_i$ as its ith row must be an orthogonal matrix.

16. Recall from Exercise 11 of Section 8-1 that $(A\mathbf{X}) \cdot \mathbf{Y} = \mathbf{X} \cdot (A^T\mathbf{Y})$, where A is an $n \times n$ matrix and $\mathbf{X}$ and $\mathbf{Y}$ are vectors in $\mathbb{R}^n$. Give an argument to show that if $T: \mathbb{R}^n \to \mathbb{R}^n$ is a linear transformation whose matrix A is orthogonal, then the length of $T(\mathbf{X})$ equals the length of $\mathbf{X}$, and the angle between $T(\mathbf{X})$ and $T(\mathbf{Y})$ equals the angle between $\mathbf{X}$ and $\mathbf{Y}$.

17. In Exercise 5 of Section 2-2 we showed that the solution set for a homogeneous system of n equations is a subspace of $\mathbb{R}^n$. Give an argument that shows that every subspace of $\mathbb{R}^n$ is the solution set for a system of equations in n unknowns. Show also that if the dimension of the subspace is k, then the subspace is the solution set for a system of $n - k$ equations in n unknowns.

SECTION 8-4
The Cross Product in $\mathbb{R}^3$

PREVIEW *In this section we introduce the cross product of two vectors in $\mathbb{R}^3$ and give some geometrical interpretations connected with it.*

There are many situations in which it is extremely convenient to find a vector which is orthogonal to two noncollinear vectors in $\mathbb{R}^3$.

For definiteness, we are dealing with two noncollinear vectors $\mathbf{X} = (x_1, x_2, x_3)$ and $\mathbf{Y} = (y_1, y_2, y_3)$. We want to find a vector $\mathbf{Z} = (z_1, z_2, z_3)$ so that $\mathbf{X} \cdot \mathbf{Z} = 0$ and $\mathbf{Y} \cdot \mathbf{Z} = 0$. This leads to the homogeneous system of two equations in unknowns z_1, z_2, and z_3.

$$x_1z_1 + x_2z_2 + x_3z_3 = 0$$
$$y_1z_1 + y_2z_2 + y_3z_3 = 0$$

You can solve this system of equations and find that the solution set can be interpreted geometrically as a line since the rank of the matrix is 2. One solution is the vector

$$\mathbf{Z} = (x_2y_3 - x_3y_2, \, x_3y_1 - x_1y_3, \, x_1y_2 - x_2y_1).$$

This particular vector is called the cross product of **X** and **Y** and is denoted by **X** × **Y**.

Definition 8-6_____

If $\mathbf{X} = (x_1, x_2, x_3)$ and $\mathbf{Y} = (y_1, y_2, y_3)$, then the *cross product* or *vector product* of **X** with **Y** is the vector

$$\mathbf{X} \times \mathbf{Y} = (x_2 y_3 - x_3 y_2,\ x_3 y_1 - x_1 y_3,\ x_1 y_2 - x_2 y_1).$$

_____

It is essential to observe the order in which the product is expressed, because, as you can check,

$$\mathbf{Y} \times \mathbf{X} = (y_2 x_3 - y_3 x_2,\ y_3 x_1 - y_1 x_3,\ y_1 x_2 - y_2 x_1) = -\mathbf{X} \times \mathbf{Y}.$$

Example Let $\mathbf{X} = (1,1,2)$ and $\mathbf{Y} = (3,-1,1)$. Then $\mathbf{X} \times \mathbf{Y} = (3,5,-4)$. You can check that, indeed, $\mathbf{X} \times \mathbf{Y}$ is orthogonal to both **X** and **Y**.

Example Let $\mathbf{X} = (2,1,-5)$ and $\mathbf{Y} = (1,8,3)$. Then $\mathbf{X} \times \mathbf{Y} = (43,-11,15)$. Again, you can check that $\mathbf{X} \times \mathbf{Y}$ is orthogonal to **X** and to **Y**.

The formulas for the components of $\mathbf{X} \times \mathbf{Y}$ are somewhat complicated. There is a way to remember how to compute $\mathbf{X} \times \mathbf{Y}$ using the **i**, **j**, **k** notation: compute the "determinant" of the "matrix"

$$\begin{pmatrix} \mathbf{i} & \mathbf{j} & \mathbf{k} \\ x_1 & x_2 & x_3 \\ y_1 & y_2 & y_3 \end{pmatrix}.$$

If you compute as though you are expanding along the first row, you get

$$\mathbf{X} \times \mathbf{Y} = \mathbf{i}(x_2 y_3 - x_3 y_2) - \mathbf{j}(x_1 y_3 - x_3 y_1) + \mathbf{k}(x_1 y_2 - x_2 y_1).$$

Example Let $\mathbf{X} = (1,1,2)$ and $\mathbf{Y} = (3,-1,1)$. Then $\mathbf{X} \times \mathbf{Y}$ can be computed by computing the "determinant" of the "matrix"

$$\begin{pmatrix} \mathbf{i} & \mathbf{j} & \mathbf{k} \\ 1 & 1 & 2 \\ 3 & -1 & 1 \end{pmatrix}.$$

Thus $\mathbf{X} \times \mathbf{Y} = 3\mathbf{i} + 5\mathbf{j} - 4\mathbf{k} = (3,5,-4)$.

As you have probably observed, $\mathbf{X} \times \mathbf{Y}$ is not the only vector orthogonal to **X** and to **Y**; any scalar multiple of $\mathbf{X} \times \mathbf{Y}$ is, too. There are at least two reasons for

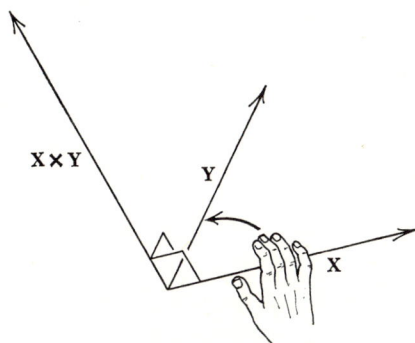

Figure 8-23

choosing this particular vector as $\mathbf{X} \times \mathbf{Y}$. One reason is that it is easy to compute $\mathbf{X} \times \mathbf{Y}$ from $\mathbf{X}$ and $\mathbf{Y}$, and the "determinant" method makes it easy to remember. The second is that $\mathbf{X} \times \mathbf{Y}$ has some interesting geometric facts connected with it. The direction of $\mathbf{X} \times \mathbf{Y}$ can be pictured by the "right-hand rule." If you picture pushing $\mathbf{X}$ into $\mathbf{Y}$ with the fingers of the right hand through the smallest angle between $\mathbf{X}$ and $\mathbf{Y}$ in the plane determined by them, then the thumb points in the general direction of $\mathbf{X} \times \mathbf{Y}$ (see Figure 8-23).

But there is more; the length of $\mathbf{X} \times \mathbf{Y}$ equals the area of the parallelogram determined by $\mathbf{X}$ and $\mathbf{Y}$ (see Figure 8-24). Let $\mathbf{X} = (x_1, x_2, x_3)$ and $\mathbf{Y} = (y_1, y_2, y_3)$. The area of the parallelogram is

$$\text{area} = |\mathbf{X}|\,|\mathbf{Y}|\,\sin\theta,$$

where θ is the acute angle between $\mathbf{X}$ and $\mathbf{Y}$. Then

$$\text{area} = |\mathbf{X}|\,|\mathbf{Y}|\,\sin\theta = |\mathbf{X}|\,|\mathbf{Y}|\,\sqrt{1 - \cos^2\theta}$$
$$= |\mathbf{X}|\,|\mathbf{Y}|\,[1 - (\mathbf{X}\cdot\mathbf{Y})^2/|\mathbf{X}|^2\,|\mathbf{Y}|^2]^{1/2}$$
$$= [|\mathbf{X}|^2\,|\mathbf{Y}|^2 - (\mathbf{X}\cdot\mathbf{Y})^2]^{1/2}$$
$$= [(x_1^2 + x_2^2 + x_3^2)(y_1^2 + y_2^2 + y_3^2) - (x_1y_1 + x_2y_2 + x_3y_3)^2]^{1/2}$$
$$= [(x_2y_3 - x_3y_2)^2 + (x_3y_1 - x_1y_3)^2 + (x_1y_2 - x_2y_1)^2]^{1/2}$$
$$= |\mathbf{X} \times \mathbf{Y}|.$$

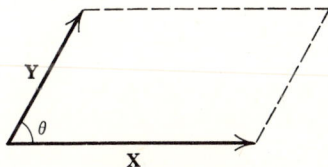

Figure 8-24

SUMMARY *The cross product* $\mathbf{X} \times \mathbf{Y}$ *of the vectors* $\mathbf{X} = (x_1, x_2, x_3)$ *and* $\mathbf{Y} = (y_1, y_2, y_3)$ *is the vector*

$$(x_2 y_3 - x_3 y_2, \ x_3 y_1 - x_1 y_3, \ x_1 y_2 - x_2 y_1).$$

It can be computed as the "determinant" of the "matrix"

$$\begin{pmatrix} \mathbf{i} & \mathbf{j} & \mathbf{k} \\ x_1 & x_2 & x_3 \\ y_1 & y_2 & y_3 \end{pmatrix}.$$

The vector $\mathbf{X} \times \mathbf{Y}$ *is orthogonal to* $\mathbf{X}$ *and to* $\mathbf{Y}$, *and its length equals the area of the parallelogram determined by* $\mathbf{X}$ *and* $\mathbf{Y}$.

Exercises for Section 8-4

1. Compute $\mathbf{X} \times \mathbf{Y}$ in the following cases.

(a) $\mathbf{X} = (2,1,3)$ and $\mathbf{Y} = (1,5,6)$ (d) $\mathbf{X} = (1,0,0)$ and $\mathbf{Y} = (0,1,0)$

(b) $\mathbf{X} = (1,5,6)$ and $\mathbf{Y} = (2,1,3)$ (e) $\mathbf{X} = (1,5,6)$ and $\mathbf{Y} = (4,2,6)$

(c) $\mathbf{X} = (3,-1,4)$ and $\mathbf{Y} = (1,1,0)$

2. In each of the parts, find a unit vector orthogonal to each of the given vectors.

(a) $(2,1,3)$, $(1,1,2)$ (c) $(2,5,1)$, $(6,-7,1)$

(b) $(-1,0,2)$, $(1,1,0)$ (d) $(1,2,1)$, $(2,1,2)$

More Challenging Exercises

3. Verify the following properties of the cross product.

(a) $\mathbf{X} \times \mathbf{Y} = -\mathbf{Y} \times \mathbf{X}$

(b) If $\mathbf{X}$ and $\mathbf{Y}$ are collinear, then $\mathbf{X} \times \mathbf{Y} = \mathbf{0}$.

(c) If $\mathbf{X} \times \mathbf{Y} = \mathbf{0}$ and neither $\mathbf{X}$ nor $\mathbf{Y}$ is the zero vector, then $\mathbf{X}$ and $\mathbf{Y}$ are collinear.

(d) $a\mathbf{X} \times \mathbf{Y} = a(\mathbf{X} \times \mathbf{Y})$

(e) $\mathbf{X} \times a\mathbf{Y} = a(\mathbf{X} \times \mathbf{Y})$

(f) $\mathbf{X} \times (\mathbf{Y} + \mathbf{Z}) = \mathbf{X} \times \mathbf{Y} + \mathbf{X} \times \mathbf{Z}$

In the following exercises we show how the construction of the cross product can be used in higher-dimensional spaces.

4. Suppose we are given three linearly independent vectors $(2,1,0,1)$, $(3,4,-1,2)$, and $(1,0,1,1)$ in $\mathbb{R}^4$. Form a 3×4 matrix using these vectors as rows.

$$A = \begin{pmatrix} 2 & 1 & 0 & 1 \\ 3 & 4 & -1 & 2 \\ 1 & 0 & 1 & 1 \end{pmatrix}$$

Construct a vector in $\mathbb{R}^4$ as follows:

$$(\det A(-\mid 1), \ -\det A(-\mid 2), \ \det A(-\mid 3), \ -\det A(-\mid 4))$$

(refer to Definition 6-3 for notation). Verify that this vector is orthogonal to the three given vectors (you should have computed the vector $(1,2,3,-4)$).

5. We state the result of Exercise 4 in a slightly different form. Let i_1, i_2, i_3, i_4 denote the four vectors in the natural basis for $\mathbb{R}^4$. Compute the "determinant" of the "matrix"

$$\begin{pmatrix} i_1 & i_2 & i_3 & i_4 \\ 2 & 1 & 0 & 1 \\ 3 & 4 & -1 & 2 \\ 1 & 0 & 1 & 1 \end{pmatrix}.$$

The resulting vector is orthogonal to the three given vectors.

6. In this exercise we give an indication why the results of Exercises 4 and 5 are true. Suppose that (a_1,a_2,a_3,a_4), (b_1,b_2,b_3,b_4), and (c_1,c_2,c_3,c_4) are linearly independent vectors in $\mathbb{R}^4$. Determine a vector by computing the "determinant" of the "matrix"

$$\begin{pmatrix} i_1 & i_2 & i_3 & i_4 \\ a_1 & a_2 & a_3 & a_4 \\ b_1 & b_2 & b_3 & b_4 \\ c_1 & c_2 & c_3 & c_4 \end{pmatrix}.$$

Show that the vector determined by this process is orthogonal to the three given vectors. (Hint: take the dot product of each given vector with the vector constructed here. Show that you obtain the first-row expansion for the determinant of a 4×4 matrix having the given vectors as second, third, and fourth rows, and one of the given vectors in the first row.)

7. Give an argument to show that if $n - 1$ linearly independent vectors in $\mathbb{R}^n$ are given, then a similar process can be used to construct a vector which is orthogonal to the given vectors.

9

Change
of
Basis,
Equivalence
of
Matrices

Ordered Bases, Coordinates of Vectors Relative to Ordered Bases, Matrix Representation of Linear Transformations Relative to Ordered Bases

PREVIEW *If we agree on an order in listing a basis, then we can use the coordinate n-tuple for the vector relative to the basis to represent the vector. Using this representation, we find that a linear transformation has a matrix representation relative to an ordered basis in its domain and an ordered basis in the vector space containing its range.*

In Chapters 4 through 6 we dealt with linear transformations and their matrix representations. We used the natural bases in $\mathbb{R}^n$ because they were sufficient for our needs then. In this chapter and in the next, however, we must consider different bases in $\mathbb{R}^n$. The reason is that we can often understand linear functions and quadratic functions much better when we use bases other than the natural bases. This matter is discussed in more detail in the next chapter.

If you refer to Proposition 3-6, you see that if $\mathbf{X}_1, \ldots, \mathbf{X}_n$ form a basis for $\mathscr{V}$ and $\mathbf{X}$ is a vector in $\mathscr{V}$, then $\mathbf{X}$ has a unique representation as a proper linear combination of the basis vectors. For example, the vector (3,4,5) has the representation

$$(3,4,5) = (-6)(1,0,0) + (4)(1,1,0) + (5)(1,0,1)$$

in terms of the vectors (1,0,0), (1,1,0), and (1,0,1) which form a basis for $\mathbb{R}^3$. Therefore, the three scalars -6, 4, and 5 determine the vector (3,4,5) relative to the basis (1,0,0), (1,1,0), and (1,0,1). We must be careful, though, to multiply the proper basis vector by the proper scalar. We do this by agreeing to list the scalars and the vectors in the basis in the same order so that the first scalar multiplies the first vector, the second scalar multiplies the second vector, and so on. In this case, for example, we say that the ordered triple $(-6,4,5)$ constitutes the coordinates of the vector (3,4,5) relative to the basis (1,0,0), (1,1,0), and (1,0,1), whereas the ordered triple $(4,5,-6)$ constitutes the coordinates of (3,4,5) relative to the basis (1,1,0), (1,0,1), and (1,0,0).

Definition 9-1_____

If $\mathbf{X}_1, \ldots, \mathbf{X}_n$ form a basis for a vector space, then we say that the n-tuple $(a_1, \ldots, a_n)$ constitutes the coordinates of the vector $\mathbf{X}$ relative to $\mathbf{X}_1, \ldots, \mathbf{X}_n$ if

$X = a_1X_1 + \cdots + a_nX_n$. This language presupposes that the vectors in the basis and the scalars in the n-tuple are written in the corresponding order. To emphasize the order, we sometimes use the phrase "ordered basis" rather than just "basis."

Example The vectors $(2,1)$ and $(1,-1)$ form a basis for $\mathbb{R}^2$. The coordinates of the vector $(1,2)$ relative to this basis are $(1,-1)$ since $(1,2) = 1(2,1) - 1(1,-1)$. The coordinates of $(1,2)$ relative to the basis $(1,-1)$ and $(2,1)$ are $(-1,1)$ since $(1,2) = -1(1,-1) + 1(2,1)$. The coordinates of $(2,1)$ relative to the basis $(2,1)$ and $(1,-1)$ are $(1,0)$ since $(2,1) = 1(2,1) + 0(1,-1)$.

There is a slight possibility of confusion because we use $(1,-1)$ and $(1,2)$ to represent the same vector. However, no confusion arises if we say that $(1,-1)$ constitutes the coordinates of a vector relative to $(2,1)$ and $(1,-1)$.

Example The vectors $(2,1,0)$, $(0,1,1)$, and $(1,0,1)$ form a basis for $\mathbb{R}^3$. Since

$$(3,2,2) = 1(2,1,0) + 1(0,1,1) + 1(1,0,1)$$

we say that $(1,1,1)$ are the coordinates of $(3,2,2)$ relative to the basis above. Similarly, $(-1,1,2)$ are the coordinates of $(0,0,3)$ relative to this basis since

$$(0,0,3) = -1(2,1,0) + 1(0,1,1) + 2(1,0,1).$$

As these examples show, we now use n-tuples to denote vectors in $\mathbb{R}^n$ and also to denote the coordinates of a vector relative to some ordered basis. Because we use the same notation in different senses, it is essential to add some modifying phrase such as "relative to a basis" when we use an n-tuple to denote the coordinates of a vector.

Brief Exercises

1. Find the coordinates of the vectors $(1,0)$, $(0,1)$, and $(-4,3)$ relative to the ordered basis $(1,1)$, $(-1,2)$.

2. Find the coordinates of $(1,3,1)$, $(5,-1,3)$, and $(1,6,2)$ relative to the ordered basis $(1,0,1)$, $(2,1,0)$, $(0,1,1)$.

3. Find the coordinates of $(2,1,1,1)$ and $(0,5,-1,2)$ relative to $(1,0,1,1)$, $(0,1,1,1)$, $(1,1,1,0)$, $(1,1,0,0)$.

Answers: **1** $(1,0)$ has coordinates $(\frac{2}{3}, -\frac{1}{3})$; $(0,1)$ has coordinates $(\frac{1}{3}, \frac{1}{3})$; $(-4,3)$ has coordinates $(-\frac{5}{3}, \frac{7}{3})$ **2** $(1,3,1)$ has coordinates $(-1,1,2)$; $(5,-1,3)$ has coordinates $(\frac{13}{3}, \frac{1}{3}, -\frac{4}{3})$; $(1,6,2)$ has coordinates $(-\frac{7}{3}, \frac{5}{3}, \frac{13}{3})$ **3** $(2,1,1,1)$ has coordinates $(1,0,0,1)$; $(0,5,-1,2)$ has coordinates $(-\frac{3}{2}, \frac{7}{2}, -\frac{6}{2}, \frac{9}{2})$

The reason we introduce coordinates of a vector relative to an ordered basis is that we need this concept to discuss the matrix representation of a linear transformation relative to bases other than the natural bases.

Example Consider the linear transformation $T: \mathbb{R}^2 \to \mathbb{R}^2$ given by the equation $T(x,y) = (2x + 3y, x - y)$. The transformation T has a domain, $\mathbb{R}^2$ in this case, and it has a vector space containing its image, also $\mathbb{R}^2$ in this case. We allow different ordered bases in the domain and in the space containing the image. For this example, we use $(1,1)$ and $(-1,1)$ as the ordered basis in the domain, and we use $(2,1)$ and $(0,1)$ as the ordered basis in the space containing the image of T.

The purpose of this example is to find a 2×2 matrix which allows us to compute the coordinates of $T(x,y)$ relative to $(2,1)$ and $(0,1)$ from the coordinates of (x,y) relative to $(1,1)$ and $(-1,1)$. To do this we find the coordinates of $T(1,1)$ and the coordinates of $T(-1,1)$ relative to $(2,1)$ and $(0,1)$.

$$T(1,1) = (5,0) = \tfrac{5}{2}(2,1) - \tfrac{5}{2}(0,1)$$

$$T(-1,1) = (1,-2) = \tfrac{1}{2}(2,1) - \tfrac{5}{2}(0,1)$$

Use the coordinates of $T(1,1)$ as the first column and the coordinates of $T(-1,1)$ as the second column of a matrix.

$$\begin{pmatrix} \tfrac{5}{2} & \tfrac{1}{2} \\ -\tfrac{5}{2} & -\tfrac{5}{2} \end{pmatrix}$$

This is the matrix for T relative to the basis consisting of $(1,1)$ and $(-1,1)$ in the domain and the basis consisting of $(2,1)$ and $(0,1)$ in the space containing the image.

We show how this matrix is used to compute the coordinates of $T(x,y)$ relative to $(2,1)$ and $(0,1)$ from the coordinates of (x,y) relative to $(1,1)$ and $(-1,1)$. For example, the coordinates of $(3,2)$ relative to $(1,1)$ and $(-1,1)$ are $(\tfrac{5}{2}, -\tfrac{1}{2})$. Now

$$\begin{pmatrix} \tfrac{5}{2} & \tfrac{1}{2} \\ -\tfrac{5}{2} & -\tfrac{5}{2} \end{pmatrix} \begin{pmatrix} \tfrac{5}{2} \\ -\tfrac{1}{2} \end{pmatrix} = \begin{pmatrix} 6 \\ -5 \end{pmatrix}.$$

The coordinates of $T(3,2)$ relative to $(2,1)$ and $(0,1)$ should be $(6,-5)$.

$$T(3,2) = (12,1) = 6(2,1) - 5(0,1)$$

To see that this works in general, suppose the coordinates of (x,y) relative to $(1,1)$ and $(-1,1)$ are (a,b), i.e.,

$$(x,y) = a(1,1) + b(-1,1).$$

Then, because T is linear,

$$T(x,y) = aT(1,1) + bT(-1,1)$$
$$= a[\tfrac{5}{2}(2,1) - \tfrac{5}{2}(0,1)] + b[\tfrac{1}{2}(2,1) - \tfrac{5}{2}(0,1)]$$
$$= (\tfrac{5}{2}a + \tfrac{1}{2}b)(2,1) + (-\tfrac{5}{2}a - \tfrac{5}{2}b)(0,1).$$

Therefore, the coordinates of $T(x,y)$ relative to $(2,1)$ and $(0,1)$ are

$$(\tfrac{5}{2}a + \tfrac{1}{2}b, \ -\tfrac{5}{2}a - \tfrac{5}{2}b),$$

the same result we would have obtained by matrix multiplication.

$$\begin{pmatrix} \tfrac{5}{2} & \tfrac{1}{2} \\ -\tfrac{5}{2} & -\tfrac{5}{2} \end{pmatrix} \begin{pmatrix} a \\ b \end{pmatrix} = \begin{pmatrix} \tfrac{5}{2}a + \tfrac{1}{2}b \\ -\tfrac{5}{2}a - \tfrac{5}{2}b \end{pmatrix}$$

Example Consider the linear transformation $T: \mathbb{R}^2 \rightarrow \mathbb{R}^3$ defined by $T(x,y) = (5x + 6y, 7x - y, 8x - 3y)$. In this example we determine the matrix for T relative to the basis consisting of $(1,1)$ and $(0,1)$ in $\mathbb{R}^2$ and the basis consisting of $(1,0,0)$, $(1,1,0)$, and $(1,1,1)$ in $\mathbb{R}^3$. The columns of the matrix consist of the coordinates of $T(1,1)$ and $T(0,1)$ relative to $(1,0,0)$, $(1,1,0)$, and $(1,1,1)$. Since

$$T(1,1) = (11,6,5) = 5(1,0,0) + 1(1,1,0) + 5(1,1,1)$$

and

$$T(0,1) = (6,-1,-3) = 7(1,0,0) + 2(1,1,0) - 3(1,1,1)$$

the matrix for T is

$$\begin{pmatrix} 5 & 7 \\ 1 & 2 \\ 5 & -3 \end{pmatrix}.$$

We can use this matrix to determine the coordinates of $T(x,y)$ relative to $(1,0,0)$, $(1,1,0)$, and $(1,1,1)$ from the coordinates of (x,y) relative to $(1,1)$ and $(0,1)$. For example, since $(4,2) = 4(1,1) - 2(0,1)$, the coordinates of $(4,2)$ relative to $(1,1)$ and $(0,1)$ are $(4,-2)$. The coordinates of $T(4,2)$ relative to $(1,0,0)$, $(1,1,0)$, and $(1,1,1)$ are

$$\begin{pmatrix} 5 & 7 \\ 1 & 2 \\ 5 & -3 \end{pmatrix} \begin{pmatrix} 4 \\ -2 \end{pmatrix} = \begin{pmatrix} 6 \\ 0 \\ 26 \end{pmatrix}.$$

This is easily checked.

$$T(4,2) = (32,26,26) = 6(1,0,0) + 0(1,1,0) + 26(1,1,1)$$

This always happens. If the coordinates of (x,y) relative to $(1,1)$ and $(0,1)$ are (a,b), i.e.,

$$(x,y) = a(1,1) + b(0,1),$$

then, because T is linear,

$$T(x,y) = aT(1,1) + bT(0,1) = a(11,6,5) + b(6,-1,-3)$$
$$= a[5(1,0,0) + 1(1,1,0) + 5(1,1,1)]$$
$$+ b[7(1,0,0) + 2(1,1,0) - 3(1,1,1)]$$
$$= (5a + 7b)(1,0,0) + (a + 2b)(1,1,0) + (5a - 3b)(1,1,1).$$

Therefore, the coordinates of $T(x,y)$ relative to the second basis are $(5a + 7b, a + 2b, 5a - 3b)$. This is the same result we obtain by matrix multiplication.

$$\begin{pmatrix} 5 & 7 \\ 1 & 2 \\ 5 & -3 \end{pmatrix} \begin{pmatrix} a \\ b \end{pmatrix} = \begin{pmatrix} 5a + 7b \\ a + 2b \\ 5a - 3b \end{pmatrix}$$

These examples illustrate the general situation. Suppose $T: \mathbb{R}^n \to \mathbb{R}^m$ is a linear transformation, and suppose we have bases $\mathbf{X}_1, \ldots, \mathbf{X}_n$ in $\mathbb{R}^n$ and $\mathbf{Y}_1, \ldots, \mathbf{Y}_m$ in $\mathbb{R}^m$. The matrix for T relative to these bases is constructed as follows: the first column consists of the coordinates of $T(\mathbf{X}_1)$ relative to the $\mathbf{Y}$'s, the second column consists of the coordinates of $T(\mathbf{X}_2)$ relative to the $\mathbf{Y}$'s, and so on. Then the components for $T(\mathbf{X})$ relative to the basis vectors in $\mathbb{R}^m$ can be computed by multiplying the matrix for T and the $n \times 1$ matrix consisting of the coordinates of $\mathbf{X}$ relative to the basis in $\mathbb{R}^n$. If

$$T(\mathbf{X}_1) = a_{11}\mathbf{Y}_1 + a_{21}\mathbf{Y}_2 + \cdots + a_{m1}\mathbf{Y}_m,$$
$$T(\mathbf{X}_2) = a_{12}\mathbf{Y}_1 + a_{22}\mathbf{Y}_2 + \cdots + a_{m2}\mathbf{Y}_m,$$
$$\vdots$$
$$T(\mathbf{X}_n) = a_{1n}\mathbf{Y}_1 + a_{2n}\mathbf{Y}_2 + \cdots + a_{mn}\mathbf{Y}_m,$$

then the matrix for T relative to the given bases is

$$A = \begin{pmatrix} a_{11} & a_{12} & \cdots & a_{1n} \\ a_{21} & a_{22} & \cdots & a_{2n} \\ \cdot & \cdot & & \cdot \\ \cdot & \cdot & & \cdot \\ \cdot & \cdot & & \cdot \\ a_{m1} & a_{m2} & \cdots & a_{mn} \end{pmatrix}$$

Then the coordinates of $T(\mathbf{X})$ relative to the second basis can be computed as the product of the $m \times n$ matrix A with the $n \times 1$ matrix consisting of the coordinates of $\mathbf{X}$ relative to the first basis, for if the coordinates of $\mathbf{X}$ relative to $\mathbf{X}_1, \ldots, \mathbf{X}_n$ are given by $(x_1, \ldots, x_n)$, then, because T is linear,

$$
\begin{aligned}
T(\mathbf{X}) &= T(x_1\mathbf{X}_1 + \cdots + x_n\mathbf{X}_n) \\
&= x_1 T(\mathbf{X}_1) + \cdots + x_n T(\mathbf{X}_n) \\
&= x_1[a_{11}\mathbf{Y}_1 + a_{21}\mathbf{Y}_2 + \cdots + a_{m1}\mathbf{Y}_m] \\
&\quad + x_2[a_{12}\mathbf{Y}_1 + a_{22}\mathbf{Y}_2 + \cdots + a_{m2}\mathbf{Y}_m] + \cdots \\
&\quad + x_n[a_{1n}\mathbf{Y}_1 + a_{2n}\mathbf{Y}_2 + \cdots + a_{mn}\mathbf{Y}_m] \\
&= (a_{11}x_1 + a_{12}x_2 + \cdots + a_{1n}x_n)\mathbf{Y}_1 \\
&\quad + (a_{21}x_1 + a_{22}x_2 + \cdots + a_{2n}x_n)\mathbf{Y}_2 + \cdots \\
&\quad + (a_{m1}x_1 + a_{m2}x_2 + \cdots + a_{mn}x_n)\mathbf{Y}_m.
\end{aligned}
$$

The coordinates of $T(\mathbf{X})$ can also be computed as the following matrix product.

$$
\begin{pmatrix}
a_{11} & a_{12} & \cdots & a_{1n} \\
a_{21} & a_{22} & \cdots & a_{2n} \\
\cdot & \cdot & & \cdot \\
\cdot & \cdot & & \cdot \\
\cdot & \cdot & & \cdot \\
a_{m1} & a_{m2} & \cdots & a_{mn}
\end{pmatrix}
\begin{pmatrix}
x_1 \\
x_2 \\
\cdot \\
\cdot \\
\cdot \\
x_n
\end{pmatrix}
$$

Definition 9-2

Suppose $\mathbf{X}_1, \ldots, \mathbf{X}_n$ is a basis for $\mathbb{R}^n$ and $\mathbf{Y}_1, \ldots, \mathbf{Y}_m$ is a basis for $\mathbb{R}^m$, and suppose $T: \mathbb{R}^n \to \mathbb{R}^m$ is a linear transformation. The *matrix for T relative to these ordered bases* is the matrix described above.

Brief Exercises

1. Suppose $T: \mathbb{R}^2 \to \mathbb{R}^2$ is defined by $T(x,y) = (x + 2y, 2x - 3y)$. Find the matrix for T relative to the following ordered bases, the first in the domain of T, the second in the space containing the image.

   (a) $(1,0)$, $(0,1)$; $(2,1)$, $(-1,0)$ (b) $(1,1)$, $(-2,3)$; $(1,1)$, $(0,2)$

2. Same exercise for the linear transformation $T: \mathbb{R}^3 \to \mathbb{R}^2$ defined by $T(x,y,z) = (x + y + z, 2x - y + 3z)$.

   (a) $(1,0,0)$, $(0,1,0)$, $(0,0,1)$; $(2,0)$, $(-1,2)$
   (b) $(1,0,0)$, $(1,1,0)$, $(1,1,1)$; $(1,0)$, $(0,1)$

3. Same exercise for the linear transformation $T: \mathbb{R}^2 \to \mathbb{R}^3$ defined by $T(x,y) = (x + y, 2x - y, x + 2y)$.

    (a) $(1,1)$, (-1.2); $(1,0,0)$, $(1,0,1)$, $(0,1,1)$
    (b) $(1,0)$, $(0,1)$; $(1,0,1)$, $(0,1,1)$, $(1,1,0)$

Answers:

1(a) $\begin{pmatrix} 2 & -3 \\ 3 & -8 \end{pmatrix}$
    1(b) $\begin{pmatrix} 3 & 4 \\ -2 & -\frac{17}{2} \end{pmatrix}$
    2(a) $\begin{pmatrix} 1 & \frac{1}{4} & \frac{5}{4} \\ 1 & -\frac{1}{2} & \frac{3}{2} \end{pmatrix}$

2(b) $\begin{pmatrix} 1 & 2 & 3 \\ 2 & 1 & 4 \end{pmatrix}$
    3(a) $\begin{pmatrix} 0 & -6 \\ 2 & 7 \\ 1 & -4 \end{pmatrix}$
    3(b) $\begin{pmatrix} 1 & 2 \\ 1 & 0 \\ 1 & -1 \end{pmatrix}$

We have seen above that a linear transformation determines a matrix relative to some given ordered bases. It is also true that a matrix determines a linear transformation.

Example   Suppose we are given the matrix

$$A = \begin{pmatrix} 2 & 1 & 0 \\ 3 & 0 & -2 \end{pmatrix},$$

and suppose we are dealing with the bases $(1,0,1)$, $(1,1,1)$, and $(0,1,1)$ in $\mathbb{R}^3$ and $(1,1)$ and $(-1,1)$ in $\mathbb{R}^2$. In this example we show that there is a linear transformation $T: \mathbb{R}^3 \to \mathbb{R}^2$ whose matrix is A relative to the given ordered bases.

If the vector $\mathbf{X}$ in $\mathbb{R}^3$ has coordinates (a,b,c) relative to the basis in $\mathbb{R}^3$, then $T(\mathbf{X})$ must have coordinates

$$\begin{pmatrix} 2 & 1 & 0 \\ 3 & 0 & -2 \end{pmatrix} \begin{pmatrix} a \\ b \\ c \end{pmatrix} = \begin{pmatrix} 2a + b \\ 3a - 2c \end{pmatrix}$$

relative to the basis in $\mathbb{R}^2$. This shows that T is a linear function because of the following properties of matrix multiplication (see Exercise 20, Section 4-4).

$$A(\mathbf{X} + \mathbf{X}') = A\mathbf{X} + A\mathbf{X}'$$

$$A(r\mathbf{X}) = r(A\mathbf{X})$$

Proposition 9-1

Suppose $\mathbf{X}_1, \ldots, \mathbf{X}_n$ and $\mathbf{Y}_1, \ldots, \mathbf{Y}_m$ are bases for $\mathbb{R}^n$ and $\mathbb{R}^m$, respectively. Then each linear transformation from $\mathbb{R}^n$ to $\mathbb{R}^m$ determines a unique $m \times n$ matrix, and each $m \times n$ matrix determines a linear transformation from $\mathbb{R}^n$ to $\mathbb{R}^m$ relative to these ordered bases.

There are some facts which we established earlier when we dealt with linear transformations relative to the natural ordered bases in $\mathbb{R}^n$ which are still true in the more general setting we are now discussing. Most of them are easy to see; we list them without any further discussion so that we can refer to them later.

Proposition 9-2

Suppose that S and T are linear transformations from $\mathbb{R}^n$ to $\mathbb{R}^m$, and suppose that the matrices A and B represent S and T, respectively, relative to some choice of ordered bases. Then $A + B$ represents $S + T$ and rA represents rS relative to the same ordered bases.

Proposition 9-3

Suppose that S is a linear transformation from $\mathbb{R}^n$ to $\mathbb{R}^m$ and that T is a linear transformation from $\mathbb{R}^m$ to $\mathbb{R}^p$. Suppose further that some ordered basis is chosen in each of the spaces $\mathbb{R}^n$, $\mathbb{R}^m$, and $\mathbb{R}^p$. If the matrix for S is A and the matrix for T is B relative to the appropriate bases, then the matrix for $T \circ S$ is BA.

Proposition 9-4

Suppose T is a linear transformation from $\mathbb{R}^n$ to $\mathbb{R}^n$, and suppose the matrix for T is A relative to some choice of ordered bases in the domain and in the range of T. The transformation T is nonsingular if and only if A is nonsingular; if T is nonsingular, then A^{-1} is the matrix for T^{-1} relative to these same bases. (The basis in the domain of T^{-1} is the one in the image of T; the basis in the image of T^{-1} is the one in the domain of T.)

SUMMARY *Coordinates of a vector relative to a basis are introduced; they are the scalars used to express the vector as a linear combination of the basis vectors. We indicate these with the n-tuple notation. Then we show how to find the matrix for a linear transformation* $T: \mathbb{R}^n \to \mathbb{R}^m$ *relative to the ordered bases* $\mathbf{X}_1, \ldots, \mathbf{X}_n$ *in* $\mathbb{R}^n$ *and* $\mathbf{Y}_1, \ldots, \mathbf{Y}_m$ *in* $\mathbb{R}^m$: *the first column consists of the coordinates of* $T(\mathbf{X}_1)$ *relative to the* $\mathbf{Y}$'s, *the second column consists of the coordinates of* $T(\mathbf{X}_2)$ *relative to the* $\mathbf{Y}$'s, *and so on.*

Many facts established earlier are still valid if one is careful about the ordered bases. Each linear transformation determines a matrix relative to a choice of ordered bases. Each matrix determines a linear transformation relative to a choice of ordered bases. The sum of linear transformations is represented by the sum of matrices; the scalar multiple of a linear transformation is represented by the scalar multiple of the matrix. The product of matrices is represented by the composition of linear transformations. The linear transformation is nonsingular if and only if the matrix is nonsingular; if that is the case, then the matrix for the inverse transformation is the inverse of the matrix.

Exercises for Section 9-1

1. Suppose $T: \mathbb{R}^2 \to \mathbb{R}^3$ is defined by $T(x,y) = (x + 2y, 3x - y, x + y)$. Find the matrix for T relative to the ordered bases

   (a) $(1,0)$, $(1,1)$ and $(1,0,0)$, $(0,1,0)$, $(0,0,1)$.
   (b) $(1,0)$, $(1,1)$ and $(1,0,0)$, $(1,1,0)$, $(1,1,1)$.
   (c) $(1,1)$, $(3,1)$ and $(1,0,1)$, $(0,1,1)$, $(1,1,0)$.

2. Suppose $T: \mathbb{R}^3 \to \mathbb{R}^3$ has the matrix

$$\begin{pmatrix} 1 & 2 & 1 \\ 0 & -3 & 2 \\ 1 & 0 & 1 \end{pmatrix}$$

relative to the natural ordered bases in both the domain and image of T. Find the matrix for T relative to the following pairs of ordered bases in the domain and image of T, respectively.

   (a) $(1,0,0)$, $(0,1,0)$, $(0,0,1)$ and $(1,1,0)$, $(1,0,1)$, $(0,1,1)$
   (b) $(1,1,0)$, $(1,0,1)$, $(0,1,1)$ and $(1,1,0)$, $(1,0,1)$, $(0,1,1)$
   (c) $(1,1,0)$, $(1,0,1)$, $(0,1,1)$ and $(1,0,0)$, $(0,1,0)$, $(0,0,1)$

3. Suppose the matrix

$$\begin{pmatrix} 1 & 2 & 1 \\ 3 & 1 & 0 \\ 0 & 2 & 1 \end{pmatrix}$$

represents a linear transformation $T: \mathbb{R}^3 \to \mathbb{R}^3$ relative to the ordered bases $(1,2,0)$, $(1,0,1)$, $(0,1,0)$ in the domain of T and $(1,1,0)$, $(1,0,1)$, $(0,1,1)$ in the image of T. Find an explicit formula for $T(x,y,z)$ and thereby verify that the matrix determines a linear function.

4. Supply the details of the argument for Proposition 9-2.

5. Supply an argument for Proposition 9-3.

6. Supply an argument for Proposition 9-4.

SECTION 9-2
Change of Basis

PREVIEW *We discuss how to find the coordinates of a vector relative to a second basis from its coordinates relative to a first basis. This is done with the transition matrix or the matrix for a change from one basis to another. Then we show that if we are given a nonsingular matrix P and a basis, then P can be the transition matrix from the given basis or to the given basis.*

As we have seen many times, a vector space may have many bases. The coordinates of a vector may be different relative to different bases. We show how to find the coordinates of a vector relative to an ordered basis from its coordinates relative to another ordered basis.

Example Suppose we consider the ordered basis consisting of $(2,1)$ and $(-1,1)$ in $\mathbb{R}^2$ and the ordered basis consisting of $(1,1)$ and $(0,3)$. In this example we show how to compute the coordinates of a vector relative to $(1,1)$ and $(0,3)$ if we know its coordinates relative to $(2,1)$ and $(-1,1)$.

For an example, the coordinates of the vector $(5,1)$ relative to $(2,1)$ and $(-1,1)$ are $(2,-1)$ since

$$(5,1) = 2(2,1) - (-1,1).$$

You can find these coordinates by solving the system of equations

$$x\begin{pmatrix}2\\1\end{pmatrix} + y\begin{pmatrix}-1\\1\end{pmatrix} = \begin{pmatrix}5\\1\end{pmatrix}.$$

We can use the same method to find the coordinates of $(5,1)$ relative to $(1,1)$ and $(0,3)$, i.e. solve the equations

$$x\begin{pmatrix}1\\1\end{pmatrix} + y\begin{pmatrix}0\\3\end{pmatrix} = \begin{pmatrix}5\\1\end{pmatrix}.$$

However, we can proceed in another way. First, we find the coordinates of $(2,1)$ and $(-1,1)$ relative to $(1,1)$ and $(0,3)$.

$$(2,1) = 2(1,1) - \tfrac{1}{3}(0,3)$$
$$(-1,1) = -(1,1) + \tfrac{2}{3}(0,3)$$

Then substitute these values in the equation

$$
\begin{aligned}
(5,1) &= 2(2,1) - (-1,1) \\
&= 2(2(1,1) - \tfrac{1}{3}(0,3)) \\
&\quad -(-(1,1) + \tfrac{2}{3}(0,3)) \\
&= 5(1,1) - \tfrac{4}{3}(0,3).
\end{aligned}
$$

This procedure can be used to determine the coordinates of any vector (x,y) relative to $(1,1)$ and $(0,3)$ once the coordinates are known relative to $(2,1)$ and $(-1,1)$. To see this, suppose the coordinates of (x,y) relative to $(2,1)$ and $(-1,1)$ are (a,b). Then

$$
\begin{aligned}
(x,y) &= a(2,1) + b(-1,1) = a(2(1,1) - \tfrac{1}{3}(0,3)) + b(-(1,1) + \tfrac{2}{3}(0,3)) \\
&= (2a - b)(1,1) + (-\tfrac{1}{3}a + \tfrac{2}{3}b)(0,3).
\end{aligned}
$$

Thus the coordinates of (x,y) relative to $(1,1)$ and $(0,3)$ are $(2a - b, -\tfrac{1}{3}a + \tfrac{2}{3}b)$. This can be obtained with the matrix product

$$
\begin{pmatrix} 2 & -1 \\ -\tfrac{1}{3} & \tfrac{2}{3} \end{pmatrix} \begin{pmatrix} a \\ b \end{pmatrix}.
$$

This suggests another way to view the problem of changing coordinates. Think of the identity function I on $\mathbb{R}^2$. It is a linear transformation which sends every vector to itself, $I(x,y) = (x,y)$. Consequently, by the considerations of Section 9-1, there is a matrix for I relative to the basis $(1,1)$ and $(0,3)$ in the domain of I and the basis $(2,1)$ and $(-1,1)$ in the space containing the image. That matrix is

$$
\begin{pmatrix} 2 & -1 \\ -\tfrac{1}{3} & \tfrac{2}{3} \end{pmatrix}.
$$

We use it to compute the coordinates of $I(x,y) = (x,y)$ relative to $(1,1)$ and $(0,3)$ from the coordinates of (x,y) relative to $(2,1)$ and $(-1,1)$ exactly what we want to do.

Example In this example we show how to compute the change of coordinates from the ordered basis $(2,1,0)$, $(1,0,1)$, and $(-1,1,1)$ to the ordered basis $(2,-1,1)$, $(3,0,1)$, and $(2,1,0)$. To do this, consider the identity function I on $\mathbb{R}^3$. It is a linear transformation sending (x,y,z) into (x,y,z). The matrix for I relative to the basis $(2,1,0)$, $(1,0,1)$, and $(-1,1,1)$ in the domain of I and the basis $(2,-1,1)$, $(3,0,1)$, and $(2,1,0)$ in the space containing the image is

$$
\begin{pmatrix} 0 & -2 & -6 \\ 0 & 3 & 7 \\ 1 & -2 & -5 \end{pmatrix}
$$

since

$$(2,1,0) = \quad 0(2,-1,1) + 0(3,0,1) + (2,1,0)$$
$$(1,0,1) = -2(2,-1,1) + 3(3,0,1) - 2(2,1,0)$$
$$(-1,1,1) = -6(2,-1,1) + 7(3,0,1) - 5(2,1,0).$$

Therefore, if (x,y,z) has coordinates (a,b,c) relative to the first basis, then its co-ordinates relative to the second basis are $(-2b - 6c, 3b + 7c, a - 2b - 5c)$ since

$$\begin{pmatrix} 0 & -2 & -6 \\ 0 & 3 & 7 \\ 1 & -2 & -5 \end{pmatrix} \begin{pmatrix} a \\ b \\ c \end{pmatrix} = \begin{pmatrix} -2b - 6c \\ 3b + 7c \\ a - 2b - 5c \end{pmatrix}.$$

The matrix

$$\begin{pmatrix} 0 & -2 & -6 \\ 0 & 3 & 7 \\ 1 & -2 & -5 \end{pmatrix}$$

is called the matrix for the change of basis from the basis $(2,1,0)$, $(1,0,1)$, and $(-1,1,1)$ to the basis $(2,-1,1)$, $(3,0,1)$, and $(2,1,0)$. Some call it the transition matrix from the first basis to the second.

Example   In the previous two examples we have found the matrix for certain changes of bases. In this example we find the matrices for the change of basis from the second to the first in each of the two previous examples. The important point to observe is that the matrices we obtain are the inverses of the ones obtained earlier (refer also to Proposition 9-4).

The matrix for the change of basis in $\mathbb{R}^2$ from $(1,1)$, and $(0,3)$ to $(2,1)$ and $(-1,1)$ is

$$\begin{pmatrix} \frac{2}{3} & 1 \\ \frac{1}{3} & 2 \end{pmatrix}$$

because

$$(1,1) = \tfrac{2}{3}(2,1) + \tfrac{1}{3}(-1,1)$$
$$(0,3) = 1(2,1) + 2(-1,1).$$

Observe that this matrix is the inverse of the matrix obtained in the first example.

$$\begin{pmatrix} 2 & -1 \\ -\frac{1}{3} & \frac{2}{3} \end{pmatrix} \begin{pmatrix} \frac{2}{3} & 1 \\ \frac{1}{3} & 2 \end{pmatrix} = \begin{pmatrix} 1 & 0 \\ 0 & 1 \end{pmatrix}$$

The matrix for the change of basis in $\mathbb{R}^3$ from $(2,-1,1)$, $(3,0,1)$, and $(2,1,0)$ to $(2,1,0)$, $(1,0,1)$, and $(-1,1,1)$ is

$$\begin{pmatrix} -\tfrac{1}{4} & \tfrac{1}{2} & 1 \\ \tfrac{7}{4} & \tfrac{3}{2} & 0 \\ -\tfrac{3}{4} & -\tfrac{1}{2} & 0 \end{pmatrix}$$

since

$$(2,-1,1) = -\tfrac{1}{4}(2,1,0) + \tfrac{7}{4}(1,0,1) - \tfrac{3}{4}(-1,1,1)$$
$$(3,0,1) = \tfrac{1}{2}(2,1,0) + \tfrac{3}{2}(1,0,1) - \tfrac{1}{2}(-1,1,1)$$
$$(2,1,0) = 1(2,1,0) + 0(1,0,1) + 0(-1,1,1).$$

Observe that this matrix is the inverse of the one we obtained in the previous example.

$$\begin{pmatrix} 0 & -2 & -6 \\ 0 & 3 & 7 \\ 1 & -2 & -5 \end{pmatrix} \begin{pmatrix} -\tfrac{1}{4} & \tfrac{1}{2} & 1 \\ \tfrac{7}{4} & \tfrac{3}{2} & 0 \\ -\tfrac{3}{4} & -\tfrac{1}{2} & 0 \end{pmatrix} = \begin{pmatrix} 1 & 0 & 0 \\ 0 & 1 & 0 \\ 0 & 0 & 1 \end{pmatrix}$$

We could have predicted that the matrices obtained here are the inverses of the ones obtained in the earlier examples. The reason lies in the fact that the matrix for the change from any basis to itself is the identity matrix. Another way of saying the same thing is that, relative to the same basis in the domain and range, the matrix for the identity function is the identity matrix. The situation for $\mathbb{R}^2$ is pictured in Figure 9-1.

These examples illustrate that we can use matrices to compute change of bases.

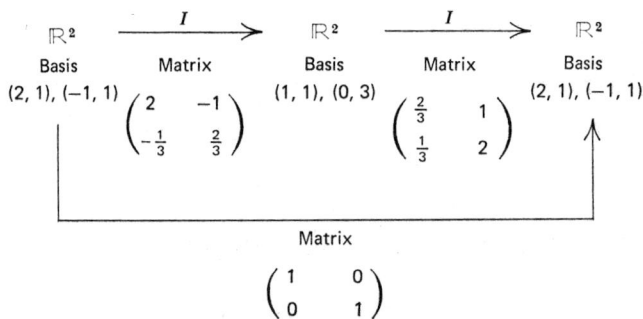

Figure 9-1

Definition 9-3_____

Suppose $X_1, \ldots, X_n$ and $Y_1, \ldots, Y_n$ are two ordered bases for $\mathbb{R}^n$. The *matrix for the change of basis from* $X_1, \ldots, X_n$ *to* $Y_1, \ldots, Y_n$ is the $n \times n$ matrix constructed as follows: the first column consists of the coordinates of X_1 relative to the Y's, the second column consists of the coordinates of X_2 relative to the Y's, and so on. This matrix is also called the *transition matrix* from the basis $X_1, \ldots, X_n$ to the basis $Y_1, \ldots, Y_n$.

_____

If A is the transition matrix from $X_1, \ldots, X_n$ to $Y_1, \ldots, Y_n$ in $\mathbb{R}^n$, and if B is the transition matrix from $Y_1, \ldots, Y_n$ to $X_1, \ldots, X_n$, then AB is the matrix for the change of basis from $Y_1, \ldots, Y_n$ to $Y_1, \ldots, Y_n$; therefore $AB = I_n$ (see Figure 9-2). This shows that a transition matrix is nonsingular.

Proposition 9-5_____

The matrix for a change of basis in $\mathbb{R}^n$ from $X_1, \ldots, X_n$ to $Y_1, \ldots, Y_n$ is the matrix for the identity function on $\mathbb{R}^n$ relative to these bases, $X_1, \ldots, X_n$ in the domain and $Y_1, \ldots, Y_n$ in the image. A transition matrix is a nonsingular matrix. The transition matrix from $Y_1, \ldots, Y_n$ to $X_1, \ldots, X_n$ is the inverse of the transition matrix from $X_1, \ldots, X_n$ to $Y_1, \ldots, Y_n$.

_____

Brief Exercises

1. Find the transition matrix from the first basis to the second.

   (a) $1,0)$, $(0,1)$; $(1,1)$, $(-1,2)$ (c) $(1,-1)$, $(2,1)$; $(1,2)$, $(-3,1)$
   (b) $(1,1)$, $(-1,2)$; $(1,0)$, $(0,1)$ (d) $(1,2)$, $(-3,1)$; $(1,-1)$, $(2,1)$

2. Verify that the matrices in 1(a) and (b) are inverses of each other. Same for the matrices in 1(c) and (d).

3. Find the transition matrix from the first basis to the second.

   (a) $(1,2)$, $(2,1)$; $(1,0)$, $(0,1)$ (c) $(1,2)$, $(2,1)$; $(3,-1)$, $(-2,1)$
   (b) $(1,0)$, $(0,1)$; $(3,-1)$, $(-2,1)$

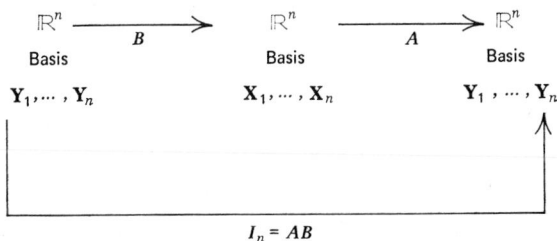

Figure 9-2

4. Explain why the matrix in 3(c) is the product of the matrices in (a) and (b) (in one order).

Selected Answers:

1(a)
$$\begin{pmatrix} \frac{2}{3} & \frac{1}{3} \\ -\frac{1}{3} & \frac{1}{3} \end{pmatrix}$$

1(b)
$$\begin{pmatrix} 1 & -1 \\ 1 & 2 \end{pmatrix}$$

1(c)
$$\begin{pmatrix} -\frac{2}{7} & \frac{5}{7} \\ -\frac{3}{7} & -\frac{3}{7} \end{pmatrix}$$

1(d)
$$\begin{pmatrix} -1 & -\frac{5}{3} \\ 1 & -\frac{2}{3} \end{pmatrix}$$

3(a)
$$\begin{pmatrix} 1 & 2 \\ 2 & 1 \end{pmatrix}$$

3(b)
$$\begin{pmatrix} 1 & 2 \\ 1 & 3 \end{pmatrix}$$

3(c)
$$\begin{pmatrix} 5 & 4 \\ 7 & 5 \end{pmatrix}$$

In Proposition 9-5 we saw that if $\mathbf{X}_1, \ldots, \mathbf{X}_n$ and $\mathbf{Y}_1, \ldots, \mathbf{Y}_n$ are bases for $\mathbb{R}^n$, then there is a nonsingular matrix P which is the matrix for the change of basis from $\mathbf{X}_1, \ldots, \mathbf{X}_n$ to $\mathbf{Y}_1, \ldots, \mathbf{Y}_n$. Thus there are three things present: the old basis $\mathbf{X}_1, \ldots, \mathbf{X}_n$, the new basis $\mathbf{Y}_1, \ldots, \mathbf{Y}_n$, and the transition matrix P. One way to interpret Proposition 9-5 is that the transition matrix P can be determined from the two bases. We now show two things:

(1) given a basis $\mathbf{X}_1, \ldots, \mathbf{X}_n$ and a nonsingular $n \times n$ matrix P, we can find a basis $\mathbf{Y}_1, \ldots, \mathbf{Y}_n$ so that P is the transition matrix from $\mathbf{X}_1, \ldots, \mathbf{X}_n$ to $\mathbf{Y}_1, \ldots, \mathbf{Y}_n$; and

(2) given a basis $\mathbf{Y}_1, \ldots, \mathbf{Y}_n$ and a nonsingular $n \times n$ matrix P, we can find a basis $\mathbf{X}_1, \ldots, \mathbf{X}_n$ so that P is the transition matrix from $\mathbf{X}_1, \ldots, \mathbf{X}_n$ to $\mathbf{Y}_1, \ldots, \mathbf{Y}_n$.

Of these two, the second is easier to show.

Example   Suppose we have the basis for $\mathbb{R}^2$ consisting of $\mathbf{Y}_1 = (1,1)$ and $\mathbf{Y}_2 = (2,-1)$, and suppose we are given the nonsingular matrix

$$P = \begin{pmatrix} 1 & -1 \\ 3 & 4 \end{pmatrix}.$$

In this example we find a basis for $\mathbb{R}^2$ consisting of $\mathbf{X}_1$ and $\mathbf{X}_2$ so that P is the transition matrix from the basis formed by $\mathbf{X}_1$ and $\mathbf{X}_2$ to the basis formed by $\mathbf{Y}_1$ and $\mathbf{Y}_2$.

Recall that the first column of P consists of the coordinates of $\mathbf{X}_1$ relative to $\mathbf{Y}_1$ and $\mathbf{Y}_2$, whereas the second column of P consists of the coordinates of $\mathbf{X}_2$ relative to $\mathbf{Y}_1$ and $\mathbf{Y}_2$. It is therefore a simple matter to find $\mathbf{X}_1$ and $\mathbf{X}_2$.

$$\mathbf{X}_1 = 1\mathbf{Y}_1 + 3\mathbf{Y}_2 = 1(1,1) + 3(2,-1) = (7,-2)$$
$$\mathbf{X}_2 = (-1)\mathbf{Y}_1 + 4\mathbf{Y}_2 = -(1,1) + 4(2,-1) = (7,-5)$$

Therefore the matrix

$$P = \begin{pmatrix} 1 & -1 \\ 3 & 4 \end{pmatrix}$$

is the transition matrix from the basis consisting of $(7,-2)$ and $(7,-5)$ to the basis consisting of $(1,1)$ and $(2,-1)$.

Example Suppose we have the basis for $\mathbb{R}^3$ consisting of $\mathbf{Y}_1 = (1,0,1)$, $\mathbf{Y}_2 = (2,-1,3)$, and $\mathbf{Y}_3 = (-1,1,1)$, and suppose we are given the nonsingular matrix

$$P = \begin{pmatrix} 1 & 2 & -1 \\ 3 & 1 & 0 \\ -1 & 2 & 4 \end{pmatrix}.$$

In this example we find vectors $\mathbf{X}_1$, $\mathbf{X}_2$, and $\mathbf{X}_3$ which form a basis for $\mathbb{R}^3$ so that P is the transition matrix from the basis consisting of $\mathbf{X}_1$, $\mathbf{X}_2$, and $\mathbf{X}_3$ to the basis formed by $\mathbf{Y}_1$, $\mathbf{Y}_2$, and $\mathbf{Y}_3$.

This is a simple matter, because the columns of P consist of the coordinates of the $\mathbf{X}$'s relative to the $\mathbf{Y}$'s. Therefore,

$$\mathbf{X}_1 = \mathbf{Y}_1 + 3\mathbf{Y}_2 - \mathbf{Y}_3 = (1,0,1) + 3(2,-1,3) - (-1,1,1) = (8,-4,9)$$

$$\mathbf{X}_2 = 2\mathbf{Y}_1 + \mathbf{Y}_2 + 2\mathbf{Y}_3 = 2(1,0,1) + (2,-1,3) + 2(-1,1,1) = (2,1,7)$$

$$\mathbf{X}_3 = -\mathbf{Y}_1 + 0\mathbf{Y}_2 + 4\mathbf{Y}_3 = -(1,0,1) + 0(2,-1,3) + 4(-1,1,1) = (-5,4,3).$$

Consequently P is the matrix for the change of basis from $(8,-4,9)$, $(2,1,7)$, and $(-5,4,3)$ to $(1,0,1)$, $(2,-1,3)$, and $(-1,1,1)$.

These two examples illustrate the general situation. If a basis for $\mathbb{R}^n$ consisting of $\mathbf{Y}_1, \ldots, \mathbf{Y}_n$ and a nonsingular $n \times n$ matrix P are given, it is easy to find the basis consisting of $\mathbf{X}_1, \ldots, \mathbf{X}_n$ so that P is the transition matrix from the $\mathbf{X}$'s to the $\mathbf{Y}$'s: the columns of P consist of the coordinates of the $\mathbf{X}$'s relative to the $\mathbf{Y}$'s.

However, if we are given a basis formed by $\mathbf{X}_1, \ldots, \mathbf{X}_n$ and a nonsingular $n \times n$ matrix P, more calculations are required to find the basis formed by $\mathbf{Y}_1, \ldots, \mathbf{Y}_n$ so that P is the transition matrix from $\mathbf{X}_1, \ldots, \mathbf{X}_n$ to $\mathbf{Y}_1, \ldots, \mathbf{Y}_n$.

Example Suppose we are given the basis for $\mathbb{R}^2$ consisting of $\mathbf{X}_1 = (1,1)$ and $\mathbf{X}_2 = (2,-1)$, and suppose we are given the nonsingular matrix

$$P = \begin{pmatrix} 1 & -1 \\ 3 & 4 \end{pmatrix}.$$

In this example we find the basis for $\mathbb{R}^2$ consisting of $\mathbf{Y}_1$ and $\mathbf{Y}_2$ so that P is the transition matrix from $\mathbf{X}_1$ and $\mathbf{X}_2$ to $\mathbf{Y}_1$ and $\mathbf{Y}_2$.

From Proposition 9-5 we know that P^{-1} is the transition matrix from $\mathbf{Y}_1$ and $\mathbf{Y}_2$ to $\mathbf{X}_1$ and $\mathbf{X}_2$. The inverse of P is

$$P^{-1} = \begin{pmatrix} \frac{4}{7} & \frac{1}{7} \\ -\frac{3}{7} & \frac{1}{7} \end{pmatrix}.$$

The columns of P^{-1} are the coordinates of $\mathbf{Y}_1$ and $\mathbf{Y}_2$ relative to $\mathbf{X}_1$ and $\mathbf{X}_2$. Therefore,

$$\mathbf{Y}_1 = \tfrac{4}{7}\mathbf{X}_1 - \tfrac{3}{7}\mathbf{X}_2$$
$$= \tfrac{4}{7}(1,1) - \tfrac{3}{7}(2,-1) = (-\tfrac{2}{7},1)$$
$$\mathbf{Y}_2 = \tfrac{1}{7}\mathbf{X}_1 + \tfrac{1}{7}\mathbf{X}_2$$
$$= \tfrac{1}{7}(1,1) + \tfrac{1}{7}(2,-1) = (\tfrac{3}{7},0).$$

Therefore, the matrix

$$P = \begin{pmatrix} 1 & -1 \\ 3 & 4 \end{pmatrix}$$

is the transition matrix from the basis formed by $(1,1)$ and $(2,-1)$ to the basis formed by $(-\tfrac{2}{7},1)$ and $(\tfrac{3}{7},0)$.

This example shows that if we are given a basis for $\mathbb{R}^n$ formed by $\mathbf{X}_1, \ldots, \mathbf{X}_n$ and a nonsingular $n \times n$ matrix P, it is possible to find a basis formed by $\mathbf{Y}_1, \ldots, \mathbf{Y}_n$ so that P is the transition matrix from $\mathbf{X}_1, \ldots, \mathbf{X}_n$ to $\mathbf{Y}_1, \ldots, \mathbf{Y}_n$. One way is to find P^{-1}, the transition matrix from $\mathbf{Y}_1, \ldots, \mathbf{Y}_n$ to $\mathbf{X}_1, \ldots, \mathbf{X}_n$. The columns of P^{-1} are the coordinates of the $\mathbf{Y}$'s relative to the $\mathbf{X}$'s, so we can then find the $\mathbf{Y}$'s.

Proposition 9-6

Suppose P is a nonsingular $n \times n$ matrix.
(1) If a basis for $\mathbb{R}^n$ consisting of $\mathbf{X}_1, \ldots, \mathbf{X}_n$ is given, there is a basis formed by $\mathbf{Y}_1, \ldots, \mathbf{Y}_n$ so that P is the transition matrix from $\mathbf{X}_1, \ldots, \mathbf{X}_n$ to $\mathbf{Y}_1, \ldots, \mathbf{Y}_n$.
(2) If a basis for $\mathbb{R}^n$ consisting of $\mathbf{Y}_1, \ldots, \mathbf{Y}_n$ is given, then there is a basis formed by $\mathbf{X}_1, \ldots, \mathbf{X}_n$ so that P is the transition matrix from $\mathbf{X}_1, \ldots, \mathbf{X}_n$ to $\mathbf{Y}_1, \ldots, \mathbf{Y}_n$.

Brief Exercises

1. Suppose P is the nonsingular matrix

$$\begin{pmatrix} 2 & 1 \\ -1 & 3 \end{pmatrix}.$$

In each of the following, find X_1 and X_2 so that P is the transition matrix from X_1 and X_2 to Y_1 and Y_2.

(a) $Y_1 = (1,-2)$, $Y_2 = (-1,5)$ (c) $Y_1 = (1,0)$, $Y_2 = (0,1)$

(b) $Y_1 = (-1,5)$, $Y_2 = (1,-2)$

2. Let P be the matrix in Exercise 1. Find Y_1 and Y_2 so that P is the transition matrix from $(2,1)$ and $(1,-1)$ to Y_1 and Y_2.

3. Let P be the nonsingular matrix
$$\begin{pmatrix} 1 & 0 & 1 \\ -1 & 2 & 0 \\ 0 & 1 & 2 \end{pmatrix}.$$

(a) If $Y_1 = (2,1,3)$, $Y_2 = (1,1,0)$, and $Y_3 = (1,-1,1)$, find X_1, X_2, and X_3 so that P is the transition matrix from the X's to the Y's.

(b) If $X_1 = (2,1,3)$, $X_2 = (1,1,0)$, and $X_3 = (1,-1,1)$, find Y_1, Y_2, and Y_3 so that P is the transition matrix from the X's to the Y's.

Answers: **1(a)** $X_1 = (3,-9)$, $X_2 = (-2,13)$ **1(b)** $X_1 = (-3,12)$, $X_2 = (2,-1)$ **1(c)** $X_1 = (2,-1)$, $X_2 = (1,3)$ **2** $Y_1 = (1,\frac{2}{7})$, $Y_2 = (0,-\frac{3}{7})$ **3(a)** $X_1 = (1,0,3)$, $X_2 = (3,1,1)$, $X_3 = (4,-1,5)$ **3(b)** $Y_1 = (3,\frac{7}{3},\frac{11}{3})$, $Y_2 = (1,\frac{4}{3},\frac{2}{3})$, $Y_3 = (-1,-\frac{5}{3},-\frac{4}{3})$

SUMMARY *If* $X_1, \ldots, X_n$ *and* $Y_1, \ldots, Y_n$ *are bases for* $\mathbb{R}^n$, *the transition matrix from* $X_1, \ldots, X_n$ *to* $Y_1, \ldots, Y_n$ *is formed by using the coordinates of the* X's *relative to the* Y's *as columns; this transition matrix is nonsingular, and its inverse is the transition matrix from the* Y's *to the* X's.
    If a nonsingular $n \times n$ *matrix* P *is given and if* $X_1, \ldots, X_n$ *is a basis, then there is a basis* $Y_1, \ldots, Y_n$ *so that* P *is the transition matrix from* $X_1, \ldots, X_n$ *to* $Y_1, \ldots, Y_n$. *Also if* P *is given and* $Y_1, \ldots, Y_n$ *is a basis, then there is a basis* $X_1, \ldots, X_n$ *so that* P *is the transition matrix from* $X_1, \ldots, X_n$ *to* $Y_1, \ldots, Y_n$.

Exercises for Section 9-2

1. Find the matrix for the change of bases indicated.

(a) In $\mathbb{R}^2$ from $(1,0)$, $(0,1)$ to $(2,3)$, $(-1,2)$
(b) In $\mathbb{R}^2$ from $(2,3)$, $(-1,2)$ to $(1,0)$, $(0,1)$
(c) In $\mathbb{R}^2$ from $(2,3)$, $(-1,2)$ to $(3,0)$, $(-1,4)$
(d) In $\mathbb{R}^3$ from $(1,0,0)$, $(0,1,0)$, $(0,0,1)$ to $(1,1,0)$, $(1,0,1)$, $(0,1,1)$
(e) In $\mathbb{R}^3$ from $(1,1,0)$, $(1,0,1)$, $(0,1,1)$ to $(1,0,0)$, $(1,1,0)$, $(1,1,1)$

2. Find an ordered basis $\mathbf{X}_1$, $\mathbf{X}_2$ in $\mathbb{R}^2$ so that the matrix

$$\begin{pmatrix} 1 & 1 \\ 2 & 3 \end{pmatrix}$$

is the matrix for the change of basis from $\mathbf{X}_1$, $\mathbf{X}_2$ to $(1,0)$, $(0,1)$.

3. Find an ordered basis $\mathbf{Y}_1$, $\mathbf{Y}_2$ in $\mathbb{R}^2$ so that the matrix

$$\begin{pmatrix} 1 & 1 \\ 2 & 3 \end{pmatrix}$$

is the matrix for the change of basis from $(1,0)$, $(0,1)$ to $\mathbf{Y}_1, \mathbf{Y}_2$.

4. Find an ordered basis $\mathbf{X}_1$, $\mathbf{X}_2$, $\mathbf{X}_3$ in $\mathbb{R}^3$ so that the matrix

$$\begin{pmatrix} 1 & 0 & 1 \\ 0 & 1 & 1 \\ 1 & 1 & 0 \end{pmatrix}$$

is the matrix for the change of basis from $\mathbf{X}_1$, $\mathbf{X}_2$, $\mathbf{X}_3$ to $(1,0,0)$, $(0,1,0)$, $(0,0,1)$.

5. Find an ordered basis $\mathbf{Y}_1$, $\mathbf{Y}_2$, $\mathbf{Y}_3$ in $\mathbb{R}^3$ so that the matrix

$$\begin{pmatrix} 1 & 0 & 1 \\ 0 & 1 & 1 \\ 1 & 1 & 0 \end{pmatrix}$$

is the matrix for the change of basis from $(1,0,0)$, $(0,1,0)$, $(0,0,1)$ to $\mathbf{Y}_1$, $\mathbf{Y}_2$, $\mathbf{Y}_3$.

More Challenging Exercises

6. A change of basis in $\mathbb{R}^n$ is called *orthogonal* if the matrix for the change of basis is an orthogonal matrix (refer to Exercises 15 and 16 in Section 8-3). Show that an orthogonal change of basis does not change distances or angles.

7. Suppose A is an orthogonal matrix. Show that A transforms any orthonormal basis into an orthonormal basis; i.e., if $\mathbf{X}_1, \ldots, \mathbf{X}_n$ form an orthonormal basis, then $A\mathbf{X}_1, \ldots, A\mathbf{X}_n$ form an orthonormal basis also.

8. Suppose A is the matrix for a change of basis from an orthonormal basis to another orthonormal basis in $\mathbb{R}_n$. Must A be orthogonal?

9. Suppose $\mathbf{X}_1, \ldots, \mathbf{X}_n$ is an ordered basis for $\mathbb{R}^n$ and A is a nonsingular $n \times n$ matrix. Proposition 9-6 states that there is an ordered basis $\mathbf{Y}_1, \ldots, \mathbf{Y}_n$ so

that A is the matrix for the change of basis from the **X**'s to the **Y**'s. Is the ordered basis $Y_1, \ldots, Y_n$ determined uniquely by this condition?

10. Suppose $X_1, \ldots, X_n$ and $Y_1, \ldots, Y_n$ are two bases for $\mathbb{R}^n$. If A is the $n \times n$ matrix having the **X**'s as columns (in the order indicated by the subscripts) and B is the $n \times n$ matrix having the **Y**'s as columns (in the same order), then show that $A^{-1}B$ is the transition matrix from the **Y**'s to the **X**'s.

SECTION 9-3
Different Matrices for the Same Linear Transformation, Equivalence of Matrices

PREVIEW *From Section 9-1 we know that most linear transformations can be represented by many matrices depending upon the ordered bases used. In this section we show that two matrices A and B of the same size represent the same linear transformation if and only if they are related by the equation $B = PAQ$, where P and Q are nonsingular matrices of the proper size.*

In Section 9-1 we explained how to find the matrix for a linear transformation $T: \mathbb{R}^n \to \mathbb{R}^m$ relative to some given ordered bases. You probably have noticed from the examples in Section 9-1 that if you change the ordered basis in $\mathbb{R}^n$ or in $\mathbb{R}^m$ (or in both), you are likely to change the matrix needed to represent T. In this section we discuss this point in detail. We show that the matrices A and B represent the same linear transformation relative to possibly different ordered bases if and only if A and B are related to each other by the equation $B = PAQ$, where P and Q are properly chosen nonsingular matrices. We illustrate this with some examples.

Example Consider the linear transformation $T: \mathbb{R}^2 \to \mathbb{R}^3$ given by $T(x,y) = (x + y, x - y, 2x + 3y)$. The matrix for T relative to the bases consisting of $(1,0)$ and $(0,1)$ in $\mathbb{R}^2$ and $(1,0,0)$, $(1,1,0)$, and $(1,1,1)$ in $\mathbb{R}^3$ is

$$A = \begin{pmatrix} 0 & 2 \\ -1 & -4 \\ 2 & 3 \end{pmatrix}.$$

The matrix for T relative to the bases consisting of $(1,1)$ and $(2,3)$ in $\mathbb{R}^2$ and $(1,0,1)$, $(1,1,0)$, and $(0,1,1)$ in $\mathbb{R}^3$ is

$$B = \begin{pmatrix} \frac{7}{2} & \frac{19}{2} \\ -\frac{3}{2} & -\frac{9}{2} \\ \frac{3}{2} & \frac{7}{2} \end{pmatrix}.$$

The matrices A and B were determined by the method outlined in Section 9-1. The goal of this example is to find the nonsingular matrices P and Q so that $B = PAQ$.

To do this, consider Figure 9-3. We can think of the transformation in two ways. One is indicated by the top arrow. The other way to think of T is to follow the

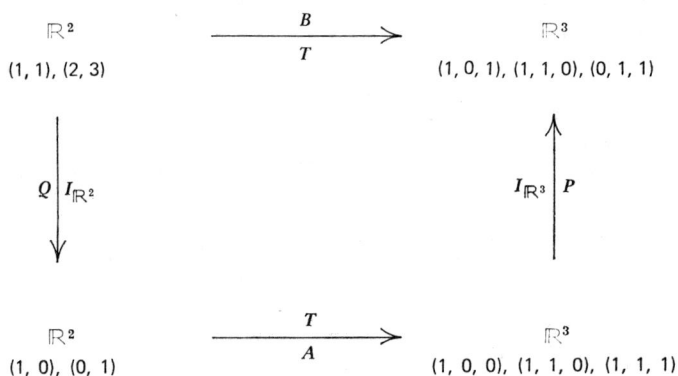

$$
\begin{array}{ccc}
\mathbb{R}^2 & \xrightarrow[\ T\ ]{B} & \mathbb{R}^3 \\
(1,1),\,(2,3) & & (1,0,1),\,(1,1,0),\,(0,1,1) \\
\Big\downarrow{\scriptstyle Q}\ {\scriptstyle I_{\mathbb{R}^2}} & & \Big\uparrow{\scriptstyle I_{\mathbb{R}^3}}\ {\scriptstyle P} \\
\mathbb{R}^2 & \xrightarrow[\ A\ ]{T} & \mathbb{R}^3 \\
(1,0),\,(0,1) & & (1,0,0),\,(1,1,0),\,(1,1,1)
\end{array}
$$

Figure 9-3

arrows down, across, and up. The longer path represents the function $T = I_{\mathbb{R}^3} \circ T \circ I_{\mathbb{R}^2}$. The matrix Q is the transition matrix from $(1,1)$ and $(2,3)$ to $(1,0)$ and $(0,1)$.

$$Q = \begin{pmatrix} 1 & 2 \\ 1 & 3 \end{pmatrix}$$

The matrix P is the transition matrix from $(1,0,0)$, $(1,1,0)$, and $(1,1,1)$ to $(1,0,1)$, $(1,1,0)$, and $(0,1,1)$.

$$P = \begin{pmatrix} \frac{1}{2} & 0 & \frac{1}{2} \\ \frac{1}{2} & 1 & \frac{1}{2} \\ -\frac{1}{2} & 0 & \frac{1}{2} \end{pmatrix}$$

You can easily check that $B = PAQ$.

Example   Consider the matrices

$$A = \begin{pmatrix} 3 & 1 \\ 0 & 2 \end{pmatrix} \qquad B = \begin{pmatrix} 2 & 14 \\ -6 & 12 \end{pmatrix}$$

$$P = \begin{pmatrix} 2 & 1 \\ 0 & 3 \end{pmatrix} \qquad Q = \begin{pmatrix} 1 & 1 \\ -1 & 2 \end{pmatrix}.$$

It is easy to verify that P and Q are nonsingular and that $B = PAQ$. The goal of this example is to show that A and B are matrices for the same linear transformation (see Figure 9-4).

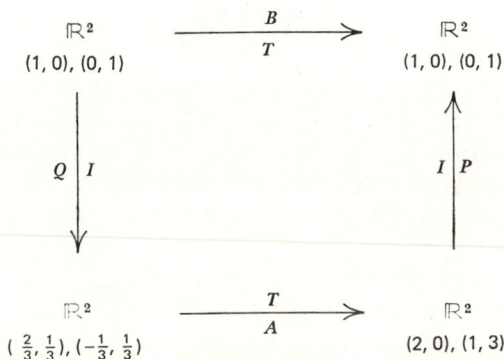

Figure 9-4

The matrix B represents the transformation given by $T(x,y) = (2x + 14y, -6x + 12y)$ relative to the natural bases as indicated by the top arrow in Figure 9-4. The matrix Q is the transition matrix from $(1,0)$ and $(0,1)$ to $(\frac{2}{3},\frac{1}{3})$ and $(-\frac{1}{3},\frac{1}{3})$; the matrix P is the transition matrix from $(2,0)$ and $(1,3)$ to $(1,0)$ and $(0,1)$ as we explained before Proposition 9-6. Then A represents T relative to the basis $(\frac{2}{3},\frac{1}{3})$ and $(-\frac{1}{3},\frac{1}{3})$ in the domain of T and $(2,0)$ and $(1,3)$ in the image of T.

These examples illustrate the fact that the matrices A and B represent the same linear transformation if and only if $B = PAQ$, where P and Q are nonsingular matrices. This can be seen by referring to Figure 9-5. Suppose we know that $B = PAQ$. Pick bases $\mathbf{X}'_1, \ldots, \mathbf{X}'_n$ for $\mathbb{R}^n$ and $\mathbf{Y}'_1, \ldots, \mathbf{Y}'_m$ for $\mathbb{R}^m$. Then by Proposition 9-1 the matrix A represents a linear transformation relative to these bases. By Proposition 9-6 there are bases $\mathbf{X}_1, \ldots, \mathbf{X}_n$ for $\mathbb{R}^n$ and $\mathbf{Y}_1, \ldots, \mathbf{Y}_m$ for $\mathbb{R}^m$ so that Q is the transition matrix from $\mathbf{X}_1, \ldots, \mathbf{X}_n$ to $\mathbf{X}'_1, \ldots, \mathbf{X}'_n$ and so that P is the transition matrix from $\mathbf{Y}'_1, \ldots, \mathbf{Y}'_m$ to $\mathbf{Y}_1, \ldots, \mathbf{Y}_m$. Then the matrix B represents T relative to $\mathbf{X}_1, \ldots, \mathbf{X}_n$ and $\mathbf{Y}_1, \ldots, \mathbf{Y}_m$, since PAQ is the matrix for $I_{\mathbb{R}^m} \circ T \circ I_{\mathbb{R}^n} = T$.

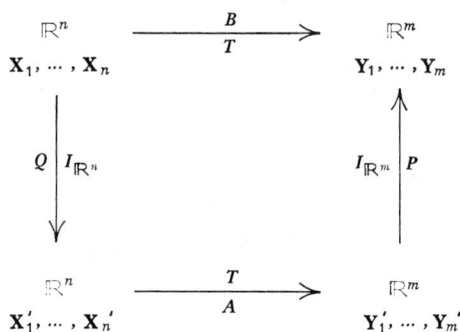

Figure 9-5

On the other hand, suppose B represents T relative to $\mathbf{X}_1, \ldots, \mathbf{X}_n$ and $\mathbf{Y}_1, \ldots,$ $\mathbf{Y}_m$, and suppose A represents T relative to $\mathbf{X}'_1, \ldots, \mathbf{X}'_n$ and $\mathbf{Y}'_1, \ldots, \mathbf{Y}'_m$. If Q is the transition matrix from $\mathbf{X}_1, \ldots, \mathbf{X}_n$ to $\mathbf{X}'_1, \ldots, \mathbf{X}'_n$ and P is the transition matrix from $\mathbf{Y}'_1, \ldots, \mathbf{Y}'_m$ to $\mathbf{Y}_1, \ldots, \mathbf{Y}_m$ then $B = PAQ$ because PAQ is the matrix for $T = I_{\mathbb{R}^m} \circ T \circ I_{\mathbb{R}^n}$ relative to $\mathbf{X}_1, \ldots, \mathbf{X}_n$ and $\mathbf{Y}_1, \ldots, \mathbf{Y}_m$.

Proposition 9-7_____
The matrices A and B represent the same linear transformation (relative to possibly different ordered bases) if and only if $B = PAQ$, where P and Q are nonsingular matrices.

Definition 9-4_____
The matrices A and B are *equivalent* if and only if there are nonsingular matrices P and Q so that $B = PAQ$.

There is another way to interpret this result. Recall from Proposition 5-6 that a nonsingular matrix is a product of elementary matrices. Furthermore, recall from Proposition 5-2 that the row operations can be effected by multiplying on the left with elementary matrices, and from Exercise 4 in Section 5-2 that the column operations can be effected by multiplying on the right by elementary matrices.

So, suppose $B = PAQ$, where P and Q are nonsingular. Then there are elementary matrices $G_1, \ldots, G_k$ and $H_1, \ldots, H_t$ so that $P = G_k \cdots G_1$ and $Q = H_1 \cdots H_t$. Thus $B = G_k \cdots G_1 A H_1 \cdots H_t$. Consequently A and B are equivalent if and only if one can be obtained from the other by a succession of row and column operations.

Practically speaking, how can we decide whether two matrices are equivalent? The answer to this lies in the observation that if A and B are equivalent to a third matrix C, then they are equivalent to each other. This fact is easy to see. If A and C

represent the same linear function, and if B and C represent the same linear function, then A and B represent the same linear function. (Or, if you prefer, you can make a matrix argument: suppose $C = P_1 A Q_1$ and $C = P_2 B Q_2$, where P_1, P_2, Q_1, and Q_2 are nonsingular matrices. Then $P_1 A Q_1 = P_2 B Q_2$ and so $B = P_2^{-1} P_1 A Q_1 Q_2^{-1}$. Since the product of nonsingular matrices is nonsingular and since the inverse of a nonsingular matrix is nonsingular, A and B are equivalent, because $P_2^{-1} P_1$ and $Q_1 Q_2^{-1}$ are nonsingular.)

We transform A and B by row and column operations to matrices of the form

$$
\begin{pmatrix}
1 & 0 & \cdots & 0 & 0 & \cdots & 0 \\
0 & 1 & \cdots & 0 & 0 & \cdots & 0 \\
\cdot & \cdot & & \cdot & \cdot & & \cdot \\
\cdot & \cdot & & \cdot & \cdot & & \cdot \\
\cdot & \cdot & & \cdot & \cdot & & \cdot \\
0 & 0 & \cdots & 1 & 0 & \cdots & 0 \\
0 & 0 & \cdots & 0 & 0 & \cdots & 0 \\
\cdot & \cdot & & \cdot & \cdot & & \cdot \\
\cdot & \cdot & & \cdot & \cdot & & \cdot \\
\cdot & \cdot & & \cdot & \cdot & & \cdot \\
0 & 0 & \cdots & 0 & 0 & \cdots & 0
\end{pmatrix}
$$

where the $r \times r$ identity is in the upper left hand corner. Thus A and B are equivalent if and only if they can be transformed to the same matrix in the form above by row and column operations. But the number of nonzero rows in the matrix above is the rank of the matrix by Proposition 7-1. So, finally, the matrices A and B are equivalent if and only if they have the same rank.

Proposition 9-8

Suppose A and B are matrices of the same size. The following statements are equivalent.

(1) The matrices A and B are equivalent.

(2) The matrices A and B represent the same linear transformation relative to possibly different ordered bases.

(3) Each matrix can be obtained from the other by a succession of row and column transformations.

(4) The matrices A and B have the same rank.

(5) The matrices A and B can be transformed by row and column operations to the same matrix of the form

$$
\begin{pmatrix} I_r & 0 \\ 0 & 0 \end{pmatrix}
$$

where I_r is the $r \times r$ identity.

SUMMARY   *This section is devoted to establishing conditions which are equivalent to the fact that two matrices A and B of the same size represent the same linear transformation. They are:*

(1) *B = PAQ, where P and Q are nonsingular,*
(2) *each can be obtained from the other by some row and column operations,*
(3) *A and B have the same rank, and*
(4) *each can be transformed with row and column operations to a matrix of the form*

$$\begin{pmatrix} I_r & 0 \\ 0 & 0 \end{pmatrix}.$$

Exercises for Section 9-3

1. In each of the following let A be the given matrix. Using row and column operations, transform A into the form

$$B = \begin{pmatrix} I_r & 0 \\ 0 & 0 \end{pmatrix}.$$

Find nonsingular matrices P and Q of the appropriate size so that $B = PAQ$.

(a) $\begin{pmatrix} 2 & 1 & 3 \\ 1 & 4 & 1 \end{pmatrix}$ (b) $\begin{pmatrix} -1 & 0 & 2 \\ 1 & 1 & 3 \\ 0 & 1 & 5 \end{pmatrix}$ (c) $\begin{pmatrix} 2 & 1 & 0 \\ 3 & 0 & 1 \\ 0 & -1 & 2 \end{pmatrix}$

2. In Exercise 1(a) transform A to the form B described there in two different ways. Thus find nonsingular matrices P, P_1, Q, and Q_1 so that $B = PAQ = P_1AQ_1$ (this shows that the matrices P and Q are not unique).

3. Determine whether or not the following matrices are equivalent. If they are, find P and Q, nonsingular matrices, so that $B = PAQ$.

$$A = \begin{pmatrix} 2 & 1 & 3 \\ 1 & 4 & 1 \end{pmatrix} \qquad B = \begin{pmatrix} -1 & 2 & 4 \\ 5 & 1 & 0 \end{pmatrix}$$

4. Same exercise as 3 for

$$A = \begin{pmatrix} -1 & 1 & 2 \\ 5 & 0 & 1 \\ 0 & -2 & 1 \end{pmatrix} \qquad B = \begin{pmatrix} 4 & 2 & 1 \\ 1 & 1 & 2 \\ -3 & 0 & 1 \end{pmatrix}.$$

More Challenging Exercises

In Exercises 5 through 8 we are considering the following situation. Suppose $\mathcal{V}$ and $\mathcal{W}$ are vector spaces over $\mathbb{R}$, that $T: \mathcal{V} \to \mathcal{W}$ is a linear transformation, that $\mathbf{X}_1, \ldots, \mathbf{X}_n$ and $\mathbf{Y}_1, \ldots, \mathbf{Y}_m$ are ordered bases for $\mathcal{V}$ and $\mathcal{W}$, respectively, and that A is the matrix for T relative to these bases.

We have seen in Proposition 9-8 that if the matrices A and B are equivalent, then they represent the same linear transformation relative to possibly different ordered bases. The situation we have in mind is this: suppose $B = PAQ$, where P and Q are nonsingular; we develop a way of using P and Q to find ordered bases so that B is the matrix for T relative to them. We do this by determining the changes in the bases corresponding to row and column operations on the matrix A.

5. Give an argument to support the following statements.

   (a) $F_{i,j}A$ is the matrix for T relative to $\mathbf{X}_1, \ldots, \mathbf{X}_n$ and $\mathbf{Y}_1, \ldots, \mathbf{Y}_j, \ldots,$ $\mathbf{Y}_i, \ldots, \mathbf{Y}_m$; i.e., the position of $\mathbf{Y}_i$ and $\mathbf{Y}_j$ has been switched.

   (b) $F_i(x)A$ is the matrix for T relative to $\mathbf{X}_1, \ldots, \mathbf{X}_n$ and $\mathbf{Y}_1, \ldots, x^{-1}\mathbf{Y}_i, \ldots,$ $\mathbf{Y}_m$; i.e., the vector $\mathbf{Y}_i$ has been multiplied by x^{-1}.

   (c) $F_{i,j}(x)A$ is the matrix for T relative to $\mathbf{X}_1, \ldots, \mathbf{X}_n$ and $\mathbf{Y}_1, \ldots, \mathbf{Y}_i, \ldots,$ $\mathbf{Y}_j - x\mathbf{Y}_i, \ldots, \mathbf{Y}_n$; i.e., $\mathbf{Y}_j$ is replaced by $\mathbf{Y}_j - x\mathbf{Y}_i$.

   (d) $AF_{i,j}$ is the matrix for T relative to $\mathbf{X}_1, \ldots, \mathbf{X}_j, \ldots, \mathbf{X}_i, \ldots, \mathbf{X}_n$ and $\mathbf{Y}_1, \ldots, \mathbf{Y}_m$; i.e., the position of $\mathbf{X}_i$ and $\mathbf{X}_j$ has been switched.

   (e) $AF_i(x)$ is the matrix for T relative to $\mathbf{X}_1, \ldots, x^{-1}\mathbf{X}_i, \ldots, \mathbf{X}_n$ and $\mathbf{Y}_1, \ldots,$ $\mathbf{Y}_m$; i.e., $\mathbf{X}_i$ has been multiplied by x^{-1}.

   (f) $AF_{i,j}(x)$ is the matrix for T relative to $\mathbf{X}_1, \ldots, \mathbf{X}_i - x\mathbf{X}_j, \ldots, \mathbf{X}_j, \ldots, \mathbf{X}_n$ and $\mathbf{Y}_1, \ldots, \mathbf{Y}_m$; i.e., $\mathbf{X}_i$ has been replaced by $\mathbf{X}_i - x\mathbf{X}_j$.

6. The facts in Exercise 5 are too complicated to be remembered in the form presented there. The purpose of this exercise is to simplify the statements found there by the introduction of some notation. Let $\mathcal{X} = (\mathbf{X}_1, \ldots, \mathbf{X}_n)$ and $\mathcal{Y} = (\mathbf{Y}_1, \ldots, \mathbf{Y}_m)$; i.e., $\mathcal{X}$ and $\mathcal{Y}$ denote *ordered* bases thought of a n-tuples of *vectors*.

   Suppose B is an $n \times n$ matrix, for example. Then it is possible to assign a natural meaning to $\mathcal{X}B$; it may be thought of as multiplication of a $1 \times m$ matrix with entries from $\mathcal{V}$ and an $m \times n$ matrix with entries from $\mathbb{R}$.

$$(\mathbf{X}_1, \ldots, \mathbf{X}_n) \begin{pmatrix} b_{11} & b_{12} & \cdots & b_{1n} \\ \cdot & \cdot & & \cdot \\ \cdot & \cdot & & \cdot \\ \cdot & \cdot & & \cdot \\ b_{n1} & b_{n2} & \cdots & b_{nn} \end{pmatrix}$$

$$= (b_{11}\mathbf{X}_1 + \cdots + b_{n1}\mathbf{X}_n, \; b_{12}\mathbf{X}_1 + \cdots + b_{n2}\mathbf{X}_n, \ldots, \; b_{1n}\mathbf{X}_1 + \cdots + b_{nn}\mathbf{X}_n)$$

If, therefore, you think of $\mathscr{X}$ and $\mathscr{Y}$ as $1 \times n$ and $1 \times m$ matrices with entries from $\mathscr{V}$, then give arguments to support the following statements.

(a) $F_{i,j}A$ is the matrix for T relative to $\mathscr{X}$ and $\mathscr{Y} F_{i,j}$.

(b) $F_i(x)A$ is the matrix for T relative to $\mathscr{X}$ and $\mathscr{Y} F_i(x^{-1})$.

(c) $F_{i,j}(x)A$ is the matrix for T relative to $\mathscr{X}$ and $\mathscr{Y} F_{i,j}(-x)$.

These three statements can be combined into one statement. (1) If F is an elementary matrix, then FA is the matrix for T relative to $\mathscr{X}$ and $\mathscr{Y} F^{-1}$.

Give arguments to support the following three statements.

(d) $AF_{i,j}$ is the matrix for T relative to $\mathscr{X}F_{i,j}$ and $\mathscr{Y}$.

(e) $AF_i(x)$ is the matrix for T relative to $\mathscr{X}F_i(x)$ and $\mathscr{Y}$.

(f) $AF_{i,j}(x)$ is the matrix for T relative to $\mathscr{X}F_{i,j}(x)$ and $\mathscr{Y}$.

These three statements can be combined into one statement. (2) If F is an elementary matrix, then AF is the matrix for T relative to $\mathscr{X}F$ and $\mathscr{Y}$.

7. Use the results of Exercise 6 to justify the following statement. If $B = PAQ$, then B is the matrix for T relative to $\mathscr{X}Q$ and $\mathscr{Y}P^{-1}$.

8. By considering Figure 9-6, give another argument for the result stated in Exercise 7.

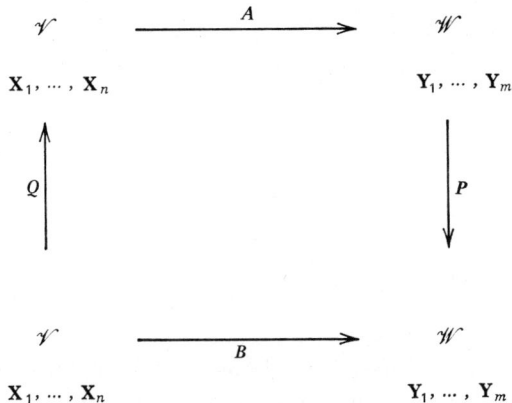

Figure 9-6

10

Eigenvalues, Eigenvectors, Quadratic Functions

Eigenvalues and Eigenvectors

PREVIEW *First we show the kind of information we can obtain about a linear transformation if its matrix is a diagonal matrix. Then we introduce the concepts of eigenvalues, eigenvectors, and characteristic polynomials for linear transformations and for matrices. Then we show that a linear transformation can be represented by a diagonal matrix if and only if there is a basis consisting of eigenvectors.*

As we mentioned at the beginning of Chapter 9, we can study linear and quadratic functions more effectively if we are not restricted to using the natural bases. We discuss quadratic functions in Section 10-2. In this section we concentrate on linear transformations.

In Section 9-1 we introduced the matrix for a linear transformation $T: \mathbb{R}^n \to \mathbb{R}^m$ relative to a basis in $\mathbb{R}^n$ and a basis in $\mathbb{R}^m$. Consequently, we needed to specify two bases, one in the domain of T and one in the space containing the image of T, even in the case that T is a linear transformation from $\mathbb{R}^n$ to $\mathbb{R}^n$. In this chapter we discuss only linear transformations from $\mathbb{R}^n$ to $\mathbb{R}^n$; therefore we deal only with square matrices. Furthermore, we now require that the same basis be used in the domain and in the space containing the image of T, except when we explicitly mention otherwise such as in a change of basis.

Definition 10-1_____

Suppose $\mathbf{X}_1, \ldots, \mathbf{X}_n$ is a basis for $\mathbb{R}^n$ and $T: \mathbb{R}^n \to \mathbb{R}^n$ is a linear transformation. We say that *A is the matrix for T relative to* $\mathbf{X}_1, \ldots, \mathbf{X}_n$ if this one basis has been used both in the domain and in the space containing the image to construct A as outlined in Definition 9-2.

Example   Suppose the matrix for $T: \mathbb{R}^2 \to \mathbb{R}^2$ relative to the natural basis is

$$\begin{pmatrix} 3 & 0 \\ 0 & 2 \end{pmatrix}.$$

This means that T stretches vectors along the x-axis by a factor of 3 since

$$\begin{pmatrix} 3 & 0 \\ 0 & 2 \end{pmatrix}\begin{pmatrix} x \\ 0 \end{pmatrix} = \begin{pmatrix} 3x \\ 0 \end{pmatrix}$$

and that it stretches vectors along the y-axis by a factor of 2 since

$$\begin{pmatrix} 3 & 0 \\ 0 & 2 \end{pmatrix}\begin{pmatrix} 0 \\ y \end{pmatrix} = \begin{pmatrix} 0 \\ 2y \end{pmatrix}.$$

Consequently T sends the x-axis into itself, magnifying every vector along the x-axis by a factor 3, and it sends the y-axis into itself, magnifying vectors along the y-axis by a factor of 2. The important thing to observe is that there are two lines which T sends back into themselves, namely the lines determined by the basis vectors. This can be detected from the matrix since it is a diagonal matrix (Definition 6-6 in Exercises for Section 6-4) (see Figure 10-1).

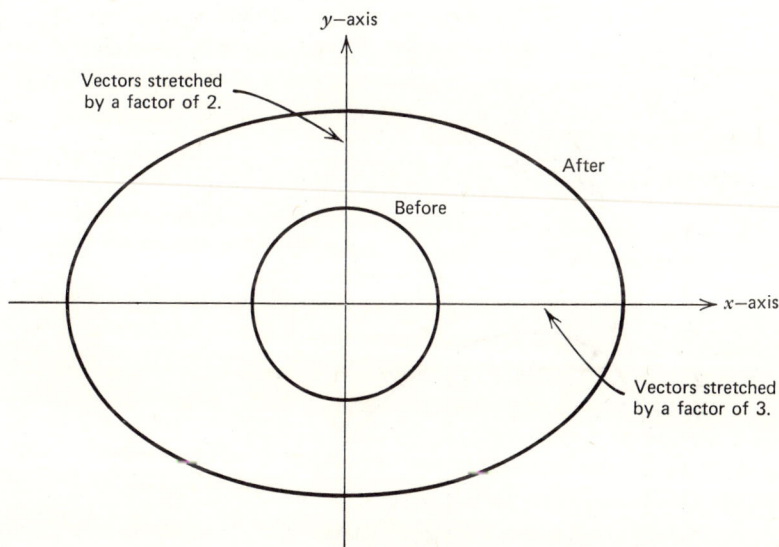

Figure 10-1

Example   Suppose $T: \mathbb{R}^2 \to \mathbb{R}^2$ is given by $T(x,y) = \frac{1}{5}(12x - y), \frac{1}{5}(-6x + 3y))$. The matrix for T relative to the natural basis is

$$\begin{pmatrix} \frac{12}{5} & -\frac{1}{5} \\ -\frac{6}{5} & \frac{13}{5} \end{pmatrix}.$$

However, the matrix for T relative to the basis consisting of $(1,2)$ and $(-1,3)$ is

$$\begin{pmatrix} 2 & 0 \\ 0 & 3 \end{pmatrix}.$$

Relative to (1,2) and (−1,3), the coordinates of a vector along the line determined by (1,2) are of the form (a,0), and the coordinates of a vector along the line determined by (−1,3) are of the form (0,b). Since

$$\begin{pmatrix} 2 & 0 \\ 0 & 3 \end{pmatrix} \begin{pmatrix} a \\ 0 \end{pmatrix} = \begin{pmatrix} 2a \\ 0 \end{pmatrix}$$

and

$$\begin{pmatrix} 2 & 0 \\ 0 & 3 \end{pmatrix} \begin{pmatrix} 0 \\ b \end{pmatrix} = \begin{pmatrix} 0 \\ 3b \end{pmatrix},$$

the transformation T stretches vectors along the line determined by (1,2) through a factor of 2 and those along the line determined by (−1,3) through a factor of 3 (see Figure 10-2).

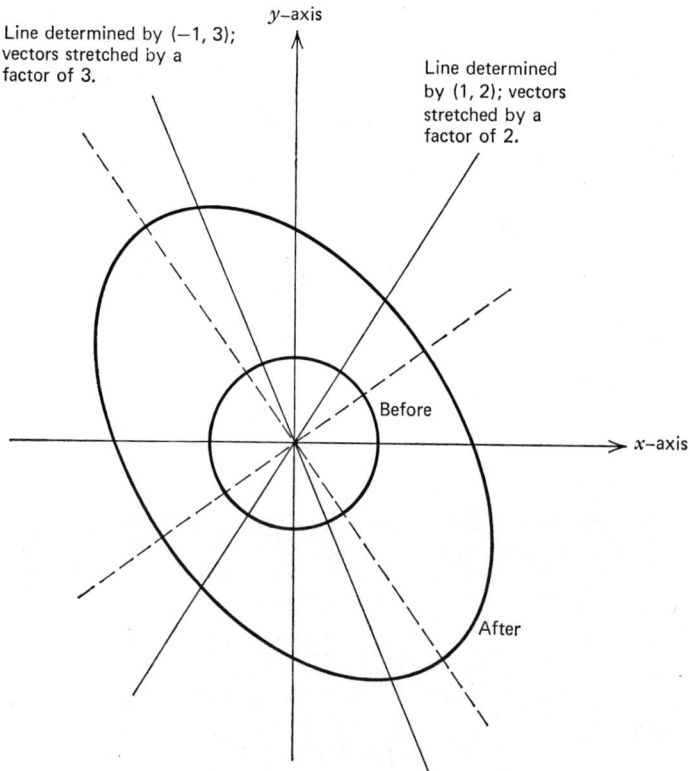

Figure 10-2

The important observation is that because the matrix for T is a diagonal matrix, the transformation maps the line determined by the first basis vector into itself and the line determined by the second basis vector into itself. The entry in the (1,1) position gives the magnification factor along the first line; the entry in the (2,2) position gives the magnification factor along the second line. From this information we can determine the effect of T on any vector (x,y). If the coordinates of (x,y) relative to (1,2) and $(-1,3)$ are (a,b), then the coordinates of $T(x,y)$ relative to the same basis are $(2a,3b)$.

These two examples give an indication of what we seek. If $T: \mathbb{R}^n \to \mathbb{R}^n$ is a linear transformation, we want to find a basis for $\mathbb{R}^n$ so that the matrix for T relative to this basis is a diagonal matrix.

Example Consider the linear transformation $T: \mathbb{R}^3 \to \mathbb{R}^3$ defined by $(2x - 3y + z, z - y, 0)$. The matrix for this transformation relative to the basis consisting of $(1,0,0)$, $(1,1,0)$, and $(1,1,1)$ is the diagonal matrix

$$\begin{pmatrix} 2 & 0 & 0 \\ 0 & -1 & 0 \\ 0 & 0 & 0 \end{pmatrix}.$$

From this matrix we immediately get the following information. The transformation magnifies by a factor of 2 the vectors along the line determined by $(1,0,0)$; it reverses the vectors along the line determined by $(1,1,0)$; and it collapses to the zero vector all vectors along the line determined by $(1,1,1)$. Furthermore if

$$(x,y,z) = a(1,0,0) + b(1,1,0) + c(1,1,1)$$

then

$$T(x,y,z) = 2a(1,0,0) - b(1,1,0) + 0(1,1,1).$$

That is, T doubles the component in the $(1,0,0)$ direction, multiplies by -1 the component in the $(1,1,0)$ direction, and nullifies the component in the $(1,1,1)$ direction.

Brief Exercises

1. Suppose the matrix for $T: \mathbb{R}^2 \to \mathbb{R}^2$ relative to the ordered basis $(2,1)$, $(1,-1)$ is

$$\begin{pmatrix} 2 & 0 \\ 0 & 3 \end{pmatrix}.$$

Describe the effect of the transformation in geometric terms.

2. Same problem for $T: \mathbb{R}^2 \to \mathbb{R}^2$ whose matrix relative to $(1,1)$, $(-3,2)$ is

$$\begin{pmatrix} 2 & 0 \\ 0 & -2 \end{pmatrix}.$$

3. Same problem for $T: \mathbb{R}^2 \to \mathbb{R}^2$ whose matrix relative to $(-1,3)$, $(4,1)$ is

$$\begin{pmatrix} -3 & 0 \\ 0 & 2 \end{pmatrix}.$$

The examples given in the early part of this section indicate our goal. Suppose $T: \mathbb{R}^n \to \mathbb{R}^n$ is a linear transformation. We want to know when there is a basis for $\mathbb{R}^n$ relative to which the matrix for T is a diagonal matrix. If there is such a basis we want a method of finding it.

Example Consider again the linear transformation $T(x,y) = (\frac{1}{5}(12x - y),$ $\frac{1}{5}(-6x + 13y))$. As we saw in the second example, the matrix for T relative to the basis consisting of $(1,2)$ and $(-1,3)$ is

$$\begin{pmatrix} 2 & 0 \\ 0 & 3 \end{pmatrix}.$$

If we use $\mathbf{X}_1 = (1,2)$ and $\mathbf{X}_2 = (-1,3)$, then we have $T(\mathbf{X}_1) = 2\mathbf{X}_1$ and $T(\mathbf{X}_2) = 3\mathbf{X}_2$. Consequently if the matrix for T is diagonal relative to the basis consisting of $\mathbf{X}_1$ and $\mathbf{X}_2$, then T sends $\mathbf{X}_1$ and $\mathbf{X}_2$ into scalar multiples of themselves.

This example indicates the kind of vectors we are looking for. They are vectors $\mathbf{X}$ for which there is a scalar r so that $T(\mathbf{X}) = r\mathbf{X}$.

Definition 10-2 _____

Suppose $T: \mathbb{R}^n \to \mathbb{R}^n$ is a linear transformation. If r is a scalar and $\mathbf{X}$ is a nonzero vector with the property that $T(\mathbf{X}) = r\mathbf{X}$, then r is called an *eigenvalue* of T and $\mathbf{X}$ is called an *eigenvector* of T corresponding to, belonging to, or *for* r. (The terms "proper value," "latent value," and "characteristic value" are also used for "eigenvalue." Similarly "proper vector," "latent vector," and "characteristic vector" are used for "eigenvector").

The same terminology is used for square matrices. If $A\mathbf{X} = r\mathbf{X}$ and $\mathbf{X}$ is nonzero, then r is an eigenvalue of A and $\mathbf{X}$ is an eigenvector of A corresponding to r.

_____

At this point we explain why we allow the zero scalar as an eigenvalue but do not allow the zero vector as an eigenvector. Since $T(\mathbf{0}) = \mathbf{0}$ it follows that if r is any scalar, then $T(\mathbf{0}) = r\mathbf{0} = \mathbf{0}$. This would mean that the zero vector is an eigenvector for any scalar. More importantly, the equation $T(0) = r0$ does not give any specific information about T since this equation holds for any linear transformation T.

We do however allow the zero scalar as an eigenvalue. If X is a nonzero vector and $T(X) = 0X = 0$, then T sends every vector along the line determined by X into 0. Refer to the third example in this section where we had $T(1,1,1) = (0,0,0)$. The transformation collapses or compresses the space along $(1,1,1)$.

In more advanced treatments you will find that complex numbers can also be eigenvalues. We are interested only in real eigenvalues in this book. However, it is important to realize that complex eigenvalues can occur.

In the next two examples we discuss a method for finding the eigenvalues and eigenvectors of a linear transformation.

Example   Consider the linear transformation $T: \mathbb{R}^2 \to \mathbb{R}^2$ whose matrix relative to the natural basis is

$$A = \begin{pmatrix} -\frac{14}{11} & \frac{9}{11} \\ -\frac{12}{11} & \frac{25}{11} \end{pmatrix}.$$

We try to find scalars r and nonzero vectors $X = (x,y)$ which have the property that $T(X) = rX$. In matrix terms this equation takes the form

$$\begin{pmatrix} -\frac{14}{11} & \frac{9}{11} \\ -\frac{12}{11} & \frac{25}{11} \end{pmatrix} \begin{pmatrix} x \\ y \end{pmatrix} = r \begin{pmatrix} x \\ y \end{pmatrix}.$$

Put all terms on the left-hand side.

$$\left[\begin{pmatrix} -\frac{14}{11} & \frac{9}{11} \\ -\frac{12}{11} & \frac{25}{11} \end{pmatrix} - \begin{pmatrix} r & 0 \\ 0 & r \end{pmatrix} \right] \begin{pmatrix} x \\ y \end{pmatrix} = \begin{pmatrix} 0 \\ 0 \end{pmatrix}$$

Since the matrix

$$\begin{pmatrix} -\frac{14}{11} - r & \frac{9}{11} \\ -\frac{12}{11} & \frac{25}{11} - r \end{pmatrix}$$

sends a nonzero vector X into 0, we know that the matrix is singular (Proposition 5-8). Therefore the determinant of the matrix is zero (Proposition 6-7). Consequently,

$$\det \begin{pmatrix} -\frac{14}{11} - r & \frac{9}{11} \\ -\frac{12}{11} & \frac{25}{11} - r \end{pmatrix} = (-\tfrac{14}{11} - r)(\tfrac{25}{11} - r) + \tfrac{108}{121}$$

$$= r^2 - r - 2 = 0.$$

The polynomial $r^2 - r - 2$ is called the characteristic polynomial of A and of T. The eigenvalues for T are the roots of the characteristic polynomial: $r^2 - r - 2 = (r - 2)(r + 1) = 0$. Therefore the eigenvalues of T are 2 and -1.

Now we find the eigenvectors for T. An eigenvector (x,y) for T corresponding to 2 must satisfy the equation

$$\left[\begin{pmatrix} -\frac{14}{11} & \frac{9}{11} \\ -\frac{12}{11} & \frac{25}{11} \end{pmatrix} - \begin{pmatrix} 2 & 0 \\ 0 & 2 \end{pmatrix}\right]\begin{pmatrix} x \\ y \end{pmatrix} = \begin{pmatrix} 0 \\ 0 \end{pmatrix}.$$

The matrix

$$\begin{pmatrix} -\frac{36}{11} & \frac{9}{11} \\ -\frac{12}{11} & \frac{3}{11} \end{pmatrix}$$

is singular, and therefore the system of equations

$$-\tfrac{36}{11}x + \tfrac{9}{11}y = 0$$
$$-\tfrac{12}{11}x + \tfrac{3}{11}y = 0$$

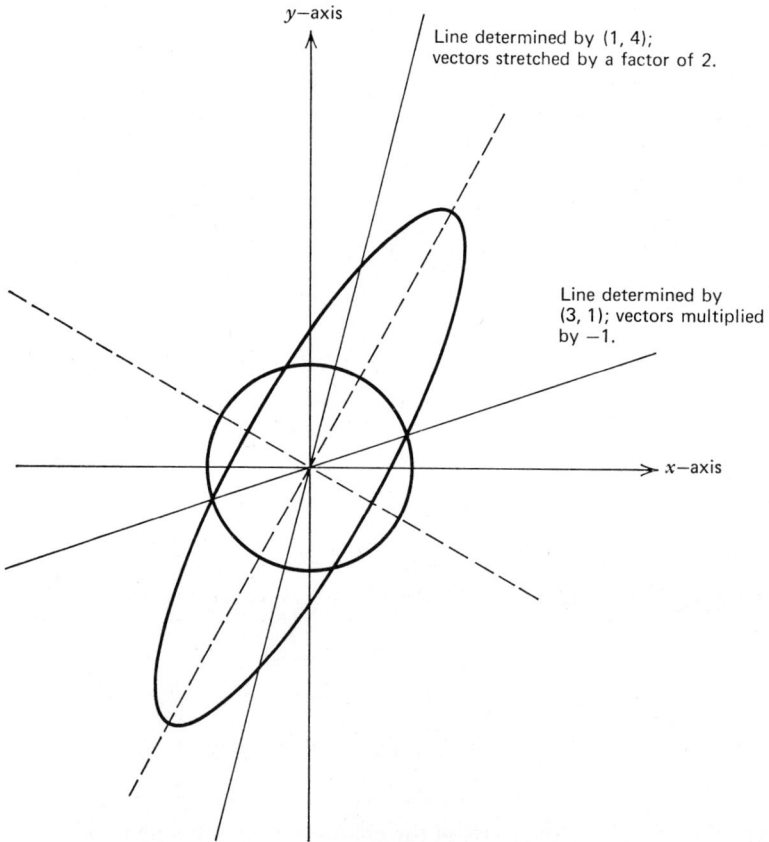

y−axis

Line determined by (1, 4); vectors stretched by a factor of 2.

Line determined by (3, 1); vectors multiplied by −1.

x−axis

Figure 10-3

has a nonzero solution, e.g., (1,4). Consequently (1,4) is an eigenvector corresponding to the eigenvalue 2.

An eigenvector (x,y) corresponding to the eigenvalue -1 must satisfy

$$\left[\begin{pmatrix} -\frac{14}{11} & \frac{9}{11} \\ -\frac{12}{11} & \frac{25}{11} \end{pmatrix} - \begin{pmatrix} -1 & 0 \\ 0 & -1 \end{pmatrix}\right]\begin{pmatrix} x \\ y \end{pmatrix} = \begin{pmatrix} 0 \\ 0 \end{pmatrix}.$$

Again the matrix

$$\begin{pmatrix} -\frac{3}{11} & \frac{9}{11} \\ -\frac{12}{11} & \frac{36}{11} \end{pmatrix}$$

is singular and thus the system of equations

$$-\tfrac{3}{11}x + \tfrac{9}{11}y = 0$$
$$-\tfrac{12}{11}x + \tfrac{36}{11}y = 0$$

has a nonzero solution, for example (3,1). Thus (3,1) is an eigenvector corresponding to the eigenvalue -1.

The transformation T stretches the plane by a factor of 2 along the line determined by (2,1), and it multiplies every vector along the line determined by (3,1) by the scalar -1 (see Figure 10-3).

Example   Consider the transformation $T: \mathbb{R}^2 \rightarrow \mathbb{R}^2$ whose matrix relative to the natural ordered basis is

$$\begin{pmatrix} 2 & 1 \\ 1 & 3 \end{pmatrix}.$$

In this example we determine whether T has any real eigenvalues and real eigenvectors.

If there is an eigenvector $\mathbf{X}$ corresponding to an eigenvalue r, then the equation $T(\mathbf{X}) = r\mathbf{X}$ must be satisfied. This equation can be rewritten as follows: using I for the identity on $\mathbb{R}^2$, $T(\mathbf{X}) - rI(\mathbf{X}) = \mathbf{0}$ or $(T - rI)(\mathbf{X}) = \mathbf{0}$. Since $\mathbf{X} \neq \mathbf{0}$, the linear function $T - rI$ must be singular, and so, by Proposition 5-8, the matrix

$$\begin{pmatrix} 2 & 1 \\ 1 & 3 \end{pmatrix} - r\begin{pmatrix} 1 & 0 \\ 0 & 1 \end{pmatrix}$$

is singular. Therefore, the determinant of

$$\begin{pmatrix} 2 - r & 1 \\ 1 & 3 - r \end{pmatrix}$$

is zero. Therefore,

$$(2 - r)(3 - r) - 1 = r^2 - 5r + 5 = 0.$$

You can solve this equation to find that the two solutions are $\frac{1}{2}(5 + \sqrt{5})$ and $\frac{1}{2}(5 - \sqrt{5})$.

An eigenvector (a,b) for $(5 + \sqrt{5})/2$ must satisfy

$$\begin{pmatrix} 2 & 1 \\ 1 & 3 \end{pmatrix} \begin{pmatrix} a \\ b \end{pmatrix} = \tfrac{1}{2}(5 + \sqrt{5}) \begin{pmatrix} a \\ b \end{pmatrix}.$$

This vector equation corresponds to two homogeneous scalar equations with coefficient matrix

$$\begin{pmatrix} -\tfrac{1}{2} - \tfrac{1}{2}\sqrt{5} & 1 \\ 1 & \tfrac{1}{2} - \tfrac{1}{2}\sqrt{5} \end{pmatrix}.$$

The eigenvalues are chosen precisely so that this matrix is singular; therefore, there is a nonzero solution. A solution is $(-1 + \sqrt{5}, 2)$. Proceeding in a similar fashion we find that an eigenvector for the eigenvalue $\frac{1}{2}(5 - \sqrt{5})$ is $(-1 - \sqrt{5}, 2)$. These two eigenvectors form a basis for $\mathbb{R}^2$, and the matrix for T relative to the ordered

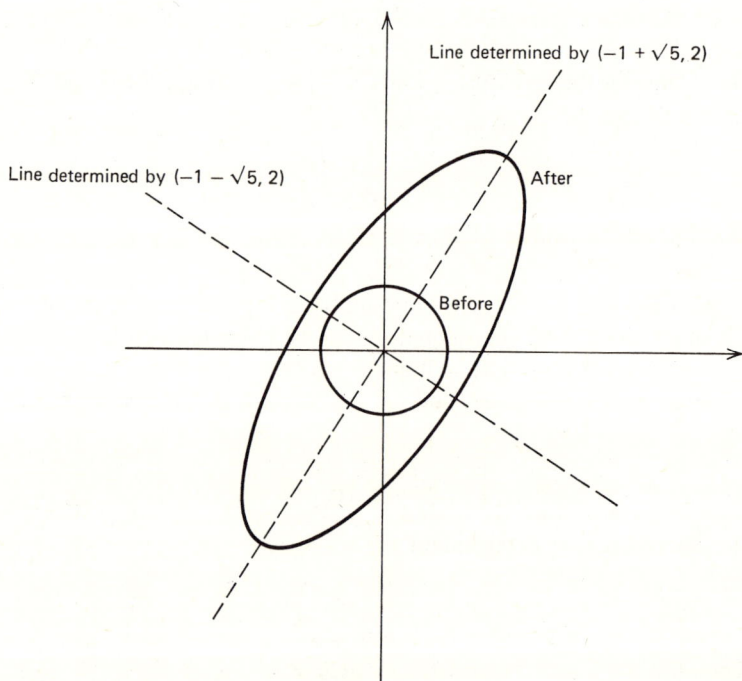

Figure 10-4

basis consisting of $(-1 + \sqrt{5}, 2)$ and $(-1 - \sqrt{5}, 2)$ is

$$\begin{pmatrix} \frac{1}{2}(5 + \sqrt{5}) & 0 \\ 0 & \frac{1}{2}(5 - \sqrt{5}) \end{pmatrix}.$$

This shows that the transformation T stretches the plane by the factor $\frac{1}{2}(5 + \sqrt{5})$—roughly 3.6—along the line determined by $(-1 + \sqrt{5}, 2)$. It also stretches the plane by the factor $\frac{1}{2}(5 - \sqrt{5})$—roughly 1.4—along the line determined by $(-1 - \sqrt{5}, 2)$. This is indicated in Figure 10-4.

These examples indicate a method for determining the eigenvalues and eigenvectors for a linear transformation $T: \mathbb{R}^n \to \mathbb{R}^n$. First, find a matrix A for T relative to some basis $\mathbf{X}_1, \ldots, \mathbf{X}_n$ in $\mathbb{R}^n$. An eigenvector $\mathbf{X}$ for T corresponding to the eigenvalue r satisfies the equation

$$T(\mathbf{X}) = r\mathbf{X}.$$

This can be rewritten as

$$(T - rI)(\mathbf{X}) = \mathbf{0}$$

where I is the identity on $\mathbb{R}^n$. Since $\mathbf{X}$ is nonzero, the transformation $T - rI$ is singular (Proposition 5-8). Thus the matrix $A - rI_n$ for $T - rI$ is singular and consequently has zero determinant (Proposition 6-7). The eigenvalues are the solutions of the equation

$$\det(A - rI_n) = 0.$$

The matrix form of the equation

$$(T - rI)(\mathbf{X}) = \mathbf{0}$$

is the equation

$$\begin{pmatrix} & & \\ & A - rI_n & \\ & & \end{pmatrix} \begin{pmatrix} x_1 \\ x_2 \\ \cdot \\ \cdot \\ \cdot \\ x_n \end{pmatrix} = 0$$

where $x_1, \ldots, x_n$ are the coordinates of the eigenvector $\mathbf{X}$ relative to the basis $\mathbf{X}_1, \ldots, \mathbf{X}_n$. Since $A - rI_n$ is singular, the system of equations has a nonzero solution. The solution gives the coordinates relative to $\mathbf{X}_1, \ldots, \mathbf{X}_n$ of eigenvectors belonging to r.

Proposition 10-1

If $T: \mathbb{R}^n \to \mathbb{R}^n$ is a linear transformation and A is a matrix for T relative to a basis for $\mathbb{R}^n$, then the following are equivalent.

(1) r is an eigenvalue for T.
(2) r is an eigenvalue for A.
(3) r is a root of the equation det $(A - rI_n) = 0$.

Definition 10-3_____

If A is a $n \times n$ matrix, then det $(A - rI_n)$ is called the *characteristic polynomial* of A, and det $(A - rI_n) = 0$ is called the *characteristic equation* of A.

_____

The roots of the characteristic equation det $(A - rI_n) = 0$ are often complex numbers. If that happens, then A has complex eigenvalues; we mentioned this possibility after Definition 10-2.

Brief Exercises

1. Find the real eigenvalues and eigenvectors of the following transformations.

(a) $T(x,y) = \frac{1}{4}(13x + 3y, 9x + 7y)$ (c) $T(x,y) = (3x - y, x + 2y)$
(b) $T(x,y) = (4x, x - y)$ (d) $T(x,y,z) = (x, x + y, x + y + z)$

2. Find the real eigenvalues and eigenvectors of the following matrices.

$$\text{(a)} \begin{pmatrix} 2 & 1 \\ 1 & 2 \end{pmatrix} \qquad \text{(b)} \begin{pmatrix} 4 & 1 \\ 3 & 0 \end{pmatrix}$$

Answers: **1(a)** Eigenvalues 4 and 1. $(1,1)$ is an eigenvector for 4; $(1,-3)$ is an eigenvector for 1 **1(b)** Eigenvalues 4 and -1. $(5,1)$ is an eigenvector for 4; $(0,1)$ is an eigenvector for -1 **1(c)** No real eigenvalues or eigenvectors **1(d)** 1 is the only eigenvalue. $(0,0,1)$ is an eigenvector for 1 **2(a)** Eigenvalues 3 and 1. $(1,1)$ is an eigenvector for 3; $(1,-1)$ is an eigenvector for 1 **2(b)** Eigenvalues are $2 + \sqrt{7}$ and $2 - \sqrt{7}$. $(1,-2 + \sqrt{7})$ is an eigenvector for $2 + \sqrt{7}$; $(1,-2 - \sqrt{7})$ is an eigenvector for $2 - \sqrt{7}$

The examples preceding Proposition 10-1 were presented to illustrate a procedure for finding eigenvalues and eigenvectors for a linear transformation or a matrix. We chose the examples so that we could find a basis of eigenvectors; however, this is not always possible. The next examples illustrate a few of the many different kinds of difficulties which can occur.

Example Consider the linear transformation $T: \mathbb{R}^2 \to \mathbb{R}^2$ determined by $T(x,y) = (2x + y, -x + 3y)$. The matrix for T relative to the natural basis is

$$A = \begin{pmatrix} 2 & 1 \\ -1 & 3 \end{pmatrix}.$$

The characteristic polynomial for A is

$$\det(A - rI_2) = \det \begin{pmatrix} 2-r & 1 \\ -1 & 3-r \end{pmatrix} = r^2 - 5r + 7.$$

The equation $r^2 - 5r + 7 = 0$ has no real numbers as roots. Therefore there are no real eigenvectors for T.

Example   Consider the linear transformation $T: \mathbb{R}^3 \to \mathbb{R}^3$ whose matrix relative to the standard basis is

$$A = \begin{pmatrix} 0 & -1 & 1 \\ 2 & 3 & 3 \\ -2 & 1 & 1 \end{pmatrix}.$$

The characteristic polynomial of A is

$$\det(A - rI_3) = \det \begin{pmatrix} -r & -1 & 1 \\ 2 & 3-r & 3 \\ -2 & 1 & 1-r \end{pmatrix}$$
$$= -r^3 + 4r^2 - 4r + 16.$$

To find the eigenvalues for T we must solve the equation

$$-r^3 + 4r^2 - 4r + 16 = 0.$$

Fortunately, this polynomial can be factored somewhat.

$$-(r-4)(r^2 + 4) = 0$$

From this you can see that 4 is the only real eigenvalue. An eigenvector for 4 is $(-1,7,3)$, as you can verify.

In this case we cannot analyze the transformation T by obtaining a diagonal matrix to represent it.

Example   Let $T: \mathbb{R}^4 \to \mathbb{R}^4$ be the linear transformation whose matrix relative to the natural basis is

$$A = \begin{pmatrix} 2 & 0 & 0 & 0 \\ 0 & 2 & 0 & 0 \\ 3 & -3 & 2 & -3 \\ 3 & -3 & 0 & -1 \end{pmatrix}.$$

In this example we find the real eigenvalues and eigenvectors of A.

The characteristic polynomial of A is

$$\det \begin{pmatrix} 2-x & 0 & 0 & 0 \\ 0 & 2-x & 0 & 0 \\ 3 & -3 & 2-x & -3 \\ 3 & -3 & 0 & -1-x \end{pmatrix} = (2-x)^3(-1-x).$$

Therefore, the eigenvalues of A are 2 and -1.

The vector (a,b,c,d) is an eigenvector of A belonging to -1 if and only if

$$\begin{pmatrix} 2 & 0 & 0 & 0 \\ 0 & 2 & 0 & 0 \\ 3 & -3 & 2 & -3 \\ 3 & -3 & 0 & -1 \end{pmatrix} \begin{pmatrix} a \\ b \\ c \\ d \end{pmatrix} = -1 \begin{pmatrix} a \\ b \\ c \\ d \end{pmatrix}.$$

This vector equation can be rewritten as a homogeneous system of four equations whose coefficient matrix is

$$\begin{pmatrix} 3 & 0 & 0 & 0 \\ 0 & 3 & 0 & 0 \\ 3 & -3 & 3 & -3 \\ 3 & -3 & 0 & 0 \end{pmatrix}.$$

The row echelon form of this matrix is

$$\begin{pmatrix} 1 & 0 & 0 & 0 \\ 0 & 1 & 0 & 0 \\ 0 & 0 & 1 & -1 \\ 0 & 0 & 0 & 0 \end{pmatrix}.$$

Consequently (a,b,c,d) is an eigenvector belonging to -1 if and only if $a = b = 0$ and $c = d$. Therefore $(0,0,1,1)$ is an eigenvector of A corresponding to the eigenvalue -1.

The vector (a,b,c,d) is an eigenvector for A belonging to 2 if and only if

$$\begin{pmatrix} 2 & 0 & 0 & 0 \\ 0 & 2 & 0 & 0 \\ 3 & -3 & 2 & -3 \\ 3 & -3 & 0 & -1 \end{pmatrix} \begin{pmatrix} a \\ b \\ c \\ d \end{pmatrix} = 2 \begin{pmatrix} a \\ b \\ c \\ d \end{pmatrix}.$$

This vector equation can be rewritten as a homogeneous system of scalar equations whose coefficient matrix is

$$\begin{pmatrix} 0 & 0 & 0 & 0 \\ 0 & 0 & 0 & 0 \\ 3 & -3 & 0 & -3 \\ 3 & -3 & 0 & -3 \end{pmatrix}.$$

The row echelon form for this matrix is

$$\begin{pmatrix} 1 & -1 & 0 & -1 \\ 0 & 0 & 0 & 0 \\ 0 & 0 & 0 & 0 \\ 0 & 0 & 0 & 0 \end{pmatrix}.$$

Therefore (a,b,c,d) is an eigenvector of A belonging to 2 if and only if $a - b - d = 0$. Solving for d we obtain the result that (a,b,c,d) is an eigenvector of A belonging to 2 if and only if

$$(a,b,c,d) = (a, b, c, a - b) = a(1,0,0,1) + b(0,1,0,-1) + c(0,0,1,0).$$

This shows that every vector in the subspace spanned by the vectors $(1,0,0,1)$, $(0,1,0,-1)$, and $(0,0,1,0)$ is an eigenvector of A corresponding to 2.

Consequently the matrix for T relative to the ordered basis $(1,0,0,1)$, $(0,1,0-1)$, $(0,0,1,0)$, and $(0,0,1,1)$ is the diagonal matrix diag $(2,2,2,-1)$.

These examples should give some insight into the possibility of changing bases so that a linear transformation $T: \mathbb{R}^n \rightarrow \mathbb{R}^n$ is represented by a diagonal matrix relative to the new basis. First, suppose there is a basis $\mathbf{X}_1, \ldots, \mathbf{X}_n$ for $\mathbb{R}^n$ consisting of eigenvectors. Then $T(\mathbf{X}_1) = r_1\mathbf{X}_1, \ldots, T(\mathbf{X}_n) = r_n\mathbf{X}_n$ (some scalars may be repeated in the list $r_1, \ldots, r_n$ of eigenvalues). The matrix for T relative to this basis is the diagonal matrix

$$\begin{pmatrix} r_1 & 0 & \cdots & 0 \\ 0 & r_2 & \cdots & 0 \\ \cdot & \cdot & & \cdot \\ \cdot & \cdot & & \cdot \\ \cdot & \cdot & & \cdot \\ 0 & 0 & \cdots & r_n \end{pmatrix}.$$

On the other hand, suppose that, relative to the basis $\mathbf{X}_1, \ldots, \mathbf{X}_n$, the transformation T can be represented by the diagonal matrix

$$\begin{pmatrix} d_1 & 0 & \cdots & 0 \\ 0 & d_2 & \cdots & 0 \\ \cdot & \cdot & & \cdot \\ \cdot & \cdot & & \cdot \\ \cdot & \cdot & & \cdot \\ 0 & 0 & \cdots & d_n \end{pmatrix}.$$

Thus $T(\mathbf{X}) = d_1\mathbf{X}_1, \ldots, T(\mathbf{X}_n) = d_n\mathbf{X}_n$, and the $\mathbf{X}$'s are eigenvectors of T and the d's are eigenvalues of T.

Proposition 10-2

The linear transformation $T: \mathbb{R}^n \to \mathbb{R}^n$ can be represented by a diagonal matrix relative to the basis $\mathbf{X}_1, \ldots, \mathbf{X}_n$ if and only if $\mathbf{X}_1, \ldots, \mathbf{X}_n$ is a basis consisting of eigenvectors of T.

SUMMARY *An eigenvector $\mathbf{X}$ of a linear transformation T for the eigenvalue r is a nonzero vector satisfying $T(\mathbf{X}) = r\mathbf{X}$, and an eigenvector $\mathbf{X}$ for a square matrix A corresponding to the eigenvalue r is a nonzero vector satisfying $A\mathbf{X} = r\mathbf{X}$. The eigenvalues for A are the roots of the characteristic polynomial of A, i.e., the scalars r satisfying*

$$\det (A - rI_n) = 0.$$

The eigenvalues for a linear transformation and its matrix are the same. Finally, a linear transformation T can be represented by a diagonal relative to the basis $\mathbf{X}_1, \ldots, \mathbf{X}_n$, used both in the domain and range, if and only if it consists of eigenvectors for T.

Exercises for Section 10-1

1. Determine whether the following matrices have real eigenvalues and eigenvectors. If they do, find them. What geometric information does this give about the transformation determined by the matrix?

(a) $\begin{pmatrix} 3 & 1 \\ 1 & 3 \end{pmatrix}$ (b) $\begin{pmatrix} 2 & 0 \\ 3 & 4 \end{pmatrix}$ (c) $\begin{pmatrix} 5 & 1 \\ 2 & -1 \end{pmatrix}$ (d) $\begin{pmatrix} 3 & -2 & -1 \\ -1 & 4 & 1 \\ 1 & 2 & 5 \end{pmatrix}$

More Challenging Exercises

2. Do A and A^T have the same eigenvalues? Do they have the same eigenvectors? Support your answers.

3. Give an example of a 2×2 matrix having no real eigenvectors.

4. Give an example of a 2×2 matrix having only one eigenvalue but two linearly independent eigenvectors.

5. Give an example of a 2×2 matrix which has only one eigenvalue and does not have two linearly independent eigenvectors.

6. If A is a nonsingular $n \times n$ matrix with eigenvalue r, does A^{-1} have r^{-1} as an eigenvalue? How do the eigenvectors of A^{-1} relate to the eigenvectors of A?

7. Give an argument to show that if A is an $n \times n$ matrix, then $\det (A - xI_n)$ is a polynomial of degree n.

8. How are the eigenvalues of A^k related to the eigenvalues of A?

9. What are the eigenvalues of a diagonal matrix?

SECTION 10-2
Quadratic Functions

PREVIEW *We introduce quadratic functions and show how symmetric matrices represent them. We find the relation between symmetric matrices for the same quadratic function relative to different bases. Finally, we show how to find a basis so that a quadratic function can be expressed as a "sum of squares," i.e., represented by a diagonal matrix.*

Throughout this book we have concentrated our attention on linear transformations and their relationship to matrices. A linear transformation from $\mathbb{R}^n$ to $\mathbb{R}$ is one of the type

$$T(x_1, \ldots, x_n) = a_1 x_1 + \cdots + a_n x_n.$$

Now we consider quadratic functions. They are functions from $\mathbb{R}^n$ to $\mathbb{R}$. A linear function from $\mathbb{R}^n$ to $\mathbb{R}$ can be described by a sum of terms of the first degree. A quadratic function from $\mathbb{R}^n$ to $\mathbb{R}$ can be described by a sum of terms each of the second degree.

Example   An example of a quadratic function from $\mathbb{R}$ to $\mathbb{R}$ is $q(x) = 5x^2$. Another is $q(x) = 7x^2$. In fact, every quadratic function from $\mathbb{R}$ to $\mathbb{R}$ is defined by the equation $q(x) = ax^2$ where a is some real number.

Example   One example of a quadratic function from $\mathbb{R}^2$ to $\mathbb{R}$ is $q(x,y) = 3x^2 - 5xy + y^2$. Another example is $q(x,y) = 4x^2 + 6xy$. Any quadratic function from $\mathbb{R}^2$ to $\mathbb{R}$ can be defined by an equation of the form $q(x,y) = ax^2 + bxy + cy^2$, where a, b, and c are real numbers.

Example   A quadratic function from $\mathbb{R}^3$ to $\mathbb{R}$ is given by an equation of the form

$$q(x_1,x_2,x_3) = a_{11}x_1^2 + a_{12}x_1x_2 + a_{13}x_1x_3 + a_{22}x_2^2 + a_{23}x_2x_3 + a_{33}x_3^2$$

where the a's are real numbers.

In each of these examples we are thinking of x, (x,y), or (x_1,x_2,x_3) as the coordinates of a vector in $\mathbb{R}$, $\mathbb{R}^2$, and $\mathbb{R}^3$, respectively, relative to the natural basis. We can use matrix multiplication to express these formulas.

Example   We can express $q(x,y) = 3x^2 - 5xy + y^2$ by the matrix product

$$\begin{pmatrix} x & y \end{pmatrix} \begin{pmatrix} 3 & -\frac{5}{2} \\ -\frac{5}{2} & 1 \end{pmatrix} \begin{pmatrix} x \\ y \end{pmatrix}.$$

The (1,1) entry and the (2,2) entry give the coefficients of x^2 and y^2, respectively. The (1,2) and (2,1) positions add to give the coefficient of the xy term (the "cross" term) since $xy = yx$, and we put half the coefficient in each position.

The matrix is to be considered as defining a function $q: \mathbb{R}^2 \to \mathbb{R}$ as follows: if $\mathbf{X}$ is a vector in $\mathbb{R}^2$, and (x,y) are its coordinates relative to the natural basis, then $q(\mathbf{X})$ can be computed by

$$q(\mathbf{X}) = \begin{pmatrix} x & y \end{pmatrix} \begin{pmatrix} 3 & -\frac{5}{2} \\ -\frac{5}{2} & 1 \end{pmatrix} \begin{pmatrix} x \\ y \end{pmatrix}.$$

Consequently, we say that this 2×2 matrix is the matrix for q relative to the standard basis.

The same function q can be expressed with a matrix product relative to a different basis. Take the basis consisting of $(1,1)$ and $(-1,1)$, for example. Let P be the transition matrix from this basis to the natural basis.

$$P = \begin{pmatrix} 1 & -1 \\ 1 & 1 \end{pmatrix}$$

So, if **X** has coordinates (x,y) relative to the standard basis and (u,v) relative to the new basis, then

$$\begin{pmatrix} x \\ y \end{pmatrix} = \begin{pmatrix} 1 & -1 \\ 1 & 1 \end{pmatrix}\begin{pmatrix} u \\ v \end{pmatrix} = \begin{pmatrix} & P & \end{pmatrix}\begin{pmatrix} u \\ v \end{pmatrix}.$$

Taking transposes (Exercise 17, Section 4-4) we obtain

$$(x,y) = \begin{matrix} (u & v) \end{matrix}\begin{pmatrix} 1 & 1 \\ -1 & 1 \end{pmatrix} = \begin{matrix} (u & v) \end{matrix}\begin{pmatrix} & P^T & \end{pmatrix}.$$

Therefore,

$$q(\mathbf{X}) = \begin{matrix} (x & y) \end{matrix}\begin{pmatrix} 3 & -\frac{5}{2} \\ -\frac{5}{2} & 1 \end{pmatrix}\begin{pmatrix} x \\ y \end{pmatrix}$$

$$= \begin{matrix} (u & v) \end{matrix}\begin{pmatrix} 1 & 1 \\ -1 & 1 \end{pmatrix}\begin{pmatrix} 3 & -\frac{5}{2} \\ -\frac{5}{2} & 1 \end{pmatrix}\begin{pmatrix} 1 & -1 \\ 1 & 1 \end{pmatrix}\begin{pmatrix} u \\ v \end{pmatrix}$$

$$= \begin{matrix} (u & v) \end{matrix}\begin{pmatrix} -1 & -2 \\ -2 & 9 \end{pmatrix}\begin{pmatrix} u \\ v \end{pmatrix}.$$

Consequently, we say that the matrix $\begin{pmatrix} -1 & -2 \\ -2 & 9 \end{pmatrix}$ represents q relative to the basis consisting of $(1,1)$ and $(-1,1)$. So, for example, if $\mathbf{X} = (2,3)$, then

$$q(\mathbf{X}) = \begin{matrix} (2 & 3) \end{matrix}\begin{pmatrix} 3 & -\frac{5}{2} \\ -\frac{5}{2} & 1 \end{pmatrix}\begin{pmatrix} 2 \\ 3 \end{pmatrix} = -9.$$

But also, since

$$(2,3) = (\tfrac{5}{2})(1,1) + (\tfrac{1}{2})(-1,1)$$

we can also compute

$$q(\mathbf{X}) = \begin{matrix} (\frac{5}{2} & \frac{1}{2}) \end{matrix}\begin{pmatrix} -1 & -2 \\ -2 & 9 \end{pmatrix}\begin{pmatrix} \frac{5}{2} \\ \frac{1}{2} \end{pmatrix} = -9.$$

Example   Consider the quadratic function from $\mathbb{R}^3$ to $\mathbb{R}$ defined by $q(x_1,x_2,x_3) = x_1^2 - 2x_1x_2 + 4x_1x_3 - x_2^2 + 4x_2x_3 + 2x_3^2$. The function q can be represented by the matrix

$$\begin{pmatrix} 1 & -1 & 2 \\ -1 & -1 & 2 \\ 2 & 2 & 2 \end{pmatrix}$$

relative to the standard basis. If $\mathbf{X}$ has the coordinates (x_1, x_2, x_3) relative to the standard basis, then

$$q(\mathbf{X}) = (x_1 \quad x_2 \quad x_3) \begin{pmatrix} 1 & -1 & 2 \\ -1 & -1 & 2 \\ 2 & 2 & 2 \end{pmatrix} \begin{pmatrix} x_1 \\ x_2 \\ x_3 \end{pmatrix}.$$

We say that the matrix

$$A = \begin{pmatrix} 1 & -1 & 2 \\ -1 & -1 & 2 \\ 2 & 2 & 2 \end{pmatrix}$$

represents the function relative to the natural basis.

As in the previous example we take the coefficient -2 of $x_1 x_2$ and put half of it in both the $(1,2)$ and $(2,1)$ positions. Similarly we put half of the coefficient 4 of $x_1 x_3$ in both the $(1,3)$ and $(3,1)$ positions. For a similar reason we use 2 in the $(2,3)$ and $(3,2)$ positions. In this way the matrix we use is symmetric, i.e., the (i,j) entry equals the (j,i) entry.

This function q can also be represented by a matrix relative to another basis, $(1,0,1)$, $(1,1,0)$, and $(0,1,1)$, for example. The matrix B for q relative to this basis should have the following property: if $\mathbf{X}$ is a vector in $\mathbb{R}^3$ having coordinates (y_1, y_2, y_3) relative to this basis, then

$$q(\mathbf{X}) = (y_1 \quad y_2 \quad y_3) \begin{pmatrix} & & \\ & B & \\ & & \end{pmatrix} \begin{pmatrix} y_1 \\ y_2 \\ y_3 \end{pmatrix}.$$

Let P be the transition matrix from the basis consisting of $(1,0,1)$, $(1,1,0)$, and $(0,1,1)$ to the standard basis.

$$P = \begin{pmatrix} 1 & 1 & 0 \\ 0 & 1 & 1 \\ 1 & 0 & 1 \end{pmatrix}$$

Therefore, if $\mathbf{X}$ has coordinates $(y_1 \quad y_2 \quad y_3)$ relative to $(1,0,1)$, $(1,1,0)$, and $(0,1,1)$ and coordinates $(x_1 \quad x_2 \quad x_3)$ relative to the standard basis, then

$$\begin{pmatrix} x_1 \\ x_2 \\ x_3 \end{pmatrix} = \begin{pmatrix} 1 & 1 & 0 \\ 0 & 1 & 1 \\ 1 & 0 & 1 \end{pmatrix} \begin{pmatrix} y_1 \\ y_2 \\ y_2 \end{pmatrix} = \begin{pmatrix} & & \\ & P & \\ & & \end{pmatrix} \begin{pmatrix} y_1 \\ y_2 \\ y_3 \end{pmatrix}$$

and, taking transposes (Exercise 17, Section 4-4),

$$(x_1 \quad x_2 \quad x_3) = \frac{(y_1 \quad y_2 \quad y_3) \begin{pmatrix} 1 & 0 & 1 \\ 1 & 1 & 0 \\ 0 & 1 & 1 \end{pmatrix}}{} = (y_1 \quad y_2 \quad y_3) \begin{pmatrix} P^T \end{pmatrix}.$$

Consequently,

$$q(\mathbf{X}) = \frac{(x_1 \quad x_2 \quad x_3) \begin{pmatrix} 1 & -1 & 2 \\ -1 & -1 & 2 \\ 2 & 2 & 2 \end{pmatrix} \begin{pmatrix} x_1 \\ x_2 \\ x_3 \end{pmatrix}}{}$$

$$= (y_1 \quad y_2 \quad y_3) \begin{pmatrix} 1 & 0 & 1 \\ 1 & 1 & 0 \\ 0 & 1 & 1 \end{pmatrix} \begin{pmatrix} 1 & -1 & 2 \\ -1 & -1 & 2 \\ 2 & 2 & 2 \end{pmatrix} \begin{pmatrix} 1 & 1 & 0 \\ 0 & 1 & 1 \\ 1 & 0 & 1 \end{pmatrix} \begin{pmatrix} y_1 \\ y_2 \\ y_3 \end{pmatrix}$$

$$= (y_1 \quad y_2 \quad y_3) \begin{pmatrix} 7 & 4 & 5 \\ 4 & -2 & 2 \\ 5 & 2 & 5 \end{pmatrix} \begin{pmatrix} y_1 \\ y_2 \\ y_3 \end{pmatrix}.$$

As an example, if $\mathbf{X} = (2,1,5)$, then

$$q(\mathbf{X}) = (2 \quad 1 \quad 5) \begin{pmatrix} 1 & -1 & 2 \\ -1 & -1 & 2 \\ 2 & 2 & 2 \end{pmatrix} \begin{pmatrix} 2 \\ 1 \\ 5 \end{pmatrix} = (11 \quad 7 \quad 16) \begin{pmatrix} 2 \\ 1 \\ 5 \end{pmatrix}$$

$$= 22 + 7 + 80 = 109.$$

The coordinates of $\mathbf{X}$ relative to $(1,0,1)$, $(1,1,0)$, and $(0,1,1)$ are $(3,-1,2)$ since

$$(2,1,5) = 3(1,0,1) - (1,1,0) + 2(0,1,1).$$

Thus

$$q(\mathbf{X}) = (3 \quad -1 \quad 2) \begin{pmatrix} 7 & 4 & 5 \\ 4 & -2 & 2 \\ 5 & 2 & 5 \end{pmatrix} \begin{pmatrix} 3 \\ -1 \\ 2 \end{pmatrix} = 109.$$

These examples show how we use square matrices to represent quadratic functions relative to different bases. The matrices we use all have a special property: the (i,j) entry equals the (j,i) entry.

Definition 10-4

A square matrix is *symmetric* if the entries in positions which are symmetric with respect to the main diagonal are equal, i.e., every (i,j) entry equals the (j,i) entry. This is the same as saying that the matrix equals its own transpose, i.e., $A = A^T$.

Definition 10-5

A *quadratic function q on* $\mathbb{R}^n$ is a function from $\mathbb{R}^n$ to $\mathbb{R}$ given by a formula of the following type: if the coordinates of **X** relative to the standard basis are given by $(x_1, \ldots, x_n)$, then

$$q(\mathbf{X}) = q(x_1, \ldots, x_n)$$

$$= a_{11}x_1^2 + a_{12}x_1x_2 + a_{13}x_1x_3 + \cdots + a_{22}x_2^2 + a_{23}x_2x_3 + \cdots + a_{nn}x_n^2.$$

As the examples indicate, we can express a quadratic function with a symmetric matrix relative to a basis for $\mathbb{R}^n$.

Definition 10-6

Suppose q is a quadratic function defined on $\mathbb{R}^n$, and suppose $\mathbf{X}_1, \ldots, \mathbf{X}_n$ is a basis for $\mathbb{R}^n$. *The symmetric matrix A represents (is the matrix for) q* relative to the basis $\mathbf{X}_1, \ldots, \mathbf{X}_n$ if the following condition is met: if **X** is in $\mathbb{R}^n$ and $(x_1, \ldots, x_n)$ are the coordinates of **X** relative to the basis, then

$$q(\mathbf{X}) = (x_1, \ldots, x_n) \begin{pmatrix} & & \\ & A & \\ & & \end{pmatrix} \begin{pmatrix} x_1 \\ \vdots \\ x_n \end{pmatrix}.$$

We have shown in the examples how to find the matrix for q relative to the standard basis; we make the matrix symmetric by placing half the coefficient of the x_ix_j term in the (i,j) position and half in the (j,i) position.

Brief Exercises

Find the matrix for the following quadratic functions relative to the standard basis.

1. $q(x,y) = x^2 - 3xy + 4y^2$

2. $q(x,y,z) = 3x^2 + 2xy - 3xz + y^2 - 5yz + z^2$

3. $q(x,y,z,w) = 4x^2 + 3y^2 + 3xz - 4z^2 + 2zw - w^2$

Answers:

1 $\begin{pmatrix} 1 & -\frac{3}{2} \\ -\frac{3}{2} & 4 \end{pmatrix}$ **2** $\begin{pmatrix} 3 & 1 & -\frac{3}{2} \\ 1 & 1 & -\frac{5}{2} \\ -\frac{3}{2} & -\frac{5}{2} & 1 \end{pmatrix}$ **3** $\begin{pmatrix} 4 & 0 & \frac{3}{2} & 0 \\ 0 & 3 & 0 & 0 \\ \frac{3}{2} & 0 & -4 & 1 \\ 0 & 0 & 1 & -1 \end{pmatrix}$

The examples earlier in this section show that a quadratic function can be repre-sented by different symmetric matrices with respect to different bases. Suppose the quadratic function q is represented by the matrix A relative to the basis $\mathbf{X}_1, \ldots, \mathbf{X}_n$ and we have another basis $\mathbf{Y}_1, \ldots, \mathbf{Y}_n$. Suppose further that the coordinates of $\mathbf{X}$ are given by $(x_1, \ldots, x_n)$ relative to the first basis and by $(y_1, \ldots, y_n)$ relative to the second. If P is the transition matrix from $\mathbf{Y}_1, \ldots, \mathbf{Y}_n$ to $\mathbf{X}_1, \ldots, \mathbf{X}_n$, then

$$\begin{pmatrix} x_1 \\ \cdot \\ \cdot \\ \cdot \\ x_n \end{pmatrix} = \begin{pmatrix} & & \\ & P & \\ & & \end{pmatrix} \begin{pmatrix} y_1 \\ \cdot \\ \cdot \\ \cdot \\ y_n \end{pmatrix}.$$

Consequently, taking transposes of both sides (Exercise 17, Section 4-4) we have

$$(x_1 \ \cdots \ x_n) = (y_1 \ \cdots \ y_n) \begin{pmatrix} & & \\ & P^T & \\ & & \end{pmatrix}.$$

Thus,

$$q(\mathbf{X}) = (x_1 \ \cdots \ x_n) \begin{pmatrix} & & \\ & A & \\ & & \end{pmatrix} \begin{pmatrix} x_1 \\ \cdot \\ \cdot \\ x_n \end{pmatrix}$$

$$= (y_1 \ \cdots \ y_n) \begin{pmatrix} & & \\ & P^T & \\ & & \end{pmatrix} \begin{pmatrix} & & \\ & A & \\ & & \end{pmatrix} \begin{pmatrix} & & \\ & P & \\ & & \end{pmatrix} \begin{pmatrix} y_1 \\ \cdot \\ \cdot \\ y_n \end{pmatrix}.$$

Proposition 10-3

If a quadratic function is represented by the matrix A relative to a basis $\mathbf{X}_1, \ldots, \mathbf{X}_n$, and if P is the transition matrix from $\mathbf{X}_1, \ldots, \mathbf{X}_n$ to another basis $\mathbf{Y}_1, \ldots, \mathbf{Y}_n$, then the quadratic function is represented by $P^T A P$ relative to $\mathbf{Y}_1, \ldots, \mathbf{Y}_n$.

We have seen that we can always use a symmetric matrix A to represent a quadratic function relative to the standard basis; i.e., $A = A^T$. Then $P^T A P$ is also symmetric since (Exercise 17, Section 4-4) $(P^T A P)^T = P^T A^T (P^T)^T = P^T A P$. Consequently, we can always use symmetric matrices to represent quadratic functions relative to any basis. This is particularly important because we can always find an orthonormal basis for $\mathbb{R}^n$ consisting of eigenvectors for A. The following proposition contains all the information we need about symmetric matrices.

Proposition 10-4

All eigenvalues of an $n \times n$ real symmetric matrix A are real numbers, and there is an orthonormal basis for $\mathbb{R}^n$ consisting of eigenvectors of A.

The next examples illustrate how we use the facts in Proposition 10-4 to analyze a quadratic function.

Example   Suppose the quadratic function q is represented by the matrix

$$A = \begin{pmatrix} \frac{9}{5} & \frac{2}{5} \\ \frac{2}{5} & \frac{6}{5} \end{pmatrix}$$

relative to the standard basis. The function is therefore $q(x,y) = \frac{9}{5}x^2 + \frac{4}{5}xy + \frac{6}{5}y^2$.

The characteristic polynomial for A is $r^2 - 3r + 2$; the eigenvalues are 2 and 1. An eigenvector for A corresponding to 2 is $(2,1)$, and an eigenvector for A corresponding to 1 is $(1,-2)$. Observe that these eigenvectors are orthogonal, as we mentioned in Proposition 10-4. By normalizing these vectors we obtain an orthonormal basis for $\mathbb{R}^2$ consisting of eigenvectors $(\frac{2}{5}\sqrt{5}, \frac{1}{5}\sqrt{5})$ and $(\frac{1}{5}\sqrt{5}, -\frac{2}{5}\sqrt{5})$.

Form a matrix P having the eigenvectors as its columns.

$$P = \begin{pmatrix} \frac{2}{5}\sqrt{5} & \frac{1}{5}\sqrt{5} \\ \frac{1}{5}\sqrt{5} & -\frac{2}{5}\sqrt{5} \end{pmatrix}$$

Now

$$P^T A P = \begin{pmatrix} \frac{2}{5}\sqrt{5} & \frac{1}{5}\sqrt{5} \\ \frac{1}{5}\sqrt{5} & -\frac{2}{5}\sqrt{5} \end{pmatrix} \begin{pmatrix} \frac{9}{5} & \frac{2}{5} \\ \frac{2}{5} & \frac{6}{5} \end{pmatrix} \begin{pmatrix} \frac{2}{5}\sqrt{5} & \frac{1}{5}\sqrt{5} \\ \frac{1}{5}\sqrt{5} & -\frac{2}{5}\sqrt{5} \end{pmatrix}$$

$$= \begin{pmatrix} 2 & 0 \\ 0 & 1 \end{pmatrix} = D.$$

The matrix P is the transition matrix from the basis consisting of $(\frac{2}{5}\sqrt{5}, \frac{1}{5}\sqrt{5})$ and $(\frac{1}{5}\sqrt{5}, -\frac{2}{5}\sqrt{5})$ to the natural basis. If a vector $\mathbf{X} = (x,y)$ has coordinates (a,b)

relative to this basis, then

$$q(\mathbf{X}) = \begin{matrix} (a & b) \end{matrix} \begin{pmatrix} 2 & 0 \\ 0 & 1 \end{pmatrix} \begin{pmatrix} a \\ b \end{pmatrix} = 2a^2 + b^2.$$

Changing basis allows us to express q as a sum of squares without any cross terms.

Example Suppose the quadratic function q is represented by the matrix

$$A = \begin{pmatrix} 3 & \sqrt{6} \\ \sqrt{6} & -2 \end{pmatrix}$$

relative to the natural basis. The quadratic function is

$$q(x,y) = 3x^2 + 2\sqrt{6}xy - 2y^2.$$

We find a basis for $\mathbb{R}^2$ relative to which the quadratic function is expressed as a sum of squares. In other words, we find a basis for $\mathbb{R}^2$ relative to which the matrix for q is a diagonal matrix.

The characteristic equation for A is

$$x^2 - x - 12 = 0.$$

The eigenvalues are 4 and -3. An eigenvector for 4 is $(\sqrt{6},1)$ and an eigenvector for -3 is $(1,-\sqrt{6})$. Observe that these vectors are orthogonal. An orthonormal basis consisting of eigenvectors can be obtained by normalizing these vectors:

$$(\tfrac{1}{7}\sqrt{42}, \tfrac{1}{7}\sqrt{7}), (\tfrac{1}{7}\sqrt{7}, -\tfrac{1}{7}\sqrt{42}).$$

Let P be the matrix having these unit vectors as columns:

$$P = \begin{pmatrix} \tfrac{1}{7}\sqrt{42} & \tfrac{1}{7}\sqrt{7} \\ \tfrac{1}{7}\sqrt{7} & -\tfrac{1}{7}\sqrt{42} \end{pmatrix}$$

The matrix P is the transition matrix from the basis consisting of $(\tfrac{1}{7}\sqrt{42}, \tfrac{1}{7}\sqrt{7})$ and $(\tfrac{1}{7}\sqrt{7}, -\tfrac{1}{7}\sqrt{42})$ to the natural basis. Relative to this new basis the matrix which represents q is

$$P^T A P = \begin{pmatrix} \tfrac{1}{7}\sqrt{42} & \tfrac{1}{7}\sqrt{7} \\ \tfrac{1}{7}\sqrt{7} & -\tfrac{1}{7}\sqrt{42} \end{pmatrix} \begin{pmatrix} 3 & \sqrt{6} \\ \sqrt{6} & -2 \end{pmatrix} \begin{pmatrix} \tfrac{1}{7}\sqrt{42} & \tfrac{1}{7}\sqrt{7} \\ \tfrac{1}{7}\sqrt{7} & -\tfrac{1}{7}\sqrt{42} \end{pmatrix}$$

$$= \begin{pmatrix} 4 & 0 \\ 0 & -3 \end{pmatrix}.$$

Consequently, if (a,b) gives the coordinates of $\mathbf{X}$ relative to this new basis, then $q(\mathbf{X}) = 4a^2 - 3b^2$.

These examples show the general situation. Suppose the quadratic function q is represented by the symmetric matrix A relative to the natural basis. Let $\mathbf{X}_1, \ldots, \mathbf{X}_n$ be an orthonormal basis consisting of the eigenvectors of A; construct the matrix P with $\mathbf{X}_1, \ldots, \mathbf{X}_n$ as columns. Then $P^T A P$ is a diagonal matrix representing q relative to $\mathbf{X}_1, \ldots, \mathbf{X}_n$; the eigenvalues of A are the entries along the diagonal.

Proposition 10-5

Every quadratic function on $\mathbb{R}^n$ can be represented by a diagonal matrix relative to an orthonormal basis.

Brief Exercises

Find the basis relative to which the following quadratic functions can be represented by a diagonal matrix.

1. $q(x,y) = x^2 + 3xy - 2y^2$ **3.** $q(x,y) = x^2 + 2xy + y^2$

2. $q(x,y) = 5x^2 - 2xy + y^2$

Answers: **1** $(-1, 1 - \sqrt{2})$, $(-1, 1 + \sqrt{2})$ **2** $(1, 2 - \sqrt{5})$, $(1, 2 + \sqrt{5})$
3 $(1,1)$, $(1,-1)$

SUMMARY   *A quadratic function on* $\mathbb{R}^n$ *is a function from* $\mathbb{R}^n$ *to* $\mathbb{R}$ *which can be expressed as a sum of second degree terms; if* $(x_1, \ldots, x_n)$ *are the coordinates of* $\mathbf{X}$ *relative to some basis for* $\mathbb{R}^n$, *then there is a symmetric matrix* A *so that*

$$q(\mathbf{X}) = (x_1, \ldots, x_n) \begin{pmatrix} & & \\ & A & \\ & & \end{pmatrix} \begin{pmatrix} x_1 \\ \cdot \\ \cdot \\ \cdot \\ x_n \end{pmatrix}.$$

In this sense the symmetric matrix A *represents the quadratic function* q *relative to this basis. If* P *is the transition matrix from a second basis to this one, then the matrix for* q *relative to the second basis is* $P^T A P$.

Since all eigenvalues of a symmetric matrix are real and since there is an orthonormal basis for $\mathbb{R}^n$ *consisting of eigenvectors for a symmetric matrix, it is possible to represent any symmetric matrix relative to an orthonormal basis with a diagonal matrix.*

Exercises for Section 10-2

1. Find the matrix for the following quadratic functions on $\mathbb{R}^2$ relative to the natural basis and relative to the basis consisting of $(1,1)$ and $(0,1)$. The formulas below are given relative to the natural basis.

(a) $q(x,y) = 5x^2 + y^2$ (b) $q(x,y) = 4x^2 + 3xy - y^2$

2. Find the matrix for the following quadratic functions on $\mathbb{R}^3$ relative to the natural basis and relative to the basis consisting of $(1,0,0)$, $(1,1,0)$, and $(1,1,1)$. The formulas below are given relative to the natural basis.

(a) $q(x,y,z) = 2x^2 + 3xy + y^2 - z^2$ (b) $q(x,y,z) = x^2 - y^2 + yz$

3. In each of the following, find a basis relative to which the following quadratic functions can be represented by a diagonal matrix. The formulas below are given relative to the natural basis.

(a) $q(x,y) = 3x^2 + 2xy - y^2$ (c) $q(x,y) = 6x^2 + 2xy$

(b) $q(x,y) = 4x^2 - 4xy + y^2$ (d) $q(x,y,z) = 4x^2 + 2y^2 - 2xz$

More Challenging Exercises

4. The matrices A and B are *similar* if and only if there is a nonsingular matrix P so that $B = P^{-1}AP$. Use the results of Section 9-3 to show that the matrices A and B are similar if and only if they represent the same linear transformation $T: \mathbb{R}^n \to \mathbb{R}^n$ relative to possibly different bases (Definition 10-1).

5. Suppose A is a symmetric matrix; by Proposition 10-4 there is an othonormal basis $\mathbf{X}_1, \ldots, \mathbf{X}_n$ consisting of eigenvectors of A. If P is the $n \times n$ matrix constructed by using $\mathbf{X}_1, \ldots, \mathbf{X}_n$ as its columns, then show that P is an orthogonal matrix, i.e., $P^T = P^{-1}$ (see Exercise 15, Section 8-3).

6. Let A be the matrix

$$A = \begin{pmatrix} 2 & 0 \\ 1 & 3 \end{pmatrix}.$$

Find a nonsingular matrix P so that $P^{-1}AP$ is a diagonal matrix.

7. Same exercise for the matrix

$$\begin{pmatrix} 1 & 2 \\ 0 & 3 \end{pmatrix}.$$

8. Let T be the linear transformation defined by $T(x,y) = (3x - y, x + 2y)$. Let A be the matrix for T relative to $(1,0)$, $(0,1)$ and B the matrix for T relative to $(1,1)$, $(0,1)$. Show that A and B are similar by producing a nonsingular matrix P so that $B = P^{-1}AP$.

9. Let T be the linear transformation defined by $T(x,y,z) = (3x - y, 2x + 2z, y - x)$. Let A be the matrix for T relative to the natural ordered basis, and let B be the matrix for T relative to $(1,1,0)$, $(1,0,1)$, and $(0,1,1)$. Show that A and B are similar by finding a nonsingular matrix P so that $B = P^{-1}AP$.

10. Suppose that A and B are similar $n \times n$ matrices. Are A^2 and B^2 similar? What about A^k and B^k?

11. In the following, for each matrix A find an orthogonal matrix P so that $P^T AP$ is a diagonal matrix.

(a) $\begin{pmatrix} 2 & \sqrt{2} \\ \sqrt{2} & 3 \end{pmatrix}$ (b) $\begin{pmatrix} 3 & 2\sqrt{2} \\ 2\sqrt{2} & -4 \end{pmatrix}$ (c) $\begin{pmatrix} \frac{11}{6} & \frac{1}{6} & \frac{4}{6} \\ \frac{1}{6} & \frac{11}{6} & -\frac{4}{6} \\ \frac{4}{6} & -\frac{4}{6} & \frac{14}{6} \end{pmatrix}$

12. A change of basis is called an orthogonal change of basis if the transition matrix is orthogonal. Find the orthogonal change of basis which transforms the given quadratic function's formula into one expressed as a sum of squares.

(a) $\frac{3}{2}x^2 + xy + \frac{3}{2}y^2$
(b) $\frac{1}{2}x^2 - 5xy + \frac{1}{2}y^2$
(c) $\frac{5}{4}x^2 - \frac{3}{2}\sqrt{3}xy - \frac{1}{4}y^2$

13. Suppose $\mathbf{X}$ and $\mathbf{Y}$ are eigenvectors of the symmetric matrix A corresponding to different eigenvalues r and s. Show that $\mathbf{X}$ and $\mathbf{Y}$ are orthogonal. (Hint: show $(A\mathbf{X}) \cdot \mathbf{Y} = r(\mathbf{X} \cdot \mathbf{Y})$. Then, using Exercise 11 of Section 8-1, show $(A\mathbf{X}) \cdot \mathbf{Y} = s(\mathbf{X} \cdot \mathbf{Y})$. Conclude that $\mathbf{X} \cdot \mathbf{Y} = 0$.)

14. Using Exercise 13, give an argument to show that if the $n \times n$ symmetric matrix A has n different eigenvalues, then there is an orthonormal basis for $\mathbb{R}^n$ consisting of eigenvectors for A.

15. Suppose that $\mathbf{X}_1, \ldots, \mathbf{X}_n$ is an orthonormal basis for $\mathbb{R}^n$. If P is the transition matrix from this basis to the standard basis, then the columns of P are given by $\mathbf{X}_1, \ldots, \mathbf{X}_n$. Suppose $\mathbf{X}$ has coordinates $(x'_1, \ldots, x'_n)$ and $\mathbf{Y}$ has coordinates $(y'_1, \ldots, y'_n)$ relative to $\mathbf{X}_1, \ldots, \mathbf{X}_n$, and suppose that $\mathbf{X}$ has coordinates $(x_1, \ldots, x_n)$ and $\mathbf{Y}$ has coordinates $(y_1, \ldots, y_n)$ relative to the standard basis. Show that

$$x_1 y_1 + \cdots + x_n y_n = x'_1 y'_1 + \cdots + x'_n y'_n.$$

Therefore the dot product of two vectors can be computed as the sum of products of corresponding coordinates relative to any orthonormal basis.

SECTION 10-3
Conic Sections

PREVIEW   *We show how to change coordinate systems so that the equation of a conic section can be simplified. Then the nature of the conic can be determined.*

In this section we discuss conic sections and their equations. Conic sections can be described geometrically as the curves which are formed by a plane cutting or intersecting a right circular cone. They are the circle, ellipse, hyperbola, parabola, two lines, one line, and one point.

The points on these curves can be described by an equation of the following form relative to some coordinate system:

$$Ax^2 + Bxy + Cy^2 + Dx + Ey + F = 0.$$

This is called the general second-degree equation. By using a different coordinate system we can simplify the form of the equation.

In this section we determine a coordinate system geometrically by specifying the point which serves as the origin and by specifying a basis of vectors, each pictured as emanating from the origin. Changing coordinate systems therefore consists in moving the origin or changing the basis vectors, or both. We also speak of coordinate axes. These are the lines determined by the vectors in the basis. Changing a coordinate system is also called changing the coordinate axes. Axes in the plane are labeled with letters such as x and y, x' and y', u and v, etc. The coordinate system is then designated as the xy-system, the $x'y'$-system, uv-system, etc. The coordinate system or axes are orthogonal or orthonormal if the basis used is orthogonal or orthonormal. In this section we deal exclusively with orthonormal systems having the same unit of measurement along every axis.

Simplifying second-degree equations is done in two steps.

(1) Rotate the axes so that the cross terms are eliminated.
(2) Translate the axes so that the first-degree terms are eliminated, if possible.

Translation of axes is simpler to describe, so we treat it first.

Example   Suppose we are given a plane with two coordinate systems, the xy-system and the $x'y'$-system as shown in Figure 10-5. Suppose that the $x'y'$-system

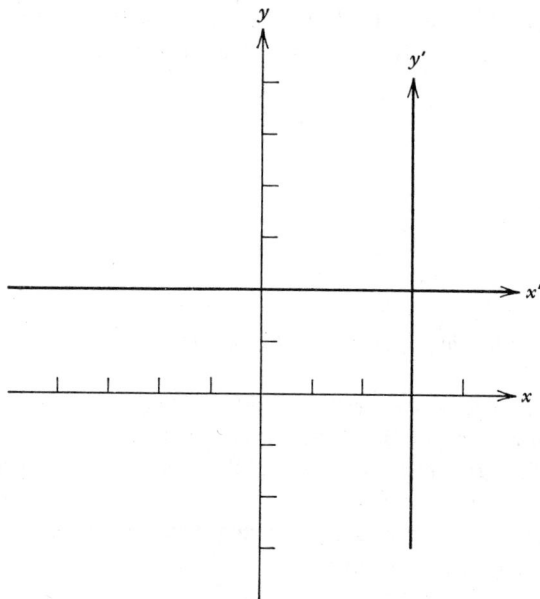

Figure 10-5

is obtained from the xy-system by translating the origin to the point $(3,2)$. This means that the x-axis is parallel to the x'-axis, the y-axis is parallel to the y'-axis, the positive direction of the x-axis and the x'-axis are the same, and the positive direction of the y-axis and the y'-axis are the same. Then the point having coordinates $(1,1)$ relative to the xy-system has coordinates $(-2,-1)$ relative to the $x'y'$-system. Similarly, the point having coordinates $(4,6)$ relative to the xy-system has coordinates $(1,4)$ relative to the $x'y$-system. In general, if a point has coordinates (x,y) relative to the xy-system and (x', y') relative to the $x'y'$-system, then the two sets of coordinates are related by the equations

$$x = x' + 3 \qquad \text{and} \qquad y = y' + 2$$

or by the equations

$$x' = x - 3 \qquad \text{and} \qquad y' = y - 2.$$

This example shows the general situation. Suppose we have two systems of orthogonal axes: call them the xy-system and the $x'y'$-system. Suppose that the x and x' lines are parallel and have the same positive direction, and that the y and y' lines are parallel and have the same positive direction. Suppose further that the origin of the $x'y'$-system occurs at the point with coordinates (h,k) relative to the xy-system. Then a point having coordinates (x,y) relative to the xy-system has

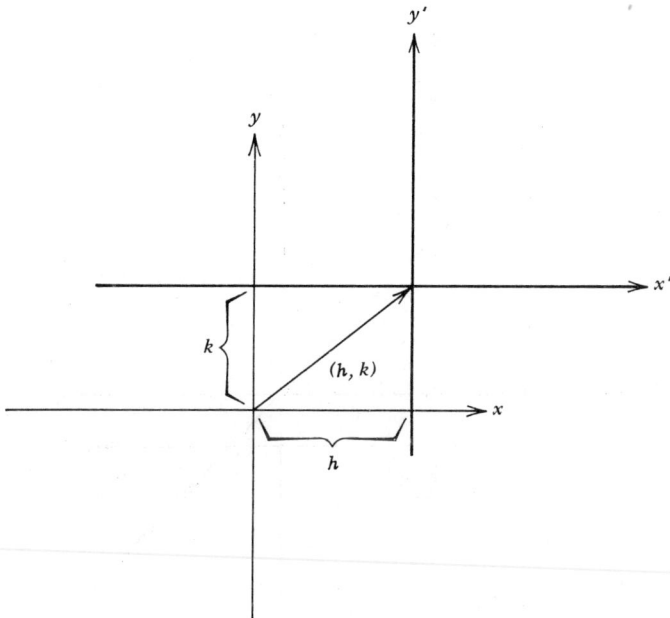

Figure 10-6

coordinates $(x - h, y - k)$ relative to the $x'y'$-system. Another way to express this is that if a point has coordinates (x,y) relative to the xy-system and (x', y') relative to the $x'y'$-system, then the coordinates are related by the equations

$$x = x' + h \quad \text{and} \quad y = y' + k$$

or by the equations

$$x' = x - h \quad \text{and} \quad y' = y - k.$$

See Figure 10-6. In this situation we say that the $x'y'$-system is obtained by translating the xy-system through (h,k)—or that the xy-system is obtained by translating the $x'y'$-system through $(-h,-k)$.

Example Suppose the $x'y'$-system has been obtained by translating the xy-system through $(3,-1)$. See Figure 10-7. Consider the line with equation $(x,y) = (2,1) + t(1,-1)$. This can be written as two equations $x = 2 + t$ and $y = 1 - t$. If a point has coordinates (x,y) relative to the xy-system and (x', y') relative to the $x'y'$-system, then $x = x' + 3$ and $y = y' - 1$. Then, relative to the $x'y'$-system, the line can be described by the equations $x = x' + 3 = 2 + t$ and $y = y' - 1 = 1 - t$ or by $x' = -1 + t$ and $y' = 2 - t$. Thus, relative to the $x'y'$-system, the

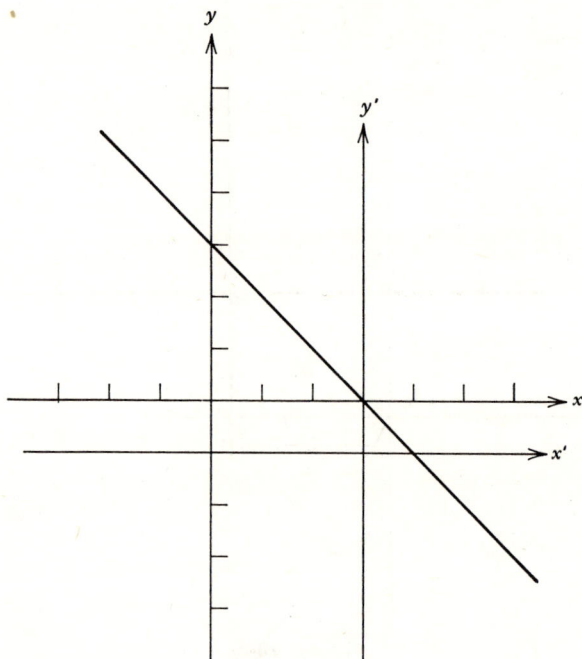

Figure 10-7

line has the vector equation

$$(x', y') = (-1 + t, 2 - t) = (-1,2) + t(1,-1).$$

This equation should also be clear from the figure.

Suppose, however, that a curve in the plane is described by the following equation.

$$x^2 + 3xy + y^2 - 5x - 2 = 0$$

Since $x = x' + 3$ and $y = y' - 1$, the equation for the curve relative to the $x'y'$-system is

$$(x' + 3)^2 + 3(x' + 3)(y' - 1) + (y' - 1)^2 - 5(x' + 3) - 2 = 0$$

or by the equation

$$x'^2 + 3x'y' + y'^2 - 2x' + 7y' - 16 = 0.$$

The example above shows how to find an equation for a curve relative to an $x'y'$-system from an equation for the same curve relative to an xy-system if the systems can be obtained from each other by a translation. If the relation between the coordinates is given by $x = x' + h$ and $y = y' + k$, then the equation for the

$x'y'$-system is obtained by substituting $x' + h$ for x and $y' + k$ for y in the equation relative to the xy-system.

Brief Exercises

1. Suppose the $x'y'$-system has been obtained from the xy-system by translating the xy-system through $(-2,4)$. The following equations describe a curve relative to the xy-system. Find an equation for the curve relative to the $x'y'$-system.

   (a) $2x + y = 4$ (b) $x^2 + y = 5$ (c) $x^2 - 2xy + 3y^2 = 1$

2. Same problem as Exercise 1 with the $x'y'$-system obtained by translating the xy-system through $(3,-2)$. (Use the equations given in Exercise 1.)

3. Suppose the $x'y'$-system has been obtained by translating the xy-system through $(2,-1)$. Find an equation for the curves relative to the xy-system if an equation for the curve relative to the $x'y'$-system is

   (a) $5x' + 2y' = 3$ (b) $4x'^2 + 2x'y' + 3y'^2 = 0$

Answers: **1(a)** $2x' + y' = 8$ **1(b)** $x'^2 - 4x' + y' = -3$ **1(c)** $x'^2 - 2x'y' + 3y'^2 - 12x' + 28y' = -67$ **2(a)** $2x' + y' = 0$ **2(b)** $x'^2 + 6x' + y' = -2$ **2(c)** $x'^2 - 2x'y' + 3y'^2 + 10x' - 18y' = -32$ **3(a)** $5x + 2y = 11$ **3(b)** $4x^2 + 2xy + 3y^2 - 14x + 2y + 15 = 0$

As we mentioned earlier in this section, the first step in simplifying the general equation for a conic is to rotate the axes so that the cross term is eliminated. We now discuss rotation of axes.

Suppose we have two systems of orthogonal axes; call them the xy-system and the $x'y'$-system. Suppose further that the $x'y'$-system is obtained by rotating the xy-system through the angle θ. We find the relation between the coordinates (x',y') of a point relative to the $x'y'$-system and its coordinates (x,y) relative to the xy-system.

One way to handle this problem is to argue geometrically as follows. Throughout this discussion, refer to Figure 10-8. Using trigonometry we obtain the following equalities.

$$AR = BQ = x' \sin \theta \qquad AB = RQ = y' \sin \theta$$

$$OB = x' \cos \theta \qquad PR = y' \cos \theta$$

Therefore we have

$$x = OA = OB - AB = x' \cos \theta - y' \sin \theta$$

$$y = AR + RP = BA + RP = x' \sin \theta + y' \cos \theta.$$

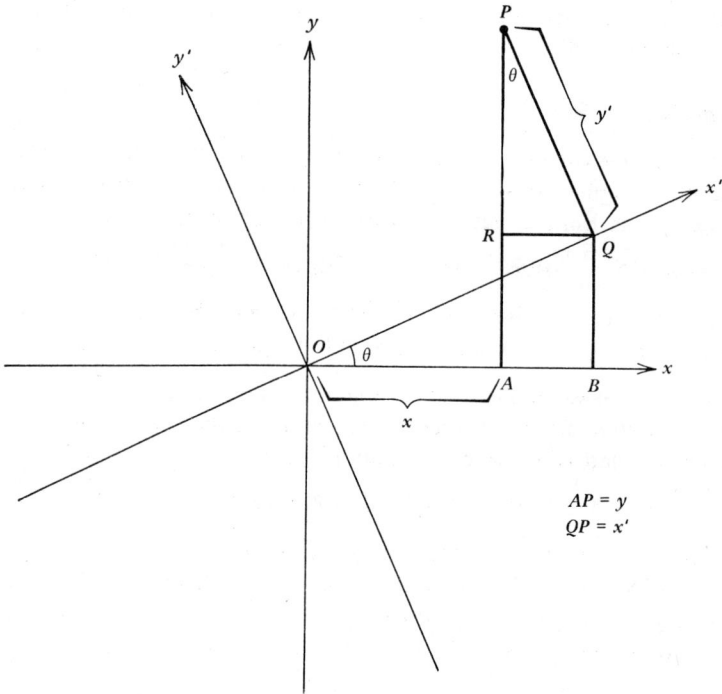

Figure 10-8

Since the xy-system is obtained from the $x'y'$-system by rotating through the angle $-\theta$, we have the corresponding equations

$$x' = \quad x \cos \theta + y \sin \theta$$

$$y' = -x \sin \theta + y \cos \theta.$$

We can write these equations as matrix equalities.

$$\begin{pmatrix} x \\ y \end{pmatrix} = \begin{pmatrix} \cos \theta & -\sin \theta \\ \sin \theta & \cos \theta \end{pmatrix} \begin{pmatrix} x' \\ y' \end{pmatrix}$$

$$\begin{pmatrix} x' \\ y' \end{pmatrix} = \begin{pmatrix} \cos \theta & \sin \theta \\ -\sin \theta & \cos \theta \end{pmatrix} \begin{pmatrix} x \\ y \end{pmatrix}$$

These matrix equalities show that we are dealing with a linear transformation, and this leads to the second way of viewing the rotation of axes. The matrix for the change from the ordered basis $(\cos \theta, \sin \theta)$ and $(-\sin \theta, \cos \theta)$ to the natural basis is

$$\begin{pmatrix} \cos\theta & -\sin\theta \\ \sin\theta & \cos\theta \end{pmatrix}.$$

Therefore

$$\begin{pmatrix} \cos\theta & -\sin\theta \\ \sin\theta & \cos\theta \end{pmatrix} \begin{pmatrix} x' \\ y' \end{pmatrix} = \begin{pmatrix} x \\ y \end{pmatrix}.$$

(Notice that the determinant of this matrix equals 1. Refer also to Exercise 4 at the end of this section.)

In the next few examples we show how to use rotations to eliminate the cross or xy term.

Example   Consider the curve described by the equation

$$2x^2 + 4xy - y^2 + 4x - 2y + 5 = 0$$

relative to the natural basis. We show how to find an orthogonal change of basis so that the equation relative to the new basis has no cross terms, i.e., no $x'y'$ term. This change of coordinate system does not change the origin.

Isolate the second degree terms $2x^2 + 4xy - y^2$. Express these in matrix form.

$$(x \quad y) \begin{pmatrix} 2 & 2 \\ 2 & -1 \end{pmatrix} \begin{pmatrix} x \\ y \end{pmatrix}$$

We now find a basis relative to which this quadratic function can be expressed as a sum of squares as in the previous section. The eigenvalues of the matrix are 3 and -2. Unit eigenvectors for these eigenvalues, respectively, are $(2/\sqrt{5}, 1/\sqrt{5})$ and $(-1/\sqrt{5}, 2/\sqrt{5})$. Thus the coordinates (x,y) of a point relative to the natural basis and (x', y') relative to the basis consisting of these eigenvectors are related by the following equations.

$$\begin{pmatrix} 2/\sqrt{5} & -1/\sqrt{5} \\ 1/\sqrt{5} & 2/\sqrt{5} \end{pmatrix} \begin{pmatrix} x' \\ y' \end{pmatrix} = \begin{pmatrix} x \\ y \end{pmatrix}$$

Relative to the new basis the quadratic function becomes $3x'^2 - 2y'^2$.

To determine the equation of the original curve relative to the new basis, we substitute $2x'/\sqrt{5} - y'/\sqrt{5}$ for x and $x'/\sqrt{5} + 2y'/\sqrt{5}$ for y in the original equation and obtain the following equation.

$$3x'^2 - 2y'^2 + 6x'/\sqrt{5} - 8y'/\sqrt{5} + 5 = 0$$

The same result can be obtained using matrix notation. Write the original equation using matrices as follows.

$$(x \quad y)\begin{pmatrix} 2 & 2 \\ 2 & -1 \end{pmatrix}\begin{pmatrix} x \\ y \end{pmatrix} + (4 \quad -2)\begin{pmatrix} x \\ y \end{pmatrix} = -5$$

The relation between the coordinates is given by the equation

$$\begin{pmatrix} 2/\sqrt{5} & -1/\sqrt{5} \\ 1/\sqrt{5} & 2/\sqrt{5} \end{pmatrix}\begin{pmatrix} x' \\ y' \end{pmatrix} = \begin{pmatrix} x \\ y \end{pmatrix}$$

Taking transposes of both sides, we obtain the equation

$$(x' \quad y')\begin{pmatrix} 2/\sqrt{5} & 1/\sqrt{5} \\ -1/\sqrt{5} & 2/\sqrt{5} \end{pmatrix} = (x \quad y)$$

Thus the equation of the curve can be expressed by the equation

$$(x' \quad y')\begin{pmatrix} 2/\sqrt{5} & 1/\sqrt{5} \\ -1/\sqrt{5} & 2/\sqrt{5} \end{pmatrix}\begin{pmatrix} 2 & 2 \\ 2 & -1 \end{pmatrix}\begin{pmatrix} 2/\sqrt{5} & -1/\sqrt{5} \\ 1/\sqrt{5} & 2/\sqrt{5} \end{pmatrix}\begin{pmatrix} x' \\ y' \end{pmatrix}$$
$$+ (4 \quad -2)\begin{pmatrix} 2/\sqrt{5} & -1/\sqrt{5} \\ 1/\sqrt{5} & 2/\sqrt{5} \end{pmatrix}\begin{pmatrix} x' \\ y' \end{pmatrix} = -5.$$

If you multiply these matrices you obtain the equation

$$3x'^2 - 2y'^2 + 6x'/\sqrt{5} - 8y'/\sqrt{5} = -5.$$

Now we are in a position to analyze equations for conic sections. The general equation for a conic is the following

$$Ax^2 + Bxy + Cy^2 + Dx + Ey + F = 0.$$

After we eliminate the cross term and translate for simplicity, we arrive at one of the following equations.

$u^2 + v^2 = r^2$ circle with center at origin and radius r

$u^2/a^2 + v^2/b^2 = 1$ ellipse with half-axes of length a along the u-axis and length b along the v-axis

$u^2/a^2 - v^2/b^2 = 1$ hyperbola opening along the u-axis with asymptotes $v = (b/a)u$ and $v = -(b/a)u$

$v = au^2$ parabola; $a > 0$: opening along the positive part of the v-axis; $a < 0$: opening along the negative part of the v-axis

$u^2/a^2 - v^2/b^2 = 0$ two lines: $v = (b/a)u$ and $v = -(b/a)u$

$u^2/a^2 + v^2/b^2 = 0$ one point

$u^2/a^2 + v^2/b^2 = -r^2$ impossible equation, so no configuration

As we mentioned earlier in this section, in simplifying a general second-degree equation we first rotate the axes to eliminate the cross term, and then we translate to remove the first-degree terms if possible. We illustrate this in the next examples.

Example Recall the curve considered in the previous example. Its equation is

$$2x^2 + 4xy - y^2 + 4x - 2y + 5 = 0.$$

Using the new basis $(2/\sqrt{5}, 1/\sqrt{5})$, $(-1/\sqrt{5}, 2/\sqrt{5})$ we find the equation to be

$$3x'^2 - 2y'^2 + 6x'/\sqrt{5} - 8y'/\sqrt{5} = -5.$$

We show how to transform this to one of the standard equations. Group the terms as follows:

$$3x'^2 + 6x'/\sqrt{5} - 2y'^2 + 8y'/\sqrt{5} = -5$$

$$3(x'^2 + 2x'/\sqrt{5}) - 2(y'^2 + 4y'/\sqrt{5}) = -5.$$

Complete the square inside both parentheses:

$$3(x'^2 + 2x'/\sqrt{5} + \tfrac{1}{5}) - 2(y'^2 + 4y'/\sqrt{5} + \tfrac{4}{5}) = -5 + \tfrac{3}{5} - \tfrac{8}{5}$$

$$3(x' + 1/\sqrt{5})^2 - 2(y' + 2/\sqrt{5})^2 = -6.$$

Let $u = x' + 1/\sqrt{5}$ and $v = y' + 2/\sqrt{5}$. The equation becomes the following.

$$3u^2 - 2v^2 = -6$$

or

$$v^2/(\sqrt{3})^2 - u^2/(\sqrt{2})^2 = 1$$

From the list of standard forms given above, we can see that the curve in question is a hyperbola opening along the v-axis with asymptotes $v = 3u/\sqrt{2}$ and $v = -3u/\sqrt{2}$. Now we review the calculations to determine what the u and v axes are. Refer to Figure 10-9.

The first change of basis was described in the last example. The $x'y'$-system was determined by the basis vectors $(2/\sqrt{5}, 1/\sqrt{5})$ and $(-1\sqrt{5}, 2/\sqrt{5})$. The next change of basis was determined by the translation

$$u = x' + 1/\sqrt{5} \qquad \text{and} \qquad v = y' + 2/\sqrt{5}.$$

Consequently the original conic is a hyperbola with center at $(0, -1)$ and opening along the line determined by the vector $(1/\sqrt{5}, -2/\sqrt{5})$.

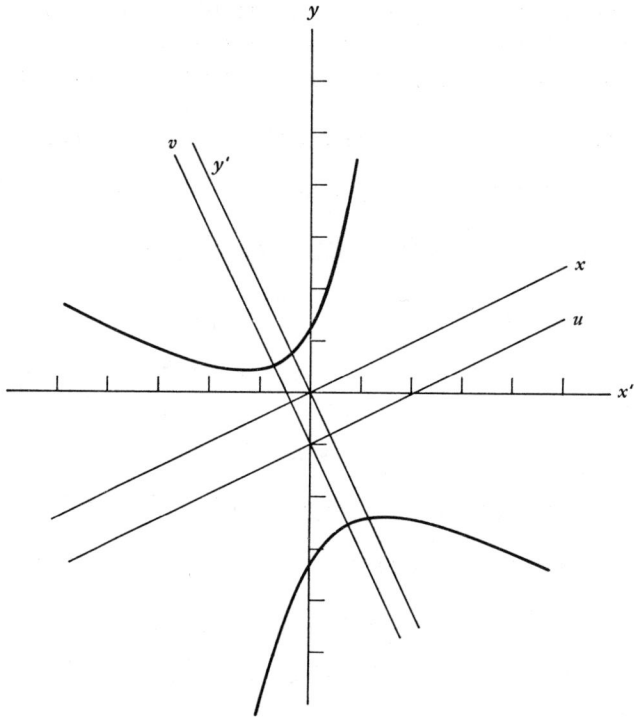

Figure 10-9

Example Consider the curve having the following equation relative to the xy-system.

$$2x^2 + 8xy + 8y^2 - 10\sqrt{5}x - 15\sqrt{5}y = -45$$

In this section we determine the geometrical nature of the curve. First consider the matrix for the quadratic part of the expression.

$$\begin{pmatrix} 2 & 4 \\ 4 & 8 \end{pmatrix}$$

We find an orthogonal change of basis relative to which the quadratic function is a sum of squares. The characteristic polynomial of this matrix is $x^2 - 10x$, and its roots are 10 and 0. A unit eigenvector for 10 is $(1/\sqrt{5}, 2/\sqrt{5})$, and a unit eigenvector for 0 is $(-2/\sqrt{5}, 1/\sqrt{5})$. The matrix

$$\begin{pmatrix} 1/\sqrt{5} & -2/\sqrt{5} \\ 2/\sqrt{5} & 1/\sqrt{5} \end{pmatrix}$$

is the matrix for the change of basis from the basis consisting of $(1/\sqrt{5}, 2/\sqrt{5})$ and $(-2/\sqrt{5}, 1/\sqrt{5})$ to the natural basis. Thus the coordinates (x,y) of a point relative to the standard basis and (x', y') relative to the new basis are related by the equation

$$\begin{pmatrix} x \\ y \end{pmatrix} = \begin{pmatrix} 1/\sqrt{5} & -2/\sqrt{5} \\ 2/\sqrt{5} & 1/\sqrt{5} \end{pmatrix} \begin{pmatrix} x' \\ y' \end{pmatrix}.$$

Consequently the equation of the curve relative to the new basis is

$$(x' \quad y') \begin{pmatrix} 1/\sqrt{5} & 2/\sqrt{5} \\ -2/\sqrt{5} & 1/\sqrt{5} \end{pmatrix} \begin{pmatrix} 2 & 4 \\ 4 & 8 \end{pmatrix} \begin{pmatrix} 1/\sqrt{5} & -2/\sqrt{5} \\ 2/\sqrt{5} & 1/\sqrt{5} \end{pmatrix} \begin{pmatrix} x' \\ y' \end{pmatrix}$$

$$+ \quad (x' \quad y') \begin{pmatrix} 1/\sqrt{5} & 2/\sqrt{5} \\ -2/\sqrt{5} & 1/\sqrt{5} \end{pmatrix} \begin{pmatrix} -10\sqrt{5} \\ -15\sqrt{5} \end{pmatrix} = -45.$$

By multiplying these matrices we obtain the equation

$$10x'^2 - 40x' + 5y' = -45.$$

Completing the square of the x' terms we obtain

$$10(x'^2 - 4x' + 4) + 5y' + 5 = 0$$
$$10(x' - 2)^2 + 5(y' + 1) = 0.$$

Make a change of coordinates by translating the axes through $(2,-1)$; this is given by the equations $u = x' - 2$ and $v = y' + 1$. This gives the equation $10u^2 + 5v = 0$ or $v = -2u^2$. Consequently we are dealing with a parabola with vertex at $(4/\sqrt{5}, 3/\sqrt{5})$ opening in the negative direction of the v-axis. See Figure 10-10.

These examples show the procedure for simplifying the form of a second-degree equation by changing coordinate systems. While the technique of diagonalizing a symmetric matrix is not essential in the case of an equation involving only two variables, it is very helpful when one wishes to simplify an equation involving first- and second-degree terms and having more than two variables. We do not treat this in any detail, but we offer the following example to show how the techniques used in the two-dimensional situation can be used in higher-dimensional situations.

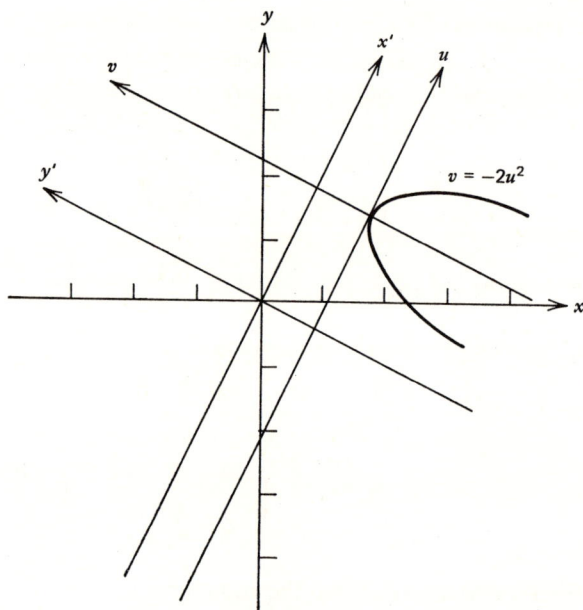

Figure 10-10

Example Consider the configuration in $\mathbb{R}^3$ having the following equation relative to the natural basis.

$$3y^2 + 4xy - 2\sqrt{2}x - 18y - 6\sqrt{2}z = -27$$

Write this in matrix notation.

$$(x \ \ y \ \ z)\begin{pmatrix} 0 & 0 & 2 \\ 0 & 3 & 0 \\ 2 & 0 & 0 \end{pmatrix}\begin{pmatrix} x \\ y \\ z \end{pmatrix} + (-2\sqrt{2} \ \ -18 \ \ -6\sqrt{2})\begin{pmatrix} x \\ y \\ z \end{pmatrix} = -27$$

Find the eigenvalues of the 3×3 matrix; they are the roots of the characteristic polynomial $-x^3 + 3x^2 + 4x - 12$. The eigenvalues are -2, 2, and 3. Unit eigenvectors corresponding to -2, 2, and 3, respectively, are $(-1/\sqrt{2}, 0, 1/\sqrt{2})$, $(1/\sqrt{2}, 0, 1/\sqrt{2})$, and $(0,1,0)$. Use these vectors as columns of a matrix for a change of basis.

$$\begin{pmatrix} x \\ y \\ z \end{pmatrix} = \begin{pmatrix} -1/\sqrt{2} & 1/\sqrt{2} & 0 \\ 0 & 0 & 1 \\ 1/\sqrt{2} & 1/\sqrt{2} & 0 \end{pmatrix}\begin{pmatrix} x' \\ y' \\ z' \end{pmatrix}$$

This gives a new equation.

$$(x' \quad y' \quad z')\begin{pmatrix} -1/\sqrt{2} & 0 & 1/\sqrt{2} \\ 1/\sqrt{2} & 0 & 1/\sqrt{2} \\ 0 & 1 & 0 \end{pmatrix}\begin{pmatrix} 0 & 0 & 2 \\ 0 & 3 & 0 \\ 2 & 0 & 0 \end{pmatrix}\begin{pmatrix} -1/\sqrt{2} & 1/\sqrt{2} & 0 \\ 0 & 0 & 1 \\ 1/\sqrt{2} & 1/\sqrt{2} & 0 \end{pmatrix}\begin{pmatrix} x' \\ y' \\ z' \end{pmatrix}$$

$$+ \quad (-2\sqrt{2} \quad -18 \quad -6\sqrt{2})\begin{pmatrix} -1/\sqrt{2} & 1/\sqrt{2} & 0 \\ 0 & 0 & 1 \\ 1/\sqrt{2} & 1/\sqrt{2} & 0 \end{pmatrix}\begin{pmatrix} x' \\ y' \\ z' \end{pmatrix} = -27$$

Multiplying these matrices we obtain the equation

$$-2x'^2 + 2y'^2 + 3z'^2 - 4x' - 8y' - 18z' = -27.$$

Completing the squares we obtain the equation

$$-2(x'^2 + 2x' + 1) + 2(y'^2 - 4y' + 4)$$
$$+ 3(z'^2 - 6z' + 9) = -27 - 2 + 8 + 27.$$

Changing coordinate axes by the translation $u = x' + 1$, $v = y' - 2$, and $w = z' - 3$, we obtain the equation

$$-2u^2 + 2v^2 + 3w^2 = 6$$
$$-u^2/3 + v^2/3 + w^2/2 = 0.$$

This shows how the techniques for simplifying the equations of conics can be employed on equations involving many unknowns provided each term is of degree two or less. A more complete treatment can be found in Brinkman and Klotz [2], pp. 431–442.

The technique of expressing a quadratic function as a sum of squares relative to some orthonormal basis is a very important one. Not only is it used in classifying conic sections and quadric surfaces, but it is also used in much advanced mathematics and in other areas of study. Chapter 12 of Noble [8] contains a discussion of the use of these methods in studying a chemical reaction system. Sections 5.4 and 5.5 of Noble [8] show how the method can be used to study electrical circuits. Campbell [3] gives some further reference. For a discussion of this method used in differential equations, see Section 6.1 of Brauer, Nohel, and Schneider [1].

SUMMARY   *The techniques for simplifying a second-degree equation are discussed. First, we find an orthogonal change of basis which eliminates the cross terms, or, equivalently, expresses the quadratic terms in the equation as a sum of* *squares. This is done by finding a basis of eigenvectors for the symmetric matrix for the quadratic terms of the equation and using these basis vectors as columns for a matrix of a change of basis. Then a translation is used to eliminate the first-degree terms, if possible.*

Exercises for Section 10-3

1. Determine the type of conic section which is described by the following equations.

   (a) $7x^2 + 6\sqrt{3}xy + 13y^2 = 4$ (d) $3 - x^2 - 12xy + 35y^2 = 6$
   (b) $5x^2 + 26xy + 5y^2 = -72$ (e) $xy = 4$
   (c) $17x^2 + 18xy - 7y^2 = 10$

2. Same problem as Exercise 1 for the following.

   (a) $2xy + 4y = 14$
   (b) $x^2 - 2xy + y^2 + 2x - 4y = -\frac{9}{4}$
   (c) $29x^2 - 6xy + 21y^2 + 32\sqrt{10}x - 24\sqrt{10}y = -80$

3. Simplify the following second-degree equations by an orthogonal change of basis and then a translation to transform the equation to a simpler form.

   (a) $z^2 - 2xy - 2\sqrt{2}x + 2y = 0$
   (b) $-3x^2 + 6\sqrt{2}xy + 2\sqrt{3}xz + 2\sqrt{6}yz + 3z^2 + \sqrt{6}x - 6\sqrt{3}y - 9\sqrt{2}z = -6$
   (c) $3x^2 + 2xz + y^2 - 8yw + 3z^2 + 2w^2 - 6\sqrt{2}x + 10\sqrt{2}y - 2\sqrt{2}z - 2\sqrt{2}w = -2$

More Challenging Exercises

4. Give an argument to show that a 2×2 orthogonal matrix having determinant equal to 1 represents a rotation of axes.

5. What geometric interpretation can you give to a 2×2 orthogonal matrix whose determinant equals -1?

6. Consider the configuration having the equation

$$Ax^2 + Bxy + Cy^2 = F$$

with $F > 0$. Give an argument to show that the configuration is an ellipse if $B^2 - 4AC < 0$, a parabola if $B^2 - 4AC = 0$, and a hyperbola if $B^2 - 4AC > 0$.

References

1. Fred Brauer, John A. Nohel, and Hans Schneider. *Linear Mathematics.* W. A. Benjamin, Inc., New York, N.Y., 1970.
2. Heinrich W. Brinkmann and Eugene A. Klotz. *Linear Algebra and Analytic Geometry.* Addison-Wesley Publishing Co., Inc., Reading, Mass., 1971.
3. Hugh C. Campbell. *Linear Algebra with Applications Including Linear Programming.* Meredith Corporation, New York, N.Y., 1971.
4. Charles W. Curtis. *Linear Algebra.*, 2nd ed. Allyn and Bacon, Inc., Boston, Mass., 1968.
5. Jimmie D. Gilbert. *Elements of Linear Algebra.* International Textbook Co., Scranton, Pa., 1970.
6. John B. Johnston, G. Baley Price, and Fred S. Van Vleck. *Linear Equations and Matrices.* Addison-Wesley Publishing Co., Inc., Reading, Mass., 1966.
7. John G. Kemeny, J. Laurie Snell, and Gerald L. Thompson. *Introduction to Finite Mathematics.* Prentice-Hall, Inc., Englewood Cliffs, N.J., 1957.
8. Ben Noble, Ed. *Applications of Undergraduate Mathematics in Engineering.* The Macmillan Co., New York, N.Y., 1967.
9. Ben Noble. *Applied Linear Algebra.* Prentice-Hall, Inc., Englewood Cliffs, N.J., 1969.
10. John H. Staib. *An Introduction to Matrices and Linear Transformations.* Addison-Wesley Publishing Co., Inc., Reading, Mass., 1969.

Answers to Selected Problems

Exercises for Section 1-1

1(a) $(\frac{1}{2},\frac{1}{2})$ **1(b)** $(\frac{3}{5} + \frac{1}{5}w, -\frac{2}{5} + z - \frac{9}{5}w, z, w)$ **1(c)** $(\frac{4}{6} + \frac{7}{6}w - \frac{9}{6}u + \frac{7}{6}v, \frac{32}{6} - \frac{28}{6}w + \frac{30}{6}u - \frac{16}{6}v, -\frac{38}{6} + \frac{13}{6}w - \frac{9}{6}u + \frac{1}{6}v, w, u,v)$ **1(d)** No solution **1(e)** $(\frac{6}{18},\frac{17}{18},\frac{23}{18})$ **1(f)** $(\frac{2}{4} + \frac{10}{4}w, \frac{6}{4} - \frac{18}{4}w + v, -\frac{1}{4} + \frac{11}{4}w - \frac{6}{4}v, w, v)$ **1(g)** $(-\frac{5}{6} - \frac{11}{6}w, \frac{4}{6} + \frac{10}{6}w, \frac{7}{6} - \frac{5}{6}w, w)$

Exercises for Section 1-2

1(a) Yes **1(b)** Yes **1(c)** Yes **1(d)** No

2(a)
$$\begin{pmatrix} 1 & 0 & 0 & \frac{12}{10} \\ 0 & 1 & 0 & \frac{3}{10} \\ 0 & 0 & 1 & -\frac{17}{10} \end{pmatrix}$$

2(b)
$$\begin{pmatrix} 1 & 0 & \frac{3}{7} & \frac{2}{7} & \frac{9}{7} \\ 0 & 1 & -\frac{13}{7} & \frac{17}{7} & \frac{8}{7} \end{pmatrix}$$

2(c)
$$\begin{pmatrix} 1 & 0 & 0 & \frac{13}{27} \\ 0 & 1 & 0 & -\frac{17}{27} \\ 0 & 0 & 1 & -\frac{15}{27} \end{pmatrix}$$

2(d)
$$\begin{pmatrix} 1 & 0 & 0 & 0 \\ 0 & 1 & 0 & 0 \\ 0 & 0 & 1 & 0 \\ 0 & 0 & 0 & 1 \end{pmatrix}$$

3(a)
$$\begin{pmatrix} 1 & 0 & 0 & 0 \\ 0 & 1 & 0 & 0 \\ 0 & 0 & 1 & 0 \\ 0 & 0 & 0 & 1 \end{pmatrix}$$

3(b)
$$\begin{pmatrix} 1 & 0 & 0 & 0 & -\frac{50}{16} \\ 0 & 1 & 0 & 0 & \frac{98}{16} \\ 0 & 0 & 1 & 0 & -\frac{4}{16} \\ 0 & 0 & 0 & 1 & \frac{50}{16} \end{pmatrix}$$

4(b) It is possible to solve for x and z in terms of y and w **4(c)** It is possible to solve for y and w in terms of x and z; "Obtain leading 1's in the second and fourth columns."

5 No for x and y. Yes for x and z. Yes for y and z **6(a)** There is no single answer. One possible answer is

$$\begin{pmatrix} 1 & 0 & 1 & 0 \\ 0 & 1 & 1 & 0 \\ 0 & 0 & 0 & 1 \end{pmatrix}$$

6(b) There is no single answer. One possible answer is

$$\begin{pmatrix} 1 & 0 & 0 & 1 \\ 0 & 1 & 1 & 0 \\ 0 & 0 & 0 & 0 \end{pmatrix}$$

9 There is no single answer to each of these. The following are possible answers.

(a) $\begin{pmatrix} -2 & 1 & 0 & 0 \\ 1 & 0 & 1 & 3 \\ 0 & 0 & 0 & 0 \\ 0 & 0 & 0 & 0 \end{pmatrix}$ **(b)** $\begin{pmatrix} 1 & 0 & 0 & 2 \\ 0 & 1 & 1 & 3 \\ 0 & 0 & 0 & 0 \\ 0 & 0 & 0 & 0 \end{pmatrix}$

(c) $\begin{pmatrix} 1 & -1 & -1 & 0 \\ 0 & 0 & 0 & 0 \\ 0 & 0 & 0 & 0 \\ 0 & 0 & 0 & 0 \end{pmatrix}$ **(d)** $\begin{pmatrix} 1 & 0 & 0 & 3 \\ 0 & 1 & 0 & 1 \\ 0 & 0 & 1 & 4 \\ 0 & 0 & 0 & 0 \end{pmatrix}$

10 Yes, no constants in the component description.   **12(a)** Yes **12(b)** Yes **12(c)** No **12(d)** Yes

13(a) $\begin{pmatrix} 1 & 0 \\ 0 & 1 \end{pmatrix}$ **13(b)** $\begin{pmatrix} 1 & 0 \\ 0 & 1 \end{pmatrix}$ **13(c)** $\begin{pmatrix} 1 & 0 & 0 \\ 0 & 1 & 0 \end{pmatrix}$

13(d) $\begin{pmatrix} 1 & 0 & 0 & 0 \\ 0 & 0 & 0 & 0 \\ 0 & 0 & 1 & 0 \end{pmatrix}$ **15(a)** $\begin{pmatrix} 1 & 1 \\ 1 & 2 \\ 2 & 3 \end{pmatrix}$ **15(b)** $\begin{pmatrix} 3 & 0 \\ 1 & 1 \end{pmatrix}$

15(c) $\begin{pmatrix} 5 & 1 & 1 \\ 1 & -4 & 1 \\ 2 & 3 & 3 \\ 3 & 2 & 1 \end{pmatrix}$

Exercises for Section 2-1

1(a) $(9,15)$ **1(b)** $(-40,-11)$ **1(c)** $(0,0)$ **1(d)** $(0,0)$ **2(a)** $(3,1)$ **2(b)** $(-\frac{1}{3},0)$ **2(c)** $(-\frac{1}{2},-\frac{1}{2})$
3(a) $(1,-5,-10)$ **3(b)** $(-2,-5,3)$ **3(c)** $(2,-43,7)$ **4(a)** $(1,\frac{1}{2},\frac{3}{2})$ **4(b)** $(-\frac{1}{2},1,-\frac{7}{2})$ **4(c)**
$(-1,-1,0)$ **5(a)** $(\frac{2}{3},-\frac{2}{3},-\frac{1}{3},0)$ **5(b)** $(\frac{2}{5},1,0,-\frac{1}{5})$ **5(c)** $(\frac{11}{6},0,\frac{39}{6},-\frac{2}{6})$ **6** $\mathbf{X} = (\frac{5}{14},\frac{4}{14})$, $\mathbf{Y} =$
$(\frac{3}{7},\frac{1}{7})$ **7** $\mathbf{X} = (2,\frac{5}{7})$, $\mathbf{Y} = (3,\frac{2}{7})$ **8** $\mathbf{X} = (\frac{3}{8},-\frac{4}{8},\frac{5}{8})$, $\mathbf{Y} = (\frac{4}{8},0,-\frac{4}{8})$, $\mathbf{Z} = (\frac{3}{8},\frac{4}{8},-\frac{3}{8})$ **9** No
solution

Exercises for Section 2-2

1(a) Yes **1(b)** Yes **1(c)** No **1(d)** Yes **1(e)** No **1(f)** Yes **2** Not a subset **7** No single answer.
One possible: all the vectors of the form $(2r,5r)$ along with those of the form $(r,-r)$,
where r is a scalar **8** No single answer. One possible: the vectors of the form (n,n), where
n is a positive integer

Exercises for Section 2-3

1(a) $(2,1) + t(1,-2)$ **1(b)** $(-1,5) + t(5,-3)$ **1(c)** $(1,1) + t(3,-4)$ **1(d)** $(7,15) + t(-7,-13)$
1(e) $(1,5) + t(0,3)$ **2(a)** $(\frac{7}{2},0)$, $(0,\frac{7}{3})$ **2(b)** $(-\frac{26}{5},0)$, $(0,26)$ **2(c)** $(10,0)$, $(0,\frac{10}{3})$ **2(d)** $(5,0)$, $(0,5)$
3(a) $(2,1,5) + t(2,0,-4)$ **3(b)** $(2,0,3) + t(-1,-1,-3)$ **3(c)** $(3,8,-2) + t(-4,-6,3)$ **4(a)**
No **4(b)** Yes **4(c)** No **5(a)** No **5(b)** No **5(c)** Yes **6(a)** No intersection **6(b)** $(3,6,5)$ **9**
$(\frac{13}{2},3)$

Exercises for Section 3-1

1(a) $(43, 59, 70, 16\sqrt{17}-15, \pi-20)$ **1(b)** $(31,339,325,224,63)$ **1(c)** $(2, \frac{9}{20})$
2(a) $6(1,3,1,4)-5(6,3,-1,0) + 2(9,0,1,3)$ **2(b)** $4(1,0,1) + 1(1,1,0)$ **3(a)** $(1,1) =$
$(-\frac{1}{5})(1,-1) + (\frac{2}{5})(3,2)$ **3(b)** $(1,1,0) = (-\frac{1}{5})(1,-1,0) + (\frac{2}{5})(3,2,0)$ **3(c)** No **3(d)** $(1,1,0)=$
$(\frac{4}{16})(1,0,1) - (\frac{2}{16})(3,4,5) + (\frac{3}{16})(6,8,2)$

Exercises for Section 3-2

1(a) Yes **1(b)** Yes **2(a)** No **2(b)** Yes **3(a)** No **3(b)** Yes **4** Yes

Exercises for Section 3-3

1(a) $(1,2,1) + s(1,-1,2) + t(3,-1,0)$ **1(b)** $(5,1,-1) + s(-5,-4,5) + t(-4,0,3)$ **2**
$(2,1,4) + s(1,-1,2) + t(3,-1,0)$ **3** $(1,2,1) + s(-5,-4,5) + t(-4,0,3)$ **4(a)** They inter-
sect in $(\frac{12}{9},\frac{3}{9},\frac{15}{9}) + v(\frac{5}{9},-\frac{1}{9},\frac{4}{9})$ **4(b)** They intersect in $(\frac{11}{3},-\frac{4}{3},\frac{2}{3}) + v(-\frac{7}{3},\frac{8}{3},\frac{2}{3})$. **5(a)** No

5(b) No **6(a)** (5,7,2) **6(b)** $(-7,4,-4)$ **7(a)** (1,0,3), (0,1,1) **7(b)** $(1,-2,0)$, $(0,-2,1)$ **7(c)** $(1,0,\frac{5}{3})$, $(0,1,\frac{4}{3})$ **8(a)** The point $(6,-9,-7)$ **8(b)** The line through $(\frac{3}{4},-\frac{5}{4},0)$ parallel to $(\frac{2}{4},\frac{2}{4},1)$ **8(c)** The plane passing through (0,1,0) and determined by $(1,-3,0)$ and (0,2,1) **10** Parallel **11** Any line through the origin not containing (1,2) **12** Any line through the origin not containing $(-2,3)$ **13** Any plane through the origin not containing (1,0,2) **14** Any plane through the origin not containing $(2,-1,3)$ **15** Any line through the origin not contained in the plane determined by (1,2,1) and $(5,-1,0)$ **16** Any line through the origin not contained in the plane determined by $(2,-1,3)$ and $(4,-5,0)$ **18** Yes **19** Yes

Exercises for Section 3-4

1 Linearly independent **2** Linearly independent **3** Linearly independent **4** Linearly independent **5** Linearly dependent **6** Linearly independent **7** Linearly dependent

Exercises for Section 3-5

1 Basis **2** Not a basis **3** Basis **4** Not a basis **5** Basis **6** Not a basis **7** Basis **8** Not a basis **9** Not a basis **10(a)** Basis **10(b)** (1,1), (2,1); (1,1), $(-5,3)$; (2,1), $(-5,3)$ **11(a)** Any three vectors **11(b)** Basis **11(c)** Any three of the given vectors **11(d)** Any three vectors except for the following: (1,0,0), (0,1,0), (1,1,0); (0,0,1), (1,1,1), (1,1,0) **14** (1,0,2) and $(0,1,-1)$ **15** $(1,0,0,\frac{1}{17})$, (0,1,0,1), $(0,0,1,-\frac{11}{17})$.

Exercises for Section 3-6

1(a) Yes **1(b)** No **1(c)** No **1(d)** No **1(e)** Yes **2(a)** No **2(b)** Yes **2(c)** No **2(d)** No **2(e)** Yes **3(a)** Yes **3(b)** Yes **4(a)** $(1,0,-2)$, $(0,1,-1)$ **4(b)** $(0,-1,1)$ **4(c)** No basis **5(a)** $(1,0,0,-1)$, $(0,1,0,-1)$, (0,0,1,0) **5(b)** $(-1,0,1,0)$, $(-\frac{1}{2},\frac{3}{2},0,1)$ **5(c)** $(1,1,-2,1)$ **5(d)** $(1,1,-2,1)$

9(a) $x - 2y \qquad = 0$
$\quad\;\; - y + z \;\;= 0$
9(b) $x - y - z \;= 0$
9(c) $x - y \qquad\;\; = 0$
$\qquad\quad z \qquad = 0$
$\;\; x \qquad\quad\;\; - w = 0$

9(d) $x - y - z \qquad = 0$
$\qquad\quad y \qquad - w = 0$
9(e) $x - y - z \qquad = 0$
9(f) Same answer as **9(e)**

Exercises for Section 4-1

1(a) $x^2 + 3x + 7$ **1(b)** $3x^2 + 6x + 16$ **1(c)** $10x^2 + 18x + 21$ **1(d)** $16x^4 + 8x^3 + x^2 + 2$ **1(e)** $4x^4 + 17x^2 + 18$ **2(a)** $(2x^2 + 9x + 10, 15x^2 + 10x - 11)$ **2(b)** $(5x^2 - 9x + 4, -15x^2 + 25x + 4)$ **2(c)** $(22x^2 + 22x - 2, 4x^4 + 48x^3 + 172x^2 + 278x + 158)$ **2(d)** $(-12x^2 - 6x + 6, 3x^4 - 18x^3 + 42x^2 - 15x - 6)$

Exercises for Section 4-2

1(a)(i) No **1(a)(ii)** No **1(a)(iii)** Yes **1(b)(i)** No **1(b)(ii)** No **1(b)(iii)** No **1(c)(i)** Yes **1(c)(ii)** Yes **2(a)** $T(x,y) = (\frac{1}{2}x + \frac{3}{2}y, -2x + 3y, -\frac{1}{2}x + \frac{7}{2}y)$ **2(b)** $T(x,y) = (x - y, \frac{8}{5}x - \frac{6}{5}y)$ **2(c)** $T(x,y) = (-5x - 2y, -10x - 4y, 0)$ **2(d)** $(x + y + 3z, 3x + 2y)$ **2(e)** T is not defined on a basis **3(a)** $(3,-1,8) + t(7,-1,18)$. A line through $(3,-1,8)$ parallel to $(7,-1,18)$ **3(b)** $(0,5) + t(-3,6)$ **3(c)** $(0,5) + t(0,0)$. One point **4** No **5** Parallel lines, two points, or one point

Exercises for Section 4-3

1(a) $\begin{pmatrix} 3 & 1 \\ 2 & -5 \end{pmatrix}$ $T(5,1) = (16,5)$, $T(-3,7) = (-2,-41)$, $T(17,-18) = (33,124)$

1(b) $\begin{pmatrix} 2 & 1 \\ 1 & -1 \end{pmatrix}$ $T(5,1) = (11,4), T(-3,7) = (1,-10), T(17,-18) = (16,35)$

1(c) $\begin{pmatrix} 4 & 2 \\ -5 & 3 \end{pmatrix}$ $T(5,1) = (22,-22), T(-3,7) = (2,36), T(17,-18) = (32,-139)$

2(a) $\begin{pmatrix} 1 & -2 & 1 \\ 1 & 1 & -3 \end{pmatrix}$ **2(b)** $\begin{pmatrix} 2 & 1 & 3 \\ 1 & -1 & 1 \end{pmatrix}$ **2(c)** $\begin{pmatrix} 5 & 2 & -1 \\ 1 & 1 & 0 \end{pmatrix}$

3(a) $\begin{pmatrix} 1 & 2 \\ 1 & -1 \\ 3 & 5 \\ 4 & -1 \end{pmatrix}$ **3(b)** $\begin{pmatrix} 1 & 1 \\ 1 & 1 \\ 2 & 1 \\ 2 & 1 \end{pmatrix}$ **4** $\begin{pmatrix} \frac{1}{2} & -\frac{7}{2} & \frac{11}{2} \\ 0 & 0 & 1 \\ -\frac{1}{2} & \frac{3}{2} & \frac{5}{2} \end{pmatrix}$

5 $\begin{pmatrix} 5 & 1 \\ 1 & 0 \\ 6 & -4 \end{pmatrix}$ **6** $\begin{pmatrix} 1 & 2 \\ -\frac{1}{3} & \frac{2}{3} \\ -\frac{2}{3} & \frac{10}{3} \end{pmatrix}$

7 $T(2,1) = (5,3,-6), T(7,8) = (22,24,-3), T(16,2) = (34,6,-72)$ **8** $T(3,1,1) = (6,-2,31)$, $T(0,-1,2) = (3,11,11), T(3,1,-6) = (-8,-23,-11)$ **9** $T(2,1) = (5,-2), T(3,4) = (10,7)$ **10** Sends it to the line $(27,26) + t(-14,-20)$ **11** T sends $X_0 + tX_1$ into $T(X_0) + tT(X_1)$; a line if $T(X_1) \neq 0$, a point if $T(X_1) = 0$ **12** Sends it into the triangle with vertices $(3,-1)$, $(8,-5)$, $(3,13)$ **13** Sends it to the parallelogram with vertices $(0,0)$, $(1,-2)$, $(6,2)$, $(5,4)$ **14** Sends it into the parallelepiped determined by $(0,0,0)$, $(2,3,2)$, $(1,1,1)$, and $(5,4,3)$

Exercises for Section 4-4

1(a) $(S - 4T)(x,y) = (-14x + 5y, -3x - 5y, -x + 6y)$, $(3S + 2T)(x,y) = (14x + y, 5x - y, 11x + 4y)$ **1(b)** $(S - 4T)(x,y,z) = (-19x - 5y + 7z, -x - 10y - 9z)$, $(3S + 2T)(x,y,z) = (13x - y - 7z, 11x + 12y + z)$ **1(c)** $(S - 4T)(x,y,z) = (-14x - 2y + 13z, -3x + 3y - 7z)$, $(3S + 2T)(x,y,z) = (14x + 8y - 17z, 5x - 5y + 7z)$

3(a) $\begin{pmatrix} 16 & -12 & 10 \\ 16 & 30 & 40 \end{pmatrix}$ **3(b)** $\begin{pmatrix} 2 & 3 \\ 1 & -1 \\ 2 & 1 \end{pmatrix}$ **3(c)** $\begin{pmatrix} 0 & 0 \\ 0 & 0 \\ 0 & 0 \end{pmatrix}$

6 $\begin{pmatrix} 57 & 1 \\ 35 & -40 \\ 54 & -38 \end{pmatrix}$ **7** $\begin{pmatrix} 23 & 45 & 60 \\ 29 & 49 & 82 \\ 37 & 69 & 49 \end{pmatrix}$ **11** $\begin{pmatrix} 45 & 67 \\ 96 & 175 \\ 25 & 28 \end{pmatrix}$

13
$$2A^2 + 3A = \begin{pmatrix} 5 & 22 \\ 0 & 27 \end{pmatrix} \qquad A^2 - 4A + 3I_2 = \begin{pmatrix} 0 & 0 \\ 0 & 0 \end{pmatrix}$$

14
$$4A^3 - 3A^2 + 2A = \begin{pmatrix} 1 & 26 & 50 \\ 62 & 14 & 88 \\ 189 & 150 & 240 \end{pmatrix}$$

Exercises for Section 5-1

1(a) $\frac{1}{12} \begin{pmatrix} -4 & 5 & 1 \\ 8 & -4 & -8 \\ 4 & -2 & 2 \end{pmatrix}$ **1(b)** $\begin{pmatrix} 1 & -2 & 1 \\ 2 & -5 & 4 \\ -1 & 3 & -2 \end{pmatrix}$

1(c) $\frac{1}{8} \begin{pmatrix} -5 & 4 & -1 & -3 \\ -11 & 4 & 1 & 3 \\ 11 & -4 & -1 & 5 \\ 2 & 0 & 2 & -2 \end{pmatrix}$ **1(d)** $\frac{1}{4} \begin{pmatrix} 0 & 1 & 3 & -1 \\ 0 & -1 & -3 & 5 \\ 0 & 1 & -1 & -1 \\ 2 & 0 & 0 & -2 \end{pmatrix}$

1(e) $\begin{pmatrix} 1 & 0 & 0 & -1 \\ 1 & -1 & -1 & 1 \\ -1 & 2 & 1 & -1 \\ -1 & 1 & 1 & 0 \end{pmatrix}$

2(a) $(\frac{10}{3}, -\frac{8}{3}, -3)$ **2(b)** $(\frac{4}{3}, -\frac{1}{3})$ **2(c)** $(\frac{7}{3}, -\frac{8}{3})$ **2(d)** $(3, -3, -1)$ **2(e)** $(0,0,1)$

Exercises for Section 5-2

6(a) $F_{1,2}F_{2,1}(-1)F_1(-\frac{1}{4})F_{1,2}(-5)F_{2,1}(-1)F_2(2)A$ **6(b)** $F_{1,2}F_{2,1}(-2)A$ **6(c)** $F_{2,1}F_{3,2} \times$
$F_{3,1}(-1)F_{3,2}(-1)F_1(\frac{1}{3})F_{1,2}(-2)F_{2,1}(1)F_{2,3}(-4)F_{1,3}(1)A$ **7(a)** $A = F_2(\frac{1}{2})F_{2,1}(1)F_{1,2}(5) \times$
$F_1(-4)F_{2,1}(1)F_{1,2}B$ **7(b)** $A = F_{2,1}(2)F_{1,2}B$ **7(c)** $A = F_{1,3}(-1)F_{2,3}(4)F_{2,1}(-1)F_{1,2}(2) \times$
$F_1(3)F_{3,2}(1)\ F_{3,1}(1)F_{3,2}F_{2,1}B$

Exercises for Section 5-3

There are many correct answers to Exercise 1. Here are some possible ones.
1(a) $A = F_{1,2}(2)F_{1,2}B$ **1(b)** $A = F_{1,3}(2)F_1(-1)F_{2,1}(3)F_{3,1}(1)F_{3,2}F_{1,2}B$ **1(c)** $A =$
$F_{1,2}(2)F_{3,2}(1)F_{3,1}(1)F_3(3)F_{1,3}(-1)F_{2,3}(1)$

Exercises for Section 6-1

1(a) -12 **1(b)** 1 **1(c)** 1 **1(d)** 18 **1(e)** 0 **1(f)** -292 **2(a)** -13 **2(b)** -21 **2(c)** 0 **3** -930
4 -202

Exercises for Section 6-2

1(a) -13 **1(b)** -21 **1(c)** 0 **1(d)** -930 **1(e)** -202 **3** -236 **4(a)** 0 **4(b)** 0 **4(c)** 1 **4(d)** -42
4(e) -1088 **4(f)** -201

Exercises for Section 6-3

1(a) $\begin{pmatrix} 2 & -1 \\ 0 & 2 \end{pmatrix}$ **1(b)** $\begin{pmatrix} 2 & -1 \\ 4 & 3 \end{pmatrix}$ **1(c)** $\begin{pmatrix} 1 & 0 & -1 \\ -3 & -1 & 3 \\ 4 & 2 & -5 \end{pmatrix}$

1(d) $\begin{pmatrix} -6 & 4 & 3 \\ 12 & -8 & -5 \\ -2 & 2 & 1 \end{pmatrix}$ **2(a)** $\begin{pmatrix} \frac{1}{2} & -\frac{1}{4} \\ 0 & \frac{1}{2} \end{pmatrix}$ **2(b)** $\begin{pmatrix} \frac{2}{10} & -\frac{1}{10} \\ \frac{4}{10} & \frac{3}{10} \end{pmatrix}$

2(c) $\begin{pmatrix} -1 & 0 & 1 \\ 3 & 1 & -3 \\ -4 & -2 & 5 \end{pmatrix}$ **2(d)** $\begin{pmatrix} -3 & 2 & \frac{3}{2} \\ 6 & -4 & -\frac{5}{2} \\ -1 & 1 & \frac{1}{2} \end{pmatrix}$

3

$$x = \frac{\det \begin{pmatrix} 5 & 1 \\ 6 & -1 \end{pmatrix}}{\det \begin{pmatrix} 2 & 1 \\ 1 & -1 \end{pmatrix}} = \frac{-11}{-3} = \frac{11}{3}$$

$$y = \frac{\det \begin{pmatrix} 2 & 5 \\ 1 & 6 \end{pmatrix}}{\det \begin{pmatrix} 2 & 1 \\ 1 & -1 \end{pmatrix}} = \frac{7}{-3} = \frac{-7}{3}$$

4

$$x = \frac{\det \begin{pmatrix} 7 & 1 & -2 \\ 3 & -1 & 2 \\ 5 & 1 & 1 \end{pmatrix}}{\det \begin{pmatrix} 1 & 1 & -2 \\ 1 & -1 & 2 \\ 2 & 1 & 1 \end{pmatrix}} = \frac{-30}{-6} = 5$$

$$y = \frac{\det \begin{pmatrix} 1 & 7 & -2 \\ 1 & 3 & 2 \\ 2 & 5 & 1 \end{pmatrix}}{\det \begin{pmatrix} 1 & 1 & -2 \\ 1 & -1 & 2 \\ 2 & 1 & 1 \end{pmatrix}} = \frac{16}{-6} = \frac{-8}{3}$$

$$z = \frac{\det \begin{pmatrix} 1 & 1 & 7 \\ 1 & -1 & 3 \\ 2 & 1 & 5 \end{pmatrix}}{\det \begin{pmatrix} 1 & 1 & -2 \\ 1 & -1 & 2 \\ 2 & 1 & 1 \end{pmatrix}} = \frac{14}{-6} = \frac{-7}{3}$$

5

$$x = \frac{\det \begin{pmatrix} 8 & -1 & 3 \\ 0 & 4 & -5 \\ 2 & -1 & 1 \end{pmatrix}}{\det \begin{pmatrix} 2 & -1 & 3 \\ 1 & 4 & -5 \\ 6 & -1 & 1 \end{pmatrix}} = \frac{-22}{-46} = \frac{11}{23}$$

$$y = \frac{\det \begin{pmatrix} 2 & 8 & 3 \\ 1 & 0 & -5 \\ 6 & 2 & 1 \end{pmatrix}}{\det \begin{pmatrix} 2 & -1 & 3 \\ 1 & 4 & -5 \\ 6 & -1 & 1 \end{pmatrix}} = \frac{-222}{-46} = \frac{111}{23}$$

$$z = \frac{\det \begin{pmatrix} 2 & -1 & 8 \\ 1 & 4 & 0 \\ 6 & -1 & 2 \end{pmatrix}}{\det \begin{pmatrix} 2 & -1 & 3 \\ 1 & 4 & -5 \\ 6 & -1 & 1 \end{pmatrix}} = \frac{-182}{-46} = \frac{91}{23}$$

Exercises for Section 7-1

1(a) 1 **1(b)** 2 **1(c)** 3 **1(d)** 4 **2** The rank of a submatrix cannot be larger than the rank of the matrix

Exercises for Section 7-2

Some of these problems have many answers. Here are some.
1 Ker T: $(-23,10,12)$; Im T: $(1,0)$, $(0,1)$ **2** Ker T: $(0,1,0,0)$; Im T: $(1,0,0)$, $(0,1,0)$, $(0,0,1)$
3(a) Ker A: $(-1,-1,1)$; Im A: $(1,0)$, $(0,1)$ **3(b)** Ker A: $(-4,-6,5,0)$, $(-4,-1,0,5)$;
Im A: $(1,0,1)$, $(0,1,0)$ **3(c)** Ker A is the zero subspace; Im A: $(1,0,0)$, $(0,1,0)$, $(0,0,1)$ **3(d)**
Ker A: $(-14,20,-19,32,0)$, $(2,-12,-19,0,16)$; Im A: $(1,0,0)$, $(0,1,0)$, $(0,0,1)$

Exercises for Section 7-3

1(a) Line through $(0,0,0)$ determined by $(\frac{1}{2},\frac{5}{2},1)$ **1(b)(i)** Line through $(1,1,0)$ parallel to $(\frac{1}{2},\frac{5}{2},1)$ **1(b)(ii)** Line through $(1,0,0)$ parallel to $(\frac{1}{2},\frac{5}{2},1)$ **1(c)** Parallel **2(a)** Yes, if $a - 2b \neq 0$ **2(b)** Plane passing through the points $(0,0,0)$, $(1,0,-4)$, and $(0,1,6)$ **2(c)(i)** Plane passing through $(0,0,2)$ determined by the vectors $(1,0,-4)$ and $(0,1,6)$ **2(c)(ii)** Plane passing through $(0,0,1)$ determined by the vectors $(1,0,-4)$ and $(0,1,6)$ **2(c)(iii)** Plane passing through $(0,0,-2)$ determined by the vectors $(1,0,-4)$ and $(0,1,6)$ **2(d)** Parallel **3(a)** No **3(b)** The one point $(0,0,0)$ **3(c)(i)** $(1,1,0)$ **3(c)(ii)** $(1,-1,0)$ **3(d)** Different points **4(a)** Dim Im $T = 2$ **4(b)** Since Im $T = \mathbb{R}^2$, there is a solution for any values of a and b **4(c)** 1 **5(a)** Dim Im $T = 1$; generated by $(2,1)$ **5(b)** Since Im $T \neq \mathbb{R}^2$, there are values of a and b for which there are no solutions **5(c)** 2 **6(a)** Dim Im $T = 3$ **6(b)** Since Im $T = \mathbb{R}^3$, there is a solution for any values of a, b, and c **6(c)** 0 **7(a)** $(\frac{3}{2},-\frac{1}{2}) + (0,0)$ **7(b)** $(1,-1,0) + z(-1,1,1)$ **8(a)** Rank $A = 3$ since det $A(-\,|4,5) = -10 \neq 0$ **8(b)** Any triples except y, w, v **9** All triples

Exercises for Section 8-1

1(a) $\sqrt{53}$ **1(b)** $\sqrt{34}$ **1(c)** $\sqrt{53}$ **1(d)** $\sqrt{22}$ **1(e)** $\sqrt{59}$ **1(f)** $\sqrt{125}$ **2(a)** $(1\sqrt{6}, 1/\sqrt{6}, 2/\sqrt{6})$ **2(b)** $5(/\sqrt{66}, 1/\sqrt{66}, 6/\sqrt{66}, 2/\sqrt{66})$ **2(c)** $(-1/\sqrt{31}, -1/\sqrt{31}, 2/\sqrt{31}, 5/\sqrt{31})$ **2(d)** $(1/\sqrt{43}, 1/\sqrt{43}, -1/\sqrt{43}, 2/\sqrt{43}, 6/\sqrt{43})$ **3(a)** $\sqrt{6}$ **3(b)** $\sqrt{35}$ **3(c)** $\sqrt{51}$ **3(d)** $\sqrt{55}$ **4(a)** $1,0,0,1,0,1$ **4(b)** 19 **4(c)** -43 **5(a)** $\sqrt{29}$ **5(b)** $\sqrt{10}$ **5(c)** $\sqrt{33}$ **5(d)** $\sqrt{41}$ **6(a)** 22 **6(b)** 4 **6(c)** -126 **7(a)** $5,2,3$ **7(b)** $8, -6,17$ **7(c)** x,y,z **8(a)** $(2,-\frac{3}{2})$ **8(b)** $(\frac{1}{3},2)$ **8(c)** $(-\frac{4}{3},3)$ **8(d)** $(\frac{3}{4}, \frac{7}{4})$ **8(e)** $(5,\frac{19}{14})$ **8(f)** $(3,-\frac{1}{2},\frac{5}{2},1)$ **8(g)** $(-\frac{1}{3},\frac{14}{3},\frac{17}{3},\frac{7}{3})$ **9** $(\frac{3}{2}-\sqrt{3}, 4 - \sqrt{3})$ and $(\frac{3}{2} + \sqrt{3}, 4 + \sqrt{3})$

Exercises for Section 8-2

1(a) $-1/\sqrt{10}$ **1(b)** $11/\sqrt{221}$ **1(c)** $-2/\sqrt{533}$ **1(d)** 0 **2(a)** 0 **2(b)** $11/6\sqrt{21}$ **2(c)** $3/2\sqrt{7}$ **2(d)** $14/\sqrt{806}$ **3(a)** Yes **3(b)** Yes **3(c)** No **3(d)** Each is orthogonal to the others **4(a)** $(\frac{3}{2},\frac{1}{2})$ **4(b)** $(\frac{4}{5},\frac{3}{5})$ **4(c)** $(\frac{33}{29},\frac{44}{29},\frac{22}{29})$ **4(d)** $(\frac{10}{11},-\frac{15}{11},\frac{10}{11},\frac{20}{11})$ **5(a)** $(-\frac{3}{13},\frac{2}{13}) + (\frac{16}{13},\frac{24}{13})$ **5(b)** $(\frac{28}{41},-\frac{35}{41}) + (\frac{95}{41},\frac{76}{41})$ **5(c)** $(-4,-2) + (-2,4)$ **5(d)** $(\frac{7}{6},-\frac{7}{6}\frac{14}{6}) + (\frac{5}{6},\frac{13}{6},\frac{4}{6})$ **5(e)** $(1,-2,6) + (4,2,0)$ **5(f)** $(\frac{4}{47},-\frac{8}{47},-\frac{4}{47},-\frac{16}{47}\frac{20}{47}) + (\frac{137}{47},\frac{55}{47},\frac{192}{47},\frac{110}{47},\frac{121}{47})$ **6(a)** $5\sqrt{2}/2$ **6(b)** $\sqrt{2}/2$ **6(c)** $3\sqrt{10}/5$

Exercises for Section 8-3

2 $(3,1) = (4/\sqrt{2})(1/\sqrt{2}, 1/\sqrt{2}) - (2/\sqrt{2})(-1/\sqrt{2}, 1/\sqrt{2}) = ((3 + \sqrt{3})/2)(1/2,\sqrt{3}/2) + ((3\sqrt{3} - 1)/2)(\sqrt{3}/2,-1/2)$. $(2,-1) = (1/\sqrt{2})(1/\sqrt{2}, 1/\sqrt{2}) - (3/\sqrt{2})(-1/\sqrt{2}, 1/\sqrt{2}) =$

$((2 - \sqrt{3})/2)(1/2, \sqrt{3}/2) + ((2\sqrt{3}+1)/2)(\sqrt{3}/2, -1/2)$ **3** $(2,1,6) = (3/\sqrt{2})(1/\sqrt{2}, 1/\sqrt{2}, 0) +$
$(5/\sqrt{3})(-1/\sqrt{3}, 1/\sqrt{3}, 1/\sqrt{3})$ + $(13/\sqrt{6})(1/\sqrt{6}, -1/\sqrt{6}, 2/\sqrt{6})$ = $(3/\sqrt{2})(1/\sqrt{2},$
$1/\sqrt{2}, 0) - (1/\sqrt{2})(-1/\sqrt{2}, 1/\sqrt{2}, 0) + 6(0,0,1)$ **4** $(1,2,-1,6) = (6/\sqrt{3})(1/\sqrt{3}, 0,$
$1/\sqrt{3}, 1/\sqrt{3}) + (-3/\sqrt{3})(1/\sqrt{3}, 1/\sqrt{3}, 0, -1/\sqrt{3}) + (9/\sqrt{3})(0, 1/\sqrt{3}, -1/\sqrt{3}, 1/\sqrt{3})$
$+ 0(1/\sqrt{3}, -1/\sqrt{3}, -1/\sqrt{3}, 0)$

In Exercises 5, 6, and 7, there is no unique answer. Be sure to check that the vectors are mutually orthogonal and in the subspace under consideration **8** $(\frac{5}{6}, \frac{5}{6}, \frac{10}{6})$ **9** $(\frac{5}{14})(2,1,3,0) -$
$(\frac{3}{238})(-4,5,1,14)$ **10(a)** $(\frac{20}{11}, \frac{19}{11}, \frac{14}{11})$ **(10b)** $(\frac{39}{14}, \frac{20}{14}, \frac{47}{14})$

13 $2x + y - z \qquad\qquad = 0$
$\quad 2x + 3y \qquad - w = 0$

14 $x + y - z \qquad\qquad = 0$
$\qquad\quad y \qquad - w \quad = 0$
$\quad x + 2y \qquad\qquad - v = 0$

Exercises for Section 8-4

1(a) $-9i - 9j + 9k$ **1(b)** $9i + 9j - 9k$ **1(c)** $-4i + 4j + 4k$ **1(d)** k **1(e)** $18i + 18j - 18k$
2(a) $(-1/\sqrt{3}, -1/\sqrt{3}, 1/\sqrt{3})$ **2(b)** $(-\frac{2}{3}, \frac{2}{3}, -\frac{1}{3})$ **2(c)** $(3/\sqrt{131}, 1/\sqrt{131}, -11/\sqrt{131})$ **2(d)**
$(1/\sqrt{2}, 0, -1/\sqrt{2})$

Exercises for Section 9-1

1(a) $\begin{pmatrix} 1 & 3 \\ 3 & 2 \\ 1 & 2 \end{pmatrix}$ **1(b)** $\begin{pmatrix} -2 & 1 \\ 2 & 0 \\ 1 & 2 \end{pmatrix}$ **1(c)** $\begin{pmatrix} \frac{3}{2} & \frac{1}{2} \\ \frac{1}{2} & \frac{7}{2} \\ \frac{3}{2} & \frac{9}{2} \end{pmatrix}$

2(a) $\begin{pmatrix} 0 & -\frac{1}{2} & 1 \\ 1 & \frac{5}{2} & 0 \\ 0 & -\frac{5}{2} & 1 \end{pmatrix}$ **2(b)** $\begin{pmatrix} -\frac{1}{2} & 1 & \frac{1}{2} \\ \frac{7}{2} & 1 & \frac{5}{2} \\ -\frac{5}{2} & 1 & -\frac{3}{2} \end{pmatrix}$ **2(c)** $\begin{pmatrix} 3 & 2 & 3 \\ -3 & 2 & -1 \\ 1 & 2 & 1 \end{pmatrix}$

3 $T(x,y,z) = (2x + y + z, -3x + 2y + 7z, x + y + 2z)$ **4** $T(x,y,z,w) = (\frac{1}{12})(-9x + 25y - 4z + 17w, 6x + 74y - 20z + 34w, 8y - 8z + 4w)$

Exercises for Section 9-2

1(a) $\begin{pmatrix} \frac{2}{7} & \frac{1}{7} \\ -\frac{3}{7} & \frac{2}{7} \end{pmatrix}$ **1(b)** $\begin{pmatrix} 2 & -1 \\ 3 & 2 \end{pmatrix}$ **1(c)** $\begin{pmatrix} \frac{11}{12} & -\frac{2}{12} \\ \frac{9}{12} & \frac{6}{12} \end{pmatrix}$

1(d)
$$\begin{pmatrix} \frac{1}{2} & \frac{1}{2} & -\frac{1}{2} \\ \frac{1}{2} & -\frac{1}{2} & \frac{1}{2} \\ -\frac{1}{2} & \frac{1}{2} & \frac{1}{2} \end{pmatrix}$$
1(e)
$$\begin{pmatrix} 0 & 1 & -1 \\ 1 & -1 & 0 \\ 0 & 1 & 1 \end{pmatrix}$$

2 $X_1 = (1,2)$, $X_2 = (1,3)$ **3** $Y_1 = (3,-2)$, $Y_2 = (-1,1)$ **4** $X_1 = (1,0,1)$, $X_2 = (0,1,1)$, $X_3 = (1,1,0)$ **5** $Y_1 = (\frac{1}{2}, -\frac{1}{2}, \frac{1}{2})$, $Y_2 = (-\frac{1}{2}, \frac{1}{2}, \frac{1}{2})$, $Y_3 = (\frac{1}{2}, \frac{1}{2}, -\frac{1}{2})$

Exercises for Section 9-3

There are many answers to Exercises 1, 3, and 4. Here are some

1(a)
$$P = \begin{pmatrix} \frac{4}{7} & -\frac{1}{7} \\ -\frac{1}{7} & \frac{2}{7} \end{pmatrix} \qquad Q = \begin{pmatrix} 1 & 0 & -\frac{11}{17} \\ 0 & 1 & \frac{1}{7} \\ 0 & 0 & 1 \end{pmatrix}$$

1(b)
$$P = \begin{pmatrix} -1 & 0 & 0 \\ 1 & 1 & 0 \\ -1 & -1 & 1 \end{pmatrix} \qquad Q = \begin{pmatrix} 1 & 0 & 2 \\ 0 & 1 & -5 \\ 0 & 0 & 1 \end{pmatrix}$$

1(c) Since A is nonsingular, we can use $P = A^{-1}$ and $Q = I_3$ or we can use $P = I_3$ and $Q = A^{-1}$

$$A^{-1} = \begin{pmatrix} -\frac{1}{4} & \frac{1}{2} & -\frac{1}{4} \\ \frac{3}{2} & -1 & \frac{1}{2} \\ \frac{3}{4} & -\frac{1}{2} & \frac{3}{4} \end{pmatrix}$$

3
$$P = \begin{pmatrix} -\frac{4}{7} & \frac{1}{7} \\ \frac{18}{7} & -\frac{1}{7} \end{pmatrix} \qquad Q = \begin{pmatrix} 1 & -\frac{39}{14} & -\frac{39}{7} \\ 0 & \frac{78}{14} & \frac{71}{7} \\ 0 & \frac{1}{2} & 1 \end{pmatrix}$$

4 $P = BA^{-1}$, $Q = I_3$

Exercises for Section 10-1

1(a) Eigenvalues 4 and 2. $(1,1)$ eigenvector for 4; $(1,-1)$ eigenvector for 2 **1(b)** Eigenvalues 4 and 2. $(0,1)$ eigenvector for 4; $(2,-3)$ eigenvector for 2 **1(c)** Eigenvalues are $2 + \sqrt{11}$ and $2 - \sqrt{11}$. $(1, -3 + \sqrt{11})$ is an eigenvector for $2 + \sqrt{11}$; $(1, -3 - \sqrt{11})$ is an eigenvector for $2 - \sqrt{11}$ **1(d)** Eigenvalues are 6, 4, 2. $(-1,1,1)$ is an eigenvector for 6; $(1, -1, 1)$ is an eigenvector for 4; $(-1, -1, 1)$ is an eigenvector for 2

Exercises for Section 10-2

In Exercises 1 and 2 we do not give the matrix relative to the natural basis.

1(a) $\begin{pmatrix} 6 & 1 \\ -1 & 1 \end{pmatrix}$ **1(b)** $\begin{pmatrix} 6 & \frac{1}{2} \\ \frac{1}{2} & -1 \end{pmatrix}$ **2(a)** $\begin{pmatrix} 2 & \frac{7}{2} & \frac{7}{2} \\ \frac{7}{2} & 6 & 6 \\ \frac{7}{2} & 6 & 5 \end{pmatrix}$

2(b) $\begin{pmatrix} 1 & 1 & 1 \\ 1 & 0 & \frac{1}{2} \\ 1 & \frac{1}{2} & 1 \end{pmatrix}$

3(a) $(1, -2 + \sqrt{5})$, $(1, -2 - \sqrt{5})$. **3(b)** $(10, -5)$, $(1,2)$ **3(c)** $(1, -3 + \sqrt{10})$, $(-1, 3 + \sqrt{10})$ **3(d)** $(0,1,0)$, $(1, 0, 2 - \sqrt{5})$, $(1, 0, 2 + \sqrt{5})$

6 $P = \begin{pmatrix} 1 & 0 \\ -1 & 1 \end{pmatrix}$ **7** $P = \begin{pmatrix} 1 & 1 \\ 0 & 1 \end{pmatrix}$ **8** $P = \begin{pmatrix} 1 & 0 \\ 1 & 1 \end{pmatrix}$

9 $P = \begin{pmatrix} 1 & 1 & 0 \\ 1 & 0 & 1 \\ 0 & 1 & 1 \end{pmatrix}$

11(a) $P = \begin{pmatrix} \frac{1}{3}\sqrt{3} & -\frac{1}{3}\sqrt{6} \\ \frac{1}{3}\sqrt{6} & \frac{1}{3}\sqrt{3} \end{pmatrix}$

11(b) $P = \begin{pmatrix} \frac{1}{3} & \frac{2}{3}\sqrt{2} \\ -\frac{2}{3}\sqrt{2} & \frac{1}{3} \end{pmatrix}$

11(c) $P = \begin{pmatrix} \frac{1}{2}\sqrt{2} & -\frac{1}{3}\sqrt{3} & \frac{1}{6}\sqrt{6} \\ \frac{1}{2}\sqrt{2} & \frac{1}{3}\sqrt{3} & -\frac{1}{6}\sqrt{6} \\ 0 & \frac{1}{3}\sqrt{3} & \frac{1}{3}\sqrt{6} \end{pmatrix}$

12(a) $P = \begin{pmatrix} \frac{1}{2}\sqrt{2} & -\frac{1}{2}\sqrt{2} \\ \frac{1}{2}\sqrt{2} & \frac{1}{2}\sqrt{2} \end{pmatrix}$ **12(b)** $P = \begin{pmatrix} \frac{1}{2}\sqrt{2} & \frac{1}{2}\sqrt{2} \\ -\frac{1}{2}\sqrt{2} & \frac{1}{2}\sqrt{2} \end{pmatrix}$

12(c) $P = \begin{pmatrix} \frac{1}{2} & \frac{1}{2}\sqrt{3} \\ \frac{1}{2}\sqrt{3} & -\frac{1}{2} \end{pmatrix}$

Exercises for Section 10-3

In these answers, you must keep in mind the possibility that the different coordinate axes can be named by different letters. **1(a)** Ellipse; $u^2/(\frac{1}{2})^2 + v^2 = 1$ **1(b)** Hyperbola; $v^2/9 - u^2/4 = 1$ **1(c)** Hyperbola; $u^2/(\frac{1}{2})^2 - v^2 = 1$ **1(d)** Ellipse; $u^2/(\sqrt{2}/\sqrt{3})^2 + v^2/(\sqrt{3}/\sqrt{13})^2 = 1$ **1(e)** Hyperbola; $u^2/(\sqrt{2})^2 + v^2/(\sqrt{2})^2 = 1$ **2(a)** Hyperbola with center at $(-2,0)$ opening along line parallel to $(1/\sqrt{2}, 1/\sqrt{2})$; simplified equation $u^2 - v^2 = 1$ **2(b)** Parabola with vertex at $(-\frac{3}{4}, \frac{3}{4})$ opening along line parallel to $(-1/\sqrt{2}, 1/\sqrt{2})$; simplified equation $v = \sqrt{2}u^2$ **2(c)** Ellipse with center at $(7/\sqrt{10}, 3/\sqrt{10})$ and axes along lines determined by $(3/\sqrt{10}, -1/\sqrt{10})$ and $(1/\sqrt{10}, 3/\sqrt{10})$; simplified equation $u^2/2 + v^2/3 = 1$ **3(a)** $u^2 + v^2 - w^2 = 1$ **3(b)** $u^2 - v^2 + w = 0$ **3(c)** $r^2 + 2s^2 - t^2 + 3u^2 = 1$

Index

Addition, associativity of, 51, 54
  closure with respect to, 51, 54, 60
  commutativity of, 51, 54
  of functions, 146
  of linear transformations, 181–183
  of matrices, 176
  of vectors in $\mathbb{R}^2$, 49
  of vectors in $\mathbb{R}^3$, 50
  of vectors in $\mathbb{R}^4$, 52
  of vectors in $\mathbb{R}^n$, 53
  stability with respect to, 51, 54, 60
Additive function, 157, 158
Adjoint of a matrix, 257
Adjugate of a matrix, 257
Arrow representation of vectors, in $\mathbb{R}^2$, 66–70
  in $\mathbb{R}^3$, 71–75
Associative property for, addition of
      vectors, 51, 54
  composition of linear transformations,
      149, 150
  multiplication of matrices, 189
Augmented matrix, 7, 40
Axes, for coordinate system, 401
  x-axis, 66
  y-axis, 66
  z-axis, 71

Basis, 123
  natural, 133
  ordered, 347
  orthogonal, 323
  orthonormal, 323
  standard (= natural), 133

$C_{i,j}$, $C_i(x)$, $C_{i,j}(x)$, 33
Change of basis, 355–363
  matrix for, 355–359
  orthogonal, 364
Characteristic equation, 384

Characteristic polynomial, 384
Characteristic value, 378
Characteristic vector, 378
Clique in sociology, 199
Closure, relative to addition, 51, 54, 60
  relative to scalar multiplication, 51, 54, 60
Coefficient matrix, 7, 39
Coefficients of unknowns, 2, 39
Cofactors, 255, 257
  matrix of, 257
Collinear vectors, 81
Column, 7, 38
  nonzero, 20
  zero, 20
Column echelon form, 34
Column operations, 33
Column rank, 273
Column space, 273
Complementary subspace, 113, 139, 140
Components, 49, 50, 53
Composition, associativity of, 150
  of functions, 147–149
  of linear transformations, 183–185
Conic sections, 401–414
Constants, 2, 39
Coordinate system, 401
Coordinates, 49, 50, 53
  of a vector relative to an ordered basis,
      346, 347
Coplanar vectors, 105
Cosine of angle between vectors, 312–315
Cosines, law of, 311–312
Cramer's rule, 259–263
Cross product, 340

Dependence, linear, 115
Determinant, 230–269
  evaluation of, 232, 234, 237–240, 242–252
  expansion along a column, 244–246

431

Determinant (*continued*)
  expansion along a row, 237
  formula for A^{-1}, 258
  of a nonsingular matrix, 267
  of a product, 267
  of a transpose, 244, 268
  of a 2 × 2 matrix, 330–232
  of a 3 × 3 matrix, 233, 234
  properties of the, 242–244, 248
Diagonal matrix, 269
Dimension, 132
Dimension theorem, 288
Distance between points, 304–306
Domain of a function, 144
Dot product, 307
  geometrical interpretation of, 311–320

Echelon, 21
  column, 34
  row, 22
Economy, model for, 43–44
Eigenvalue, 378
Eigenvector, 378
Electrical circuits, 40–41, 200–201
Elementary matrix, 213
  determinant of, 266
  inverse of, 218–219
Empty set, 55
Entry of a matrix, 18, 37, 39
Equations, homogeneous, 3
  linear, 2–6, 35–37
  of a circle, 408
  of an ellipse, 408
  of a hyperbola, 408
  of a line, 76–81
  of a parabola, 408
  of a plane, 100–105
  of a subspace, 338, 339
  systems of, 2–15, 35–40, 292–299
Equivalent matrices, 368–369

$F_{i,j}$, $F_i(x)$, $F_{i,j}(x)$, 213
Finitely generated, 95
Function, 144
  addition, 146
  additive, 158
  composition, 149, 150
  domain, 144
  equality, 144
  homogeneous, 158

Function (*continued*)
  image, 144
  range, 144
  scalar multiple, 146
  value at **X**, 144

Generating set (= spanning set), 95
Gram–Schmidt process, 332–336

Homogeneous function, 158
Homogeneous system of linear
    equations, 3, 14, 63, 119, 130

Identity, function, 191
  matrix, 191
Image, of a function, 144
  of a linear transformation, 283
  of a matrix, 285
Independence, linear, 115
Initial 1, of a column, 34
  of a row, 20
Inner product, 307
Inverse, method for computing, 208–211
  of a linear transformation, 207
  of a matrix, 206, 258
Invertible, linear transformation, 207
  matrix, 206

Kernel, of a linear transformation, 283
  of a matrix, 284, 285
Kirchhoff's law of current, 40
  law of voltages, 40

Length of a vector, 306
Leontief, 44
Line in $\mathbb{R}^2$, 76–78
  in $\mathbb{R}^3$, 79–81
Linear combination, 87
  nonzero, 87
  proper, 87
  zero, 87
Linear dependence, 115
Linear equation, 2–6, 35–37
  definition of, 39
  homogeneous, 3, 14, 63, 119, 130
  solution, 3
  system of, 2–15, 35–40, 292–299
Linear independence, 115
Linear span, 95
Linear transformation, 152*ff*

Linear transformation (*continued*)
  addition of, 181–183
  composition, 185
  coordinate-free description, 158
  definition, 153, 155
  equality, 160
  identity, 191
  image, 283
  inverse, 207
  invertible, 207
  kernel, 283
  matrix for, 167–170, 350–351
  negative, 188
  noninvertible, 207
  nonsingular, 207
  nullity, 286
  nullspace, 286
  of a matrix, 171–172, 352–353
  rank, 286
  scalar multiple, 182
  singular, 207
  zero, 188
Lower triangular matrix, 269

Main diagonal, 269
Map, mapping (= function), 144
Matrices, addition, 176
  equivalent, 368, 369
  multiplication, 180
  row equivalent, 18
  scalar multiple, 176
  similar, 399
Matrix, adjoint, 257
  adjugate, 257
  augmented, 7, 40
  coefficient, 7, 39
  cofactor, 257
  column, 7, 38
  column echelon form, 34
  column rank, 273
  column space, 273
  elementary, 213
  entry, 18, 37, 39
  for a change of basis, 355–359
  for a linear transformation, 167–170,
    350–351
  for a quadratic function, 394
  identity, 191
  image of, 285
  inverse, 206, 258

Matrix (*continued*)
  invertible, 206
  kernel, 284, 285
  lower triangular, 269
  main diagonal, 269
  negative, 188
  noninvertible, 206
  nonsingular, 206
  nullity, 286
  nullspace, 284
  of cofactors, 257
  orthogonal, 339
  rank of, 278, 286
  row, 38
  row echelon form, 22
  row rank, 273
  row space, 273
  singular, 206
  submatrix, 235
  symmetric, 394
  transition, 359
  transpose, 35
  upper triangular, 269
  zero, 188

Natural basis, 133
Negative, of linear transformation, 188
  of matrix, 188
  of vector, 51, 54
Noncollinear vectors, 81
Noncoplanar vectors, 105
Nonempty set, 55
Noninvertible, linear transformation, 207
  matrix, 206
Nonnull set, 55
Nonsingular, linear transformation, 207
  matrix, 206
Nonvoid set, 55
Nonzero linear combination, 87
Normalize a vector, 328
n-tuple, 53
Nullity, of linear transformation, 286
  of matrix, 286
Null set, 55
Nullspace, of linear transformation, 283
  of matrix, 284

Ohm's law, 40
Ordered basis, 347
Ordered n-tuple, 53

Ordered pair, 3, 49
Ordered triple, 3, 50
Orthogonal basis, 323
Orthogonal change of basis, 364
Orthogonal complement, 338
Orthogonal matrix, 339
Orthogonal projection, of a vector on
    a subspace, 333
  of one vector on another, 320
Orthogonal vectors, 315
Orthonormal basis, 323

Pair (ordered), 3, 49
Parallelogram law, 311
Perpendicular vectors, 315
Planes, 100–105
Position in a matrix, 37
Product, cross, 340
  dot, 307
  inner, 307
  matrix, 180
  vector, 340
Proper linear combination, 87
Pythagorean theorem, 311

Quadratic function, 389–394

$R_{i,j}$, $R_i(x)$, $R_{i,j}(x)$, 10
$\mathbb{R}$, 47
$\mathbb{R}^2$, 49
$\mathbb{R}^3$, 50
$\mathbb{R}^4$, 52
$\mathbb{R}^n$, 53
Range of a function, 144
Rank, column, 273
  of a linear transformation, 286
  of a matrix, 278, 286
  row, 273
Row echelon form, 22
Row equivalent matrices, 18
Row of a matrix, 7, 38
  nonzero row, 20
  zero row, 20
Row operations on a matrix, 10
Row rank, 273
Row space, 273

Scalar, 47
Scalar multiplication, closure or stability
    with respect to, 51, 54, 60

Scalar multiplication (*continued*)
  of a function, 146
  of a linear transformation, 182
  of a matrix, 176
  of vectors, 49, 50, 52, 53
Similar matrices, 399
Singular, linear transformation, 207
  matrix, 206
Solution of a system of linear equations, 3
Span, 95
Spanning set, 95
Stability, with respect to addition, 51,
    54, 60
  with respect to scalar multiplication, 51,
    54, 60
Standard basis (= natural basis), 133
Submatrix, 235
Subspace, basis for, 135
  complementary, 113, 139, 140
  dimension of, 135
  of $\mathbb{R}^2$ and $\mathbb{R}^3$, 109–111
  of a vector space, 59, 60
  zero, 63
Sum, of functions, 146
  of linear transformations, 181–183
  of matrices, 176
  of vectors, 49, 50, 52, 53
Symmetric matrix, 394

Transformation (= map, mapping, function), 144
  linear, 152*ff*
Transition matrix, 359
Transpose, 35
Triangular matrix, lower, 269
  upper, 269
Triple, 3, 50

Unit vector, 306
Upper triangular matrix, 269

Vector, 47–55
  addition, 49, 50, 52, 53
  components, 49, 50, 52, 53
  length, 306
  negative, 51, 54, 55
  product, 340
  scalar multiplication, 49, 50, 52, 53
  unit, 306
  zero, 51, 54, 55
Vector equation, of a line in $\mathbb{R}^2$, 76–78
  of a line in $\mathbb{R}^3$, 79
  of a plane in $\mathbb{R}^3$, 100–105

Vectors, collinear, 81
  coplanar, 105
  noncollinear, 81
  noncoplanar, 105
Vector space, 46
  dimension, 132
  finitely generated, 95
  over $\mathbb{R}$, 46
  subspace, 59, 60

Void set, 55

Zero column, 20
Zero linear combination, 87
Zero linear transformation, 188
Zero matrix, 188
Zero row, 20
Zero subspace, 63
Zero vector, 51, 54, 55